T0291011

The Biodemography of Ageing and Longevity

Why and how we age are probably two of science's oldest questions, echoing personal beliefs and concerns about our own finitude. From the earliest musings of ancient philosophers to recent pharmacological trials aimed at slowing ageing and prolonging longevity, these questions have fascinated scientists across time and fields of research. Taking advantage of the natural diversity of ageing trajectories, within and across species, this interdisciplinary volume provides a comprehensive view of the recent advances in ageing and longevity through a biodemographic approach. It includes the key facts, theories, ongoing fields of investigation, big questions and new avenues for research in ageing and longevity, as well as considerations on how extending longevity integrates into the social and environmental challenges that our society faces. This is a useful resource for students and researchers curious to unravel the mysteries of longevity and ageing, from their origins to their consequences, across species, space and time.

Jean-François Lemaître is Research Director at the French National Centre for Scientific Research (CNRS), based at Claude Bernard Lyon 1 University. He is an evolutionary biologist who studies ageing in wild mammals through a multidisciplinary approach combining evolutionary demography, ecology and ecophysiology. His research currently focuses on identifying factors shaping the diversity of reproductive ageing patterns observed across mammal species in both females and males.

Samuel Pavard is Professor of Eco-anthropology at the National Museum of Natural History in Paris. He is a demographer with broad interests in understanding the biological, environmental, cultural and social determinants of ageing and longevity in humans and, in particular, in investigating the joint evolution of actuarial and reproductive senescence with cognitive and social capabilities in our species. He is currently investigating the role played by evolutionary trade-offs between mortality components in shaping mortality at old age, in mammals in general and in humans in particular.

The Biodemography of Ageing and Longevity

Edited by

JEAN-FRANÇOIS LEMAÎTRE

French National Centre for Scientific Research (CNRS)

SAMUEL PAVARD

National Museum of Natural History, Paris

French Institute for Demographic Studies (INED), Paris

Shaftesbury Road, Cambridge CB2 8EA, United Kingdom

One Liberty Plaza, 20th Floor, New York, NY 10006, USA

477 Williamstown Road, Port Melbourne, VIC 3207, Australia

314–321, 3rd Floor, Plot 3, Splendor Forum, Jasola District Centre, New Delhi – 110025, India

103 Penang Road, #05–06/07, Visioncrest Commercial, Singapore 238467

Cambridge University Press is part of Cambridge University Press & Assessment, a department of the University of Cambridge.

We share the University's mission to contribute to society through the pursuit of education, learning and research at the highest international levels of excellence.

www.cambridge.org
Information on this title: www.cambridge.org/9781316519196

DOI: 10.1017/9781009007245

When citing this work, please include a reference to the DOI 10.1017/9781009007245

First published 2025

A catalogue record for this publication is available from the British Library

Library of Congress Cataloging-in-Publication Data
Names: Lemaître, Jean-François, editor. | Pavard, Samuel, editor.
Title: The biodemography of ageing and longevity / edited by Jean-François Lemaître, French National Centre for Scientific Research (CNRS), Samuel Pavard National Museum of Natural History, Paris.
Description: Cambridge, United Kingdom ; New York, NY, USA : Cambridge University Press, 2025. | Includes bibliographical references and index.
Identifiers: LCCN 2024001603 | ISBN 9781316519196 (hardback) | ISBN 9781009007245 (ebook)
Subjects: LCSH: Aging. | Longevity. | Population biology.
Classification: LCC QH529 .B56 2025 | DDC 571.8/78–dc23/eng/20240510
LC record available at https://lccn.loc.gov/2024001603

ISBN 978-1-316-51919-6 Hardback

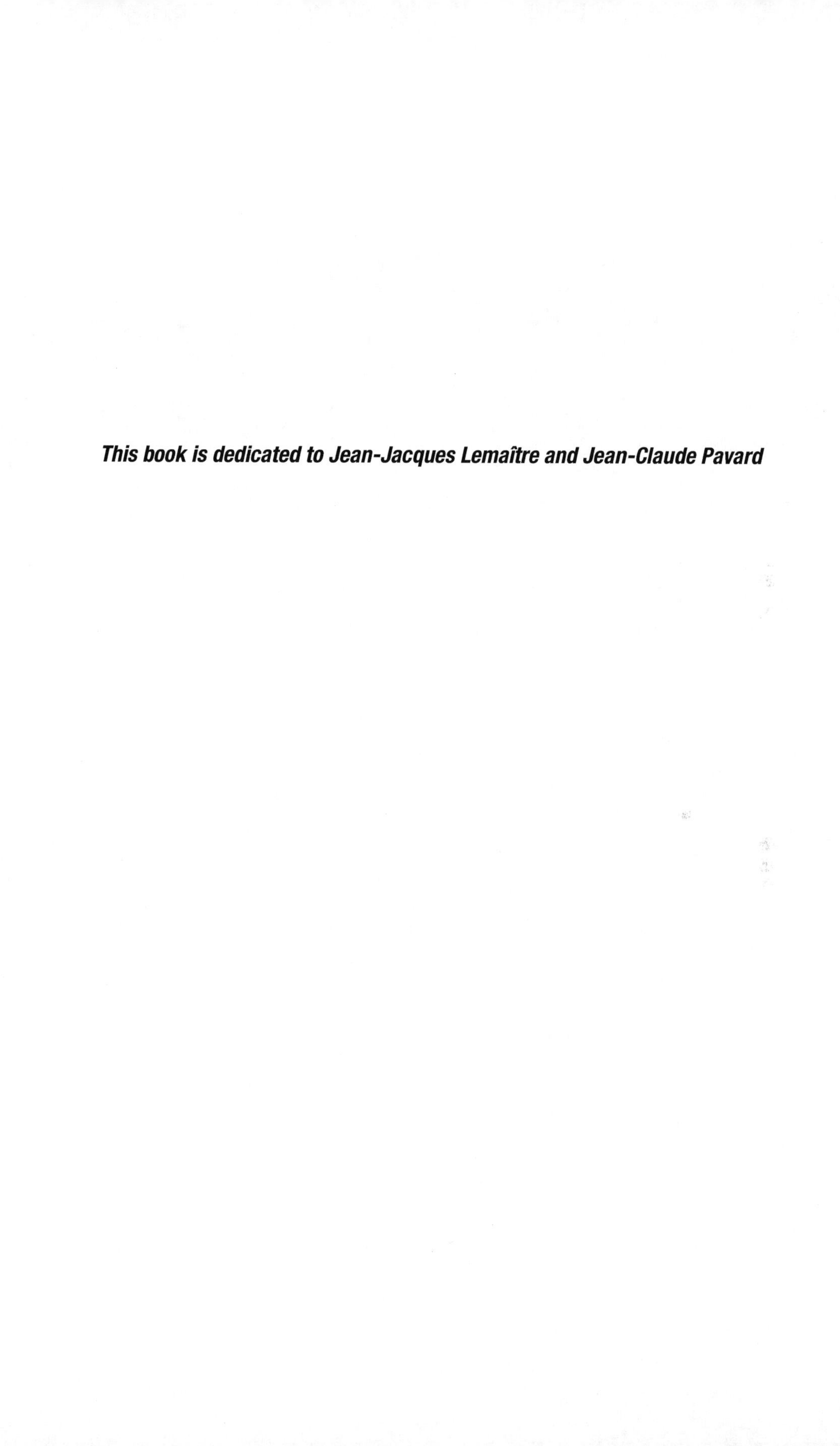

This book is dedicated to Jean-Jacques Lemaître and Jean-Claude Pavard

Contents

Contributors

Michael J. Adkesson
Chicago Zoological Society and Brookfield Zoo, Brookfield, IL, USA

Annette Baudisch
Interdisciplinary Centre on Population Dynamics, University of Southern Denmark, Odense, Denmark

Fabrice Bertile
University of Strasbourg, French National Centre for Scientific Research (CNRS), Hubert Curien Pluridisciplinary Institute (IPHC) UMR 7178, Strasbourg, France

Jelle Boonekamp
Institute of Biodiversity and Animal Health, University of Glasgow, Glasgow, UK

Carlo Giovanni Camarda
French Institute for Demographic Studies (INED), Paris, France

Hugo Cayuela
Laboratory of Biometry and Evolutionary Biology, UMR CNRS 5558, Claude Bernard University Lyon 1, Villeurbanne, France

Julie Choquette
Department of Demography, University of Montreal, Canada

Fernando Colchero
Department of Primate Behavior and Evolution Max Planck Institute for Evolutionary Anthropology, Leipzig, Germany

Dalia A. Conde
Species360 Conservation Science Alliance, Bloomington, MN, USA
Interdisciplinary Centre on Population Dynamics, University of Southern Denmark, Odense, Denmark
Institute of Biology, University of Southern Denmark, Odense, Denmark

François Criscuolo
University of Strasbourg, French National Centre for Scientific Research (CNRS), Hubert Curien Pluridisciplinary Institute (IPHC) UMR 7178, Strasbourg, France

Sarah Cubaynes
Center for Functional and Evolutionary Ecology, University of Montpellier, CNRS, EPHE, UMR5175, Montpellier, France

Linh Hoang Khanh Dang
French Institute for Demographic Studies (INED), Paris, France

Joris Deelen
Max Planck Institute for Biology of Ageing, Cologne, Germany
Cologne Excellence Cluster on Cellular Stress Responses in Ageing-Associated Diseases (CECAD), University of Cologne, Cologne, Germany

Géraldine Duthé
French Institute for Demographic Studies (INED), Paris, France

Jean-Michel Gaillard
Laboratory of Biometry and Evolutionary Biology, UMR CNRS 5558, Claude Bernard University Lyon 1, Villeurbanne, France

Michael D. Gurven
Integrative Anthropological Sciences, Department of Anthropology, Leonard and Gretchan Broom Center for Demography, University of California, Santa Barbara, CA, USA
Broom Center for Demography, University of California, Santa Barbara, Santa Barbara, CA, USA

Thomas B.L. Kirkwood
Population Health Sciences Institute, Newcastle University, Newcastle upon Tyne, UK

Jean-François Lemaître
Laboratory of Biometry and Evolutionary Biology, UMR CNRS 5558, Claude Bernard University Lyon 1, Villeurbanne, France

Gabriel A. Marais
Centro de Investigação em Biodiversidade e Recursos Genéticos (CIBIO), InBIO Laboratório Associado, Vairão Campus, University of Porto, Vairão, Portugal
Department of Biology, Faculty of Sciences, University of Porto, Porto, Portugal
BIOPOLIS Program in Genomics, Biodiversity and Land Planning, CIBIO, Vairão Campus, Vairão, Portugal

Gilles Maurer
Center for Functional and Evolutionary Ecology, Univ. Montpellier, CNRS, EPHE, UMR5175, Montpellier, France
Zooparc de Beauval & Beauval Nature, Saint-Aignan, France

France Meslé
French Institute for Demographic Studies (INED), Paris, France

Pat Monaghan
Institute of Biodiversity and Animal Health, University of Glasgow, Glasgow, UK

Soumaila Ouedraogo
New York University Abu Dhabi, United Arab Emirates

Nadine Ouellette
Department of Demography, University of Montreal, Canada

Samuel Pavard
UMR7206 Eco-anthropology, National Museum of Natural History, CNRS, UPC, Paris, France
French Institute for Demographic Studies (INED), Paris, France

Gianni Pes
Department of Biomedical Science, University of Sassari, Italy

Gilles Pison
UMR7206 Eco-anthropology, National Museum of Natural History, CNRS, UPC, Paris, France
French Institute for Demographic Studies (INED), Paris, France

Dominique Pontier
Laboratory of Biometry and Evolutionary Biology, UMR CNRS 5558, Claude Bernard University Lyon 1, Villeurbanne, France

Michel Poulain
Catholic University of Louvain, Belgium
Estonian Institute for Population Studies, Tallinn University, Estonia

Alexia Prskawetz
Wittgenstein Centre of Demography and Global Human Capital (University of Vienna), Vienna, Austria
Institute of Statistics and Mathematical Methods in Economics, Research Unit Economics, Vienna University of Technology
International Institute for Applied Systems Analysis, Laxenburg, Austria

Jean-Marie Robine
EPHE-PSL University
French Institute for Demographic Studies (INED), Paris, France

Victor Ronget
University of Mainz iomE (Institute of organismic and molecular evolution), Mainz, Germany

Miguel Sánchez-Romero
Wittgenstein Centre of Demography and Global Human Capital (University of Vienna), Vienna, Austria
International Institute for Applied Systems Analysis, Laxenburg, Austria

Larissa Smulders
Max Planck Institute for Biology of Ageing, Cologne, Germany
CECAD, University of Cologne, Cologne, Germany

Johanna Staerk
Department of Primate Behavior and Evolution Max Planck Institute for Evolutionary Anthropology, Leipzig, Germany

Morgane Tidière
Species360 Conservation Science Alliance, Bloomington, MN, USA
Interdisciplinary Centre on Population Dynamics, University of Southern Denmark, Odense, Denmark
Institute of Biology, University of Southern Denmark, Odense, Denmark

Lucie Vanhoutte
UMR7206 Eco-anthropology, National Museum of Natural History, CNRS, UPC, Paris, France
French Institute for Demographic Studies (INED), Paris, France

Cristina Vieira
Laboratory of Biometry and Evolutionary Biology, UMR CNRS 5558, Claude Bernard University Lyon 1, Villeurbanne, France

Nigel Gilles Yoccoz
Department of Arctic and Marine Biology, The Arctic University of Norway, Tromsø, Norway

Foreword

Lemaître and Pavard explore a vibrant field in their book *The Biodemography of Ageing and Longevity*.

Since the dawn of human consciousness humanity has pondered the mysteries of life's duration, its inevitable decline and the philosophical and practical implications of ageing. Today, these questions are underscored by the societal, economic and health-related challenges they raise.

This book arrives as a timely resource, providing a rich, cross-disciplinary examination of ageing. It's a must-read for those interested in biodemography, which integrates biological, social and mathematical perspectives on ageing.

The authors offer a fresh appraisal of longevity research through evolutionary biology, illuminate the promises of this research for contemporary and future societal challenges and advocate for a multidisciplinary research philosophy.

From the theoretical underpinnings of ageing to practical measures of longevity, the book delves into evolutionary biology, demography and biogerontology. It highlights the diversity of ageing patterns across the tree of life and the importance of distinguishing between chronological and biological age.

Addressing sex differences in longevity and ageing, the authors extend the discussion to human populations, offering insights into the evolutionary roots of these differences and their implications for healthspan and lifespan.

The latter chapters pivot towards human ageing, exploring historical and modern factors that shape mortality, and discussing how societies might adapt to an increasing senior demographic through integrated approaches in labour, education, innovation and intergenerational solidarity.

In summary, *The Biodemography of Ageing and Longevity* stands as an important scholarly contribution to our understanding of life's twilight journey, enlightening readers on the multifaceted process of ageing and its societal implications.

Steve Horvath, inventor of the epigenetic clock and a principal investigator at Altos Labs

Acknowledgements

Editing this book has been a long but extremely pleasant journey. Along the way, we were able to count on unfailing support from Cambridge University Press. We are particularly grateful to Aleksandra Serocka and Dominic Lewis.

The structure and content of this book have been fuelled by our multiple exchanges with many students, colleagues and friends over the last few years. For these (sometimes very lively) discussions on the evolution of the ageing process, its causes and consequences, we are deeply indebted to Margaux Bieuville, Hugo Cayuela, Louise Cheynel, Alan Cohen, Fernando Colchero, Christophe Coste, François Criscuolo, Jean-Michel Gaillard, Mike Garratt, Eric Gilson, Steve Horvath, Edward Ivimey-Cook, Alexei Maklakov, Jessica Metcalf, Jacob Moorad, Dan Nussey, Daniel Promislow, Florentin Remot, Victor Ronget, Roberto Salguero-Gómez, Florence Solari, Tamás Székely and Morgane Tidière.

We are extremely grateful to all the researchers who devoted a substantial amount of their time to commenting on the book chapters and thus contributing to the quality of the book's scientific content. Thank you to Hippolyte d'Albis, Elisabetta Barbi, Annette Baudisch, Marie-Pier Bergeron Boucher, Barry Bogin, Anne Bronikowski, Rochelle Buffenstein, Kaare Christensen, Tim H. Clutton-Brock, François Criscuolo, Philip Dammann, Thomas B.L. Kirkwood, Barthelemy Kuate Defo, Jean-Michel Gaillard, Kim R. Hill, Owen Jones, Sang-Hyop Lee, Peter Lenárt, Audrey Maille, Alexei Maklakov, Jacob Moorad, Alexey A. Moskalev, Alyson van Raalte, Jean-Marie Robine, Richard G. Rogers, Gustavo De Santis, Alexander Scheuerlein, Rebecca Sear, Mauro Serafini, Thomas Spoorenberg and Gerald Wilkinson.

We are grateful to our academic institutions and current labs, both for the stimulating intellectual environment they offer us and for the freedom they give us to develop our independent research programmes: the French National Centre for Scientific Research (CNRS) and the Laboratory of Biometry and Evolutionary Biology at the University Lyon 1 for Jean-François; the Eco-anthropology unit, the National Museum of Natural History, Paris, and the CNRS for Samuel.

Diverse grants that have supported our research during these last years have also been very helpful for this book. We thus thank, for their financial support, the French National Research Agency (ANR-15-CE32-0002-01; ANR-20-CE02-0015; ANR-22-CE02-0021), the CNRS Ecology & Environment, and the Biometrics and Evolutionary Biology Laboratory for Jean-François; and the French National Research Agency (ANR-18-CE02-0011) for Samuel.

Finally, we thank all the authors, who come from so many different fields of research, from biology to the human sciences, from cells to societies, from theory to empirical studies and fieldwork. They, together, made it possible to create this book through their innovative and enthusiastic contributions, their curiosity and their sense of dialogue. They all collectively contribute to the field of biodemography of ageing.

1 The Eternal Youth of Ageing Research

Jean-François Lemaître and Samuel Pavard

1.1 Human Ageing at the Crossroads between North and South, Past and Future, Worry and Hope

Individuals are increasingly likely to live longer in contemporary societies. In Western societies, this is the result of about two centuries of (dis)continuous increase in life expectancy from about 40 years in the middle of the eighteenth century to approximatively 85 nowadays ([1] but see [2]). At the population level, this increase in life expectancy (associated with a decline in fertility) results in a growing proportion of seniors in these populations (i.e. population ageing). It also leads to the appearance of new age classes such as supercentenarians (i.e. individuals who reach 110 years of age or more).

Such population ageing poses crucial societal issues and challenges regarding health policies and the redistribution of wealth between generations [3]. But there is much more at play. Building a society in which a growing number of individuals have an equal chance of living a long and healthy life is, above all, a social and political challenge. This book is a project that legitimizes and makes the case for massive investment in many research areas in order to better understand the biological, social and environmental determinants of ageing, at both individual and population levels.

The tremendous social implications also explain why the question of whether or not there is a biological limit to human longevity arouses such intense and conflicting debate among scientists (see [4] and the multiple commentaries on this article for an illustrative example of the debate). Moreover, the debate around biological limits crystallizes a societal debate on the limits of human progress [5]. Since the second half of the twentieth century, societal and medical progress has allowed people to expect to live longer, but will this continue in the future? Will it be accompanied by a parallel increase in healthspan? And if not, what does this really tell us about the future of our societies?

At the international level, the population ageing of wealthy societies is also changing the relationships they maintain. For instance, in an increasing number of Northern countries, the ageing of the populations has become an important driver of labour migration from the South to the North. In developing societies, on the other hand, seniors are largely forgotten by public policies. For example, the United Nations 2030 Agenda for Sustainable Development promotes a reduction in infant and maternal mortality but includes only a few considerations on health, gender disparities and the survival of elderly people [6].[1] However, seniors are subject to many forms of health, socio-economic and environmental vulnerabilities in countries with little or no health-care system favourable to these age groups [7].

At a global scale, there is no doubt that the increase in the life expectancy of a growing number of individuals in developed countries has come at the cost of a large drain on the resources of the planet. Yet globalization, as well as environmental and climatic deterioration, generates new risks for older people, who are more vulnerable to environmental fluctuations, such as climatic changes or epidemics [8]. More rational, equal and sustainable development, as well as the fight against global warming and biodiversity loss, will become major issues in the aim to improve future life expectancy (or even maintain the current level).

But, by contrast, does favouring living long and healthy lives conflict with sustainable development? Fortunately not. First, because actual research on ageing opens new avenues towards late-onset disease prevention and health – for instance in terms of environmental conditions, such as lifestyle and diet, or in terms of socio-cultural practices, such as socialization through family ties and solidarity – that align with environmentally sustainable behaviours. Second, an increasing community of researchers argue that our improved understanding of the evolutionary roots of ageing and its variations between species and populations will lead to findings that allow us to modulate both the onset and rate of ageing in our species [9]. Indeed, a significant part of the recent increase in lifespan results from biomedical progress that allows individuals to survive – at various levels of health – formerly lethal diseases (such as cardiometabolic diseases, cancer or kidney diseases). For many researchers, however, the next step will be to extend lifespan by lengthening healthspan thanks to the development of treatments against ageing (also denoted the fifth epidemiological transition, see [10]). If proved true, this could lead to equal access to low-cost healthspan extension rather than extending life by costly treatments of chronic fatal age-related diseases. Finally, population ageing is the culmination of a demographic transition that, once completed, heralds population demographic stationarity or even decline. Although no biological link between fertility decrease and lifespan increase has been observed during demographic transitions, these issues are nevertheless socio-culturally negatively correlated. It is therefore likely that securing equal access to a long healthy life, within and between countries, is a part of desirable change of the worldwide demography, and will be a factor contributing to slowing global population growth (planned for the next decades [11]) and a decreasing drain on environmental resources.

Living healthier – and perhaps longer – lives thus remains a valid social and political objective of modern democracies, and, far from being in conflict with other

[1] Compared with United Nations Millennium Development Goals, the 2030 Agenda for Sustainable Development states, as its third goal, to 'ensure healthy lives and promote well-being for all at all ages' and, as a target, to 'reduce by one third premature mortality from non-communicable diseases through prevention and treatment and promote mental health and well-being'; these can be seen as mostly aiming to improve adult health. However, it is noticeable that policies dedicated to old age are not specifically mentioned as a target, in contrast to those concerning children and reproductive women.

contemporary issues, this objective embraces other crucial objectives, whether these involve reducing inequalities between humans or buffering the impacts of human societies on their environments.

1.2 Taking Advantage of the Multidimensional Variability of the Ageing Process

Based on the multiple challenges that ageing societies will meet in the near future (see Section 1.1), one may easily recognize the fundamental importance as well as the timely nature of studying ageing. Given the many facets of the ageing process (e.g. biological, social, mathematical), we believe that the flourishing field of biodemography offers a relevant framework for building integrative research programmes.

In their recent textbook, James Carey and Deborah Roach define biodemography as the 'science concerned with identifying a set of population principles' (see [12], p. 1) through an integrative approach bridging conceptual and methodological advances from life sciences and demography. Focused on ageing, the multidisciplinary nature of the biodemographic approach takes on its full meaning. For instance, upstream of the ageing process, the study of the biological, behavioural and environmental drivers of survival or reproductive performance through the life course is the scientific goal of many research areas (e.g. biogerontology, evolutionary ecology), and downstream, the demographic consequences of ageing at both population and individual levels modulate the expression of age-specific selective forces, leading to complex eco-evolutionary feedback.

Moreover, biodemography allows us, by its essence, to fully embrace the diversity of longevity and ageing trajectories offered by the natural world and thus to expand our horizons in terms of biological models [12]. It allows us to correlate this diversity with several biological, behavioural and environmental factors, putatively shaping ageing patterns, as well as understanding their evolution.

This is why, in the 2010s, we started to see a rise in the number of studies exploring the mechanisms of the ageing process in non-model organisms [13, 14]. The starting point of this increase is rooted in the influential papers or books that have clearly demonstrated that actuarial senescence (the increase in mortality risk with increasing age, also called survival ageing) is widespread across the tree of life, but also highly variable in shape between and within taxonomic groups [13, 15, 16]. For instance, some species display early and rapid increase in mortality risk, while others display intriguing non-continuous patterns of mortality increase with age. Some, even, are characterized by a low and relatively stable mortality risk over the life course [13, 17]. We are only scratching the surface in our understanding of the ecological and biological factors shaping this diversity of actuarial senescence patterns, but the current burst in the field of evolutionary biology of ageing will undoubtedly shed light on the evolutionary roots, as well as the interplay between the evolutionary and demographic consequences of actuarial senescence [18].

Importantly, this inter-specific diversity constitutes an untapped resource for biogerontologists aiming to identify cues for improving healthspan and lifespan, thus calling for integrative research projects. Indeed, even if some ageing mechanisms are shared across species (e.g. insulin/insulin-like signalling and target of rapamycin pathways [19]), the set of cellular mechanisms (or their relative importance) allowing some species to slow down their somatic deterioration and thus to delay and/or buffer the increase in mortality risk over the life course are partly taxa or species-specific [20, 21]. Moreover, in response to some specific sources of mortality, inherent to their Bauplan or ecosystem, some species have evolved sophisticated mechanisms allowing them to limit or even escape the risk of contracting some specific late-onset diseases (e.g. cardiovascular diseases or cancers [22, 23]). The natural world thus constitutes a field of possibilities that likely seize opportunities for the improvement of human healthspan. Yet identifying the most relevant species and candidate mechanisms is a complex task, which requires bridging methodological and theoretical advances from several scientific fields. This will be highlighted throughout the book's chapters.

A first challenge to achieving this is the thorough assessment of the diversity of the ageing process across space and time for a given species. Indeed, the variability in ageing patterns is not limited to the species level and, for a given species, populations often differ in both longevity and actuarial senescence patterns (e.g. [24]). This can be largely explained by some differences in environmental conditions, even if the relative contributions of the diverse ecological traits that can modulate ageing patterns (e.g. temperature, pathogen prevalence and availability of resources) remain to be quantified.

In this context, climate change will undoubtedly have profound consequences on population-specific ageing trajectories. In animals, it has been recently highlighted that global warming can lead lizard populations to extinction through an accelerated rate of telomere erosion [25]. Rapid shifts in ageing trajectories can thus be an accelerator of the biodiversity crisis [26]. Moreover, in the specific case of human populations, global warming can concomitantly hasten ageing trajectories by altering some key cellular and physiological processes underpinning the ageing process and also by directly impairing the health and living conditions of the elderly [27]. Therefore, there is a crucial need for theoretical studies that explore the possible consequences of climate change on population ageing, as well as empirical studies investigating how the multiple ecological traits characterizing a population can influence ageing on the cellular, physiological and demographic levels.

Studying the change in longevity and ageing trajectories through time can also offer relevant insights, especially when it comes to forecasting the future of ageing. Such investigations can be performed at an evolutionary time scale – even if assessing the longevity or ageing trajectories of extinct and fossil species is particularly difficult due to a lack of data (e.g. [28]) and rather rely on phylogenetic reconstructions (e.g. [29]) – or at much shorter time scale (e.g. centuries or decades). For the latter, historical demographic records of human populations across the world constitute invaluable resources that reveal how the modifications of our pathogenic, social and economic landscapes since the nineteenth century have shaped ageing trajectories.

Similar investigations in animals would be highly valuable as these could provide additional information on how environmental conditions can drive the ageing process. As many long-term field studies monitoring individuals from birth to death emerged in the 1960s and 1970s [30], such studies will be soon achievable in both short-lived and long-lived species.

Understanding and predicting population differences along longevity and ageing trajectories is mandatory for establishing public health policies. However, many insights from studies exploring the variability of the ageing process at the individual level can contribute to healthspan improvement, notably through personalized medicine. Within populations, individuals largely differ in terms of physiological conditions, health and longevity prospects. This variability largely depends on (epi) genetic, environmental factors (as well as their interactions) and luck. It is thus important to notice that the study of large cohorts through 'omic' approaches have started to provide key information on the epigenomic and genomic determinants of longevity and health in human populations (e.g. [31, 32]). Social factors (in a broad sense, the quantity and quality of the social bonds, the social level) can also largely contribute to individual differences in health and longevity prospects, as the influence of favourable social environment on hallmarks of ageing (sensu [33]) is increasingly documented (e.g. [34, 35]).

1.3 Overview of the Book's Contents

The present book has multiple objectives. It aims to provide a comprehensive reappraisal of current longevity and ageing research through evolutionary biology and evolutionary demography lenses, and to highlight the promise of this research for dealing with current and future challenges that our societies will face. These book chapters demonstrate that a multidisciplinary philosophy organized by research in biodemography is mandatory for the future of ageing research. This aspect is illustrated in Chapter 2, where Thomas B.L. Kirkwood provides a synthesis of the various theories that have been elaborated so far with regard to the evolution of ageing and demonstrates that theoretical frameworks merging evolutionary biology and biogerontology advances are now badly needed.

Chapters 3 and 4 then focus on the critical question of how to measure longevity (or lifespan). While the concept of 'longevity' is, by its essence, at the core of ageing research, its measurement remains far from trivial and the various metrics used to quantify the duration of life can lead to various (mis)interpretations. More specifically, Victor Ronget, Gilles Maurer and Sarah Cubaynes (Chapter 3) provide the reader with a thorough review of the various metrics that have been proposed to quantify the duration of life between and within species, highlight potential biases intimately linked to each metric and give recommendations for their interpretation. Next, Jean-Michel Gaillard and Nigel Gilles Yoccoz (Chapter 4) specifically focus on the concept of 'exceptional longevity', a term that is widely used in science and media but that lacks a statistically anchored definition. Here, the authors review old and recent approaches that can be used to identify species or individuals who display longevities outwith the range of expected values, and discuss the biological implications of each metric. Taken together, Chapters 3 and 4 offer novel insights on the use and misuse of longevity in evolutionary biology, demography and biogerontology.

In Chapter 5, Annette Baudisch goes beyond the concept of longevity and focuses on full age-specific trajectories. While the occurrence of demographic ageing in survival (i.e. actuarial senescence, the increase in mortality risk with age) across species has already been largely reviewed (e.g. [16]), the author of this chapter specifically focuses on species that show negligible (or an absence of) actuarial senescence. In this carefully organized piece of work, she reviews the current evidence for such an absence of survival ageing and emphasizes the questions raised by these patterns, from both theoretical and mechanistic points of view. Taken together, Chapters 4 and 5 provide new tools and insights for identifying – on the basis of demographic properties – promising model species for ageing research (e.g. species that have potentially evolved cellular mechanisms that enable them to buffer against the increase in mortality with age and to reach a striking longevity). Unfortunately, the current set of wild animal populations for which demographic data to estimate accurate ageing parameters and longevity metrics (i.e. age-specific estimation of mortality trough life) is available remain limited. Since longitudinal follow-up of known-age individuals cannot be reasonably conducted on all species on earth, the rise of demographic (but also physiological) data accumulated in zoo populations appears particularly relevant. In Chapter 6, Morgane Tidière, Johanna Staerk, Michael J. Adkesson, Dalia A. Conde and Fernando Colchero thus demonstrate how zoos and aquariums constitute untapped resources to study ageing at different levels of organization (species, populations, individuals) by implementing approaches combing demography, ecology and biogerontology.

After this set of chapters focused on the ageing process at the demographic level, Chapters 7 and 8 then focus on the genetic, cellular and physiological bases that could underlie the variability observed in longevity and actuarial senescence. These chapters tackle this salient problematic in ageing research through two complementary angles, by focusing predominantly on the inter-specific level (Chapter 7) and intra-specific level (Chapter 8). In the former, Jean-François Lemaître, Jean-Michel Gaillard, Samuel Pavard, François Criscuolo and Fabrice Bertile highlight future avenues of research in the comparative biology of ageing. More specifically, they argue that bridging methodological and conceptual advances in both biological sciences (in particular 'omic' approaches) and demography would provide new insights on the (epi)genetic and cellular mechanisms shaping the diversity of actuarial senescence patterns across the tree of life. This chapter is closely linked to Chapters 4 and 5 as it highlights relevant approaches to identify the mechanistic basis of the evolution of 'exceptional longevity', as well as delayed or slow actuarial senescence across species. In Chapter 8, Pat Monaghan and Jelle Boonekamp focus on biological age, a trait that is gaining increasing attention in both medical and ecological research, as it provides indirect insights on individuals' health and future survival prospects [36]. More specifically, the authors synthetize the various approaches that can be used to assess biological age in wild populations of animals and emphasize the importance of teasing apart chronological and biological age in evolutionary ecology of ageing studies.

Chapter 9 focuses on a long-standing topic that is attracting the attention of scientists from many different research fields: sex differences in longevity and ageing. In this chapter, Jean-François Lemaître, Jean-Michel Gaillard, Dominique Pontier, Hugo Cayuela, Cristina Vieira and Gabriel A. Marais first review the current support for the diverse evolutionary theories that have been proposed to explain the sex differences in longevity and actuarial senescence regularly documented in the animal kingdom. They then extent the current debates on the origin of sex differences on mortality patterns to the sex differences in reproductive and physiological ageing, two aspects that now need to be tackled using a multidisciplinary approach. Given that differences in ageing, healthspan and longevity between sexes are pervasive in human populations, this chapter also embraces the evolutionary roots of sex differences in human disease aetiology (e.g. infectious diseases). It thus sets the scene for the subsequent chapters, which will predominantly focus on human ageing.

Chapters 10–13 indeed turn more particularly to human ageing and longevity and take the reader on a journey through time and space to better understand the factors that have shaped adult mortality throughout evolutionary time, more recently from the Neolithic to historical times, and which are currently at work in societies across the world. More precisely, in Chapter 10, Samuel Pavard and Michael D. Gurven review and organize the literature on the joint evolution between the human life cycle and human cognitive and social capabilities, and discuss the biocultural mechanisms that have decoupled the extended longevity and the short reproductive period in our species, making humans an outlier on most ecological continuums. Michael D. Gurven then extends this discussion (Chapter 11) by a comprehensive review of the scarce and precious data on mortality by age and cause in human forager populations. By doing so, he deconstructs many preconceived ideas about adult mortality and the dynamics of past populations, as well as emphasizing the current challenges faced by these populations, who are sometimes brutally confronted with unchosen environmental and cultural changes. In the pivotal chapter, Chapter 12, Nadine Ouellette and Julie Choquette guide the reader through the changes in longevity resulting from the transition from hunting and gathering to farming, and through more recent demographic and epidemiological transitions. They then introduce the crucial notions of compression of mortality and rectangularization of the human survival curve, the subjacent trade-offs between healthspan and lifespan, and evidence the emergence of a new statistically relevant age class: the oldest olds. The authors thus pave the way to Chapters 14 and 15, which aim at better describing the mortality of these new age classes, and Chapters 16 and 17, which aim at elucidating the factors that enabled these particular individuals to live so long. They conclude by discussing the current disparities on levels of demographic transitions in the Global South and the epidemiological mechanisms behind these. Running with this idea, Géraldine Duthé, Lucie Vanhoutte, Soumaila Ouedraogo and Gilles Pison

(Chapter 13) then discuss how such demographic and epidemiologic transitions theory has to be revised to encompass the current challenges that face countries of the Global South. In many of these countries, adult mortality is indeed the product of the coexistence of communicable and non-communicable or chronic diseases across a complex spatial, cultural and environmental landscape, shaped by urbanization, nutrition and socio-economic conditions.

Chapters 14 and 15 will then turn to the future by questioning – through an epistemological and statistical discussion – the existence and the reachability of a limit to human longevity. Carlo Giovanni Camarda and Jean-Marie Robine (Chapter 14) provide the reader with a comprehensive review of the history of the ideas and the current debate on the limit of longevity, and show how these debates were anchored, first, in natural history and scientific methods and, second, in the development of statistics and data. This discussion is continued in Chapter 15, by Linh Hoang Khanh Dang, Nadine Ouellette, Carlo Giovanni Camarda and France Meslé, who provide a history of the statistical models proposed for exploring mortality trajectories at old ages and these models' respective ability in detecting mortality plateaus and hidden heterogeneity in individuals' frailty – both key to addressing the question of the limit of longevity.

The two following chapters aim at deciphering why some individuals live long lives and why others do not; they thus aim to provide a better understanding of the process behind ageing in our species and also to identify potential levers that could enable future progress in improving lifespan and healthspan. Larissa Smulders and Joris Deelen, in Chapter 16, take a pedagogical approach to explaining the methods used in such phenotypic and genetic comparison, employing data on long-lived individuals and families. They then shed light on how such studies have allowed the identification of key features of healthy long life, such as having a favourable immune-metabolic profile, but have also unravelled deeper questions, such as the absence of clear effects of protective genetic variants (beyond a few ones) and susceptibility alleles to late-onset diseases in modulating survival to very old age. Finally, the authors identify new avenues using 'omic' data on such specific data to investigate ageing from a system biology perspective, as proposed Chapter 7 through comparative biology lenses. By contrast, Gianni Pes and Michel Poulain (Chapter 17) are interested in the overall biological, environmental, nutritional and cultural context that allows certain populations around the world (in the so-called blue zones, BZ) to exhibit lower adult mortality and a higher number of survivors beyond 100 years of age. After a discussion on genetic and hormonal potential explanations, they provide a strong argument for the role played by one's way of life on long healthy ageing, from physical activity and nutrition to social ties and purpose.

Finally, how can modern societies cope with this growing number of seniors, that is, population ageing? This is the question posed by Miguel Sánchez-Romero and Alexia Prskawetz (Chapter 18) in the final chapter of this book. To answer it, the authors apply demo-economic models to country-level data to analyse the impact of increased longevity on the sustainability of intergenerational solidarity. Without omitting the tensions that such increased longevity causes to the economy, the authors also show that when demography, labour market, education, innovation and solidarity are taken into account, modern democracies have all the cards in hand to face these changes. In consequence, increases in healthspan and lifespan may remain indicators of global societies' progress in the future, as much as they have been in the past.

1.4 Conclusion

From the oldest human stories, such as the epic of Gilgamesh (written around 2100–1200 BC), to the works of ancient Greek philosophers such as Aristotle and down to the present day, questions about the length of life, its limits and the deterioration of the body throughout life have been at the heart of philosophy, the arts and the life sciences. Today these issues are more topical than ever, given the social, economic and public health challenges to which they are intimately linked, and require innovative and dynamic scientific research at the crossroads of many scientific disciplines. In recent years, several books have brilliantly summarized major advances in the study of longevity or senescence, whether from the perspective of evolutionary ecology (e.g. [16]) or medical sciences (e.g. [37]). Here, our book builds on the multidisciplinary nature of biodemography to concomitantly summarize the large amount of knowledge accumulated over the past few years in ageing research, as well as to stimulate the development of integrative research projects on ageing. We hope that this book will be an inspiring resource for established academics and, perhaps more importantly, for students, who will be at the forefront of the coming profound changes linked to ageing populations.

References

1. Oeppen, J., Vaupel, J.W. 2002. Broken limits to life expectancy. *Science* **296**, 1029–1031 (doi:10.1126/science.1069675).
2. Vallin, J., Meslé, F. 2009. The segmented trend line of highest life expectancies. *Popul. Dev. Rev.* **35**, 159–187 (doi:10.1111/j.1728-4457.2009.00264.x).
3. UNECE (United Nations Economic Commission for Europe). 2006. What UNECE does for you: challenges and opportunities of population ageing. https://unece.org/info/Population/pub/2775.
4. Dong, X., Milholland, B., Vijg, J. 2016. Evidence for a limit to human lifespan. *Nature* **538**, 257–259.
5. Marck, A., Antero, J., Berthelot, G., Saulière, G., Jancovici, J.-M., Masson-Delmotte, V., Boeuf, G., Spedding, M., Le Bourg, É., Toussaint, J.-F. 2017. Are we reaching the limits of *Homo sapiens*? *Front. Physiol.* **8**, 812.
6. United Nations. 2023. Transforming our world: the 2030 Agenda for Sustainable Development. Sustainable Development Knowledge Platform. https://sustainabledevelopment.un.org/post2015/transformingourworld/publication.
7. Alejandria-Ganzales, M.C.P., Ghosh, S., Sacco, N. 2019. *Aging in the Global South: Challenges and Opportunities*. Lexington Books.

8. WHO (World Health Organization). 2022. The decade in a climate-changing world. Decade of Healthy Ageing Connection Series No. 3. www.who.int/publications/m/item/decade-of-healthy-ageing-connection-series-no3.
9. Olshansky, S.J. 2021. Aging like Struldbruggs, Dorian Gray or Peter Pan. *Nat. Aging* **1**, 576–578.
10. Horiuchi, S. 1999. Epidemiological transitions in human history. *Health Mortal. Issues Glob. Concern*, 54–71. (Reprinted in Jonathan Watson, Pavel Ovseiko, eds. 2005. *Health Care Systems: Major Themes in Health and Social Welfare. Volume 4. Rethinking Health Care Systems*, pp. 109–135. Routledge.)
11. United Nations, Department of Economic and Social Affairs, Population Division. 2022. *World Population Prospects*. Online Edition. https://population.un.org/wpp/.
12. Carey, J.R., Roach, D.A. 2020. *Biodemography: An Introduction to Concepts and Methods*. Princeton University Press.
13. Jones, O.R., Scheuerlein, A., Salguero-Gómez, R., Camarda, C.G., Schaible, R., Casper, B.B., Dahlgren, J.P., Ehrlén, J., García, M.B., Menges, E.S. 2014. Diversity of ageing across the tree of life. *Nature* **505**, 169.
14. Lu, A.T. et al. 2023. Universal DNA methylation age across mammalian tissues. *Nat. Aging* **3**, 1144–1166 (doi:10.1038/s43587-023-00462-6).
15. Nussey, D.H., Froy, H., Lemaître, J.-F., Gaillard, J.-M., Austad, S.N. 2013. Senescence in natural populations of animals: widespread evidence and its implications for bio-gerontology. *Ageing Res. Rev.* **12**, 214–225.
16. Shefferson, R.P., Jones, O.R., Salguero-Gómez, R. 2017. *The Evolution of Senescence in the Tree of Life*. Cambridge University Press.
17. Reinke, B.A., Cayuela, H., Janzen, F.J., Lemaître, J.-F., Gaillard, J.-M., Lawing, A.M., Iverson, J.B., Christiansen, D.G., Martínez-Solano, I., Sánchez-Montes, G. 2022. Diverse aging rates in ectothermic tetrapods provide insights for the evolution of aging and longevity. *Science* **376**, 1459–1466.
18. Gaillard, J.-M., Lemaître, J.-F. 2020. An integrative view of senescence in nature. *Funct. Ecol.* **34**, 4–16.
19. Flatt, T., Partridge, L. 2018. Horizons in the evolution of aging. *BMC Biol.* **16**, 93 (doi:10.1186/s12915-018-0562-z).
20. Hoekstra, L.A., Schwartz, T.S., Sparkman, A.M., Miller, D.A.W., Bronikowski, A.M. 2020. The untapped potential of reptile biodiversity for understanding how and why animals age. *Funct. Ecol.* **34**, 38–54 (doi:10.1111/1365-2435.13450).
21. Voituron, Y., Guillaume, O., Dumet, A., Zahn, S., Criscuolo, F. 2023. Temperature-independent telomere lengthening with age in the long-lived human fish (*Proteus anguinus*). *Proc. R. Soc. B* **290**, 20230503.
22. Natterson-Horowitz, B., Aktipis, A., Fox, M., Gluckman, P.D., Low, F.M., Mace, R., Read, A., Turner, P.E., Blumstein, D.T. 2023. The future of evolutionary medicine: sparking innovation in biomedicine and public health. *Front. Sci.* **1**, 997136.
23. Gorbunova, V., Seluanov, A., Zhang, Z., Gladyshev, V.N., Vijg, J. 2014. Comparative genetics of longevity and cancer: insights from long-lived rodents. *Nat. Rev. Genet.* **15**, 531.
24. Cayuela, H., Lemaître, J.-F., Muths, E., McCaffery, R.M., Frétey, T., Le Garff, B., Schmidt, B.R., Grossenbacher, K., Lenzi, O., Hossack, B.R. 2021. Thermal conditions predict intraspecific variation in senescence rate in frogs and toads. *Proc. Natl. Acad. Sci.* **118**, e2112235118.

25. Dupoué, A. et al. 2022. Lizards from warm and declining populations are born with extremely short telomeres. *Proc. Natl. Acad. Sci.* **119**, e2201371119.
26. Robert, A., Chantepie, S., Pavard, S., Sarrazin, F., Teplitsky, C. 2015. Actuarial senescence can increase the risk of extinction of mammal populations. *Ecol. Appl.* **25**, 116–124.
27. Leyva, E.W.A., Beaman, A., Davidson, P.M. 2017. Health impact of climate change in older people: an integrative review and implications for nursing. *J. Nurs. Scholarsh.* **49**, 670–678 (doi:10.1111/jnu.12346).
28. Weon, B.M. 2016. Tyrannosaurs as long-lived species. *Sci. Rep.* **6**, 19554.
29. Romiguier, J., Ranwez, V., Douzery, E.J., Galtier, N. 2013. Genomic evidence for large, long-lived ancestors to placental mammals. *Mol. Biol. Evol.* **30**, 5–13.
30. Clutton-Brock, T., Sheldon, B.C. 2010. Individuals and populations: the role of long-term, individual-based studies of animals in ecology and evolutionary biology. *Trends Ecol. Evol.* **25**, 562–573.
31. Deelen, J., Evans, D.S., Arking, D.E., Tesi, N., Nygaard, M., Liu, X., Wojczynski, M.K., Biggs, M.L., van Der Spek, A., Atzmon, G. 2019. A meta-analysis of genome-wide association studies identifies multiple longevity genes. *Nat. Commun.* **10**, 3669.
32. Hillary, R.F., Stevenson, A.J., Cox, S.R., McCartney, D.L., Harris, S.E., Seeboth, A., Higham, J., Sproul, D., Taylor, A.M., Redmond, P. 2021. An epigenetic predictor of death captures multi-modal measures of brain health. *Mol. Psychiatry* **26**, 3806–3816.
33. López-Otín, C., Blasco, M.A., Partridge, L., Serrano, M., Kroemer, G. 2023. Hallmarks of aging: an expanding universe. *Cell* **186**, 243–278 (doi:10.1016/j.cell.2022.11.001).
34. Epel, E.S., Blackburn, E.H., Lin, J., Dhabhar, F.S., Adler, N.E., Morrow, J.D., Cawthon, R.M. 2004. Accelerated telomere shortening in response to life stress. *Proc. Natl. Acad. Sci. USA* **101**, 17312–17315.
35. Fiorito, G., Polidoro, S., Dugué, P.-A., Kivimaki, M., Ponzi, E., Matullo, G., Guarrera, S., Assumma, M.B., Georgiadis, P., Kyrtopoulos, S.A. 2017. Social adversity and epigenetic aging: a multi-cohort study on socioeconomic differences in peripheral blood DNA methylation. *Sci. Rep.* **7**, 16266.
36. Ferrucci, L., Gonzalez-Freire, M., Fabbri, E., Simonsick, E., Tanaka, T., Moore, Z., Salimi, S., Sierra, F., de Cabo, R. 2020. Measuring biological aging in humans: a quest. *Aging Cell* **19**, e13080 (doi:10.1111/acel.13080).
37. Musi, N., Hornsby, P. 2021. *Handbook of the Biology of Aging*. Academic Press.

2 Theories of Ageing across Ages

Thomas B.L. Kirkwood

2.1 Introduction

What ageing is, and why it occurs, are questions that have long engaged human interest. The anthropological literature contains many accounts of myths and beliefs about the origins of ageing and death. Scientific study of ageing is generally thought to have begun with Aristotle who in *De Longitudine et Brevitae Vitae* (*On Length and Shortness of Life*), written 2 300 years ago, observed: ‘It is not clear whether in animals or plants universally it is a single or diverse cause that makes some to be long-lived, others short-lived’ (see [1]). Aristotle sought to connect his extensive observation of longevity in the natural world, where he observed patterns consistent with present-day life history research (such as relationships between body size, gestation period and fecundity) with his theory of elements, namely, that the body consists of the hot and the cold, the dry and the wet. He concluded that old age, which he contrasted with illness and disease, results in the death of the organism by the gradual exhaustion of ‘vital heat’, particularly in relation to the heart and lungs.

Many centuries would elapse before major scientific enquiry into the nature and causes of ageing would occur again. In 1881, a lecture by the distinguished naturalist August Weismann to the Association of German Naturalists on the subject *Über die Dauer des Lebens* (*The Duration of Life*) began the first substantive attempt to explain ageing within the joint framework of the modern cell theory of life and the then fairly recent theory of evolution by natural selection. Central to Weismann’s thinking was the fundamental distinction he drew between *germ-line*, the reproductive lineage of cells through which life is transmitted to the next generation, and *soma*, which constitutes the rest of the cells that make up the individual organism. Germ-line must be at least capable of immortality, since otherwise life would necessarily die out. Soma has no such requirement, and Weismann developed – both in his lecture and in a series of subsequent essays – his thoughts about what causes the soma to age and die [2, 3]. The strands of Weismann’s thinking were complicated, undergoing significant modification, not always identified as such, in his successive works. This led a contemporary commentator to complain that to follow his lines of thought was ‘a matter of no small difficulty, inasmuch as the argument has to be traced through a number of essays’ and that ‘his statements of opinion are so fluctuating that it is difficult to determine what his opinion actually is’ [4]. Nevertheless, there can be no doubt about the richness of Weismann’s endeavour to develop an understanding of ageing [5], although a number of factors conspired to cause his works soon to be overlooked. Chief among these was the fact that a central proposal – namely, that the physiological decline, which occurs in senescent animals, was due to limitation in the proliferative potential of somatic cells – appeared soon to be invalidated by the empirical claim that somatic chicken cells could be grown indefinitely outside the donor organism [6]. By the time it was shown, half a decade later and against considerable resistance based on the widespread but mistaken acceptance of immortality of somatic cells in culture, that somatic cells do in fact undergo replicative senescence [7], Weismann's theoretical prediction was largely forgotten.

The early decades of the twentieth century saw a variety of ideas about ageing being explored. These included suggestions: (1) that ageing was caused by noxious effects of gut microbiota [8]; (2) that the rate of ageing was strongly connected with metabolism (the 'rate of living' theory) [9]; and (3) that ageing and mortality were closely linked with determinacy of growth, such that fish growing indefinitely through life exhibited negligible ageing [10]. However, it was not until the 1950s and 1960s that significant development of theories of ageing began to occur. The subsequent proliferation of theories was rapid, so that by 1990 it could be claimed in a review titled 'An Attempt at a Rational Classification of Theories of Ageing' that there were then more than 300 theories of ageing, with the number continuing to grow [11]. In the light of the very substantial increase in both the volume and diversity of research on the biology of ageing that has occurred since 1990, it would be possible in principle to expand this classification by cataloguing the many ideas about ageing that are currently in play. Such a project might have its uses, but it is not what this chapter aims to do. Indubitably, there is a need for rationality in thinking about theories of ageing and there is value in classification. However, whereas the 1990 classification was based on collecting the diverse array of so-called theories that populated the pre-1990 biogerontological literature and organizing them into categories, the aim here is to examine the conceptual principles that underpin the theoretical framework on ageing into which fresh data can be directed and which can thereby be challenged and modified as required.

It should be emphasized that this chapter has been written to provide access to development of theories of ageing, examining earlier ideas that maybe less familiar to present-day researchers but which are nevertheless important as conceptual background. As the field continues to grow, many excellent reviews provide access to the current state of the art. To capture the full richness of these reviews would not only have been beyond the scope of such a chapter but would also quickly become dated; only a few are cited here. The organization of the chapter is, first, to consider how the concept of ageing is understood; second, to look at ideas about why ageing occurs; third, to address mechanisms; lastly, to make some observations about the need to explore how the network of theories can be better integrated and causality predicted and tested.

2.2 Theories of Ageing: Basics

We easily understand, in general terms, what we mean by the words 'theory of ageing', but the term is broad and it will help to be as precise about meaning as possible. Both 'theory' and 'ageing' are concepts of some complexity.

As regards 'theory', an essential feature of a genuine scientific theory is that its predictions should be amenable to test. Karl Popper introduced the concept of a 'basic statement' of a theoretical proposition, which he defined as a premise that could in principle be falsified by empirical observation [12]. While Popper's emphasis on the logic of falsifiability has since been downplayed by others, notably through the impact of the work of Thomas Kuhn [13], it is widely held that a concept that does not lead to specific, testable predictions is at best of limited value. This requires that any theory should be clear about exactly what it proposes and to which systems it applies.

As regards 'ageing', the term has a breadth of meaning that also requires careful consideration. In his authoritative compendium *Longevity, Senescence and the Genome*, Caleb Finch quoted Peter Medawar as stating that 'ageing' is used to describe virtually all time-dependent changes to which biological entities from molecules to ecosystems are subject, though the mechanisms and consequences for function may be vastly different [14, 15]. Hence, Finch concluded: 'Because many age-related changes in adult organisms have little or no adverse effect on vitality or lifespan, I avoid the word ageing in this book' [14, p. 5]. The problem with simply avoiding the term 'ageing' altogether, however, is that the various shades of meaning associated with it all have some validity, albeit within different spheres. The approach that will be taken here is therefore to try to be explicit about which shade of meaning is being used within the current context.

The defining feature of ageing, as most commonly accepted, is its increasing impact on the age-specific death rate in a population of individuals living under conditions where extrinsic causes of mortality – being generally those imposed by the environment (such as cold, famine, infection, predation, trauma) – are absent or significantly diminished. Mortality rates in humans show an approximately exponential rise with age, noted first by Gompertz [16]. Similar patterns were observed in other mammals [17], as well as in a variety of other species ([18], see also Chapter 7). This led to suggestions that the Gompertz model might be regarded almost as a 'law' of mortality; however, more extensive comparative studies reveal considerable diversity in patterns of age-related mortality across the tree of life ([14, 19], see also Chapters 5 and 7). Ultimately, the diversity of life history patterns provides a challenge, to which theories of ageing will need to rise. In the first instance, however, it has been found useful to define ageing as being present when there is a progressive, generalized impairment of function, resulting in a loss of adaptive response to stress and an increasing probability of death [20]. This contrasts with the scenario where mortality rates remain constant with age, as appears to be the case for *Hydra* [21], which constitutes the non-ageing condition. Age-related decline in fertility is quite often included also as an indicator of ageing (e.g. [22]), although declining fertility might be a consequential feature of more generalized impairment.

2.3 Theories on Why Ageing Occurs

Before examining suggestions for why ageing occurs at all, we briefly consider whether such explanation is even necessary. Could it be that deterioration is somehow inevitable? In support of this, it might be argued that the second law of thermodynamics, which mandates an irreversible increase in entropy, is sufficient. However, the second law applies to closed systems, whereas living organisms are open. They take in energy that might be used to counteract the inevitability of simply sliding into entropic meltdown. Furthermore, as Weismann observed, germ-line is immortal. This falsifies any claim for inescapability of functional collapse, as does any instance of non-ageing species, such as *Hydra*.

The first explicit theory about why ageing occurs was the suggestion by Weismann that: 'Individuals are injured by the operation of external forces and for this reason alone it is necessary that new and perfect individuals should commonly arise and take their place, and this necessity would remain even if the individuals possessed the power of living eternally' [2, p. 24]. The point was elaborated with the statement that 'Worn out individuals are not only valueless to the species, but they are even harmful, for they take the place of those that are sound' [2, p. 24]. The suggestion is undermined by unfortunate circularity in assuming what it seeks to explain – namely, that deterioration occurs with age – but it does encapsulate for the first time the idea that an explanation for why ageing occurs may be discovered in the actions of natural selection.

The idea that, despite its obvious drawbacks, ageing somehow confers direct adaptive benefits underpins the idea that it is programmed in much the same way as development. The existence of programmed ageing often seems to be implicit in thinking about the topic, its attraction being that it conforms to the way many people most easily think about evolution, namely that new traits are selected to adapt the genotype in ways that are evidently fitter for survival. However, despite commanding a considerable measure of popular support, the idea of programmed ageing faces significant obstacles [23]. First, there was long held to be scant evidence that ageing is a major contributor to mortality in the wild, for the simple reason that the majority of individuals die before senescence (the process and manifestations of ageing) becomes strongly apparent (see e.g. [24, 25]). This is not to say that senescence in the wild is not observed. Comprehensive analysis of field data from a wide range of species indicates detectable senescence in many of them [26, 27]. Whether the numbers reaching the age of senescence are large enough to open the door to potential evolution of programmed ageing is open to question [28], but there is a second obstacle that becomes only more difficult to surmount as ageing in the wild becomes more frequent. This is that ageing presents a direct fitness cost through its impact not only on survivorship but also generally on fecundity. The case for programmed ageing – namely, that it might either help to prevent overcrowding and thereby lessen the risk of severe depletion of resources (e.g. [29]), or that it might assist adaptation to a changing environment by accelerating the turnover of generations [30] – rests heavily on an assumption of group selection [31]. If ageing had evolved, for example, as a programmed means to regulate population size, then a mutant in which the programme was inactivated would enjoy a selective advantage, and the mutant genotype would spread. Notwithstanding the difficulties facing theories of programmed ageing, there have in the last decade appeared a number of publications that have sought to substantiate the case for programmed ageing, based on arguments such as enhanced evolvability or actions of group selection in spatially structured environments. A review of these claims, and of the mathematical models on which they are based, has shown, however, that none of them withstands close scrutiny [28].

We will return to ideas about programmed ageing later in this section but for now we turn to a group of theories that share in the recognition that, since ageing occurs late in the life history, its evolution most likely results from more indirect effects of natural selection. An important feature of any life cycle in which there is repeated opportunity for reproduction (iteroparity) is that the force of natural selection weakens with increasing age [15, 32, 33]. This is because selection operates through differential effects of genes on reproductive success, and so its force – that is, its ability to discriminate between alternative genotypes – declines in proportion to the decline in the remaining fraction of the organism's total lifetime expectation of reproduction [34, 35]. If a gene effect is manifest early in life, it will be able to influence the overall fitness of individuals bearing that gene better than if it expressed relatively late, when many will have died already. It should be noted that this is true whether or not the species exhibits ageing, since even if individuals do not age they are nevertheless subject to environmental causes of mortality. Thus, the principle is relevant in seeking to explain the evolutionary origin of ageing.

Recognition of how the force of selection attenuates with age was at the core of a seminal proposition by Peter Medawar that has come to be known as the mutation accumulation theory of ageing [15]. If a new mutation arises that affects the age of expression of a gene's effects, and if the mutation is favourable either to survival or fecundity, then, other things being equal, natural selection can be expected to favour bringing forward its time of expression to ever earlier ages, so that greater advantage can be gained. Conversely, if the mutation is deleterious, then postponing its expression to a later age will lessen its overall negative impact. If expression is postponed for long enough, so that in the wild environment a great majority of individuals would have died already, it is effectively beyond the reach of any further opposing selection. Medawar thus suggested there might accumulate a miscellaneous collection of late-acting, deleterious genes [15]. In the wild these would have little or no consequence, but in a protected environment, where individuals survived longer, their collective ill effects would become manifest as ageing. The mutation accumulation theory remains of considerable interest, and more recent work has clarified not only how selection might act upon late-acting deleterious genes and their resulting effects on mortality [36, 37], but also their influence on human ageing [38].

A pivotal step in explaining why ageing occurs was the proposal by George C. Williams that ageing is an indirect effect of evolution, being a by-product of selection for genes that in themselves are beneficial, but which come at a cost [33]. Williams' theory used a premise similar to that of the mutation accumulation theory except for one important difference. The genes in question were assumed to be pleiotropic, the same genes having both beneficial effects early in life but bad effects late. In this scenario, selection could be expected to favour retention of the early benefits, even at the cost of their potential for harm. Because of the decline in the force of natural selection with age, even small early benefits could outweigh seriously negative side-effects, provided the bad effects occurred late enough that their impacts of fitness would be minimal. The theory has come to be known as the antagonistic pleiotropy theory of ageing and its compelling logic commands widespread acceptance [39], although it is sometimes overlooked that it is specifically based on the idea of particular genes having the necessary qualities. Williams used the hypothetical case of a gene that regulates calcium metabolism to give substance to the idea, the gene promoting bone development in early life but causing calcification of arteries later [33]. This example illustrates how there might be scope to nullify the later-life deleterious effects if the gene in question were simply to be silenced. However, the decline in the force of selection might not deliver sufficient pressure for this to occur. The antagonistic pleiotropy theory continues to attract considerable attention and there is growing evidence for its importance in the biology of ageing [40, 41].

While the mutation accumulation and antagonistic pleiotropy theories are primarily genetic, based on how selection can be expected to act on genes with age-specific properties, a more physiological explanation of why ageing occurs was offered by asking how an organism can be expected to optimize its use of metabolic resources to promote fitness [42, 43]. This is the disposable soma theory of ageing, whose origins lay in considering mechanisms of cellular ageing and asking why somatic cells experience progressive deterioration [44]. The disposable soma theory notes that it is important for the organism to invest sufficiently in the energetic costs of somatic maintenance to ensure that the body does not fall apart too soon, but that it would not be optimal to invest in maintaining the body well enough for it to have the potential to last indefinitely, since most individuals in wild environments die relatively young from extrinsic hazards. The significance of this is that it leads to an explanation not only for why ageing occurs but also for how it is caused, the primary mechanisms of ageing being predicted to involve the accumulation of molecular and cellular defects. Damage accumulation is thereby predicted to occur at rates that are held in check by selection for adequate maintenance and repair via what have been termed 'longevity assurance mechanisms', thereby reflecting the ecology and biodemographics of survival; hence, it is a simple extension of the idea to explain the evolution of longevity.

It will be clear already that that the evolutionary theories just considered (mutation accumulation, antagonistic pleiotropy, disposable soma) are complementary, not mutually exclusive. Furthermore, there is rich scope for further developments. One important challenge is to better understand the connections between evolutionary theory and life history trade-offs, such as those between lifespan, reproduction and growth. Antagonistic pleiotropy and disposable soma theory are commonly viewed chiefly as proposing that trade-offs are the key force structuring much life history variation and diversity of ageing rates (for review see [45]). One factor not yet sufficiently considered is the possibility that simple energetic strategies to reduce the environmental risk of dying might have important and very general consequences [46]. Ecological factors mean that energy acquisition is unavoidably associated with mortality risk. The need to acquire energy is dictated by the organism's physiology. Because somatic maintenance accounts for a substantial part of the lifetime need of acquired energy, restraining allocation to maintenance from early on might reduce mortality risk but it incurs a penalty in terms of increasing somatic damage. This suggests that allocation to somatic maintenance is primarily tuned to expected lifespan by stabilizing selection and is not necessarily traded against reproductive effort, growth or other traits, a conclusion that is relevant across a very wide variety of life histories.

Earlier in this chapter a focus was made on considering ageing within the context of a life cycle in which there is repeated opportunity for reproduction. This was because it is in such cases that the enigma of ageing presents itself most clearly. If there is repeated opportunity for reproduction, that is, iteroparity, why do these opportunities not continue without limit? However, it is now time to consider cases of semelparity, such as is seen in Pacific salmon, *Oncorhynchus nerka*, where there is only one opportunity for reproduction, following which death occurs rather rapidly. Semelparity is sometimes offered as evidence for the existence of programmed ageing, but this is a mistake. The post-reproductive death of a semelparous organism is the consequence of a life history strategy wherein an extended period of growth is followed by endocrine-mediated transition to a phase of intense reproductive effort, during which the individual is entirely focused on maximizing fecundity during the one and only opportunity to breed, even if this quickly results in its demise. What is programmed is the exclusive focus on the success of the organism's 'big bang' reproduction. Post-reproductive death is incidental.

One further idea concerning evolution of ageing is the suggestion that there may be developmental programmes that continue to run throughout life, causing disruptive effects at higher ages ([47, 48]; an early precursor to such thinking may be found in [49]). Such mechanisms might appear to give rise to a form of programmed ageing, but actually they would reflect simply the absence of sufficient force of selection to shut them down. Thus, they are consistent with, and complementary to, the framework of evolutionary ideas considered earlier. However, while it is certainly true that developmental processes can have important consequences for ageing, lending some credence to developmental programme theories, they do not easily account for the pronounced inter- and intra-individual variability that are commonly seen in outcomes of ageing [50].

As a coda to examining theories about why ageing occurs, we conclude this section by briefly inverting the question of why soma is mortal and asking how germ-line secures its immortality. Theories on the immortality of germ-line began with Weismann [2], and were reviewed also by Medvedev [11]. Of course, the immortality of the germ-line does not require that individual germ cells live forever – they can and do die. However, the lineage continues, for which three theoretical explanations are possible and are probably all required: better maintenance than in somatic cells [e.g. 51, 52, 53], stringent cell selection in the germ cell lineage, and selection against defective progeny.

2.4 Theories on Mechanisms of Ageing

The great majority of the theories of ageing considered in 1990 classification cited earlier addressed molecular, cellular or physiological causes of ageing [11]. These theories mostly arose during or after the birth of modern molecular biology in the 1950s and 1960s. Although the science of ageing, as we now know it, was then in its infancy, this was a period of intense interest in the possibility of understanding the causes of ageing. For example, the Ciba Foundation in London convened a series of five Colloquia on Ageing between 1955 and 1959 that included many leading scientists of the time. The contents of the Colloquia – each resulted in a volume published by J. & A. Churchill, London – contains material of historical interest, but there is little that has much relevance today. However, three mechanistic theories of that time remain highly significant, albeit in greatly updated form.

In 1956, Harman proposed what has proved to be one of the most influential ideas in biogerontology, the free radical theory of ageing [54]. The idea originated from emerging evidence that irradiation, a hugely topical focus of interest at that time, contributed to mutation, cancer and ageing. Harman's research was performed under the auspices of the US Atomic Energy Commission. After observing that one mechanism by which irradiation affected cells was through the liberation of the hydroxyl (OH) and superoxide (HO_2) radicals, Harman noted that even in the absence of irradiation, OH and HO_2 radicals were generated from the interaction of the respiratory enzymes involved in the direct utilization of molecular oxygen. While highly reactive radicals such as OH would be expected mostly to react near where they were produced, Harman foresaw that they could also be expected to react with other cellular constituents, including nucleoproteins and nucleic acids. Thus, the functional efficiency of cells would be impaired. In addition, there would be a slow oxidation of the connective tissue as well as a contribution to somatic mutations and cancer.

A very extensive literature has examined the free radical theory of ageing from many perspectives, showing: (1) that oxidative damage of the kinds predicted by the theory does increase with age; (2) that long-lived mutants of animal models, including worms, flies and mice, generally display increased resistance to oxidative damage; and (3) that transgenic animals created to produce enhanced protection against free radicals show an extension of lifespan (for review, see [55]). However, later results revealed a more complicated picture increasingly at variance with the initial predictions of the free radical theory, leading to a spate of articles suggesting that the theory might be false (e.g. [56, 57]). The reality is, however, that the challenges to the free radical theory of ageing signal not the death of the theory but the need to examine its predictions more carefully [58, 59]. The complexities of free radical generation within cells, the molecular defences against them, the spatial and localization aspects of free radical generation and removal within the cell, and the consequences of radical-induced damage all need to be better understood. Lastly, and this is a point to which we shall return in Section 2.5, it is now clear that any contribution to ageing from free radical damage is just part of a broader spectrum of molecular and cellular damage contributing to age-related declines. It may well be important, but it is not the only player in the game.

Closely following the free radical theory, in terms of chronology, another radiation-related theory was the suggestion by Leo Szilard that a build-up of somatic mutations could be the cause of ageing [60]. Szilard, a physicist who had played a key role in the genesis of the atomic bomb, noted that ionizing radiation, a potent mutagen, can shorten the lifespan. Mutations cause changes in somatic cells that are essentially irreversible other than through rare back-mutation. Thus, over time somatic mutations would be expected to build up, which could cause loss of functional integrity. Szilard couched his suggestion in terms of a quantitative model whereby mutational hits in a diploid organism would at first be silent, inactivating only the affected gene copy, unless the other copy was already faulty through carrying an inherited defect. These silent hits would accumulate until the second copy was also inactivated by a further mutation, when the cell would die. The precision of Szilard's initial predictions led to an immediate challenge to his hypothesis [61], which was soon followed up by seemingly negative results obtained from irradiation studies on insect models of varying ploidy [62, 63]. If, as Szilard supposed, the majority of mutations are recessive, increasing ploidy should result in a longer lifespan since there are then more copies of each gene that must be hit before the last copy is inactivated, and vice versa.

Although Szilard's particular formulation of the somatic mutation theory of ageing did not stand up well to experimental challenge, the suggestion that somatic mutations and DNA damage play causative roles in ageing came to be widely regarded as plausible. This was in the light of greatly advanced understanding of the molecular mechanisms that are involved, assisted by rapid progress in developing the necessary experimental techniques to study the extent and consequences of the mutations in question (e.g. [64]). A central role for DNA damage and somatic mutations is strongly argued in recent reviews [65, 66].

A third early theory on mechanisms of ageing was the protein error theory proposed by Orgel [67]. Whereas the free radical and somatic mutation theories put forward suggestions of mechanisms leading to straightforward accumulation of damage, the error theory was more kinetic in that it proposed an essential instability in the processes of genetic information transfer within cells. In order for the cell to function, the information encoded in DNA must be transcribed first into RNA and then translated into protein. Each step of this process is mediated by a highly specific set of information-handling molecules, most of which are proteins. Despite an impressive level of specificity in their functions, levels of errors are detectable in the operation of each of these systems, resulting in a small but by no means negligible error rate of approximately one mistake in every 10^3 to 10^4 amino acids [68]. Orgel pointed out that such errors, while transient in themselves, might initiate a cyclic feedback leading to what he termed error catastrophe [67]. Such a process would be as relevant to post-mitotic cells (in which somatic mutations would be less likely to arise) as to dividing cells, and it might therefore have general relevance for ageing.

The protein error theory generated considerable interest, both in the interpretation of experiments designed to test it and in the modelling of its predictions (for review see [69], p. 4). The idea that error feedback could occur was not in serious dispute, and the debate centred on whether such feedback happened at a level high enough to have biological significance. The main difficulty lay in the fact that it proved difficult to measure low rates of random amino acid misincorporation, a difficulty that remains challenging even now. In the case of DNA and RNA, novel genomic techniques soon made it much easier to detect fine-grained sequence variation. However, for proteins, despite improvements in the resolving power of proteomic analyses, the lack of a protein equivalent to the polymerase chain reaction means that random amino acid substitutions result in a diversity of molecules that are individually rare and therefore elusive. Interest in the protein error theory waned, and it seemed that the balance of experimental evidence did not support it, although it remains an open question whether the results that appeared to be negative were in fact really capable of deciding the issue. In the last decade a resurgence of interest in the contribution of protein errors to cellular ageing has occurred, which recognizes that defective proteins accumulate so as to overwhelm the capacity of the chaperone and turnover systems to maintain cellular health, resulting in breakdown of cellular proteostasis. This breakdown may be caused either by an increase in the production of abnormal proteins or by a decline in the efficiency of the relevant maintenance systems. Most likely, both factors are at play. The term 'proteostasis collapse', which is used to describe such an outcome [70, 71], is somewhat reminiscent of Orgel's term 'error catastrophe', at least in the sense that progressive accumulation of defective proteins appears to be an important contributor to ageing, and it may be that some kinds of feedback are involved.

Each of the three mechanistic theories so far considered has the potential to offer on its own an explanation of ageing. However, it can readily be seen that these early theories were not mutually exclusive, and it is indeed likely that all three processes contribute to functional decline. To these three foundational concepts we can now add many further suggestions as to which mechanisms are involved in ageing. Space does not permit enumeration of all these alternatives, but some of the more significant among them include the effects of mitochondrial mutation, telomere erosion, and epigenetic dysregulation, while even the more niche concepts such as collagen cross-linkage or mobilization of mobile genetic elements may well have parts to play. This multiplicity of candidate mechanisms calls for integration of the various mechanistic theories of ageing, which will be considered next.

2.5 Integration of Theories

Ageing is complex, and this complexity is most likely reflected in its causality. Not only might multiple mechanisms be at work simultaneously, but they may very well interact synergistically. This possibility has been explored theoretically through the development of network models combining different mechanisms such as oxidative stress, loss of proteostasis and mitochondrial dysfunction [72]. Such models can not only provide proof-of-principle for exploring interactions between different mechanisms, but they have also demonstrated their utility in predicting interactions between processes, hitherto treated separately, that can be tested by experiments. An example may be seen in the resolution of the long-standing enigma of the substantial stochastic variation that has been observed in the division potential of human diploid fibroblasts, an important model of cellular ageing [73]. Theoretical modelling showed that such variation was predicted if interactions were considered between telomere erosion, oxidative stress and somatic mutations in nuclear and mitochondrial DNA [74]. Subsequent experimental studies, motivated by the modelling predictions, confirmed that mitochondrial dysfunction accounts for stochastic heterogeneity in telomere-dependent cellular senescence [75].

Just as with the mechanistic theories, the major evolutionary theories discussed in Section 2.3 are not mutually exclusive. They share a common principle in the decline, with age, in the power of natural selection and thus could be considered as a network.

Integration of the evolutionary and mechanistic theories is also desirable, in order to provide a more unified understanding of why and how ageing occurs. In this regard, integration arises most naturally in the case of the disposable soma theory, which predicts that ageing is driven through an accumulation across the life course of a variety of kinds of damage [76]. These kinds of damage represent all of the various mechanistic hypotheses described in Section 2.4, including those now commonly referred to as hallmarks of ageing ([77], but see also [78]). This tendency for damage to accumulate is countered by the action of an extensive array of error-preventing and error-correcting systems. Thus, there is direct connection between the evolution of ageing and longevity and the mechanisms through which these traits are realized. The most promising direction of travel for integration of theories of ageing would appear to be through their positioning within the development of frameworks for the 'systems biology' of ageing [79, 80].

2.6 Theories, Models and Causality

It was noted in Section 2.2 that scientific theories are expected to make clear and specific predictions and thus to be open to being falsified. Theories of ageing should be about causality; they should suggest a reason for why and/or how ageing occurs. The complexity of ageing often makes this requirement hard to meet. For the evolutionary theories, the issue is relatively straightforward if it is made clear, for example, that the purpose of the theory is to describe why selection should cause ageing to evolve within an iteroparous organism that initially shows no deterioration with the passage of time. Of course, it might be argued that to suppose the existence of such an immortal organism is itself questionable, but the point is that unless this kind of assumption is made, even as a thought experiment, there is risk of creating a circular argument. It is helpful, therefore, if the premise on which the theory is based is made clear from the outset. Evolutionary theories of ageing face important challenges in explaining life histories that are more complicated than the simple iteroparous case. This makes it all the more important that the purpose of the theory and the grounds on which it is based should be spelled out rather carefully. It is also important to recognize that natural selection is essentially quantitative. An evolutionary theory should therefore be presented in a way that can be justified by numerical analysis. Purely verbal statements run the risk of being mere storytelling, of which we should be properly cautious (see [81] and consequent debates).

In the case of mechanistic theories, there is an obvious challenge that is often insufficiently addressed, namely, to state which specific aspects of the ageing phenotype the theory seeks to explain. For example, the free radical theory has the potential to account for age changes very generally, whereas the somatic mutation theory is likely to be more relevant to ageing in proliferative tissues rather than post-mitotic ones. One might therefore consider the former a 'global theory' and the latter an 'aspect theory'. The intended scope of the theory is of some importance in considering how it might be tested. If the idea is merely to suggest that a particular kind of fault should be an increasingly prevalent feature of the ageing body, this is easy enough to determine. But if the suggestion is of a causal nature, then what would constitute a reasonable test of the prediction? One possibility might be to increase the frequency of the particular fault and seek evidence that lifespan was thereby shortened. However, there are many ways to shorten a life that do not necessarily relate to what occurs during normal ageing. The more appealing approach is to decrease the frequency of the fault and show that lifespan was increased. Positive results of this nature are highly informative, but in the case of negative results it should be remembered that even if the mechanism is truly a contributor to ageing it may be only one of many, so the potential effect on lifespan might remain masked.

The challenges of examining theories on causes of ageing, whether evolutionary or mechanistic, are made all the more difficult when it is necessary to rely on predictive models that may not fully capture the actions of the process in question. Not only does the agreement between empirical data and the model's predictions need to be tested, but also whether the model truly captures the essence of the theory. Lessons might be drawn from epidemiology, where methodologies have been well established to unpick cases of complex multi-causality [82, 83].

References

1. King, R.A.H. 2001. *Aristotle on Life and Death*. Duckworth.
2. Weismann, A. 1889. *Essays upon Heredity and Kindred Biological Problems*. Clarendon Press.
3. Weismann, A. 1891. *Essays upon Heredity and Kindred Biological Problems*. Clarendon Press.
4. Vines, S.H. 1889. An examination of some points in Prof. Weismann's theory of heredity. *Nature* **40**, 621–626.

5. Kirkwood, T.B.L., Cremer, T. 1982. Cytogerontology since 1881: a reappraisal of August Weismann and a review of modern progress. *Hum. Genet.* **60**, 101–121.
6. Carrel, A. 1912. On the permanent life of tissues outside of the organism. *J. Exp. Med.* **15**, 516–527.
7. Hayflick, L., Moorhead, P.S. 1961. The serial cultivation of human diploid cell strains. *Exp. Cell Res.* **25**, 585–621.
8. Metchnikoff, E. 1907. *The Prolongation of Life: Optimistic Studies*. Heinemann.
9. Pearl, R. 1928. *The Rate of Living*. Knopf.
10. Bidder, G.P. 1932. Senescence. *BMJ* **115**, 5831.
11. Medvedev, Z.A. 1990. An attempt at a rational classification of theories of ageing. *Biol. Rev.* **65**, 375–398.
12. Popper, K. 1959. *The Logic of Scientific Discovery*. Hutchinson.
13. Kuhn, T.S. 1962. *The Structure of Scientific Revolutions*. University of Chicago Press.
14. Finch, C.E. 1990. *Longevity, Senescence, and the Genome*. University of Chicago Press.
15. Medawar, P.B. 1952. *An Unsolved Problem of Biology*. College.
16. Gompertz, B. 1825. On the nature of the function expressive of the law of human mortality, and on a new mode of determining the value of life contingencies. *Philos. Trans. R. Soc. Lond.* **115**, 513–583.
17. Sacher, G.A. 1978. Evolution of longevity and survival characteristics in mammals. In *The Genetics of Aging* (ed. E.L. Schneider), pp. 151–168. Plenum Press.
18. Ronget, V., Lemaître, J.-F., Tidière, M., Gaillard, J.-M. 2020. Assessing the diversity of the form of age-specific changes in adult mortality from captive mammalian populations. *Diversity* **12**, 354.
19. Jones, O.R. et al. 2014. Diversity of ageing across the tree of life. *Nature* **505**, 169.
20. Maynard Smith, J. 1962. Review lectures on senescence. I. The causes of ageing. *Proc. R. Soc. Lond. Ser. B* **157**, 115–127.
21. Martinez, D.E. 1998. Mortality patterns suggest lack of senescence in hydra. *Exp. Gerontol.* **33**, 217–225.
22. Gaillard, J.M., Lemaître J.-F. 2020. An integrative view of senescence in nature. *Funct. Ecol.* **34**, 4–16.
23. Kirkwood, T.B.L., Melov, S. 2011. On the programmed/non-programmed nature of ageing within the life history. *Curr. Biol.* **21**, 701–707.
24. Lack, D. 1954. *The Natural Regulation of Animal Numbers*. Clarendon Press.
25. Berry, R.J., Jakobson, M.E. 1971. Life and death in an island population of the house mouse. *Exp. Gerontol.* **6**, 187–197.
26. Brunet-Rossinni, A.K., Austad, S.N. 2006. Senescence in wild populations of mammals and birds. In *Handbook of the Biology of Aging* (6th ed.) (eds E.J. Masoro, S.N. Austad), pp. 243–266. Elsevier.
27. Nussey, D.H., Froy, H., Lemaître, J.-F., Gaillard, J.-M., Austad, S.N. 2013. Senescence in natural populations of animals: widespread evidence and its implications for bio-gerontology. *Ageing Res. Rev.* **12**, 214–225.
28. Kowald, A., Kirkwood, T.B.L. 2016. Can aging be programmed? A critical literature review. *Aging Cell* **15**, 986–998.
29. Wynne-Edwards, V.C. 1962. *Animal Behaviour in Relation to Social Behaviour*. Oliver and Boyd.
30. Libertini, G. 1988. An adaptive theory of the increasing mortality with chronological age in populations in the wild. *J. Theor. Biol.* **132**, 145–162.

31. Maynard Smith, J. 1976. Group selection. *Q. Rev. Biol.* **51**, 277–283.
32. Haldane, J.B.S. 1941. *New Paths in Genetics*. Allen & Unwin.
33. Williams, G.C. 1957. Pleiotropy, natural selection, and the evolution of senescence. *Evolution* **11**, 398–411.
34. Hamilton, W.D. 1966. The moulding of senescence by natural selection. *J. Theor. Biol.* **12**, 12–45.
35. Charlesworth, B. 1980. *Evolution in Age-Structured Populations*. Cambridge University Press.
36. Wachter, K.W., Evans, S.N., Steinsaltz, D. 2013. The age-specific force of natural selection and biodemographic walls of death. *Proc. Natl. Acad. Sci. USA* **110**, 10141–10146.
37. Wachter, K.W., Steinsaltz, D., Evans, S.N. 2014. Evolutionary shaping of demographic schedules. *Proc. Natl. Acad. Sci. USA* **111**, 10846–10853.
38. Rodriguez, J.A., Marigorta, U.M., Hughes, D.A., Spataro, N., Bosch, E., Navarro, A. 2017. Antagonistic pleiotropy and mutation accumulation influence human senescence and disease. *Nat. Ecol. Evol.* **1**, 0055.
39. Gaillard, J.-M., Lemaître, J.-F. 2017. The Williams' legacy: a critical reappraisal of his nine predictions about the evolution of senescence. *Evolution* **71**, 2768–2785 (doi:10.1111/evo.13379).
40. Austad, S.N., Hoffman, J.M. 2018. Is antagonistic pleiotropy ubiquitous in aging biology? *Evol. Med. Public Health* **2018**, 287–294.
41. Flatt, T., Partridge, L. 2018. Horizons in the evolution of aging. *BMC Biol.* **16**, 93 (doi:10.1186/s12915-018-0562-z).
42. Kirkwood, T.B. 1977. Evolution of ageing. *Nature* **270**, 301.
43. Kirkwood, T.B., Holliday, R. 1979. The evolution of ageing and longevity. *Proc. R. Soc. Lond. B* **205**, 531–546.
44. Kirkwood, T.B. 2017. The disposable soma theory. In *The Evolution of Senescence in the Tree of Life* (eds R.P. Shefferson, O.R. Jones, R. Salguero-Gómez), pp. 23–39. Cambridge University Press.
45. Cohen, A.A., Coste, C., Li, X.-Y., Bourg, S., Pavard, S. In press. Are trade-offs really the key drivers of aging and lifespan? *Funct. Ecol.* **34**, 153–166.
46. Omholt, S.W., Kirkwood, T.B.L. 2021. Ageing as a consequence of selection to reduce the environmental risk of dying. *Proc. Natl. Acad. Sci. USA* **118**, e2102088118.
47. Blagosklonny, M.V. 2009. TOR-driven aging: speeding car with no brakes. *Cell Cycle* **8**, 4055–4059.
48. Magalhaes, J.P. 2012. Programmatic features of aging originating in development: aging mechanisms beyond molecular damage? *FASEB J.* **26**, 4821–4826.
49. Comfort, A. 1956. *The Biology of Senescence*. Routledge and Kegan Paul.
50. Finch, C.E., Kirkwood, T.B.L. 2000. *Chance, Development and Aging*. Oxford University Press.
51. Saretzki, G., Armstrong, L., Leake, A., Lako, M., Zglinicki, T. 2004. Stress defense in murine embryonic stem cells is superior to that of various differentiated murine cells. *Stem Cells* **22**, 962–971.
52. Saretzki, G. et al. 2008. Downregulation of multiple stress defense mechanisms during differentiation of human embryonic stem cells. *Stem Cells* **26**, 455–464.
53. Moore, L. et al. 2021. The mutational landscape of human somatic and germline cells. *Nature* **597**, 381–386.
54. Harman, D. 1956. Aging: a theory based on free radical and radiation chemistry. *J. Gerontol.* **11**, 298–300.

55. Beckman, K.B., Ames, B.N. 1998. The free radical theory of aging matures. *Physiol. Rev.* **78**, 547–581.
56. Perez, V.I., Bokov, A., Remmen, H., Mele, J., Ran, Q., Ikeno, Y., Richardson, A. 2009. Is the oxidative stress theory of aging dead? *Biochim. Biophys. Acta* **1790**, 1005–1014.
57. Lapointe, J., Hekimi, S. 2010. When a theory of aging ages badly. *Cell. Mol. Life Sci.* **67**, 1–8.
58. Murphy, M.P. et al. 2011. Unraveling the biological roles of reactive oxygen species. *Cell Metab.* **13**, 361–366.
59. Kirkwood, T.B.L., Kowald, A. 2012. The free-radical theory of ageing: older, wiser and still alive. *Bioessays* **34**, 692–700.
60. Szilard, L. 1959. A theory of ageing. *Nature* **184**, 957–958.
61. Maynard Smith, J. 1959. A theory of ageing. *Nature* **184**, 956–957.
62. Clark, A.M., Rubin, M.A. 1961. The modification by X-irradiation of the life span of haploids and diploids of the wasp *Habrobracon sp. Radiat. Res.* **15**, 244–253.
63. Lamb, M.J. 1965. The effects of X-irradiation on the longevity of triploid and diploid female *Drosophila melanogaster. Exp. Gerontol.* **1**, 181–187.
64. Burnet, F.M. 1974. *Intrinsic Mutagenesis: A Genetic Approach to Aging*. John Wiley & Sons.
65. Schumacher, B., Pothof, J., Vijg, J., Hoeijmakers, J.H.J. 2021. The central role of DNA damage in the ageing process. *Nature* **592**, 695–703.
66. Vijg, J. 2021. From DNA damage to mutations: all roads lead to aging. *Ageing Res. Rev.* **68**, 101316.
67. Orgel, L.E. 1963. The maintenance of the accuracy of protein synthesis and its relevance to ageing. *Proc. Natl. Acad. Sci. USA* **49**, 517–521 (doi:10.1073/pnas.49.4.517).
68. Loftfield, R.B., Vanderjagt, D. 1972. The frequency of errors in protein biosynthesis. *Biochem. J.* **89**, 82–92.
69. Kirkwood, T.B.L., Holliday, R., Rosenberger, R.F. 1984. Stability of the cellular translation apparatus. *Int. Rev. Cytol.* **92**, 93–132.
70. Taylor, R., Dillin, A. 2011. Aging as an event of proteostasis collapse. *Cold Spring Harb. Perspect. Biol.* **3**, 004440.
71. Santra, M., Dill, K.A., Graff, A.M.R. 2019. Proteostasis collapse is a driver of cell aging and death. *Proc. Natl. Acad. Sci. USA* **116**, 22173–22178.
72. Kowald, A., Kirkwood, T.B.L. 1996. A network theory of ageing: the interactions of defective mitochondria, aberrant proteins, free radicals and scavengers in the ageing process. *Mutat. Res.* **316**, 209–236.
73. Smith, J.R., Whitney, R.G. 1980. Intraclonal variation in proliferative potential of human diploid fibroblasts: stochastic mechanism for cellular aging. *Science* **207**, 82–84.
74. Sozou, P.D., Kirkwood, T.B.L. 2001. A stochastic model of cell replicative senescence based on telomere shortening, oxidative stress, and somatic mutations in nuclear and mitochondrial DNA. *J. Theor. Biol.* **213**, 573–586.
75. Passos, J.F. et al. 2007. Mitochondrial dysfunction accounts for the stochastic heterogeneity in telomere-dependent senescence. *PLoS Biol.* **5**, e110.
76. Kirkwood, T.B. 2005. Understanding the odd science of aging. *Cell* **120**, 437–447.
77. Lopez-Otin, C., Blasco, M.A., Partridge, L., Serrano, M., Kroemer, G. 2013. The hallmarks of aging. *Cell* **153**, 1194–1217.
78. Gems, D., Magalhaes, J.P. 2021. The hoverfly and the wasp: a critique of the hallmarks of aging as a paradigm. *Ageing Res. Rev.* **70**, 101407.

79. Kirkwood, T.B.L. 2011. Systems biology of ageing and longevity. *Philos. Trans. R. Soc. B* **366**, 64–70.
80. Chauhan, A., Liebal, U.W., Vera, J., Baltrusch, S., Junghanß, C., Tiedge, M., Fuellen, G., Wolkenhauer, O., Köhling, R. 2015. Systems biology approaches in aging research. *Interdiscip. Top. Gerontol.* **40**, 155–176.
81. Gould, S.J., Lewontin, R.C. 1979. The spandrels of San Marco and the Panglossian paradigm: a critique of the adaptationist programme. *Proc. R. Soc. B* **205**, 581–598.
82. Rothman, K.J. 1976. Causes. *Am. J. Epidemiol.* **104**, 587–592.
83. Wensink, M., Westendorp, R.G.J., Baudisch, A. 2014. The causal pie model: an epidemiological method applied to evolutionary biology and ecology. *Ecol. Evol.* **4**, 1924–1930.

3 The Diversity of Longevity Metrics

Statistical Considerations, Potential Biases and Biological Implications

Victor Ronget, Gilles Maurer and Sarah Cubaynes

3.1 Introduction

Why do certain organisms live longer than others? For decades, researchers have attempted to answer this question and unravel the determinants of longevity using experimental or observational studies [1]. For a given living individual, its longevity simply corresponds to the duration of its lifespan, which is equal to the difference between its date of birth and date of death. Differences in age at death among individuals result from a combination of genetic and environmental factors [2], but also occur among homogenous individuals in standardized laboratory conditions due to stochastic processes [3, 4]. Most studies on longevity aim to identify the relative contribution of various intrinsic and extrinsic (i.e. environmentally driven) factors in determining longevity.

In search of the determinants of longevity, researchers are often interested in comparing longevity measures among groups of individuals sharing common characteristics within a population [5] (see also Chapter 16), between populations of the same species [6] (see also Chapter 17) or even across species [7, 8] (see also Chapter 7). When considering a group of individuals, defining longevity is less trivial because not all group members die at the same age, which results in a distribution of ages at death instead of a single longevity value or lifespan measurement.

Choosing an appropriate metric to summarize the longevity of a group of individuals is far from an easy task, as illustrated by the large number of metrics available to demographers [9]. Several indicators can be used to describe any statistical distribution (e.g. mean, median, quantiles, maximum). Likewise, various metrics of longevity are available to summarize the distribution of ages at death of a group of individuals sampled in a population. The question of how to measure longevity is becoming more and more important as studies of longevity spans very diverse areas of research [10], from epidemiology to biomedical study and evolutionary biology [11, 12, 13]. Recently, demographers have questioned the use of these different metrics and their potential impacts on the outcomes of studies [14].

In this chapter, our goal is to raise awareness about the importance of selecting an appropriate longevity metric when studying the determinants of longevity. Our objective is not to provide strict rules about which metric should be preferred, but more to encourage any researcher interested in studying longevity to consider and question the possible alternatives when devising new studies. Importantly, we will be solely focused on longevity metrics and will not discuss the diversity and biological implications of the actuarial ageing metrics (e.g. [15]), which is beyond the scope of this chapter (but see Chapters 5 and 7). In this chapter, we ask three major questions to illustrate the diversity of longevity metrics used across the scientific literature: (1) What are the different metrics of longevity? (2) In what contexts are specific longevity metrics used? (3) Could the choice of the longevity metric influence the result and the interpretations of a study? To answer those questions, we first define the most common longevity metrics, and discuss their characteristics and limitations. Second, using a topic-modelling approach [16], we map the use of the various longevity metrics in the literature according to their scientific areas and research topics. Finally, using two examples, we illustrate how the choice of longevity metrics can affect the outcome of an analysis. Based on this, we provide a list of recommendations aimed at helping researchers to carefully think about the choice of metrics when conducting a study that uses longevity.

3.2 Overview of the Most Common Longevity Metrics

In this section we briefly define and present the pros and cons associated with the most common longevity metrics (see [14] for a more in-depth discussion of these metrics and Figure 3.1 for a graphical representation of the different longevity metrics in an extinct human cohort in France). Note that, in the scientific literature, the term 'lifespan' is frequently employed as a synonymous of longevity (e.g. [7, 17]). Lifespan refers to the duration of life of an individual, it is not a statistical index summarizing the longevity of a group of individuals, and therefore it does not strictly refer to a longevity metric (see the results in Section 3.3 for additional information regarding the use of lifespan across scientific disciplines).

A longevity metric is single value characterizing the longevity of a group of individuals. Its aim is to summarize the distribution of the ages at death within a sample of individuals. The most common longevity metrics correspond to common statistical indicators used to summarize any distribution:

Life expectancy is defined as the mean of the age-at-death distribution or the average age at death in a population. Life expectancy is one of the most widely used metrics in demography because it is easy to calculate and not too sensitive to sample size. However, because the calculation of life expectancy includes all individuals, even those dying in early life, it might not be the most appropriate measurement to study the longevity of old individuals. For instance, in Figure 3.1 life expectancy from birth is quite low (53.6 years) because of the high proportion of children (14.1%) dying in their first year.

Maximum longevity is defined as the maximum age at death or the age at death of the oldest individual. This is also one of the most commonly used metric, especially for comparing longevity among species (e.g. [18, 19]; see also Chapter 4). Using maximum longevity is convenient because it requires only the age of the oldest individual and not the full age-at-death distribution, which is often not available for many species. However, there is one main statistical issue associated with the use of maximum longevity. It was demonstrated that this metric is sensitive to sample size [14]. A higher maximum longevity is most likely to be found as more individuals are sampled in a population but the magnitude of this effect changes depending on the form of age-at-death distribution. The maximum longevity in populations where the mortality follows a Gompertz law with age is moderately sensitive to sample size while maximum longevity is highly sensitive to sample size in population with a constant mortality rate across age classes [20].

The *q*th *percentile longevity* is defined as the age at which q% of the individuals are dead. Most researchers refer to the 50th percentile, also called *median longevity*. Median longevity (or median lifespan) corresponds to the age at which half of the population died. Median longevity can significantly differ from life expectancy for asymmetric age-at-death distributions with a large proportion of individuals dying young or old (e.g. Figure 3.1). For this reason median longevity is sometimes preferred to life expectancy as a longevity metric describing the central tendencies of the age-at-death distributions [21, 22]. More recently, authors have also advocated for the use of the 80th or 90th percentile, also called *80% or 90% longevity*. This metric represent the age at which 80% or 90% of the individuals are dead, it is presented as an alternative to maximum longevity when the biological question is focused on older individuals [23]. Indeed, 80% or 90% longevity metrics are more dependent on elder age classes (higher quantiles) than life expectancy and less sensitive to sample size than maximum longevity. Sufficient data to compute this metric might still be hard to get for many wild species [24]; however, detailed demographic datasets are becoming more and more available for species across the tree of life, meaning 80% or 90% longevity should be preferred to maximum longevity for comparative analyses.

As we have seen, several metrics are available, but they have different objectives, as well as specific requirements to compute them. It is essential to choose the appropriate longevity metric depending on both the research question and also the data available. Regarding the research question, studies that aim to evaluate temporal trends or compare different populations will focus their analysis on the central tendency of the ages-at-death distribution. In this case, mean or median statistical indicators, such as life expectancy or median longevity, are better descriptors of the distribution's central tendency [25]. However, these metrics will often differ because they are more or less sensitive to the tails of the distribution. For example, median longevity exceeds life expectancy when juvenile mortality increases. Studies focusing on elderly populations or extreme longevity will investigate the right tail of the distribution and therefore favour 90% longevity or maximum longevity metrics [26].

So far, we considered the full age-at-death distribution in order to compute the different longevities metrics starting from birth; however, it might be more appropriate to compute the metrics from later ages depending on the species and the question. For instance, for the cohort born in 1900 shown in Figure 3.1, if we exclude the first five

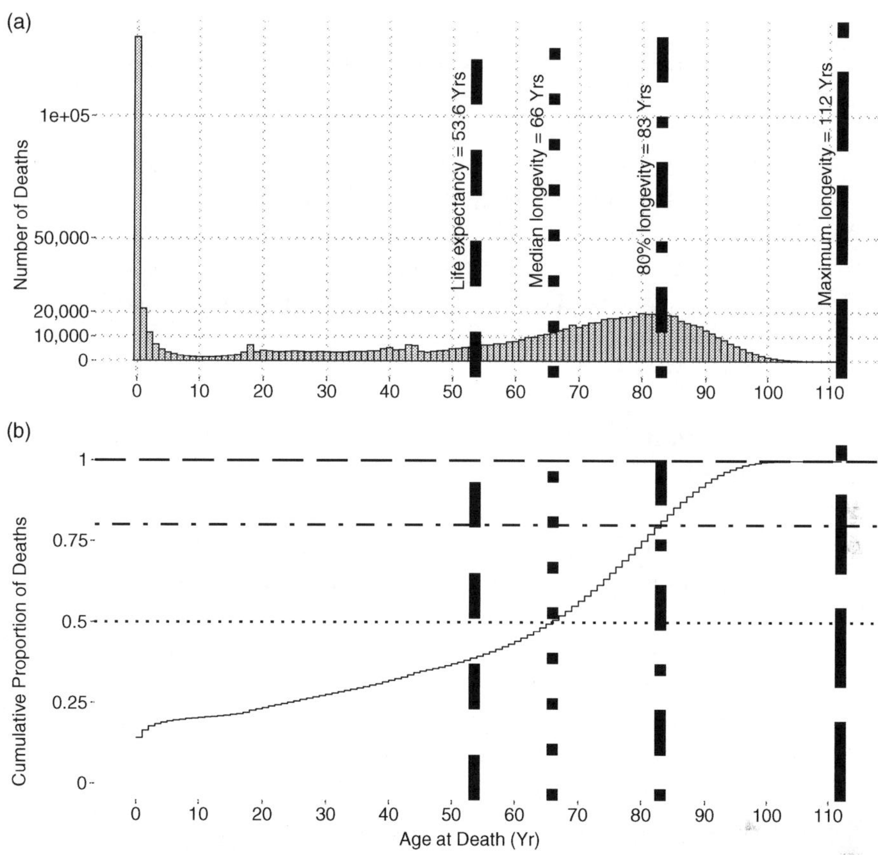

Figure 3.1 (a) Age-at-death distribution for a human cohort in France born in 1900 (data from Human Mortality Database (HMD), men and women combined), and (b) cumulative percentage of deaths per age for the same cohort. Vertical dashed and dotted lines represent different metrics of longevity calculated from birth.

years of life when most juvenile mortality happened, the life expectancy at 5 years old is 61.3 years, which is significantly higher than the life expectancy at birth (53.6 years). Computing life expectancy for only oldest ages-at-death individuals is also a common practice for studies on centenarians [27]. Moreover, in most wild species, it is difficult to keep track of the death of young individuals, so computing adult longevity metric after the age at maturity or the age at first reproduction might be more suitable [17].

In addition to the research question, the properties of the dataset – namely data volume and completeness – are also key factors in the choice of a longevity metric. If the sample size is small, maximum longevity, which is particularly sensitive to low sample size, should be avoided and life expectancy or percentile longevity should be preferred. Metrics such as life expectancy and percentile longevities can only be computed for samples in which the full age-at-death distribution is known; this is typically the case for humans or species in laboratory conditions where it is easy to get accurate ages at death. On the contrary, for wild species with imperfect monitoring this distribution is barely known and the exact age at death is recorded for a small fraction of the population, meaning that the only choice is to use maximum longevity. Therefore, the use of maximum longevity should be more common in comparative studies including a high number of species across the tree of life because high-quality demographic data are still rare [24]. As the research question and the dataset properties have various implications in the choice of longevity metric, we expect a more balanced use of these metrics in laboratory and human studies. In Section 3.3, we will illustrate how research questions and the nature of the available demographic data have favoured one or both metrics through an analysis of the scientific literature across scientific disciplines and time.

3.3 Exploring the Use of Longevity Metrics in Time and across Scientific Disciplines

In this section, we explore the extensive body of literature published on longevity since the 1990s to map the use of various metrics through time and across scientific disciplines. We choose to restrain our analysis to after 1990 because, prior to this date, many articles were not digitized more easily accessible online. We performed a topic-modelling analysis on a selection of review articles focusing on longevity. We explored the relative use of longevity metrics and how this has changed across time, scientific disciplines and main research themes or topics.

3.3.1 Data and Methods

3.3.1.1 Data Collection and Pre-processing

We aimed to select all review articles that focused on the longevity of living organisms. We chose to restrict our search to review articles to allow a broad selection of articles that can be manually screened and to exclude articles that were irrelevant (e.g. dealing with the longevity of batteries, cells or seeds) or when there were no results directly associated with longevity or life expectancy terms. Among the selected review articles, we distinguished articles associated with the term 'longevity' versus 'life expectancy' versus 'both longevity and life expectancy'. As we did not perform topic-modelling analyses on the full text articles, we choose to look generally for the term longevity, expecting that maximum longevity and quantile longevity would be coined as 'longevity' in the abstract. Thus, by comparing longevity and life expectancy, we are in fact comparing analyses of '*extreme longevity*' vs '*average longevity*'.

To do so, we conducted search on titles, abstracts and keywords of review articles from the Web of Science core collection. We used the following search code:

```
TS=((longevity NOT "life expectanc*") AND (species OR population
OR vertebrate OR animal OR plant OR human OR mammal OR insect OR
amphibian OR fish OR bird OR reptile))) AND (LA==("ENGLISH") AND
DT==("REVIEW")).
```

The 'TS' code standing for 'Topic' is given to search within Title, abstract, keywords and keywords Plus. The 'LA' code allowed us to restrict the search for articles in a specified language and 'DT' defines document-type restrictions, here reviews. We repeated the search with ("life expectanc*" NOT longevity) and (longevity AND "life expectanc*") to constitute three distinct lists. Each article was therefore assigned to one of the three mutually exclusive categories depending on the type of metric used: longevity, life expectancy or both metrics. Abstracts or full articles were checked by two authors to exclude non-relevant ones.

This procedure led to a first selection of 3 115 review articles. After manual screening and exclusion of irrelevant articles, we retained a total of $n = 1\,857$ reviews focused on longevity of living organisms, including 1 297 assigned to the longevity metric, 488 to the life expectancy metric and 72 assigned to both metrics. Our corpus was composed of the 1 857 review articles that were imported as independent text files into the program R [28], using the stm package [29]. We removed English stop words and punctuation, and applied a stemming process, reducing words to a common root.

3.3.1.2 Bibliometric Analysis

Based on the complete corpus, we constructed a data frame summarizing the unique identifying number (UT or accession number from the Web of Science), year of publication, abstract, title, the journal name, the metric category ('longevity', 'life expectancy', 'both metrics', see Section 3.3.1.1) and the Web of Science research areas for each review article. Review articles could be associated with one or several research areas. We first regrouped the 127 research areas in 12 broader disciplines [30]. Based on this data, we could calculate the number and percentage of review articles published that included each type of metric per year and per discipline.

3.3.1.3 Topic-Modelling Analysis

We then investigated the diversity of topics that longevity and life expectancy studies have focused on, and we explored potential relationships between the use of a metric and the subject(s) of the study.

To do so, we used unsupervised natural language processing, specifically a topic-modelling approach, to identify the various topics addressed in our corpus over time and research areas. The topic-modelling approach allows the identification of hidden semantic structures or patterns in a large corpus of articles. It is based on a statistical approach where topics constitute clusters of specific words or stems that co-occurred with unusual frequency [16]. Topic modelling is performed via the Latent Dirichlet Allocation (LDA) algorithm [31]. LDA is a generative statistical model that postulates that each document is a mixture of a small number of topics and that each topic is derived from a set of words. Articles usually comprise several topics with variable proportions, with topics being distributions of words. The goal of LDA is thus to model the corpus as a distribution of topics, referred to as topic prevalence, and topics as a distribution of words, called topic content.

The structural topic model (STM) is a form of topic modelling that allows us to incorporate metadata into the model and uncover how different documents might talk about the same underlying topic using different word choices. The goal of the STM is to allow researchers to identify topics and estimate their relationship to document metadata. Metadata that explain topic contents and prevalence are referred to as covariates. In our case we included three covariates: (1) the year of publication, (2) the research discipline, and (3) the metric used (longevity, life expectancy or both).

We used the stm R package to identify the different topics and investigate their prevalence within the corpus [29]. We set the number of topics to $k = 20$ after fitting models from 10 to 60 topics and inspecting perplexity [32]. While increasing the number of topics identifies more anecdotal themes, it may also discriminate topics with very similar content, making the global analysis less informative and coherent.

Model outputs included the probability of each word in our corpus being associated with a topic (estimated parameter β) and the probability that each topic was associated with a given document (estimated parameter θ). For more details on β and θ equations, see the stm R package documentation [29]. Based on these estimated probabilities, we obtained a list of the 10 words most associated with each topic. We also calculated the expected proportion of the corpus that belonged to each topic, that is, topic prevalence.

Finally, we selected the articles most representative of each topic, that is, $\theta > 0.5$. We extracted the research areas and metric category of the selected articles. For each topic, we calculated the proportion of articles associated with either longevity, life expectancy or both metrics.

3.3.2 Results

3.3.2.1 Effect of the Year of Publication

Since 1990, we have observed a steady rise in the number of reviews focused on longevity or life expectancy (Figure 3.2(a)). Overall, 70% of review articles were classified in the longevity category. The first review article using life expectancy was published in 1992. Note that less than 4% of all review articles selected in our corpus were associated with both terms, and none using both terms were published before 1997.

3.3.2.2 Variation across Research Disciplines

Various disciplines are represented in our corpus, but the great majority of review articles come from the fields of medical and biological sciences (Figure 3.2(b)). More than 90% of review articles associated with the fields of agricultural sciences and geosciences use the longevity term rather than the life expectancy term. Most review articles also use the longevity term in biological sciences (89%), chemistry (78%), multidisciplinary sciences (73%), psychology (64%), physics (60%) and medical

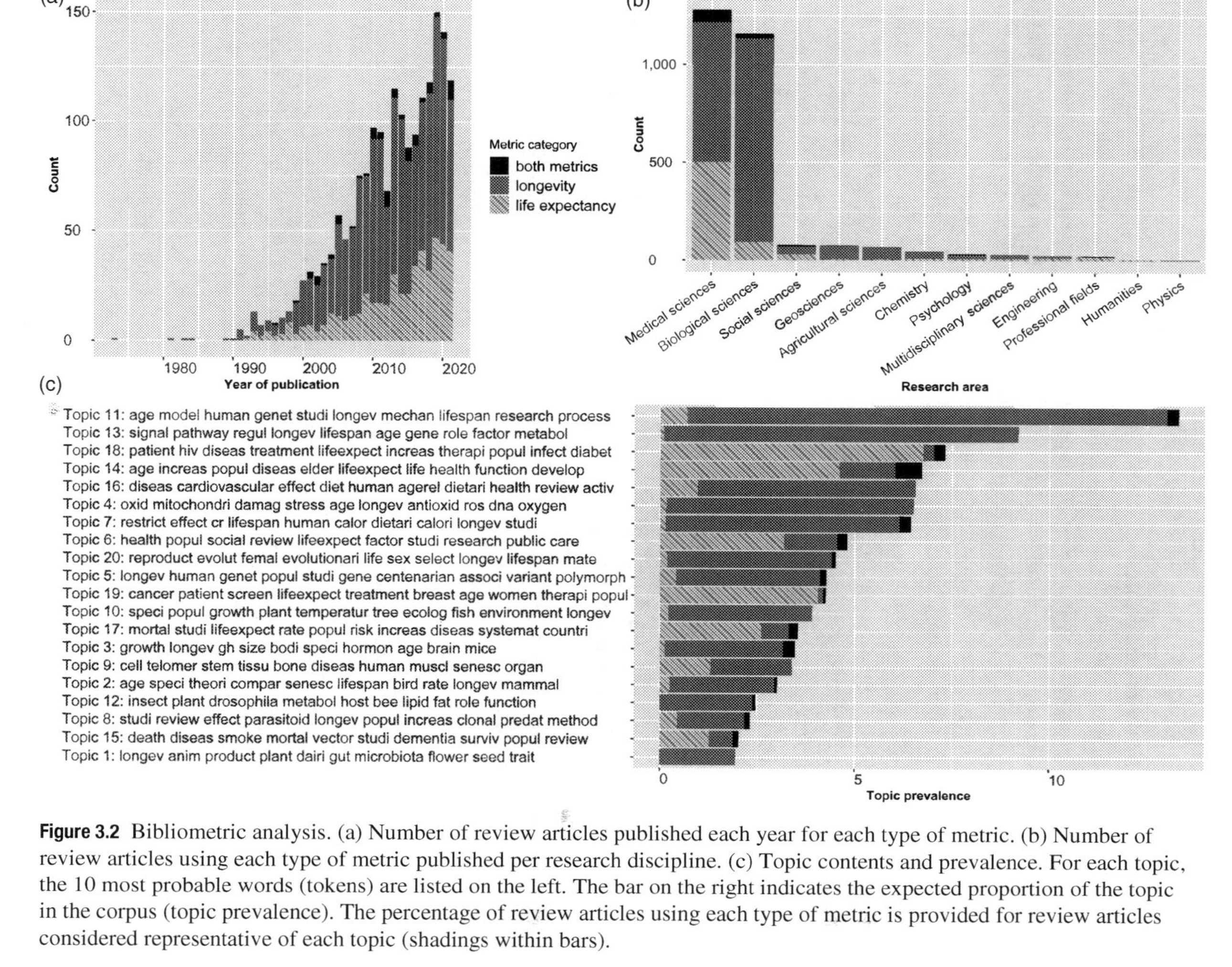

Figure 3.2 Bibliometric analysis. (a) Number of review articles published each year for each type of metric. (b) Number of review articles using each type of metric published per research discipline. (c) Topic contents and prevalence. For each topic, the 10 most probable words (tokens) are listed on the left. The bar on the right indicates the expected proportion of the topic in the corpus (topic prevalence). The percentage of review articles using each type of metric is provided for review articles considered representative of each topic (shadings within bars).

sciences (55%). By contrast, the use of longevity or life expectancy terms appears to be balanced in social sciences, and the use of life expectancy is more frequent in humanities and professional fields. Overall, only few review articles use both terms, mostly in psychology (4 of 33, ~12%), social sciences (9 of 79, ~10%), medical sciences (64 of 1 279, ~5%) and biological sciences (24 of 1 157, ~2%).

3.3.2.3 Variation across Topics

The 10 most probable words characterizing each of the 20 topics identified in our corpus are shown in Figure 3.2(c). The diversity of topics is impressive and covers: physiological, molecular and genetic studies investigating the determinant of ageing (topics 11, 13); studies exploring the impact of specific risk or protection factors such as diet or caloric restriction (topics 16 and 7), stress and oxidation (topic 4), temperature (topic 10) or smoking behaviours (topic 15) on lifespan; studies focusing on specific populations such as centenarian studies and the elderly population (topics 5 and 14), patients with cancer, diabetes or HIV (topics 18 and 19); experimental studies on model organisms (topic 12); studies on host–parasite systems (topic 8); studies on dairy products, domestic animals and cultivated plants (topic 1); studies investigating the proximal causes of discrepancies in lifespan and mortality risks across countries and social systems (topics 6 and 17); and studies investigating the evolutionary determinants of differences in lifespan across species (topic 2) or between males and females (topic 20).

The preferential use of longevity or life expectancy terms differs across topics (Figure 3.2(c)). The two most prevalent topics in the corpus are related to the genetic, molecular and physiologic determinants of ageing and lifespan (topics 11 and 13). Both topics are strongly associated with the longevity term (90% of the review articles). Several other topics also show a strong association with the use of longevity (>90%) rather than life expectancy or both terms. These latter topics deal with the role of oxidative stress (topic 4), the impact of diet (topic 7) and the effect of environmental factors on lifespan (topic 10), as well as evolutionary studies on gender differences and menopause (topic 20), metabolic studies on insects and plants (topic 12), and studies on domestic animals and plants and microbiota (topic 1).

By contrast, only two topics, related to cancer and patients with HIV or diabetes (topics 19 and 18) show a very strong association with the use of life expectancy (more than 90% of reviews), and reviews addressing the comparison of mortality rates and health conditions between human populations (topics 15, 6, 14 and 17) predominantly use the life expectancy term (60%–75%).

Overall, the use of life expectancy appears restricted to studies involving the comparison of lifespan among human populations, while the use of longevity is common when working on animals or plants, and in experimental studies evaluating the genetic and physiological determinants of lifespan. The percentage of review articles associated with both terms was near zero for most topics, peaking only at 10% among studies related to health conditions and diseases in the elderly population (topic 14).

Note that we also conducted an additional search using the keyword 'lifespan'. It led to 771 additional articles spread across all topics and research areas. The term 'lifespan' was mostly employed as a synonym of the generic term longevity. It mostly appeared in studies conducted on lab organisms, for example, studies focused on extending the longevity of model organisms such as drosophila, mole rats or *C. elegans*. Overall, when including these additional review articles, the topics identified in our analysis were quite similar. Because 'lifespan' is not a longevity metric (see Section 3.1), and the term 'lifespan' did not appear to be associated to one metric category or the other, the results presented in this section do not include these additional articles.

3.3.3 Discussion

How to measure the general notion of longevity remains a crucial debate, reinforced by the steady rise in publications referring to these metrics and spanning very diverse topics. The use of the life expectancy term versus the longevity term is far from homogeneous across research areas and even more varied across topics. Life expectancy and longevity terms are used more evenly in medical sciences, while biological sciences are skewed towards longevity. We hypothesize that this discrepancy could partly be data driven: complete demographic data available for many human populations allow us to calculate all possible metrics, while the scarcity of demographic records available for some species could be an incentive to rely on maximum longevity measures that are available more easily in the literature. The preferential use of longevity versus life expectancy terms within certain disciplines might also reflect the historical compartmentalization existing across scientific disciplines. For example, life expectancy was first proposed by human demographers, and its use still appears to be largely restricted to scientists working on human data. It might take some time for a given metric, independently of its performance, to navigate its way across disciplines.

Interestingly, the low number of studies focusing on both life expectancy and longevity terms reflect the fact that the use of different longevity metrics is not yet considered as a common practice. We encourage researchers to replicate their analyses using different longevity metrics. This practice seems perfectly feasible for human-related questions where several metrics can easily be calculated from high-quality demographic records. We hope that in the upcoming years the proportion of studies using multiple longevity metrics will increase and lead to new groundbreaking discoveries, for instance testing when factors influencing extreme longevities are similar to or different from those influencing life expectancy. For example, it was demonstrated that castration in mice influenced maximum lifespan but not median lifespan [33]. In this case, only considering one longevity metric would have led to oversimplifying the reality about the relationship between castration and longevity. This example demonstrates that the determinants of extreme and mean longevities might be different and could also point towards different biological mechanisms responsible for extreme longevities. We argue here that the different meanings of longevity should be considered more carefully in the future. We will hereafter detail two examples showing how the choice of one longevity metric could potentially affect the outcome of a study.

3.4 Does the Choice of Metrics Matter?

Age at death can be highly variable among individuals of the same species, especially in the wild, for example due to varying environmental conditions among populations [34]. A given environmental factor typically explains a low percentage of variance in longevity because the distribution of age at death often results from the interaction of multiple factors including genetics, environmental and stochastic processes.

In this section, we explore how the choice of longevity metric can influence the conclusions of a study regarding the ecological and biological drivers of longevity. We consider two contrasting examples: an example of simulated data about temperature-driven longevity across *C. elegans* populations, and a second example about comparisons of longevity metrics across mammals.

3.4.1 Simulation Study: Temperature-Driven Longevity across *C. elegans* Populations

Figure 3.3 depicts simulated age at death data for 1 000 *C. elegans* individuals sampled from 10 populations ($n = 100$ individuals per population) located (virtually) along a gradient of increasing temperature. Within each population, ages at death were simulated from a Weibull distribution with parameters chosen from the literature to mimic the negative effect of higher temperature on *C. elegans* longevity [35, 36]. Mean age at death was 20 days and simulated values ranged from 7 to 36 days among the 1 000 individuals.

We aggregated the data at the population level to calculate maximum longevity, 80% longevity, median longevity, life expectancy and mean annual temperature for each population. We then fitted linear models to estimate the relationship between ages at death and mean annual temperature, first making use of the complete data at the individual level and then using the four longevity metrics aggregated at the population level. Results about the influence of temperature on longevity varied depending on the metric used to carry out the population-level analysis (Figure 3.3).

We found a negative influence of increasing temperatures on longevity for all metrics. However, the slope of the linear relationship between temperature and longevity, as well as the uncertainty around the slope and the percentage of deviance explained by the model varied depending on the metric used. When directly analysing individual age at death, the estimated slope of the relationship between temperature and longevity (mean of –0.54 with 95% confidence interval (CI) [–0.58; –0.51]) was smaller and more precise compared to using metrics aggregated at the population level. The estimated slope was more biased and less precise

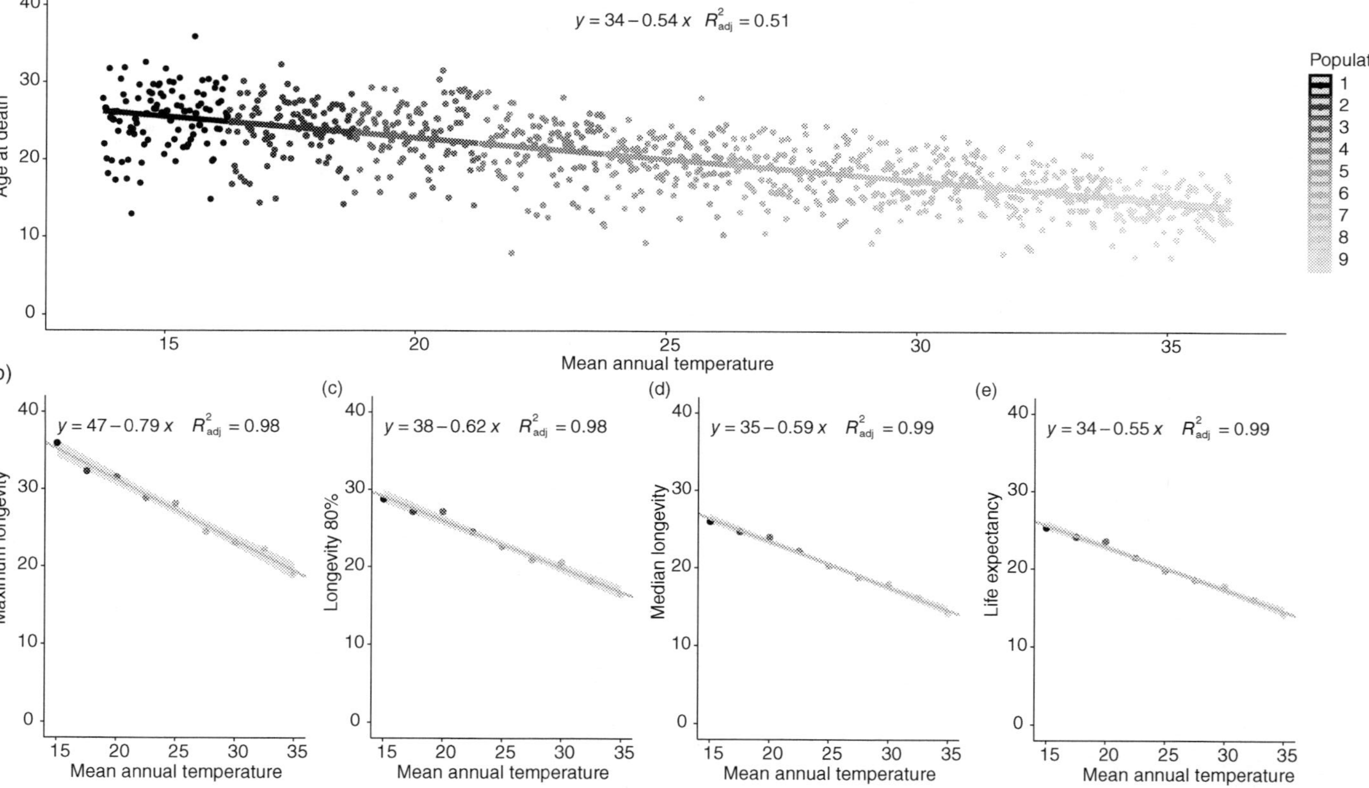

Figure 3.3 (a) Dots represent simulated data of age at death (in days) as a function of mean annual temperature (in °C) for 1 000 individuals belonging to 10 populations located along a gradient of increasing mean temperature. Regression lines from the linear model of age at death as a function of temperature (straight line) are presented with 95% confidence interval (grey area). Estimated relationships between longevity and mean annual temperature are displayed for (b) maximum longevity, (c) 80% longevity, (c) median longevity and (e) life expectancy. On the four bottom panels, each dot represents the metric value averaged over all individuals of the corresponding population. For each metric, estimated intercept and slope values, as well as adjusted R^2 values, are displayed if significant.

when using maximum longevity (mean of −0.79 with 95% CI [−0.88; −0.69]) compared to using 80% longevity (mean of −0.62 with 95% CI [−0.69; −0.54]), median longevity (mean of −0.59 with 95% CI [−0.64; −0.54]) and life expectancy (mean of −0.55 with 95% CI [−0.60; −0.50]). The percentage of deviance explained by temperature was also higher when using the four metrics aggregated at the population level compared to using the complete individual age-at-death data (see R^2 values in Figure 3.3).

These results highlight the influence of the choice of metric on the estimated relationship between a given ecological or biological factor and longevity. This simulation shows that metrics aggregated at the population level provide biased information about the underlying relationship between individual longevity and a given environmental determinant. It is therefore important to remember that longevity metrics are always summarizing the age-at-death distribution and by doing so always convey less information. Results might be particularly sensitive to the choice of metrics when working with small datasets, data with large variances and determinants explaining a small percentage of variance in longevity. In such cases, it might be more appropriate to use a longevity metric that is less sensitive to sample size, normally avoiding using maximum longevity if possible. Additionally, it points to the importance of paying attention to the metric used when comparing results obtained from several studies. Estimated coefficients obtained using different longevity metrics will not be comparable in many cases, which might be an issue when performing reviews or meta-analyses.

In this simulation, we considered only a single species with multiple populations and thus – by minimizing genetic variation among individuals and standardizing lab conditions – we could consider that there were few sources of longevity variations between populations. We might wonder how those differences across metrics change across a wider set of species in the wild with substantial variations in both genetic backgrounds and environmental conditions.

3.4.2 A Comparative Analysis: Relationships between Different Longevity Metrics in Mammals

Comparing longevities across species has always been of great interest in evolutionary sciences [37, 38]. Longevity as a species life history trait is expected to be shaped by natural and sexual selection and to be – at least to some extent – structured by the phylogenetic relatedness between species [39]. For a given population, alternative longevity metrics will always differ, for instance maximum longevity will always be higher than life expectancy. At a comparative level, what matters is the magnitude of those differences across species. One way to compare differences among longevity metrics is to assess whether metrics are proportional to each other. If there is a proportional relationship between longevity metrics, it means that we can predict one longevity metric from another one, and consequently know that the choice of longevity metric will have no major effect on the results of comparative analyses. With the following example, we focus on that hypothesis by comparing longevity metrics extracted for 48 populations of wild mammals.

All population-level data for wild mammals were extracted from work by Lemaître et al. [17]. Species were included in the analysis when data for the full age-at-death distribution was available so that maximum longevity, life expectancy and 80% longevity from the age at first reproduction could be calculated. For the sake of simplicity, only females were considered in the analysis. We also included the maximum longevity provided by the online database AnAge [40]. AnAge is an open database recording maximum longevity, mostly from captive animals. It spans a large variety of species and is arguably the most frequently used database for comparative biology of ageing (e.g. [41, 42]). Here, we wanted to compare maximum longevity measures extracted from AnAge to actual records from the wild.

First, we assessed the relationship between the various longevity metrics (maximum longevity, life expectancy, 80% longevity and maximum longevity from the AnAge database). For each relationship ($n = 6$ in total), we fitted a PGLS (phylogenetic generalized least squares) regression to account for the phylogenetic relatedness between species using the phylolm R package in R [43, 28]. Each longevity metric was log-transformed prior to the analysis. If the slope of the linear relationship with the 95% confidence interval value was not statistically different from 1, it indicated a relation of isometry between the two metrics. In other words, we can consider that the two metrics are proportional. The strength of each relationship was also assessed from the coefficient of determination R^2 (Figure 3.4).

As expected, 80% longevity was always higher than life expectancy for each species (all species are above the dashed line). We found that the relationship between 80% longevity and life expectancy is isometric, meaning that the differences between those two metrics are of comparable magnitudes across species. The correlation coefficient was also very high ($R = 0.99$). We could thus consider that these two metrics are equivalent across our 48 species of mammals.

Maximum longevity was consistently higher than life expectancy and 80% longevity as expected but the slope of the relationship was lower than 1 for both life expectancy and 80% longevity, meaning that maximum longevity was relatively closer to life expectancy/80% longevity for long-lived species compared to short-lived species. Interestingly, the correlation coefficient was also much lower. Maximum longevity is therefore likely to add more uncertainty to the relationship because it is sensitive to sample size. Depending on the population and on the species studied, the number of individuals used to assess longevities may vary a lot from few individuals in poorly studied species to millions of individuals in humans. These results show that maximum longevity is not equivalent to life expectancy/80% longevity, and that choosing maximum longevities over life expectancy/80% longevity could potentially influence results in a comparative analysis. Differences between metrics were even more striking when using maximum longevity measures coming from the AnAge database. The slopes of the relationship between maximum longevity from AnAge and life expectancy/80% longevity in the wild were much lower than 1 and the correlation coefficients were even lower. We can also note that in this example we focused only on mammal species that are known to follow a Gompertz increase of mortality with age [44]. Maximum longevity is not very sensitive to sample size

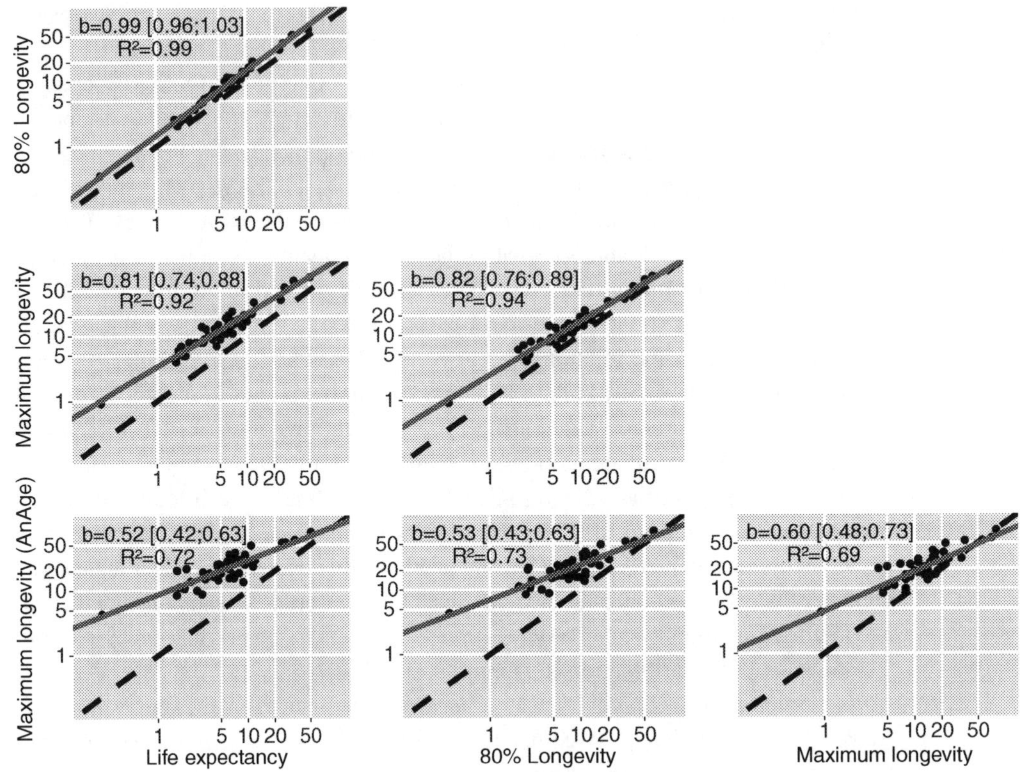

Figure 3.4 Linear relationships between longevity metrics in years for 48 mammalian species ranging from the short-lived four-striped grass mouse, *Rhabdomys pumilio* to the long-lived killer whale, *Orcinus orca* (life expectancy of 50.7 years), the linear relationship fitted is represented with a solid line, the slope value b associated with the 95% confidence interval and the R^2 value is presented. The dashed line corresponds to the line with an intercept of 0 and a slope of 1.

when the mortality follows a Gompertz law [20], so it is likely that those results could change depending on the group of species studied (e.g. in birds, age-related mortality increases following a Weibull law [45] or in plants mortality rate can remain constant independently of age [46]) and could even increase the discrepancies between maximum longevities and other metrics of longevity, especially when considering age-at-death distributions highly sensitive to sample size, such as constant mortality rate with age.

Maximum longevity estimates collected from wild populations and from databases such as AnAge could be supposed to be equivalent because they are the same metric estimated from different populations. However, we found strong differences between these two metrics. First, maximum longevity estimated from wild samples was lower than that from AnAge for most species. Moreover, the slope of the relationship between these two measures was lower than 1 and associated with a low correlation coefficient. It reflected a surprising absence of isometry between maximum longevity estimated from wild populations versus coming from the AnAge database. Maximum longevity measures from the AnAge database come from captive individuals living in a protective environment with no predation and resources always available ad libitum, which can explain the higher values compared to estimates obtained from wild populations [47]. We want to highlight here the importance of the records used to compute longevity metrics. It was demonstrated for instance that there are large variations in maximum longevities between different online databases [48]. Depending both on the sample size and environment, there can be some strong differences between populations even in the same longevity metric [47].

Second, we performed a simple comparative analysis looking at the linear relationship between body mass and each longevity metric independently. A strong allometric relationship between body mass and longevity has been demonstrated across mammals [49, 50]. Small species are expected to have reduced longevity metrics associated with fast life history strategies and large species are expected to have higher longevity metrics associated with slow life history strategies [51]. This expectation does not apply to certain outlier groups of species, such as bats, characterized by low body mass and high longevity [52]. However, when considering the whole diversity of mammals this expectation is generally true. Most of these studies have focused on the relationship between maximum longevity and body mass to include a large number of species in the analysis [53, 54]. Our objective was to highlight if the use of different longevity metrics for this specific comparative analysis could lead to contradictory results or not. We fitted a PGLS regression to each of the four relationships between metrics using the phylolm R Package in R [43, 28]. For each relationship between body mass and longevity estimated from a given metric, we reported the slope of the relationship with the 95% confidence interval and Pagel's lambda as a measure of phylogenetic inertia commonly used in comparative analyses. This measure is bounded between 0 and 1, with higher values meaning a higher phylogenetic structuration for the trait studied [55]. Results of these analyses are presented in Table 3.1.

We found a strong positive relationship between body mass and all four longevity metrics (Table 3.1). Here, the choice of longevity metric did not alter our conclusions. The slope of the estimated relationship between body mass and longevity decreased slightly when using maximum longevity but the relationship remained

Table 3.1 Slope (with 95% confidence interval) and Pagel's lambda of the linear relationships between body mass and each longevity metric

	Slope	95% CI	Lambda
Life expectancy	0.25	[0.18; 0.32]	0.49
80% longevity	0.25	[0.17; 0.32]	0.51
Maximum longevity	0.19	[0.13; 0.26]	0.5
Maximum longevity (AnAge)	0.14	[0.10; 0.19]	0.51

significant. Using one or the other longevity metric did not affect the results because body mass is strongly related to longevity across mammals. In addition, the various longevity metrics were highly correlated across species; this is also true for maximum longevity. We might be tempted to say, based on this example, that the choice of longevity metric does not really matter for inter-specific comparisons, but we were considering a highly structured relationship across all mammals. The outcome might change when investigating other group of species or biological or ecological factors less strongly associated with longevity (e.g. in [56, 57]). There is no easy answer to assess the potential bias of using maximum longevity in comparative analyses; however, it is unlikely that there will be major changes to the conclusion of a study. As a recommendation we advise researchers to perform those analyses with alternative metrics of longevity and, if possible, to test for potential biases, especially if the effect found is weak.

3.5 Conclusion

In this chapter, we discussed the use of longevity metrics and the potential caveats associated with each longevity metric. As a final point, we formulate two recommendations for researchers interested in longevity questions. First, we highlight that there is no unique and correct way to summarize the longevity of a group of individuals using a single measure. We recommend justifying the choice of metric depending on the research question and on which feature of the age-at-death distribution the study is focused. For instance, 'am I interested in the longevity of all the individuals even the ones dying young?' or 'am I interested in the longevity of individuals with an extreme longevity?' Second, we urge researchers to be extremely cautious about the data used to calculate a given longevity metric because variability in sample size and quality could result in major biases in the outcomes of studies. As a last word, the simultaneous use of several longevity metrics is a conservative practice that can minimize potential bias.

References

1. Christensen, K., Vaupel, J.W. 1996. Determinants of longevity: genetic, environmental and medical factors. *J. Inter. Med.* **240**, 333–341 (doi:10.1046/j.1365-2796.1996.d01-2853.x).
2. Deelen, J. et al. 2019. A meta-analysis of genome-wide association studies identifies multiple longevity genes. *Nat. Commun.* **10**, 1–14.
3. Hartemink, N., Caswell, H. 2018. Variance in animal longevity: contributions of heterogeneity and stochasticity. *Popul. Ecol.* **60**, 89–99 (doi:10.1007/s10144-018-0616-7).
4. Jouvet, L., Rodríguez-Rojas, A., Steiner, U.K. 2018. Demographic variability and heterogeneity among individuals within and among clonal bacteria strains. *Oikos* **127**, 728–737.
5. Shadyab, A.H., LaCroix, A.Z. 2015. Genetic factors associated with longevity: a review of recent findings. *Ageing Res. Rev.* **19**, 1–7.

6. Colchero, F. et al. 2016. The emergence of longevous populations. *PNAS* **113**, E7681–E7690 (doi:10.1073/pnas.1612191113).
7. Healy, K. et al. 2014. Ecology and mode-of-life explain lifespan variation in birds and mammals. *Proc. R. Soc. Lond. B: Biol. Sci.* **281**, 20140298 (doi:10.1098/rspb.2014.0298).
8. Mayne, B., Berry, O., Davies, C., Farley, J., Jarman, S. 2019. A genomic predictor of lifespan in vertebrates. *Sci. Rep.* **9**, 17866 (doi:10.1038/s41598-019-54447-w).
9. Wrycza, T., Baudisch, A. 2014. The pace of aging: intrinsic time scales in demography. *Demo. Res.* **30**, 1571–1590 (doi:10.4054/DemRes.2014.30.57).
10. Yang, Y.C., Boen, C., Gerken, K., Li, T., Schorpp, K., Harris, K.M. 2016. Social relationships and physiological determinants of longevity across the human life span. *Proc. Nat. Acad. Sci.* **113**, 578–583 (doi:10.1073/pnas.1511085112).
11. Calabrese, V., Cornelius, C., Cuzzocrea, S., Iavicoli, I., Rizzarelli, E., Calabrese, E.J. 2011. Hormesis, cellular stress response and vitagenes as critical determinants in aging and longevity. *Mol. Aspects Med.* **32**, 279–304.
12. Gorbunova, V., Seluanov, A., Zhang, Z., Gladyshev, V.N., Vijg, J. 2014. Comparative genetics of longevity and cancer: insights from long-lived rodents. *Nat. Rev. Genet.* **15**, 531–540.
13. Minias, P., Podlaszczuk, P. 2017. Longevity is associated with relative brain size in birds. *Ecol. Evol.* **7**, 3558–3566 (doi:10.1002/ece3.2961).
14. Ronget, V., Gaillard, J. 2020. Assessing ageing patterns for comparative analyses of mortality curves: going beyond the use of maximum longevity. *Func. Ecol.* **34**, 65–75.
15. Colchero, F. et al. 2021. The long lives of primates and the 'invariant rate of ageing' hypothesis. *Nat. Commun.* **12**, 3666 (doi:10.1038/s41467-021-23894-3).
16. Blei, D., Carin, L., Dunson, D. 2010. Probabilistic topic models. *IEEE Sig. Process. Mag.* **27**, 55–65 (doi:10.1109/MSP.2010.938079).
17. Lemaître, J.-F. et al. 2020. Sex differences in adult lifespan and aging rates of mortality across wild mammals. *Proc. Nat. Acad. Sci.* 117, 8546–8553.
18. de Magalhães, J.P., Costa, J., Church, G.M. 2007. An analysis of the relationship between metabolism, developmental schedules, and longevity using phylogenetic independent contrasts. *J. Gerontol. Ser. A* **62**, 149–160 (doi:10.1093/gerona/62.2.149).
19. Romiguier, J. et al. 2014. Comparative population genomics in animals uncovers the determinants of genetic diversity. *Nature* **515**, 261–263.
20. Vaupel, J.W. 2003. Post-Darwinian longevity. *Pop. Dev. Rev.* **29**, 258–269.
21. Bross, T.G., Rogina, B., Helfand, S.L. 2005. Behavioral, physical, and demographic changes in Drosophila populations through dietary restriction. *Aging Cell* **4**, 309–317 (doi:10.1111/j.1474-9726.2005.00181.x).
22. Lee, W.-S., Monaghan, P., Metcalfe, N.B. 2013. Experimental demonstration of the growth rate–lifespan trade-off. *Proc. R. Soc. B: Biol. Sci.* **280**, 20122370 (doi:10.1098/rspb.2012.2370).
23. Moorad, J.A., Promislow, D.E.L., Flesness, N., Miller, R.A. 2012. A comparative assessment of univariate longevity measures using zoological animal records. *Aging Cell* **11**, 940–948 (doi:10.1111/j.1474-9726.2012.00861.x).
24. Conde, D.A. et al. 2019. Data gaps and opportunities for comparative and conservation biology. *Proc. Nat. Acad. Sci.* **116**, 9658–9664.
25. Ho, J.Y., Hendi, A.S. 2018. Recent trends in life expectancy across high income countries: retrospective observational study. *BMJ* **362**, k2562.
26. Gavrilova, N.S., Gavrilov, L.A. 2020. Are we approaching a biological limit to human longevity? *J. Gerontol.: Ser. A* **75**, 1061–1067.

27. Modig, K., Andersson, T., Vaupel, J., Rau, R., Ahlbom, A. 2017. How long do centenarians survive? Life expectancy and maximum lifespan. *J. Inter. Med.* **282**, 156–163 (doi:10.1111/joim.12627).
28. R Core Team. 2017. R: A Language and Environment for Statistical Computing. R Foundation for Statistical Computing, Vienna, Austria. www.R-project.org/.
29. Roberts, M.E., Stewart, B.M., Tingley, D. 2019. Stm: an R package for structural topic models. *J. Stat. Soft.* **91**, 1-40 (doi:10.18637/jss.v091.i02).
30. Milojević, S. 2020. Practical method to reclassify Web of Science articles into unique subject categories and broad disciplines. *Quan. Sci. Stud.* **1**, 183–206.
31. Blei, D.M., Ng, A.Y., Jordan, M. 2003. Latent Dirichlet Allocation Michael I. Jordan. *J. Mach. Learn. Res.* **3**, 993–1022.
32. Grün, B., Hornik, K. 2011. Topicmodels: an R package for fitting topic models. *J. Stat. Software* **40**, 1–30 (doi:10.18637/jss.v040.i13).
33. Garratt, M., Try, H., Brooks, R.C. 2021. Access to females and early life castration individually extend maximal but not median lifespan in male mice. *GeroScience* **43**, 1437–1446 (doi:10.1007/s11357-020-00308-8).
34. Norry, F.M., Loeschcke, V. 2002. Temperature-induced shifts in associations of longevity with body size in Drosophila melanogaster. *Evolution* **56**, 299–306.
35. Vanfleteren, J.R., De Vreese, A., Braeckman, B.P. 1998. Two-parameter logistic and Weibull equations provide better fits to survival data from isogenic populations of Caenorhabditis elegans in axenic culture than does the Gompertz model. *J. Gerontol. Ser. A: Biol. Sci. Med. Sci.* **53**, B393–B403.
36. Stroustrup, N., Ulmschneider, B.E., Nash, Z.M., López-Moyado, I.F., Apfeld, J., Fontana, W. 2013. The *Caenorhabditis elegans* lifespan machine. *Nat. Meth.* **10**, 665–670.
37. Sacher, G.A. 1978. Evolution of longevity and survival characteristics in mammals. In *Genet. Aging* (ed. E.L. Schneider), pp. 151–168. Plenum Press.
38. Ricklefs, R.E. 2010. Insights from comparative analyses of aging in birds and mammals. *Aging Cell* **9**, 273–284.
39. Gaillard, J.-M., Pontier, D., Allaine, D., Lebreton, J.D., Trouvilliez, J., Clobert, J. 1989. An analysis of demographic tactics in birds and mammals. *Oikos*, **56**, 59–76.
40. De Magalhaes, J.P., Costa, J. 2009. A database of vertebrate longevity records and their relation to other life-history traits. *J. Evol. Biol.* **22**, 1770–1774.
41. Farré, X. et al. 2021. Comparative analysis of mammal genomes unveils key genomic variability for human life span. *Mol. Biol. Evol.* **38**, 4948–4961 (doi:10.1093/molbev/msab219).
42. Collard, M.K., Bardin, J., Laurin, M., Ogier-Denis, E. 2021. The cecal appendix is correlated with greater maximal longevity in mammals. *J. Anat.* **239**, 1157–1169.
43. Tung Ho, L., Ané, C. 2014. A linear-time algorithm for Gaussian and non-Gaussian trait evolution models. *Syst. Biol.* **63**, 397–408.
44. Ronget, V., Lemaître, J.-F., Tidière, M., Gaillard, J.-M. 2020. Assessing the diversity of the form of age-specific changes in adult mortality from captive mammalian populations. *Diversity* **12**, 354.
45. Ricklefs, R.E. 2000. Intrinsic aging-related mortality in birds. *J. Avian Biol.* **31**, 103–111.
46. Roach, D.A., Smith, E.F. 2020. Life-history trade-offs and senescence in plants. *Func. Ecol.* **34**, 17–25.
47. Tidière, M., Gaillard, J.-M., Berger, V., Müller, D.W., Lackey, L.B., Gimenez, O., Clauss, M., Lemaître, J.-F. 2016. Comparative analyses of longevity and senescence reveal variable survival benefits of living in zoos across mammals. *Sci. Rep.* **6**, 36361.

48. Lemaître, J.-F., Müller, D.W., Clauss, M. 2014. A test of the metabolic theory of ecology with two longevity data sets reveals no common cause of scaling in biological times. *Mammal Rev.* **44**, 204–214.
49. Lindstedt, S.L., Calder III, W.A. 1981. Body size, physiological time, and longevity of homeothermic animals. *Q. Rev. Biol.* **56**, 1–16.
50. Calder, W.A. 1983. Body size, mortality, and longevity. *J. Theor. Biol.* **102**, 135–144 (doi:10.1016/0022-5193(83)90266-7).
51. Gaillard, J.-M., Yoccoz, N.G., Lebreton, J.-D., Bonenfant, C., Devillard, S., Loison, A., Pontier, D., Allaine, D. 2005. Generation time: a reliable metric to measure life-history variation among mammalian populations. *Am. Nat.* **166**, 119–123 (doi:10.1086/430330).
52. Brunet-Rossinni, A.K., Austad, S.N. 2004. Ageing studies on bats: a review. *Biogerontology* **5**, 211–222 (doi:10.1023/B:BGEN.0000038022.65024.d8).
53. Hofman, M.A. 1993. Encephalization and the evolution of longevity in mammals. *J. Evol. Biol.* **6**, 209–227.
54. González-Lagos, C., Sol, D., Reader, S.M. 2010. Large-brained mammals live longer. *J. Evol. Bio.* **23**, 1064–1074.
55. Pagel, M. 1994. Detecting correlated evolution on phylogenies: a general method for the comparative analysis of discrete characters. *Proc. R. Soc. Lond. B* **255**, 37–45 (doi:10.1098/rspb.1994.0006).
56. Sohal, R.S., Sohal, B.H., Brunk, U.T. 1990. Relationship between antioxidant defenses and longevity in different mammalian species. *Mech. Ageing Dev.* **53**, 217–227.
57. Cooper, N., Kamilar, J.M., Nunn, C.L. 2012. Host longevity and parasite species richness in mammals. *PLoS ONE* **7**, e42190 (doi:10.1371/journal.pone.0042190).

4 The Meaning of 'Exceptional Longevity'

A Critical Reappraisal

Jean-Michel Gaillard and Nigel Gilles Yoccoz

4.1 Introduction

The longevity (or lifespan) of an organism corresponds to the time elapsed between its birth and its death. This life history trait (sensu [1]) has attracted a lot of attention, likely in relation to its huge variation both among and within species. It is far beyond the scope of this chapter to provide a comprehensive review of comparative analyses of longevity (but see Chapters 7 and 17 for specific discussions on variation in longevity at inter- and intra-specific levels, respectively). Here, we will only briefly mention some salient points we think illustrate well our current understanding of variation in longevity that takes place among organisms both within and among populations, and across species. Thus, across animal species, longevity spans several orders of magnitude. For instance, adult mayflies (*Ephemera simulans*) often live for a few days at the most [2], whereas the bowhead whale (*Balaena mysticetus*) can live up to 211 years of age [3], and the Greenland shark (*Somniosus microcephalus*) might live for five centuries [4]. Within species, and even within populations, individuals also display a large range of ages at death, with some individuals outliving most conspecifics by a large amount. For instance, while 50% of yellow-bellied marmot (*Marmota flaviventris*) females have died at around 4 years of age, one individual lived up to 14 years of age [5]. Of course, such a tremendous diversity of longevity among and within species is not totally random and several drivers of variation in longevity have been identified so far. Across species, the phylogeny of the focal species [6], its body size [7, 8], and its lifestyle are the main structuring factors shaping variation in longevity (see [9] for the positive effect of arboreality on longevity across mammals and [10, 11] for the positive effect of sociality on longevity among mole-rat species and mammals, respectively, [12] for the positive effect of hibernation on longevity among bat species, [13] for the positive effect of flying, arboreality and fossoriality on longevity across bird and mammal species, and [14] for the positive effect of protective phenotypes (e.g. armour, venom) on longevity across ectotherm species). Thus, within mammals, bats of the order Chiroptera markedly outlive shrews from the family Soricidae; large species

We warmly thank Rochelle Buffenstein, Philip Dammann and an anonymous reviewer for insightful comments on a previous draft of this chapter.

such as elephants or whales markedly outlive small species such as small rodents; and herbivorous, fossorial and social species such as Alpine marmots (*Marmota marmota*) markedly outlive carnivorous, non-fossorial and non-social species such as coyotes (*Canis latrans*). Within species, differences in terms of sex [15], body mass or condition [16, 17], level of heterozygosity [18], dominance status [19], movement activity [20], and home range location have all been reported to generate variation in longevity among conspecifics [21].

This complex pattern of variation in longevity makes the identification of species or individuals displaying exceptional longevity difficult, especially in the wild where experiments aiming to extend (e.g. via dietary restriction [22]) or reduce (e.g. via exposure to stressors [23]) longevity are virtually impossible to perform. In this chapter, we aim to demonstrate the strong context-dependence of exceptional longevity using mostly mammals as a case study. The choice of mammals is justified by the huge life history variation, with both body size and pace of life varying over several orders of magnitude across species in this Vertebrate class, and also by the availability of accurate data on age-specific mortality from a large number of mammalian populations monitored in the wild (see e.g. [15]). To identify species with 'exceptional longevities', we will first perform allometric analyses based on a simple longevity metric as repeatedly done during the last decades (see e.g. the pioneer comparative analysis of maximum longevity across mammals performed by Sacher and Wolstenholme [24]). However, instead of only focusing on size allometry, we will also consider the time scaling (sensu [25]). We will estimate expected longevity based either on species-specific size or on species-specific pace of life and we will scale the predicted longevity to the observed longevity using the concept of longevity quotient (LQ) first proposed by Palmore [26], and increasingly used in comparative analyses of longevity (e.g. [12, 27, 28, 29, 30]). LQ will be used to identify species displaying exceptional longevity and we will discuss similarities and differences in the two sets of exceptionally long-lived species. Second, we will take advantage of the Finch and Pike model to estimate species-specific expected maximum longevity [31], based on the assumption that age-specific mortality from the age at first reproduction onwards is satisfactorily described by a Gompertz model [31] (see [32] for the current empirical evidence supporting the overall suitability of the Gompertz model to describe age-specific mortality in mammals). A strong positive residual value from the regression of observed maximal longevity on maximal longevity predicted by the Finch and Pike model will thus indicate the occurrence of exceptional longevity. Lastly, to identify individuals with 'exceptional longevities' among individuals within species we will apply the life-table modelling approach based on extreme value theory recently proposed by Huang et al. to three especially detailed long-term monitoring of individual fates [33]: the medflies studied in laboratory conditions by Carey et al. [34], the rhesus macaques (*Macaca mulatta*) studied on Cayo Santiago by Lee et al. [35], and the Ansell's mole-rat (*Fukomys anselli*) studied by Dammann et al. under laboratory conditions [10].

4.2 Identifying Mammalian Species with Exceptional Longevity from Evolutionary Allometries of Body Size and Pace of Life

We retrieved information on adult female body mass, age at first reproduction and longevity 80% (corresponding to the age at which 80% of adults have died and providing a much more reliable longevity metric than the most commonly used maximum longevity [36, 37]) for the set of mammalian species recently analysed for sex differences in mortality patterns [15]. We only retained one population per species, leading to retrieving information for 100 species. We extracted life history estimates from Lemaître et al.'s supplementary information that includes female species-specific values for all required traits [15]. In addition, we included data from three species of mole-rats whose survival patterns have been intensively studied in recent years: *Fukomys anselli* with a female age at first reproduction of 270 days [38], Mechow's mole-rat (*Fukomys mechowii*) with a female age at first reproduction of 547 days [39], and naked mole-rat (*Heterocephalus glaber*) with a female age at first reproduction of 183 days [40]. For these species, we pooled males and females to estimate longevity 80%, as done in original survival analyses of these species because males and females display similar age-specific survival patterns [39]. Lastly, adult body mass of these species was retrieved from Dammann et al. [10]. We then estimated the allometric relationship between body mass and longevity across these 103 mammalian species by fitting a linear regression model between the two traits on a log-log scale. Although data points were not independent due to phylogenetic relatedness among species, we did not correct for phylogenetic inertia, in order to keep the models as simple as possible. As our mammalian sample was representative of the main orders of this vertebrate class, neglecting phylogenetic relatedness across species should not matter for our predictive and illustrative purposes [41].

As expected, we found a strong hypo-allometric relationship between the longevity metric and adult body mass among mammalian species ($R^2 = 0.43$, slope of 0.21 ± 0.02 (1 standard error (SE))). The confidence interval of the slope included the expected scaling value for a biological time such as longevity (0.25, e.g. [8]). We used the species-specific longevities predicted by the allometric relationship to estimate expected longevities and we calculated LQ as the ratio between observed and expected longevity. However, LQ as previously measured (see e.g. [27]) provides a biased metric because the expected longevity corresponds to the exponential of the residuals from the allometric relationship between body mass and observed longevity, and they are not 0 on average. The log-transformed LQ corresponds to the residuals and provides an unbiased measure of the species-specific longevity deviation from the allometric model based on body mass (called LQS for size-based longevity quotient hereafter). When so measured, LQS varied from −1.68 for least weasel (*Mustela nivalis*) to 1.90 for naked mole-rat and averaged 0 by construction. The quantiles of LQS were −0.43 (25%), −0.04 (median) and 0.42 (75%). Defining exceptional longevity as species belonging to the top 5% of LQS (i.e. the five out of 103 species with highest LQS) led to including naked mole-rat with an LQS of 1.90, Ansell's mole-rat with an LQS of 1.79, brown long-eared bat (*Plecotus auritus*) with an LQS of 1.75, greater mouse-eared bat (*Myotis myotis*) with an LQS of 1.64, and Verreaux's sifaka (*Propithecus verreauxi*) with an LQS of 1.42 as exceptionally long-lived. Unsurprisingly, these species only included mole-rats (two out of the three species included in the dataset), bats (two out of the four species included in the dataset) and primates (one out of the eight species included in the dataset). On the other hand, the top 5% of the shortest-lived species included least weasel (LQS of −1.68), four-striped grass mouse (*Rhabdomys pumilio*, LQS of −1.67), bank vole (*Myodes glareolus*, LQS of −1.54), Eurasian lynx (*Lynx lynx*, LQS of −1.53) and European hare (*Lepus europaeus*, LQS of −1.46).

Using time scaling instead of size scaling to identify mammalian species with exceptional longevity led to measuring another LQ metric (called LQT for time-based longevity quotient hereafter). Using the age at first reproduction as a measure of the pace of life [42], we retrieved a strong positive proportional association ($R^2 = 0.61$, slope of 0.96 ± 0.07 (1 SE)), as previously reported [43], and as expected for biological times [25, 44]. LQT predicted from species-specific pace of life varied from −2.16 for least weasel to 2.01 for naked mole-rat, and averaged 0 by construction. The quantiles of LQT were −0.36 (25%), 0.01 (median) and 0.29 (75%). Using the distribution of LQT values and defining as exceptional the top 5% led us to retain the five species with the highest LQT. Thus, naked mole-rat, with an LQT of 2.01, Ansell's mole-rat with an LQT of 1.67, European hare with an LQT of 1.32, agile wallaby (*Macropus agilis*) with an LQT of 1.07 and Verreaux's sifaka with an LQT of 1.06 were classified as displaying exceptional longevity using this criterion. On the other hand, weasel (LQT of −2.16), Dall's porpoise (*Phocoenoides dalli*, LQT of −1.44), Balkan snow vole (*Dinaromys bogdanovi*, LQT of −1.16), stoat (*Mustela erminea*, LQT of −1.04) and topi (*Damaliscus lunatus*, LQT of −2.16) corresponded to the top 5% of the shortest-lived species.

Overall, these analyses highlight that the weasel was especially short-lived and mole-rats were especially long-lived relative to both their size and their pace of life. On the other hand, bat species were only among the longest-lived species for their body mass and not for their pace of life. A positive association linked LQS and LQT, but less than half the variation in LQS was accounted for by observed variation in LQT ($R^2 = 0.41$, slope of 0.83 ± 0.01 (1 SE), Figure 4.1, see also Table 4.1 for species-specific values of LQS and LQT). The slightly hypo-allometric nature of the relationship between LQS and LQT indicates a slower increase of LQS for a given increase of LQT. Mammalian species displayed a much higher amount of variation in LQS (variance of 0.61, Figure 4.2(a)) than in LQT (variance of 0.36, Figure 4.2(b)), which indicates that the longevity of a given species is much more constrained by its pace of life than by its body mass. As longevity is a biological time, a closer association is indeed expected between longevity and age at first reproduction (the time scale we used to assess the pace of life), both measured in time units, than between longevity and body mass, measured in different currencies, even if size markedly shapes all biological times [8]. Among the mammalian species we studied, age at first reproduction was indeed positively associated with body mass, with an allometric exponent slightly below the value of 0.25 expected for biological times ($R^2 = 0.49$, slope of 0.19 ± 0.02 (1 SE)).

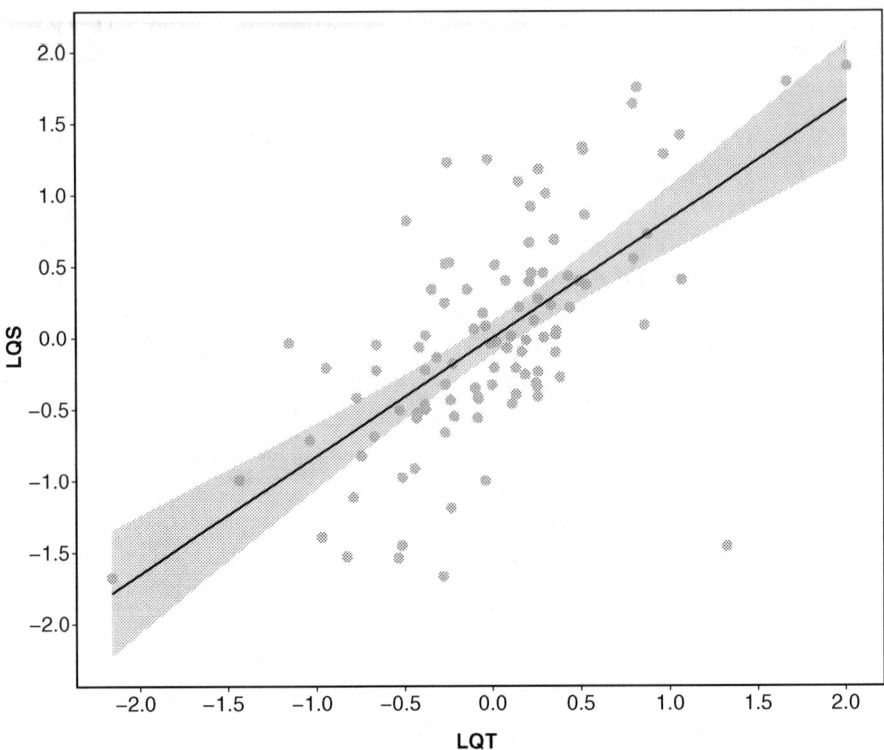

Figure 4.1 Relationship between the longevity quotient based on body mass (LQS) and the longevity quotient based on the pace of life (LQT). The slope slightly lower than 1 (0.826) indicates that LQS increases slower than LQT. The species deviating the most from the relationship (i.e. low LQS and high LQT) corresponds to the European hare (*Lepus europaeus*).

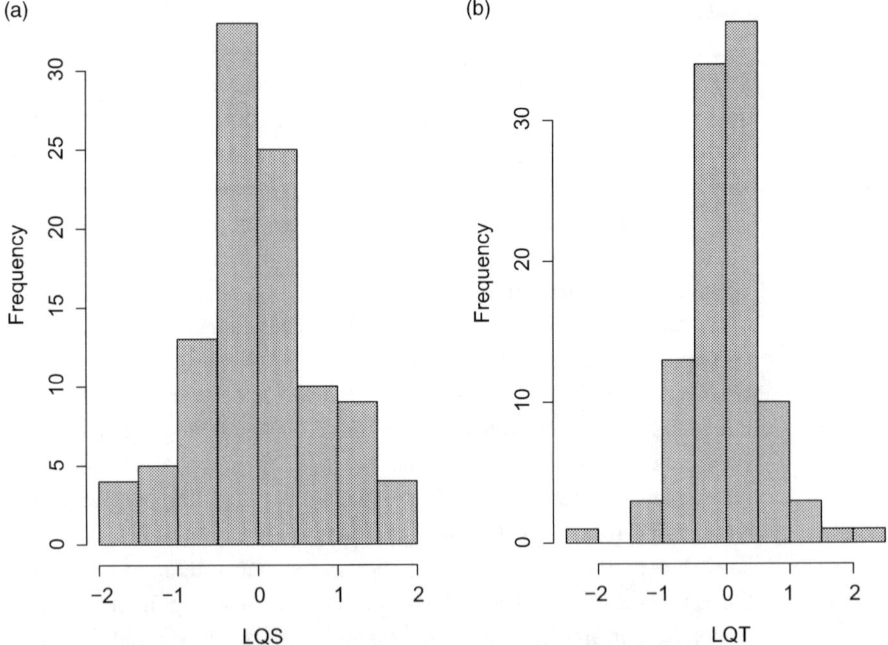

Figure 4.2 (a) Distribution of species-specific longevity quotients based on body mass (LQS) and (b) on pace of life (LQT). Longevity quotients correspond to the log-scaled ratio between species-specific observed and expected longevities (i.e. residuals from the respective allometric relationships).

Table 4.1 LQ of 103 mammalian species

Species	LQS	LQT	Species	LQS	LQT
Aepyceros melampus	**−0.434**	−0.239	*Mungos mungo*	0.372	*0.531*
Alces alces	−0.275	*0.383*	*Muscardinus avellanarius*	**−0.460**	0.111
Antilocapra americana	0.211	*0.440*	*Mustela erminea*	**−0.715**	**−1.039**
Bison bison	−0.041	−0.015	*Mustela nivalis*	**−1.676**	**−2.161**
Bison bonasus	−0.211	0.133	*Mustela putorius*	**−0.689**	**−0.673**
Capra ibex	*0.553*	*0.799*	*Myodes glareolus*	**−1.539**	**−0.536**
Capreolus capreolus	0.229	*0.331*	*Myotis daubentonii*	*0.818*	**−0.488**
Castor canadensis	**−0.498**	**−0.381**	*Myotis lucifugus*	*1.179*	0.263
Cebus capucinus	*1.248*	−0.026	*Myotis myotis*	*1.635*	*0.796*
Cervus elaphus	0.274	0.259	*Nyctereutes procyonoides*	**−0.560**	-0.084
Cervus elaphus canadensis	−0.342	0.254	*Odocoileus virginianus*	**−0.998**	−0.041
Cervus eldii hainanus	**−0.548**	−0.218	*Orcinus orca*	*1.331*	*0.511*
Cervus nippon	*0.433*	*0.432*	*Oreamnos americanus*	0.216	0.151
Cynomys ludovicianus	−0.027	0.023	*Oryctolagus cuniculus*	**−1.190**	−0.236
Damaliscus lunatus	**−1.393**	**−0.972**	*Otaria flavescens*	0.174	−0.055
Dinaromys bogdanovi	−0.035	**−1.156**	*Ovis aries*	0.089	*0.860*
Dipodomys spectabilis	−0.209	**−0.945**	*Ovis canadensis*	0.022	*0.357*
Elephas maximus	*1.009*	*0.303*	*Ovis canadensis nelsoni*	0.048	*0.364*
Enhydra lutris	0.012	0.106	*Ovis dalli*	−0.098	0.167
Equus asinus	−0.315	0.246	***Pan troglodytes***	*1.227*	−0.255
Equus hemionus	−0.072	0.084	***Panthera leo***	−0.102	*0.357*
Equus quagga antiquorum	−0.209	0.010	*Papio cynocephalus*	*0.916*	0.217
Equus quagga boehmi	−0.016	0.191	*Pecari tajacu*	−0.350	−0.099
Erinaceus europaeus	−0.325	−0.267	*Peromyscus leucopus*	**−1.453**	**−0.516**
Fukomys aselli	*1.788*	*1.670*	*Petrogale assimilis*	*0.454*	0.287
Fukomys mechowii	*0.509*	0.013	*Phoca vitulina*	−0.419	**−0.774**
Globicephala macrorhynchus	*1.283*	*0.969*	*Phoca vitulina richardsi*	0.400	0.075
Globicephala melas	*0.685*	*0.351*	*Phocarctos hookeri*	*0.452*	0.222
Gorilla beringei	*1.093*	0.148	*Phocoenoides dalli*	**−0.991**	**−1.436**
Helogale parvula	*0.517*	−0.268	*Plecotus auritus*	*1.751*	*0.818*

(continued)

Table 4.1 (cont.)

Species	LQS	LQT	Species	LQS	LQT
Heterocephalus glaber	*1.899*	*2.012*	*Procapra gutturosa*	**–0.525**	**–0.428**
Hydropotes inermis	–0.410	0.256	*Propithecus verreauxi verreauxi*	*1.417*	*1.061*
Kobus ellipsiprymnus defassa	–0.394	0.134	*Pusa hispida*	*0.666*	0.209
Kobus leche	–0.426	-0.080	*Rangifer tarandus*	–0.257	0.186
Lepus europaeus	**–1.458**	*1.320*	*Rhabdomys pumilio*	**–1.668**	–0.282
Lepus timidus	**–1.115**	**–0.793**	*Rupicapra pyrenaica*	0.397	*0.495*
Lobodon carcinophagus	*0.725*	*0.877*	*Rupicapra rupicapra*	0.122	0.237
Loxodonta africana	*0.859*	*0.524*	*Scalopus aquaticus*	–0.228	**–0.662**
Lutra lutra	–0.226	**–0.383**	*Sciurus carolinensis*	–0.182	–0.225
Lycaon pictus	**–0.466**	**–0.385**	*Spermophilus beldingi*	–0.138	–0.319
Lynx lynx	**–1.529**	**–0.830**	*Spermophilus brunneus*	–0.066	**–0.418**
Lynx rufus pallenscens	0.063	–0.103	*Sus scrofa*	–0.329	–0.002
Macaca fuscata	*1.314*	*0.517*	*Syncerus caffer*	–0.413	–0.083
Macropus agilis	0.406	*1.069*	*Tamias striatus*	0.017	**–0.381**
Mandrillus sphinx	*0.525*	–0.243	*Tamiasciurus hudsonicus*	0.394	0.209
Marmota flaviventris	**–0.561**	**–0.432**	*Tragelaphus strepsiceros*	–0.235	0.258
Marmota marmota	0.339	–0.347	*Trichosurus vulpecula*	0.003	*0.292*
Meles meles	0.084	–0.041	*Tupaia glis*	**–0.824**	**–0.744**
Mephitis mephitis	**–0.912**	**–0.443**	*Ursus americanus*	–0.049	**–0.661**
Microcebus rufus	0.339	–0.143	*Vulpes vulpes*	**–0.975**	**–0.514**
Mirounga angustirostris	**–0.505**	**–0.528**	*Zalophus californianus*	0.244	–0.271
Mirounga leonina	**–0.661**	–0.270			

LQs have been calculated from either expected longevity based on species-specific body mass (LQS) or expected longevity based on species-specific age at first reproduction (LQT). LQs are calculated as the species-specific difference between observed and expected longevities on a log scale, which correspond to the residual values from the allometric relationships linking either longevity (with lifespan 80% used as a metric) and body mass (LQS) or longevity and age at first reproduction (LQT). The LQs of the shortest-lived species (identified as belonging to the lowest quartile of LQS and LQT) are highlighted in bold while the LQs of longest-lived species (identified as belonging to the highest quartile of LQS and LQT) are in italics and underlined. The two species (*Lepus europaeus* and *Myotis daubentonii*) with opposite LQ ranking based on body mass vs age at first reproduction are highlighted in dark grey. Species highlighted in light grey have opposite LQ signs, meaning that they are rather short-lived vs long-lived depending on whether body mass or age at first reproduction is used. Among species, those highlighted in bold were markedly short- or long-lived species based on either body mass or age at first reproduction.

Whether a mammalian species can be identified as having an exceptional longevity or not may thus strongly depend on the criterion we consider for assessing the expected longevity. When using body size, as almost consistently done so far, mole-rats, bats and, to a lesser extent, primates clearly show up as being exceptionally long-lived species among mammals. All these groups have been repeatedly identified as the longest-lived species among mammals (e.g. [45] for mole-rats, [12] for bats and [28] for primates). On the other hand, when the time scale opposing slow species with a long life cycle (i.e. combining a long developmental period, a late age at first reproduction, low annual fecundity and a long lifespan, [44, 46]) to species with a short life cycle (i.e. combining a short developmental period, an early age at first reproduction, high annual fecundity and a short lifespan, [44, 46]) is considered (using age at first reproduction as a proxy), the set of the top 5% species differs and the deviation of species-specific observed longevity from expected longevity is much less. These findings thus suggest that bats do not display exceptional longevity for their pace of life but are rather especially slow-living species for their size. However, the low number of bat species included in our analyses (only four), all of small size, reproducing from 1 year of age onwards, living in temperate areas and belonging to a single family (Vespertilionidae), prevents a firm conclusion and further analyses of larger datasets will be required. Even more strikingly, the European hare was the fifth shortest-lived mammal for its size but the third longest-lived for its pace of life, indicating that hares have evolved faster life histories than expected from both their size and their lifespan. On the other hand, mole-rats, which are well-known examples of extraordinary long-lived animals [47], were identified as especially long-lived mammals relative to both their size and their pace of life, while the weasel was a short-lived species for both its body mass and pace of life.

4.3 Identifying Mammalian Species with Exceptional Longevity from Finch and Pike's Model

Almost three decades ago, Finch and Pike provided a way to estimate maximum longevity when changes in age-specific mortality follow a Gompertz model [31]. The Finch and Pike approximation of maximum longevity (t_{max}) is a direct function of the exponential rate of mortality increase with age (α, the Gompertz rate), of the age-independent mortality rate coefficient (*A*, the baseline mortality rate) and of the number of individuals monitored to estimate the parameters of the Gompertz model (*N*, the sample size). It stands as (see Eq. (3) in Finch and Pike [31]):

$$t_{max} = \mathrm{Ln}(1+\alpha \mathrm{Ln}(\mathrm{N})/\mathrm{A})/\alpha \qquad (4.1)$$

We used Eq. (4.1) to estimate expected maximum longevity in both sexes of 45 mammalian species studied in Lemaître et al. [15], for which: (1) more than 50 individuals of a given sex had reached the age at first reproduction; and (2) a positive exponential rate of annual mortality with age was obtained (i.e. the Gompertz model provided a good fit for the age-specific changes of mortality of the studied population). We retrieved from Lemaître et al. the estimates of sex-specific maximum longevity of each of the 45 mammalian species [15] (Table 4.2). We then assessed the relationship between expected and observed maximum longevity across species in both sexes. A strong relationship occurred between expected and observed maximum longevity, which was remarkably similar in both sexes (males: $R^2 = 0.84$, slope of 1.40 ± 0.10 (1 SE), Figure 4.3(a); females: $R^2 = 0.82$, slope of 1.34 ± 0.11 (1 SE), Figure 4.3(b)). In both sexes, the intercept did not differ from 0 (-0.41 ± 0.23 and -0.26 ± 0.24 in males and females, respectively), indicating that for very short-lived species, the expected maximum longevity matched with the observed one. On the other hand, the slope consistently higher than 1 indicates that the expected maximum longevity increases at a much faster rate than observed maximum longevity. The increasing overestimation of expected maximum longevity seems to indicate that the Gompertz model better describes age-specific mortality patterns of short- than long-lived mammals. We used extremely long-lived mole-rats to check whether the Finch and Pike model led to a marked overestimation of maximal longevity. As the Gompertz model did not satisfactorily fit the survival data of naked mole-rat (a rodent that displays a negligible actuarial senescence pattern [48], see also Chapter 5), and since data on Mechow's mole-rat (see [39]) were too limited to allow accurate estimates, we only used the Finch and Pike model, employing the data published by Dammann et al. on Ansell's mole-rat [38]. The Gompertz model accounting for censored individuals provided a satisfactory fit (see Figure 4.6 and Section 4.4) and led to estimate an expected

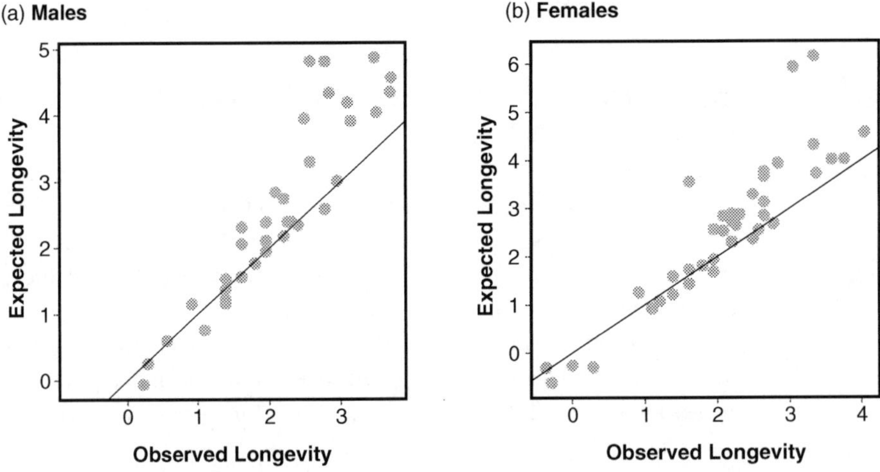

Figure 4.3 Relationship between species-specific observed maximum longevity (on a log scale) and species-specific maximum longevity (on a log scale) expected under a Gompertz model of age-specific changes in mortality in (a) males and (b) females across a set of 45 mammalian species for which a Gompertz model satisfactorily fitted age-specific mortality data in both sexes. The expected longevity increasingly overestimates observed longevity from shortest- to longest-lived species. In both (a) and (b), the regression lines correspond to the equation y = x.

Table 4.2 Sex-specific maximum longevity of a set of mammalian species

Species	Sex	Observed	Expected	Difference	Species	Sex	Observed	Expected	Difference
Alces alces	Female	16	14.57	0.09	*Myotis myotis*	Female	21	376.23	−16.92
Alces alces	Male	16	13.22	0.17	*Myotis myotis*	Male	16	122.62	−6.66
Castor canadensis	Female	8	17.06	−1.13	*Nyctereutes procyonoides*	Female	7	12.96	−0.85
Castor canadensis	Male	7	10.85	−0.55	*Nyctereutes procyonoides*	Male	6	NA	NA
Cervus elaphus canadensis	Female	8	12.53	−0.57	*Odocoileus virginianus*	Female	12	26.91	−1.24
Cervus elaphus canadensis	Female	17	51.33	−2.02	*Odocoileus virginianus*	Male	12	NA	NA
Cervus elaphus canadensis	Male	8	17.02	−1.13	***Oryctolagus cuniculus***	**Female**	**2.5**	**3.51**	**−0.4**
Cervus elaphus canadensis	Male	12	51.91	−3.33	***Oryctolagus cuniculus***	**Male**	**2.5**	**3.17**	**−0.27**
Dinaromys bogdanovi	**Female**	**5**	**4.19**	**0.16**	*Pecari tajacu*	Female	9	17.89	−0.99
Dinaromys bogdanovi	Male	5	4.75	0.05	*Pecari tajacu*	Female	15	NA	NA
Equus asinus	Female	9.5	14.2	−0.49	*Pecari tajacu*	Male	9	15.48	−0.72
Eauus asinus	Male	9.5	10.93	−0.15	*Pecari tajacu*	Male	14	NA	NA
Globicephala macrorhynchus	Female	57	97.34	−0.71	***Peromyscus leucopus***	**Female**	**0.69**	**0.74**	**−0.06**
Globicephala macrorhynchus	Male	40.5	95.73	−1.36	***Peromyscus leucopus***	**Male**	**0.46**	**NA**	**NA**
Globicephala melas	Female	43	56.03	−0.3	*Phoca vitulina richardsi*	Female	28	75.17	−1.68
Globicephala melas	Male	40	77.14	−0.93	*Phoca vitulina richardsi*	Male	23	49.97	−1.17
Hydropotes inermis	Female	7	6.97	0	*Phocoenoides dalli*	Female	10	17.67	−0.77
Hydropotes inermis	Male	7	6.96	0.01	*Phocoenoides dalli*	Male	10	10.92	−0.09
Lepus europaeus	**Female**	**3.34**	**2.95**	**0.12**	*Procapra gutturosa*	Female	6	6.12	−0.02
Lepus europaeus	**Male**	**3**	**2.13**	**0.29**	*Procapra gutturosa*	Male	6	5.85	0.02
Lepus timidus	**Female**	**4**	**3.37**	**0.16**	*Pusa hispida*	Female	29	41.56	−0.43
Lepus timidus	**Male**	**4**	**3.21**	**0.2**	*Pusa hispida*	Male	32	129.68	−3.05

(continued)

Table 4.2 (cont.)

Species	Sex	Observed	Expected	Difference	Species	Sex	Observed	Expected	Difference
Lobodon carcinophagus	Female	36	55.61	−0.54	*Rangifer tarandus*	Female	12	10.64	0.11
Lobodon carcinophagus	Male	33	56.79	−0.72	*Rangifer tarandus*	Female	14	17.21	−0.23
Lynx lynx	Female	13	NA	NA	*Rangifer tarandus*	Male	6	5.81	0.03
Lynx lynx	Male	8	NA	NA	*Rangifer tarandus*	Male	11	10.39	0.06
Lynx rufus pallenscens	Female	9	15.25	−0.69	***Scalopus aquaticus***	**Female**	**3**	**2.51**	**0.16**
Lynx rufus pallenscens	Male	13	27	−1.08	***Scalopus aquaticus***	**Female**	**4**	**4.9**	**−0.22**
Macropus agilis	Female	14	43.13	−2.08	***Scalopus aquaticus***	**Male**	**4**	**3.36**	**0.16**
Macropus agilis	Male	10	10.94	−0.09	***Scalopus aquaticus***	**Male**	**4**	**4.62**	**−0.16**
Mephitis mephitis	Female	5	5.63	−0.13	*Sus scrofa*	Female	5	35.01	−6
Mephitis mephitis	**Male**	**4**	**3.94**	**0.01**	*Sus scrofa*	Male	5	7.82	−0.56
Microcebus murinus	Female	8	NA	NA	*Syncerus caffer*	Female	13	12.83	0.01
Microcebus murinus	Male	7	8.21	−0.17	*Syncerus caffer*	Male	19	20.27	−0.07
Mustela erminea	**Female**	**1**	**0.77**	**0.23**	*Trichosurus vulpecula*	Female	9	9.98	−0.11
Mustela erminea	**Male**	**1.25**	**0.94**	**0.25**	*Trichosurus vulpecula*	Female	11	NA	NA
Mustela nivalis	**Female**	**0.75**	**0.54**	**0.28**	*Trichosurus vulpecula*	Female	14	22.87	−0.63
Mustela nivalis	**Male**	**1.75**	**1.81**	**−0.04**	*Trichosurus vulpecula*	Male	9	8.79	0.02
Mustela putorius	Female	6	NA	NA	*Trichosurus vulpecula*	Male	11	NA	NA
Mustela putorius	Male	5	10.06	−1.01	*Trichosurus vulpecula*	Male	13	123.11	−8.47
Myodes glareolus	**Female**	**1.33**	**0.74**	**0.44**	*Ursus americanus*	Female	28	473.48	−15.91
Myodes glareolus	**Male**	**1.33**	**1.28**	**0.04**	*Ursus americanus*	Male	17	76.3	−3.49
Myotis daubentonii	Female	14	39.05	−1.79	*Vulpes vulpes*	Female	7	5.4	0.23
Myotis daubentonii	Male	22	65.97	−2	*Vulpes vulpes*	Male	5	NA	NA

Observed maximum longevity (*Observed*) has been calculated in Lemaître et al. [15] from the l_x series of the life table. Expected values (*Expected*) have been estimated using the Finch and Pike model from data on sex-specific number of individuals monitored, Gompertz rate of actuarial senescence and baseline mortality rate presented in 'Supplementary Information' [15]. *Difference* corresponds to the proportional difference between observed and expected values and is calculated as: (Observed maximum longevity – Expected maximum longevity)/Observed maximum longevity. Species in bold are the shortest-lived ones (i.e. 10 shortest-lived males and 10 shortest-lived females), while species highlighted in light grey are the longest-lived ones (i.e. 10 longest-lived males and 10 longest-lived females). As discussed in the text and displayed in Figure 4.3, expected longevity is close to observed longevity in shortest-lived species but markedly overestimates longevity in longest-lived species.

maximum longevity of 26.1 years. This expected value indeed largely overestimated (by over 20%) the observed maximum longevity of 20.7 years registered in Dammann et al.'s (2019) data [38]. Therefore, in long-lived mammals, using the Gompertz rate of mortality increase with age from the age at first reproduction onwards leads to an underestimation of the true rate of mortality increase at old age, causing expected maximum longevity to exceed widely observed maximum longevity. This pattern is likely caused by the selective pressure against mortality risk taking place during early adulthood [49], which leads to a delay in the onset of actuarial senescence well beyond the age at first reproduction, as clearly identified in large herbivores and in *Fukomys anselli* [50] (see Section 4.4).

The Finch and Pike model does not allow us to identify any species with 'extreme longevity'. Rather, it provides a relevant way to check the fit quality of the Gompertz model. For species with age-specific mortality satisfactorily fitted by the Gompertz model (i.e. short-lived species in mammals), the deviation between observed and predicted longevity is low. All long-lived species displayed shorter longevity than predicted by the Finch and Pike model. This does not mean that they have extremely short longevity, but rather that the rate of senescence is underestimated in those species often displaying a delayed actuarial senescence.

4.4 Identifying Individuals with Exceptional Longevity within a Population Using Extreme Value Theory

In a recent work, Huang et al. reviewed models that have been used to model mortality rates at extreme ages [33], and proposed a new model called the smoothed threshold life table (STLT) model to allow identifying the age threshold between non-extreme and extreme ages. They applied this model to human data from the Human Mortality Database. The model of Huang et al. is composed of two parts that transition smoothly [33]: (1) a Gompertz model up to age N, and (2) a Generalized Pareto Distribution used to model extreme values. We used this model to assess the threshold age at which individuals display an exceptional longevity in three species with exceptionally detailed individual monitoring: medflies, for which a life table based on a huge sample size (over 1 million flies monitored from birth to death) was published by Carey and colleagues [34], rhesus macaques intensively monitored on Cayo Santiago (data on the fate of several hundreds of females from [35]) and the Ansell's mole-rat studied by Dammann and colleagues [38]. Based on the procedure proposed by Huang and colleagues, we first fitted a Gompertz model as the standard model for describing age-specific mortality patterns in mammals (see [50]). We then fitted the STLT model that included the potential occurrence of exceptionally long-lived individuals.

4.4.1 Assessing the Age at Which Exceptional Longevity Occurs in Medflies

The Gompertz model provided a very poor fit of the data (Figure 4.4(a), black dashed line). Applying the Finch and Pike model to these Gompertz estimates would result

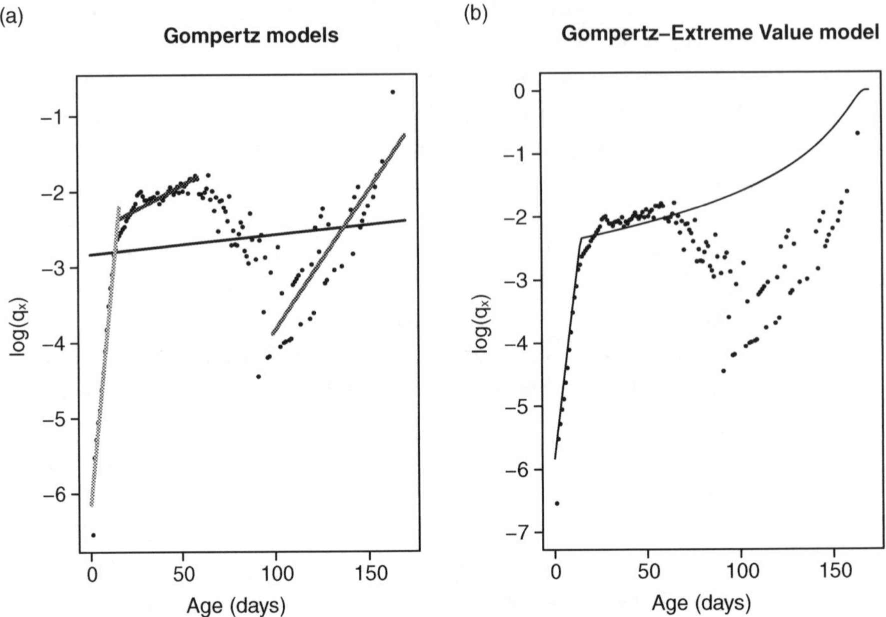

Figure 4.4 Modelling age-specific changes in mortality in medflies using either (a) a Gompertz or (b) a two-component (Gompertz–Extreme Value) STLT model. The Gompertz model provides a poor fit of age-specific mortality of flies throughout their lifespan (black dashed line in (a)) but provides an apparent good fit of early mortality (up to 17 days of age – grey dashed line, or from 17 to 60 days – pale grey continuous line) while strongly underestimating maximum longevity in both cases. The Gompertz model provides a satisfactory fit of age-specific mortality of the longest-lived flies (from 100 days of age onwards, dark grey continuous line in (a)). The STLT model provides a much better fit of mortality data than the Gompertz model, from birth to 60 days of age when 99.9% of flies have died (black line in (b)). However, the STLT model does a poor job modelling for the longest-lived individuals (i.e. from 60 to 170 days of age – see (b)).

in a maximum longevity of 186 days [31], close to the observed value of 171 days, but this is a coincidence given how poor the Gompertz model fits. When we restricted the fit to the first 17 days of life, the Gompertz model fitted better (Figure 4.4(a), grey dashed line), but would lead to a maximum longevity of 31 days as this model did not reproduce the lower mortality rates after 17 days compared to the Gompertz model fitted over the entire trajectory. When restricting the fit of the model to the period 17–60 days (Figure 4.4(a), pale grey line), the fit of the Gompertz model was somewhat better and gave a maximum longevity of $82+17=99$ days, still much lower than the observed 171 days. The STLT model provided a much better fit than the Gompertz model, although not for the longest-lived flies (Figure 4.4(b), best-fitting line of the STLT model in black). The STLT model was able to approximate the two piecewise Gompertz models detailed, but the second part corresponded to the extreme value part of the model. The model accounting for extreme longevity values provided a good fit for up to 60 days, leading us to conclude that the age at which exceptional longevity (relative to the expected longevity from a Gompertz model) occurs in medflies is 16 days, with no less than one third of flies displaying exceptional longevities, which looks a little paradoxical. The STLT model, however, strongly overestimated age-specific mortality rates after 60 days (Figure 4.4(b)). However, at 60 days, only 0.17% of flies survived, meaning that the model accounting for extreme values provided a satisfactory fit for 99.83% of flies, but only 35% of the entire age range of the fly mortality trajectory. We can then conclude that flies dying after 60 days display exceptional longevity relative to the expected longevity from the STLT model that accounts for the existence of extreme longevity values. Interestingly, a Gompertz model provided a very good fit of the age-specific mortality curve for the most long-lived flies (from 100 days onwards, dark grey line in Figure 4.4(a)), which had a baseline daily rate of mortality of 0.020 and an exponential rate of daily mortality increase of 0.037. This would result in an (additional) longevity of 59 days (i.e. 159 days in total), not far from the observed 171 days.

4.4.2 Assessing the Age at Which Exceptional Longevity Occurs in Rhesus Macaques

The Gompertz model provided a relatively good fit for the age-specific mortality patterns of female rhesus macaques (when starting at 5 years, which corresponds to the age at first reproduction, as expected from current evolutionary theories on actuarial senescence [51, 52, 53]; see Figure 4.5, best-fitting line of the Gompertz model in black). From this model, the baseline annual mortality was 0.06 and the exponential rate of annual mortality increase was 0.075. This would result in a longevity value from the Finch and Pike model of 28.8 years.

Fitting the STLT model indicated that the Gompertz model satisfactorily predicted age-specific mortality until 28 years of age (Figure 4.5, dotted line), when only 1.4% of macaques are still alive. The model accounting for the existence of extreme values provided a good fit until the end. From 28 years onwards the model indicates that mortality rate increases instead of decreasing, as observed in the medfly data.

However, it is noteworthy that the Gompertz or STLT model did not fit the early changes in mortality (mortality rates did not start increasing before an age of 10–12 years, as observed by [35]). We then refitted the STLT model to individuals older than 12 years. This led to a marked change in the shape of the predicted mortality rates, with a much better fit of the Gompertz model before the threshold age of 22 years, and a reduced increase in mortality rates after 22 years (Figure 4.5, dashed line). The occurrence of exceptional longevity in female macaques is thus restricted to a low proportion of individuals (about 6%) and a limited proportion of the entire mortality trajectory (the eight last years of life corresponding to 26.7% of the lifespan of the longest-lived females).

4.4.3 Assessing the Age at Which Exceptional Longevity Occurs in the Mole-Rat *Fukomys anselli*

The Gompertz model provided a relatively good fit for the age-specific mortality patterns of mole-rats (Figure 4.6, best-fitting line of the Gompertz model is the solid line).

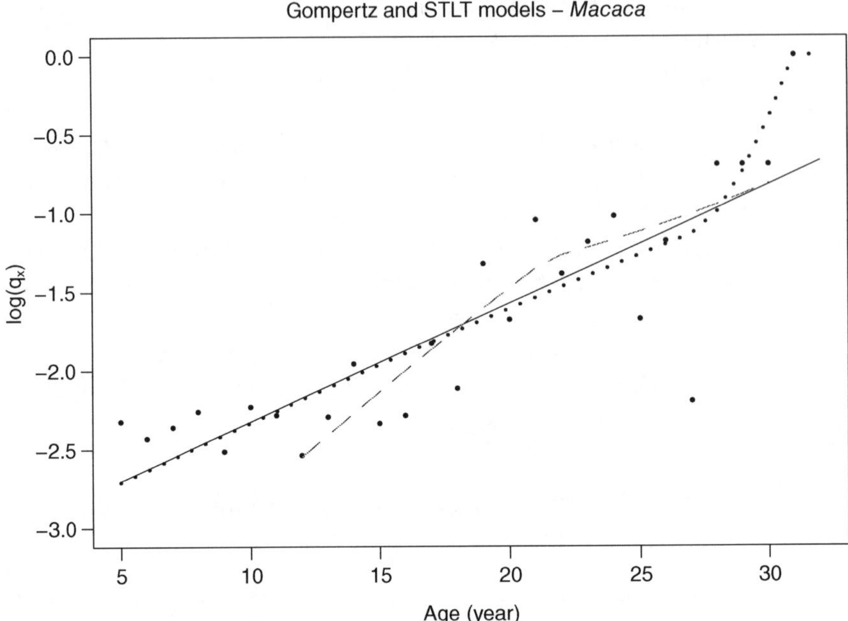

Figure 4.5 Modelling age-specific changes in mortality in female rhesus macaques using either Gompertz or STLT models. The Gompertz model provides a good fit of mortality data from 5 years onwards (solid black line) but for the longest-lived females. The STLT model from five years onwards (dotted black line) allows detecting exceptional longevity at very old ages (i.e. 28 years), corresponding to the age at which only a few female macaques are still alive. Neither model allows capturing the low mortality stage at 10–12 years of age. An STLT model with a delayed onset of actuarial senescence (starting from 12 years of age) provided the best fit of the data of old macaque females (dashed black line). According to this model, mortality first increases from 12 to 22 years of age following a Gompertz model and exceptional longevities occur in individuals older than 22 years of age.

From this model, the baseline annual mortality was 0.05 and the exponential rate of annual mortality increase was 0.096. This would result in a maximum longevity value from the Finch and Pike model of 26.1 years, well overestimated relative to the observed maximum longevity record of 20.7 years.

Fitting the STLT model indicated that the Gompertz model satisfactorily predicted age-specific mortality until 16 years of age (Figure 4.6, dotted line), when as many as 16% of mole-rats are still alive. The model accounting for the existence of extreme values provided a very good fit until the end. From 16 years onwards the model indicates that the mortality rate increases drastically, which illustrates a clear case of 'rectangularization' of the survival curve (sensu [53]). In other words, the low age-specific mortality rate spanning about 80% of the life trajectory allows a substantial proportion of individuals to reach relatively very old ages, but then mortality sharply increases and becomes extremely high within a few additional years.

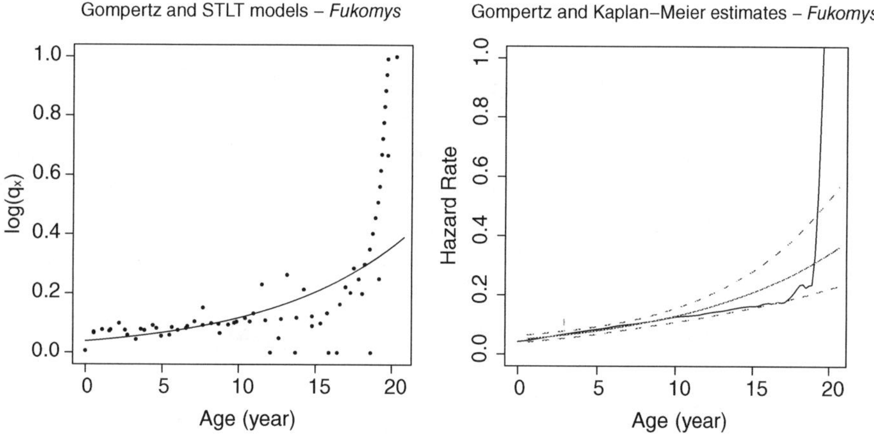

Figure 4.6 Modelling age-specific changes in mortality in the mole-rat *Fukomys anselli* using either Gompertz (solid line) and STLT (dotted line on the left panel) models, and Gompertz (grey line with associated confidence intervals in dashed lines) and Kaplan-Maier models (black line) on the right panel. The Gompertz model provides a satisfactory fit of mortality data from the age at first reproduction to about 10 years of age. However, the Gompertz model leads to a marked overestimation of age-specific mortality between 10 and ca. 16 years of age. The STLT model from the age at first reproduction onwards provides an appropriate fit to the entire mortality trajectory. In particular, it allows detecting a shift in the mortality rate at an age threshold of about 16 years of age. Prior to the age threshold, the mortality rate hardly increases with age, in line with the concept of negligible actuarial senescence, whereas from 16 years onwards, the mortality rate increases very rapidly with age. The STLT model clearly shows that the onset of actuarial senescence is substantially delayed and accounting for this delay allows estimating reliably maximum longevity (20.8 years of age estimated using the STLT model vs 20.7 observed).

4.4.4 Comparing Patterns of Exceptional Longevity among Medflies, Female Macaques and Ansell's Mole-Rats

Comparing mortality patterns among medflies, female rhesus macaques and mole-rats brings out striking differences, especially with regard to the prevalence of exceptional longevity. In medflies, as many as one third of individuals were classified as displaying exceptional longevity relative to the Gompertz model of an exponential mortality increase at a constant rate. Moreover, using the Huang et al. model to account for the occurrence of flies with extreme longevity values [33], we identified a second order of exceptional longevity displayed by some rare flies (less than 0.2%), including some flies that were able to live almost three times longer than the longest-lived 'normal' flies (i.e. corresponding to the 99.83% of flies that die before 60 days of age). On the contrary, in macaques, only 1.4% females displayed exceptional longevity relative to the Gompertz model, with the model accounting for the occurrence of female macaques with extreme longevity values nicely fitting the entire mortality trajectory from birth to the oldest age (i.e. no second order of exceptional longevity). The longest-lived females only outlived 'normal' female macaques by about a quarter of the maximum longevity (i.e. 8 years compared to 30 years). Mole-rats displayed a different mortality trajectory, about halfway between medflies and macaques. In mole-rats, there were no really exceptionally long-lived individuals. Rather, the key feature of mole-rat mortality patterns was a markedly delayed senescence (i.e. starting at about 16 years of age while starting to breed at less than 1 year) and a negligible actuarial senescence over about 80% of the life course of oldest individuals. When reaching 16 years of age, mole-rats were subjected to an especially fast mortality increase with age that led the quite high proportion of individuals reaching 16 years of age to all die within the next four years.

As the number of individuals monitored strongly influences maximum longevity, as repeatedly demonstrated in previous studies (see e.g. [37, 54]), it is tempting to link the complex mortality patterns displayed by the 0.2% longest-lived flies to the unusually high sample size and assume that similar patterns might exist in macaques or mole-rats but remain undetected due to a too low number of individuals monitored (i.e. 0.2% of longest-lived individuals correspond to about 2400 flies but less than two macaques and one mole-rat). However, considering mortality patterns in humans, for which millions of individuals have been monitored, suggests that the striking differences in patterns of exceptional longevity between flies and macaques or mole-rats cannot be fully explained by differences in sample size. Huang et al.'s analysis of human mortality data from the Netherlands reported a threshold age at which exceptional mortality occurs of about 97 years for females (see [33], table 5.1). So far, the longest-lived woman is Jeanne Calment, who died at 122 years of age (see also Chapter 14). Human exceptional longevity relative to a Gompertz model thus ranges over 20.5% of the entire mortality trajectory, with the longest-lived individual with an exceptional longevity outliving the longest-lived 'normal' individual by 1.25 times, values remarkably similar to those we obtained in both female rhesus macaques (26.7% and 1.36 times, respectively, when using the STLT model based on an onset of actuarial senescence at 12 years) and Ansell's mole-rats (15.9% and 1.29 times, respectively). In medflies, the corresponding figure is markedly different, with 90.7% of coverage and a tenfold longer life. Further work is badly needed to understand what drives and limits the magnitude of the extension of lifetime by individuals with exceptional longevity and the possibility of longest-lived individuals with exceptional longevity to outlive longest-lived 'normal' individuals.

4.5 Conclusion

These analyses provide two major outcomes about exceptional longevity patterns across species and among individuals within populations.

First, the assessment of exceptional longevity is strongly dependent on the life history descriptor used as a reference to identify expected (i.e. 'normal') longevity. While between-species comparisons are either performed using allometry analyses based on body size or sometimes even using simple analyses of distributions that ignore any structural constraint (see e.g. [55] for a recent example), we demonstrate that the pace of life markedly drives inter-specific variation in longevity and has to be considered as an alternative descriptor, or in addition to body size, when looking for identifying species with exceptional longevities. Including the pace-of-life dimension will allow researchers to tease apart species that are exceptionally long-lived from species that have exceptionally slow life histories. For instance, our preliminary analyses suggest that currently available data indicate that bats have exceptionally slow life histories rather than that they are exceptionally long-lived.

Second, both using the Finch and Pike approach to estimate species-specific maximum longevity from a Gompertz model [31], and assessing the threshold age at which individuals are exceptionally long-lived compared to their conspecifics within a given population, demonstrates the key role played by the onset of actuarial senescence. Restricting analyses of actuarial senescence to the rate of senescence is clearly inappropriate and would generally lead to flawed interpretation. For instance, our analyses of the mole-rat mortality trajectory using the STLT model pointed out the existence of a two-phase actuarial senescence pattern, with, first, a negligible senescence from the age at first reproduction to about 16 years of age, and then a steep mortality increase from 16 years up to the oldest ages. This case study on mole-rats, as well as the accumulation of high-quality data on age-specific mortality both within and across species, demonstrates that we cannot longer assume that actuarial senescence starts at the age at first reproduction. Assessing specifically an age of onset of senescence should allow both better-predicting maximum longevities using the Finch and Pike method and more accurate identification of the threshold age at which exceptional longevity occurs [31].

Biogerontological studies aiming to identify the molecular mechanisms underpinning the ageing process currently focus on species labelled as 'exceptionally long-lived' [56, p. 492] – the naked mole-rat being an iconic example of this set of species [45]. They also focus on human populations through the use of various omics approach in order to identify some genomic, proteomic and metabolomic signatures of extreme longevities, generally by targeting centenarians or supercentenarians (see Chapter 16). In both cases, the ultimate goal of these studies is to develop relevant anti-ageing interventions for humans. However, these two approaches rely on an accurate identification of species or individuals displaying extreme longevities, which appear much more complex than generally considered. In that context, our chapter offers new perspectives for a much more accurate identification of exceptional longevities, which might open new research avenues and model species in medical sciences.

References

1. Stearns, S.C. 1976. Life-history tactics: a review of the ideas. *Q. Rev. Biol.* **51**, 3–47.
2. Carey, J.R. 2002. Longevity minimalists: life table studies of two species of northern Michigan adult mayflies. *Exp. Gerontol.* **37**, 567–570.
3. George, J.C., Bada, J., Zeh, J., Scott, L., Brown, S.E., O'Hara, T., Suydam, R. 1999. Age and growth estimates of bowhead whales (*Balaena mysticetus*) via aspartic acid racemization. *Can. J. Zool.* **77**, 571–580.

4. Nielsen, J., Hedeholm, R.B., Heinemeier, J., Bushnell, P.G., Christiansen, J.S., Olsen, J., Ramsey, C.B., Brill, R.W., Simon, M., Steffensen, K.F., et al. 2016. Eye lens radiocarbon reveals centuries of longevity in the Greenland shark (*Somniosus microcephalus*). *Science* **353**, 702–704.
5. Armitage, K.B. 2014. *Marmot Biology: Sociality Individual Fitness and Population Dynamics*. Cambridge University Press.
6. Speakman, J.R. 2005. Correlations between physiology and lifespan: two widely ignored problems with comparative studies. *Aging Cell* **4**, 167–175.
7. Peters, R.H. 2012. *The Ecological Implications of Body Size*. Cambridge University Press. Available online: /core/books/ecological-implications-of-body-size/4D86337571D7F26E76F885B2548FCBFB.
8. Calder, W.A. 1984. *Size, Function, and Life History*. Harvard University Press.
9. Shattuck, M.R., Williams, S.A. 2010. Arboreality has allowed for the evolution of increased longevity in mammals. *Proc. Natl. Acad. Sci.* **107**, 4635–4639 (doi:10.1073/pnas.0911439107).
10. Dammann, P., Šaffa, G., Šumbera, R. 2022. Longevity of a solitary mole-rat species and its implications for the assumed link between sociality and longevity in African mole-rats (Bathyergidae). *Biol. Lett.* **18**, 20220243.
11. Zhu, P., Liu, W., Zhang, X., Li, M., Liu, G., Yu, Y., Zhou, X. 2023. Correlated evolution of social organization and lifespan in mammals. *Nat. Commun.* **14**, 372.
12. Wilkinson, G.S., Adams, D.M. 2019. Recurrent evolution of extreme longevity in bats. *Biol. Lett.* **15**, 20180860.
13. Healy, K., Guillerme, T., Finlay, S., Kane, A., Kelly, S.B., McClean, D., Kelly, D.J., Donohue, I., Jackson, A.L., Cooper, N. 2014. Ecology and mode-of-life explain lifespan variation in birds and mammals. *Proc. R. Soc. Lond. B Biol. Sci.* **281**, 20140298.
14. Reinke, B.A., Cayuela, H., Janzen, F.J., Lemaître, J.-F., Gaillard, J.-M., Lawing, A.M., Iverson, J.B., Christiansen, D.G., Martínez-Solano, I., Sánchez-Montes, G. 2022. Diverse aging rates in ectothermic tetrapods provide insights for the evolution of aging and longevity. *Science* **376**, 1459–1466.
15. Lemaître, J.-F., Ronget, V., Tidière, M., Allainé, D., Berger, V., Cohas, A., Colchero, F., Conde, D.A., Garratt, M., Liker, A. 2020. Sex differences in adult lifespan and aging rates of mortality across wild mammals. *Proc. Natl. Acad. Sci.* **117**, 8546–8553.
16. Gaillard, J.-M., Festa-Bianchet, M., Delorme, D., Jorgenson, J. 2000. Body mass and individual fitness in female ungulates: bigger is not always better. *Proc. R. Soc. Lond. B Biol. Sci.* **267**, 471–477.
17. Zedrosser, A., Pelletier, F., Bischof, R., Festa-Bianchet, M., Swenson, J.E. 2013. Determinants of lifetime reproduction in female brown bears: early body mass, longevity, and hunting regulations. *Ecology* **94**, 231–240.
18. Yunis, E.J., Watson, A.L., Gelman, R.S., Sylvia, S.J., Bronson, R., Dore, M.E. 1984. Traits that influence longevity in mice. *Genetics* **108**, 999–1011.
19. Schmidt, C.M., Jarvis, J.U., Bennett, N.C. 2013. The long-lived queen: reproduction and longevity in female eusocial Damaraland mole-rats (*Fukomys damarensis*). *Afr. Zool.* **48**, 193–196.
20. Gibbs, M., Dyck, H. 2010. Butterfly flight activity affects reproductive performance and longevity relative to landscape structure. *Oecologia* **163**, 341–350.
21. Tatar, M., Gray, D.W., Carey, J.R. 1997. Altitudinal variation for senescence in melanoplus grasshoppers. *Oecologia* **111**, 357–364.

22. Green, C.L., Lamming, D.W., Fontana, L. 2022. Molecular mechanisms of dietary restriction promoting health and longevity. *Nat. Rev. Mol. Cell Biol.* **23**, 56–73.
23. Monaghan, P., Heidinger, B.J., D'Alba, L., Evans, N.P., Spencer, K.A. 2012. For better or worse: reduced adult lifespan following early-life stress is transmitted to breeding partners. *Proc. R. Soc. B Biol. Sci.* **279**, 709–714.
24. Sacher, G.A. 1959. Relation of lifespan to brain weight and body weight in mammals. In *The Lifespan of Animals: CIBA Foundation Colloquia on Ageing*, vol. 5 (eds G.E.W. Wolstenholme and M. O'Connor), pp. 115–141. Churchill.
25. West, G.B. 2017. *Scale. The Universal Laws of Growth, Innovation, Sustainability, and the Pace of Life in Organisms, Cities, Economies, and Companies*. Penguin Press.
26. Palmore, E. 1969. Predicting longevity: a follow-up controlling for age. *The Gerontologist* **9**, 247–250.
27. Austad, S.N., Fischer, K.E. 1991. Mammalian aging, metabolism, and ecology: evidence from the bats and marsupials. *J. Gerontol.* **46**, 47–53.
28. Austad, S.N., Fischer, K.E. 1992. Primate longevity: its place in the mammalian scheme. *Am. J. Primatol.* **28**, 251–261.
29. Yu, Z., Seim, I., Yin, M., Tian, R., Sun, D., Ren, W., Xu, S. 2021. Comparative analyses of aging-related genes in long-lived mammals provide insights into natural longevity. *The Innovation* **2**, 100108.
30. Shilovsky, G.A., Putyatina, T.S., Markov, A.V. 2022. Evolution of longevity as a species-specific trait in mammals. *Biochem. Mosc.* **87**, 1579–1599.
31. Finch, C.E., Pike, M.C. 1996. Maximum life span predictions from the Gompertz mortality model. *J. Gerontol. A. Biol. Sci. Med. Sci.* **51**, 183–194.
32. Ronget, V., Lemaître, J.-F., Tidière, M., Gaillard, J.-M. 2020. Assessing the diversity of the form of age-specific changes in adult mortality from captive mammalian populations. *Diversity* **12**, 354.
33. Huang, F., Maller, R., Ning, X. 2020. Modelling life tables with advanced ages: an extreme value theory approach. *Insur. Math. Econ.* **93**, 95–115.
34. Carey, J.R., Liedo, P., Orozco, D., Vaupel, J.W. 1992. Slowing of mortality rates at older ages in large medfly cohorts. *Science* **258**, 457–461.
35. Lee, D.S., Kang, Y.H., Ruiz-Lambides, A.V., Higham, J.P. 2021. The observed pattern and hidden process of female reproductive trajectories across the life span in a non-human primate. *J. Anim. Ecol.* **90**, 2901–2914.
36. Moorad, J.A., Promislow, D.E., Flesness, N., Miller, R.A. 2012. A comparative assessment of univariate longevity measures using zoological animal records. *Aging Cell* **11**, 940–948.
37. Ronget, V., Gaillard, J.-M. 2020. Assessing ageing patterns for comparative analyses of mortality curves: going beyond the use of maximum longevity. *Funct. Ecol.* **34**, 65–75.
38. Dammann, P., Scherag, A., Zak, N., Szafranski, K., Holtze, S., Begall, S., Platzer, M. 2019. Comment on 'Naked mole-rat mortality rates defy Gompertzian laws by not increasing with age'. *eLife* **8**, 45415.
39. Begall, S., Nappe, R., Hohrenk, L., Schmidt, T.C., Burda, H., Sahm, A., Henning, Y. 2021. Life expectancy, family constellation and stress in giant mole-rats (*Fukomys mechowii*). *Philos. Trans. R. Soc. B* **376**, 20200207.
40. Ruby, J.G., Smith, M., Buffenstein, R. 2023. Five years later, with double the demographic data, naked mole-rat mortality rates continue to defy Gompertzian laws by not increasing with age. *eLife*. Available online: https://elifesciences.org/reviewed-preprints/88057.

41. Warton, D.I. 2022. *Eco-Stats: Data Analysis in Ecology: From t-Tests to Multivariate Abundances*. Springer International Publishing.
42. Gaillard, J.-M., Yoccoz, N.G., Lebreton, J.-D., Bonenfant, C., Devillard, S., Loison, A., Pontier, D., Allaine, D. 2005. Generation time: a reliable metric to measure life-history variation among mammalian populations. *Am. Nat.* **166**, 119–123.
43. Magalhães, J.P.D., Costa, J., Church, G.M. 2007. An analysis of the relationship between metabolism, developmental schedules, and longevity using phylogenetic independent contrasts. *J. Gerontol. A. Biol. Sci. Med. Sci.* **62**, 149–160.
44. Gaillard, J.-M., Lemaitre, J.F., Berger, V., Bonenfant, C., Devillard, S., Douhard, M., Gamelon, M., Plard, F., Lebreton, J.D. 2016. Axes of variation in life histories. In *Encycl. Evol. Biol. Elsevier* vol. 2 (ed. R.M. Kliman), pp. 312–323. Academic Press.
45. Buffenstein, R., Park, T.J., Holmes, M.M., eds. 2021. *The Extraordinary Biology of the Naked Mole-Rat*. Springer.
46. Stearns, S.C. 1983. The influence of size and phylogeny on patterns of covariation among life-history traits in the mammals. *Oikos* **41**, 173–187.
47. Stenvinkel, P., Shiels, P.G. 2019. Long-lived animals with negligible senescence: clues for ageing research. *Biochem. Soc. Trans.* **47**, 1157–1164.
48. Ruby, J.G., Smith, M., Buffenstein, R. 2018. Naked mole-rat mortality rates defy Gompertzian laws by not increasing with age. *eLife* **7**, e31157.
49. Gaillard, J.M., Loison, A., Festa-Bianchet, M., Yoccoz, N.G., Solberg, E. 2003. Ecological correlates of life span in populations of large herbivorous mammals. *Popul. Dev. Rev.* **29**, 39–56.
50. Gaillard, J.-M., Garratt, M., Lemaître, J.-F. 2017. Senescence in mammalian life history traits. In *The Evolution of Senescence in the Tree of Life* (eds R.P. Shefferson, O.R. Jones, R. Salguero-Gomez), pp. 126–155. Cambridge University Press.
51. Williams, G.C. 1957. Pleiotropy, natural selection, and the evolution of senescence. *Evolution* **11**, 398–411.
52. Hamilton, W.D. 1966. The moulding of senescence by natural selection. *J. Theor. Biol.* **12**, 12–45.
53. Fries, J.F. 1980. Aging, natural death, and the compression of morbidity. *N. Engl. J. Med.* **303**, 130–135.
54. Krementz, D.G., Sauer, J.R., Nichols, J.D. 1989. Model-based estimates of annual survival rate are preferable to observed maximum lifespan statistics for use in comparative life-history studies. *Oikos* **70**, 203–208.
55. Berkel, C., Cacan, E. 2021. Analysis of longevity in chordata identifies species with exceptional longevity among taxa and points to the evolution of longer lifespans. *Biogerontology* **22**, 329–343.
56. Brunet, A. 2020. Old and new models for the study of human ageing. *Nat. Rev. Mol. Cell Biol.* **21**, 491–493.

5 The Inevitability of Senescence

Annette Baudisch

5.1 Preamble

Out of the blue, my 6-year-old son broke into tears. It was the moment when the insight of his own finitude, his irreversible growth, ageing and death hit him – hard and suddenly. He said he wanted to remain small, he did not want to grow up and die, nor did he want me to die.

Do I have to grow old, or can I avoid it, and why? These questions come to all of us, sooner or later, to scholars and laypeople alike. The thought of growing old and dying is painful; hence it seems natural that people have been dreaming of a fountain of youth for thousands of years. The wish to remain young forever is easily sympathized with, for who wants to lose their strength and wits that life so naturally and generously builds up during two decades of childhood and youth? Yet, after a certain age, everyone experiences growing older, with no exceptions.

This chapter briefly journeys through the empirical evidence and theoretical underpinnings of the fundamental question 'to age or not to age' across the vast richness of life forms in nature. This chapter is not about 'to die or not to die'. Because all that lives will die, eventually. This is a basic law of nature. Whether for a life form, however, death will strike most often at younger or older ages, or may be blind to age entirely, is a question worth of scientific enquiry.

5.2 Empirical Evidence

5.2.1 Initial Evidence

For most of history, evidence of ageing was not recorded. Ageing was entirely up to personal experience and observation. People saw people growing old. Domestic animals also grew old. Ageing in other taxa, such as plants, fungi or algae, were hard to relate to, given their radically different body plans and ways of living. Ageing appeared to be inevitable.

The scientific study of ageing required a scientific term to avoid the inherent bias of personal experience and multifaceted associations. Initially used to mean 'wearing out', Pearl [1], and later Finch [2], defined the term *senescence*. This term describes age-related changes in an organism that adversely affect physiological functioning

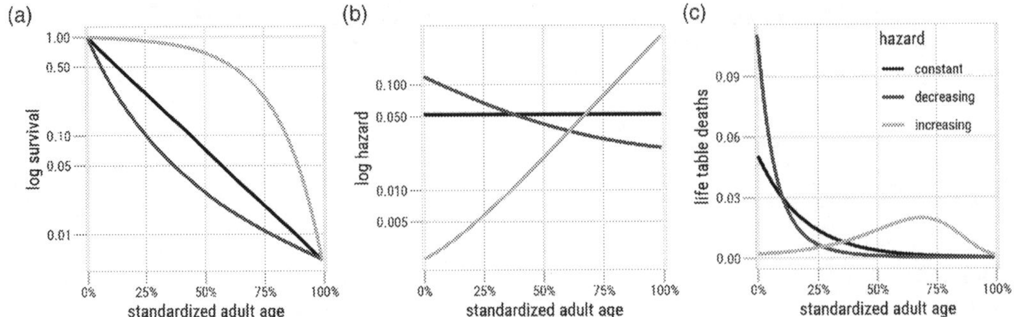

Figure 5.1 Stylized types of survival patterns (a) of Type I (light grey line), Type II (black line) and Type III (dark grey line); (b) the associated underlying increasing, constant and decreasing mortality patterns over standardized adult age; and (c) the respective distribution of death. As in Pearl [1], patterns are compared along a standardized age scale. In contrast to Pearl, here the patterns pertain to adult ages only and do not include infant mortality. Simulated data based on the Siler function.

associated with an increase in the risk of death with age. The definition cuts across levels of observation. It highlights that senescence is a multi-scale process. Changes at the genetic and physiological level eventually become apparent at the demographic level – the level of birth and death patterns. This chapter addresses actuarial senescence, the increase in the risk of death over age.

Evidence for actuarial senescence (or the absence of it) requires data on estimates of survival or death counts in a population. The first written account of data on death for males and females by age in England goes back to John Graunt in 1662 [3]. His pioneering work marked the birth of demography. Written records of vital rates emerged in the nineteenth century under William Farr, who is considered the father of vital statistics. In 1928, Raymond Pearl reviewed the existing data on life duration [1], which covered humans, mice, beetles, flies, moths, slugs, mealworms, tubeworms, rotifers and the freshwater polyp *Hydra*.

Pearl experimented with the lifespan of *Drosophila* for different treatments and strains of male and female flies and compared their survival patterns with those of other taxa. With only a few species at hand, Pearl [1] recognized that survival patterns should, in theory, broadly fall into three categories, classified by three types. These types are illustrated with stylized examples in Figure 5.1. Type I captures an upward-bending pattern, which may approach a roughly rectangular shape (Figure 5.1(a), light grey line). Type I survivorship corresponds to a population in which, initially, deaths rates are low and then rise with age (Figure 5.1(b), light grey line; mortality patterns are also known as the hazard function). Many individuals initially survive and then die at rather similar ages (Figure 5.1(c), distribution of death). The hump in the death distribution frames a typical or 'determinate' lifespan for that population. Type II captures (on a log scale) a diagonal survivorship curve (Figure 5.1(a), black line). It characterizes a population in which death rates are the same at all ages (Figure 5.1(b)). Ages at death spread widely over the age range

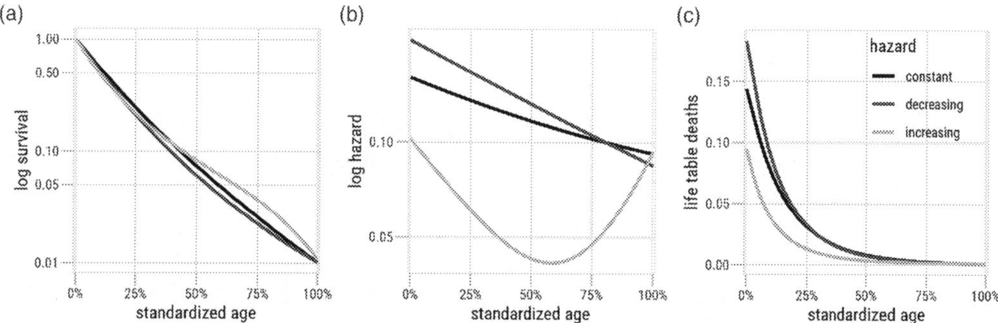

Figure 5.2 Stylized types of survival patterns (as Figure 5.1), but including premature as well as mature ages. A similarly high level of infant mortality is added to each of the three cases of increasing, constant and decreasing mortality patterns over age. The emerging patterns in (a) make it hard to differentiate between Type I, II and III survivorship; while (b) disguises negligible senescence as apparent negative senescence (black line), because infant mortality is so high that it dominates the constant mortality over adult ages. Distributions of death in (c) do not allow types of senescence to be distinguished. The hump of death that is typical for patterns of senescence disappears due to the many infant deaths. The specific examples were chosen to highlight possible problems arising from including infant mortality. These problems may be less dramatic in real-life examples.

and individual lifespans do not cluster around a common length of life. Instead, the distribution of death peters out and does not show a hump (Figure 5.1(c)). Species of Type II therefore are said to exhibit an 'indeterminate lifespan'. Type III captures (on a log scale) a downward-bending survivorship curve (Figure 5.1(a), dark grey line). It corresponds to a population in which death rates are initially high and then fall with age (Figure 5.1(b)). All but a few individuals survive the initially high hazard of death. The remaining few, however, can expect to live much longer, with no determinate lifespan (Figure 5.1(c)). Compared to the other two types, inequality in ages at death for Type III is most extreme, with most individuals experiencing short lifespans, but a few individuals enjoying vastly longer lives.

Note that Figure 5.1 depicts patterns over adult ages only and does not include the premature age range. Pearl did not explicitly distinguish between premature and adult age range when introducing his classification [1]. This is important to recognize, because including higher or lower infant mortality rates in analysing survivorship patterns can substantially change the apparent survivorship type and corresponding changes in the patterns of mortality and in the distribution of death, as demonstrated in Figure 5.2.

With life-table data on people, flies and moths, Pearl had evidence for Type I survivorship. A few years later, together with John Miner [4], he added rotifers, roaches, mice, as well as automobiles (!) to the Type I category. Pearl and Miner extended their analysis by reporting first evidence also for Type II survivorship [4], the freshwater polyp *Hydra* and the slug *Agriolimax*. These appeared to be organisms that reveal no marked change in the risk of death over their life course – *negligible senescence*. Studies published in German language on *Hydra* by Hase [5], and on *Agriolimax* by

Szabó and Szabó [6], may well be considered the first evidence for negligible senescence, long before the term was defined by Finch [2].

Pearl and Miner presented no evidence for Type III [4], but they suggested that fish and tree species show decreasing mortality with age, arguing that most of the millions of eggs or seeds would die, and only few juveniles would make it into adulthood. Their observation, however, only concerns premature ages. What this may imply for adult patterns of mortality, the authors did not speculate. It seems natural that death rates should initially fall from conception to reproductive maturity, not just for fish and tree, but for most, if not all, species as organisms develop and grow (Type III). As demonstrated in Figure 5.2, this could obscure the type of survivorship across adult ages.

With their work, Pearl and Miner pioneered a fundamental typology of survival trajectories [4], which is taught in textbooks of biology up until today. Around the same time as Pearl and Miner, in 1932 George Parker Bidder proposed the hypothesis that senescence was not observed in species that could grow without limit [7], a hypothesis that was later refuted [8]. Briefly, Bidder argued that a limit would be imposed by gravity for species on land, but size and growth were less constrained under water. Because fish could grow indefinitely, Bidder thought that fish could avoid senescence. Indeed, as reported by Medawar [9, p. 15], Bidder could not 'remember any evidence of a marine animal dying a natural death'.

In sum, the initial evidence on death rates across species was meagre. It was dominated by observations in humans, domestic animals and a few peculiar species that were picked out from among the diversity of life forms for convenient and feasible experimentation in the laboratory, such as fruit flies. One may argue from experience that humans and domestic animals were expected to senesce without even considering life-table data. And other species with available life-table data were not the best candidates to escape senescence. All cells are post-mitotic in *Drosophila* (except for their gonads), and the fragile body of a moth invites wear and tear that is hardly repairable. Yet, even among just seven species, Pearl and Miner detected two species that seemed to avoid the dictate of senescence [4]. Would not it be tempting to speculate that, with an increasing number of different sample species, further cases of negligible senescence could emerge?

5.2.2 Widespread Evidence of Negligible Senescence across Phyla

Caleb E. Finch marks a milestone of evidence on longevity and ageing [2]. In a great effort, Finch compiled existing evidence across the diversity of life forms. He found that – rather than being an exceptional phenomenon within some peculiar branch of the tree of life – widespread evidence for negligible senescence emerged across phyla [10, 11, 12].

Finch classified types of senescence into *rapid*, *gradual* and *negligible senescence*, highlighting the existence of a continuum of patterns across types of senescence. In his classification:

- *Rapid senescence* describes organisms that experience an onset of pathophysiologic changes shortly after reproductive maturity. After maturity, rapid senescence occurs over a short phase of adult life (no longer than a year). It may also occur in long-lived species that spend many years as juveniles, such as bamboo, before they reproduce and die within a short time interval. In general, Finch classifies many short-lived species as rapid senescent, which include all annual plants; short and long-lived semelparous species, such as octopus, Pacific salmon, or marsupial mice; invertebrates, such as rotifers, nematodes or flies, but also a few short-lived mammals, such as many small rodents.
- *Gradual senescence* describes adverse age-specific changes that happen more gradually and are not synchronized for adults by age. This category includes humans and all placental mammals with available data on mortality, including shorter-lived rodents with a larger spread of ages at death over one or several years.
- *Negligible* or extremely gradual *senescence* describe organisms where dysfunctional changes have so far eluded detection [2]. Negligible senescence is characterized by no detectable increase in mortality over age and is associated with indeterminate lifespan.

As an indication of indeterminate lifespan, most of the evidence on negligible senescence compiled by Finch is based on reports of exceptionally long-lived individuals and pertains to long-lived species [2], such as bristlecone pine, lobster, quahog, rockfish and tortoises. Indeterminate lifespan, however, need not be an exclusive trait of long-lived species alone. Finch proposed that short-lived species might exhibit indeterminate lifespan, and hence negligible senescence, as well. A high environmental hazard of death may keep lifespan short, even in the absence of internal deterioration. As evidence for such species, Finch and colleagues report rhesus monkeys, pipistrelle bats and five feral birds with high initial mortality rates (IMR) but slow mortality rate doubling times (MRDT) [12]. This suggested that negligible senescence might occur in both long- and short-lived species. Among the short-lived species, data on hydra represent the best-confirmed evidence of negligible senescence to date [13, 14].

Notably, the idea that short-lived species could exhibit negligible senescence appeared to be a 'radical' proposal [2, p. 222] at the end of the last century, and presenting evidence for negligible senescence across a wide range of phyla was eye-opening in the field of gerontology. Most of the emerging evidence for negligible senescence, however, was hardly reliable. It was derived from indirect measures of indeterminate lifespan or slow MRDT and mainly included species of extremely long-lived conifers, and deep-dwelling fish, birds and reptiles, as summarized by Finch [11].

A deeper search for evidence on negligible senescence was challenged by the absence of age-specific information on survival or mortality. Beyond a few cases for which life-table data were published, most such data were in the hands of individual researchers. No openly available databases on mortality patterns supported comparative research on ageing across species to fuel joint efforts that can answer big questions of interest to a whole research field. Instead, Finch and colleagues stated that they 'extracted average survival by age group' [12, p. 903] from graphs of the fraction surviving as a function of age that were depicted in published reports, or they inferred MRDT from information on reported average levels of mortality and maximum lifespan using formal demographic methods. To verify the apparent evidence on negligible senescence, especially at older ages when sample size diminishes due to the dwindling number of survivors, larger studies with age-specific information, preferably from monitoring survival individually throughout time, were greatly needed.

5.2.3 Expanding the Scope of Senescence towards 'Negative Senescence'

With the turn of the century, newly emerging evidence further shook the prevailing understanding of the limits of senescence.

At first sight unspectacularly encountered along pavements, in school playgrounds or any other place with the slightest chance to put down roots, a common weed earned its place in the history of biodemographic advances. On the lawn of the plantation where Thomas Jefferson was buried, Deborah Roach undertook a major longitudinal study on the perennial *Plantago lanceolata*. Planting individual plants and following them throughout life revealed astounding patterns that she presented at a workshop in 2002. Roach found that larger adult plants had a higher chance of surviving than smaller adult plants. As it takes time to grow large, she reported that the older and thus larger plants survived the best. Following her presentation, James W. Vaupel got up from his seat and exclaimed that this was the most exciting result of the day and that these results were evidence for 'negative senescence'.

This was a blunt statement because negligible senescence had been the most extreme pattern imaginable among gerontologists across the diversity of forms [10]. Hence, several people in the audience retorted that negative senescence did not make sense, because – as they put it – William D. Hamilton had proved in 1966 that senescence was inevitable. Only days later, I had a printed version of Hamilton's work on my desk, [15] with an image of a grumpy-looking chimpanzee on the cover page, and a note by Vaupel that read 'Please disprove! Cordially Jim'. This inspired the theoretical developments to be described later in this chapter.

Roach and Gampe observed declining mortality over adult ages in the plant [16], with lowest levels of mortality observed for the highest age groups. As a mechanism, Roach and Vaupel proposed that growth after reproductive maturity may increase reproductive potential and could reduce mortality [16, 17], which might facilitate an escape from senescence. Implementing this hypothesis in an evolutionary demographic model, we made the case that negative senescence can be an evolutionary optimal life history strategy [17].

This evidence shook Hamilton's paradigm of inevitable senescence [15]. It marked the beginning of now two decades of biodemographic research on negligible and negative senescence and the diversity of senescence patterns across the tree of life.

The possibility of negative senescence extended the theoretical continuum of types of senescence described by Finch [2]. *Negative senescence* captures life histories that are marked by declining mortality after the onset of reproduction before, eventually, mortality levels off. As candidates for negative senescence, Vaupel and colleagues listed coral species [17], clonal clusters of quaking aspens, wild leek, brown algae, neo-tropical and forest trees, cushion plants, molluscs and sea urchins. Their speculation that hydra may show falling mortality at younger ages, before entering a phase of negligible senescence, has been disproven [14].

A bottleneck for research on the extremes of senescence remained: the access to actual data on age-specific mortality or survival across species, which are the realm and expertise of evolutionary ecology and life history biology.

5.2.4 Diversity of Senescence across the Tree of Life

A strong line of research within evolutionary ecology (e.g. [18, 19, 20, 21, 22, 23]) investigated senescence patterns around the turn of the century, coming from – one might say – the opposite end of the question that is posed in this chapter.

Instead of asking whether senescence could be avoided, the accepted ecological paradigm had been that senescence was absent in the wild. Individuals should simply not live long enough to express any signs of innate deterioration in the form of reduced survival or reproduction [24]. This paradigm could prevail because exposing senescence patterns in the wild is challenging. The field took on the challenge, summarized in a special issue of *Functional Ecology* on the *Evolutionary Ecology of Senescence* [25]. With datasets of terrestrial vertebrates from 20 mark–recapture studies, Nussey and colleagues revealed that senescence in the wild is common [26]. It affects both survival and reproduction among mammals and birds [27].

Capture–recapture analyses allowed estimating survival patterns from monitoring alive individuals without the need to observe mortality events. Such analyses have provided most of the empirical evidence of actuarial senescence in wild vertebrates. The large collection of evidence for senescence in the wild [28], and evidence for the diversity of ageing across the tree of life [29], has ignited an explosion of studies documenting the presence or absence of senescence across species and environmental conditions [1, 30, 31, 32, 33, 34, 35, 36, 37, 38].

Integrating empirical and theoretical knowledge about senescence patterns, Jones and colleagues compared age trajectories of mortality and fertility based on data from life-tables and population projection matrices for 46 species across the tree of life [29], including representative species for the major clades of vertebrates and invertebrates, as well as different types of plants and a green algae. Data were of different type and quality, including of wild and captive populations as well as period and cohort data. This study, for the first time, revealed the full spectrum of senescence, including positive, negligible and negative senescence. Steeply increasing human mortality over adult ages (positive senescence) in Japan 2009 (102 years) contrasted with constant mortality (negligible senescence) in hydra (1 400 years) and the armed saltbush (9 years). U-shaped adult mortality (mixed type) was observed in roe deer (13 years) and Soay sheep (12 years), whereas decreasing mortality throughout adult age (negative senescence) marked the life courses of oarweed (*Laminaria digitata*) (8 years), netleaf oak (*Quercus rugosa*) (177 years) and the desert tortoise (*Gopherus agassizii*) (64 years).

The numbers in brackets refer to the age by which 95% of adults are expected to have died. Note that the 1 400 years for hydra is not a typo. The value was extrapolated from the extremely low mortality of hydra in the sheltered condition of a laboratory [14]. Though it is a bold stretch to extrapolate lifespan up to thousand years ahead from just a few years of observation, it highlights the remarkable survival feat of a creature as tiny as hydra, which should merely live weeks, maybe months, but not years, given the scaling law of size and lifespan across the tree of life [39, 40, 41].

5.2.5 Principle Limits to Shifting Patterns of Senescence in Primates

Jones and colleagues established another empirical milestone of research on ageing and longevity [29]. Yet, they provide only snapshots of populations at a certain time and place. Mortality, however, depends on prevailing conditions. Observed patterns change, depending on the environment. Recent evidence for intra-species variation across environmental conditions reveals macro-level constraints on how ageing can be altered in humans and primates.

Across historical time and across countries at different levels of development, Colchero and colleagues found a strong regularity in the way that human lifespan may increase when conditions become more favourable [42]. As people live longer, they tend to die at more similar ages. This link between longer lifespan and higher lifespan equality is constrained to follow a tightly linear relationship (as illustrated here by the stylized examples in Figure 5.3). The strong linear pattern is striking and remains to be explained. But the positive relationship between longer lifespan and higher lifespan equality in humans can be understood intuitively.

Lifespan extended, because lives were saved mainly at earlier ages, that is, infant, childhood and young adult ages. As lives were saved due to, for example, better hygiene and advances in medicine, survivors had to die at a later age. Thereby, deaths were removed from younger ages and added to the most common pile of death (the hump in Figure 5.1(c), light grey line), making the lifespans of two random individuals in the population more equal.

Under the hood of these dynamics lies a hidden constraint on how mortality may change for humans when conditions improve. Rather than affecting the rate of ageing – which remained relatively unchanged across time and place – the emergence of longevous populations was brought about by reducing the level of mortality [43]. A lower level of mortality led to a postponement of senescence to higher ages. This postponement is in line with the common saying that '70 is the new 60' [44]. It is also in line with the observation that there is an MRDT that is typical for human populations and does not change much across a range of conditions [11]. No change in MRDT implies no change in ageing rates.

Colchero and colleagues recently discovered similarly linear relationships for different primate population [45], such as chimpanzees (*Pan troglodytes*), gorillas (*Gorilla spp*) or baboons (*Papio spp*), that fan off from the human line. For each species, populations under wild and captive conditions aligned along a species-specific linear relationship between length of life and lifespan equality (in the stylized example

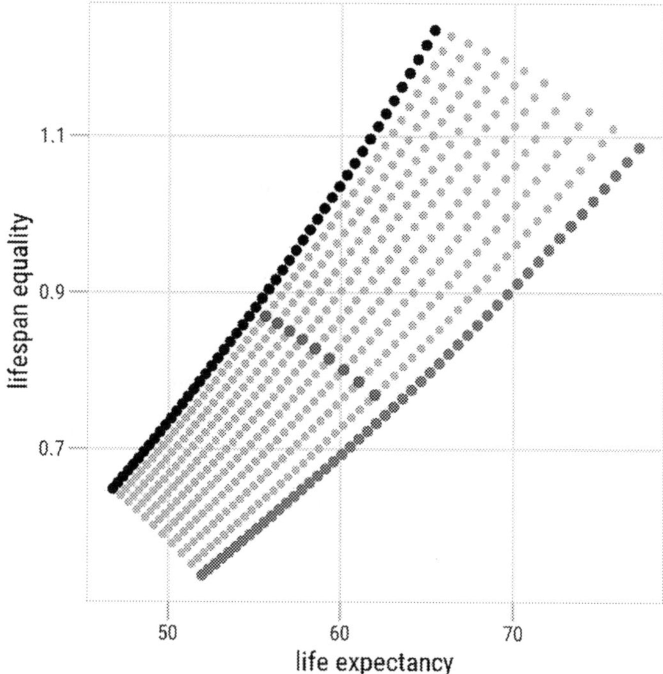

Figure 5.3 Illustration of the finding by Colchero and colleagues for stylized species (Siler model) [45, 111]. Intercept parameters for the exponential Siler terms are held constant in all species. Two species are characterized, respectively, by a high (black dotted line) and a low (grey dotted line) species-specific, fixed rate of ageing. Changing age-independent mortality (the constant additive term in the Siler model, which captures background mortality set by the environment) simulates a range of different environmental conditions (one dot per parameter combination), which align along linear patterns from lower left to upper right corner. Light grey dotted lines illustrate cases for intermediate ageing rates. The dark grey dots crossing over from the top black dotted line to the bottom grey dotted line correspond to simulated populations where the rate of ageing is successively changed from high to low, whereas age-independent mortality along the dark grey dotted line is fixed at an intermediate level.

of Figure 5.3, these different species are illustrated by the grey, black, and dark grey dotted lines). Colchero and colleagues demonstrated that only changes in the level or intercept parameters (of the Siler mortality model) could be consistent with the observed pattern [45]. If instead they modelled a change in the slope parameter – the rate of ageing – they would end up shifting a population off its own species line towards the line of another species (in Figure 5.3 the dark grey dots crossing over from the grey to the black line). Hence, one may hypothesize that the species-specific linear relationships of primates reflect a natural constraint on the plasticity of primate senescence.

These results suggest that improvements in medicine, nutrition, hygiene and any other change in the environment will most likely not alter the typical pace of senescence in human and non-human primates. Instead, Colchero and colleagues [45] provide evidence for the invariant rate of ageing hypothesis [43], which implies that biological constraints limit slowing the rate of ageing. Vaupel highlights that although ageing has been postponed to higher ages by lowering the level of mortality [43], the rate of increase in mortality has remained relatively constant.

5.2.6 Potential for Shifting Patterns of Senescence in Other Clades

Primate senescence is bound to toe the species line [45]. But how much does this depend on 'primate-ness'? Is senescence in other clades similarly constrained? Are there species that could slow the rate of ageing, or even escape from positive senescence and shift towards negligible or negative senescence?

Laboratory experiments demonstrated that constraints in flies and worms deviate from those observed for primates. For example, changing environmental conditions altered the pace of senescence in *C. elegans*. Senescence in the worm happened faster under bad, and slower under good, conditions [46], the latter known to sometimes cause dramatic extensions of lifespan [47]. Changes in diet for *Drosophila* altered the intercept but not the slope of mortality. Changes in temperature, instead, altered the slope but not the level of mortality [48]. Timing and type of feeding altered medfly mortality quite flexibly. The timing of feeding protein to the flies would enable reproduction and kick-started a steep increase in mortality. A sugar-only diet, instead, prevented reproduction and came with a slower increase in mortality with age [49], again an effect on the rate of ageing, which is not observed in primates.

Despite long-standing success in substantially manipulating ageing and longevity for species in the laboratory, none of these treatments caused a shift of senescence patterns out of type. That is, no treatment could change positive into negligible, or negligible into negative senescence. No experiment, to my knowledge, switched off a senescent increase in mortality completely.

In the wild, Finch argues that such a shift may occur in certain fish [11], such as trout [50]. While most individuals remain small, feed on insects or algae, live relatively short lives and exhibit positive senescence, a few individuals overcome a certain threshold in size beyond which they may grow into gigantic individuals (called ferox trout) that prey upon smaller fish, can live considerably longer than their conspecifics and may show negligible senescence. Similar within-species differences in lifespan are known between queens and workers in eusocial insects. For honeybees, Finch classifies these differences as a shift from rapid to gradual, rather than negligible, senescence [11]. In naked mole-rats, by contrast, all individuals, independent of breeding status, have been reported to show negligible senescence [51, 52, 53]. Note, however, that Ruby and colleagues [52] analysed survival data from a captive colony of naked mole-rats (*Heterocephalus glaber*). Given that patterns of senescence can be largely buffered in captive conditions, it is possible that naked mole-rats might suffer positive senescence in the wild.

Direct evidence for a full type-shift in senescence patterns recently emerged. Mark–recapture data on age-specific mortality in snakes suggests that the meadow viper (*Vipera ursinii*) [54] may shift between negligible and positive senescence when living in an optimal versus suboptimal environment. By contrast, hydra, the prime example of negligible senescence [5, 13, 14, 55, 56], exhibits a shift from negligible to positive senescence when it changes from asexual to sexual reproduction [57].

Such switches are not to be expected in primates [45], and probably not for mammals or birds in general. But why is that so? What is it in the mammal-ness or bird-ness that constrains whether a species can slow the rate of ageing, or shift it from positive to zero, or to negative, values? We still understand very little about the key factors that determine the type of senescence of a species. Indeed, for most species, we do not know what type of senescence they are following, and what constraints limit the plasticity of senescence in different phyla.

Conde and colleagues reveal the landscape of demographic knowledge across 97% of tetrapod species [58]. Only 1.3% of those species have data on age-specific mortality and fertility, mostly covering studies of mammals and birds. Much less is known for reptiles and almost nothing for amphibians. As Chapter 6 will elaborate, data from zoos and aquariums hold great potential to fill existing data gaps and support research on understanding the principal constraints that limit changes in mortality in response to environmental changes.

Age-specific data are needed on mortality or survival from life-tables or population projection matrices, preferably from longitudinal studies across different environmental conditions, to make progress on understanding when and why senescence is inevitable. Efforts in the community of biodemographers over recent years strongly support this direction. Open access databases for animals (DATLifeDatabase [59]; COMADRE [60]) and plants (COMPADRE [61]), and data on captive populations with millions of records for 21 000 species [62], now provide a much stronger basis for research on senesce patterns than was available to Caleb E. Finch in the early 1990s [2]. Still, even within the available knowledge base of demography, to date, researchers face a large taxon bias to study senescence in non-human species.

Notably, even among the best available data for human patterns of senescence provided by the human mortality and fertility databases (Human Fertility Database, HFD; Human Mortality Database, HMD), researchers face a bias towards countries with reliable birth and death registration, covering 41 out of 194, that is about one fifth of the recognized countries in the world.

Beyond data, unequivocal classification of senescence is important. For example, Wang and colleagues report Type III survivorship for 83 species of Yangtze River fish species [65], which would suggest negative senescence in all these fishes, that is, falling mortality over adult ages. Yet, this is not the case. Their exemplary life-table, given in Table 5.1 [65], reveals that remaining life expectancy begins to fall before the age of reproductive maturity, which implies that mortality rises with adult age in that fish species.

This mismatch demonstrates a pitfall of the original classification [4], because the authors apparently define Type III as falling mortality from birth onwards, and name fish, with high juvenile mortality, as one example for Type III survivorship (as Wang and colleagues have used it here [65]). In recent decades, however, Type III survivorship is increasingly used interchangeably with negative senescence [66].

This requires caution because the concept of negative senescence does not concern mortality at premature ages. The classification of senescence into positive, negligible and negative types concerns adult age patterns [67]. It is only equivalent to the classification of Type I, II and III survivorship when it is measured from maturity onwards. This distinction is important to prevent misreporting of apparent negative senescence.

5.2.7 Future Empirical Evidence

As more data becomes available, empirical evidence for negligible or negative senescence is expected to both increase and decrease in the future due to two different mechanisms. On the one hand, evidence may increase as data for under-represented or absent taxa emerge. On the other hand, evidence is likely to decrease, because senescence has a history of being hard to reveal [25, 68].

Regarding the former mechanism, Pearl and Miner found two out of seven species (a freshwater polyp, *Hydra fusca*; and a slug, *Agriolimax agrestis*), and Jones and colleagues found 11 out of 46 species show negligible senescence (hydra, *Hydra magnipapillata*; hermit crab, *Pagurus longicarpus*; armed saltbush, *Atriplex acanthocarpa*; red abalone, *Haliotis rufescens*) or negative senescence (red-legged frog, *Rana aurora*; red gorgonian, *Paramuricea clavata*; viburnum, *Viburnum furcatum*; oarweed, *Laminaria digitata*; netleaf oak, *Quercus rugosa*; desert tortoise, *Gopherus agassizii*; white mangrove, *Avicennia marina*) [4, 29]. This means there is an absence of evidence for senescence in about one quarter of the species in these studies. By contrast, the database AnAge [69] lists only a tiny fraction of seven species to show negligible senescence: olm ('human fish', *Proteus anguinus*), Blanding's turtle (*Emydoidea blandingii*), eastern box turtle (*Terrapene carolina carolina*), rougheye fish (*Sebastes aleutianus*), red sea urchin (*Mesocentrotus franciscanus*), ocean quahog clam (*Arctica islandica*) and Great Basin bristlecone pine (*Pinus longaeva*).[1] This reflects the strong taxon bias towards mammals and birds, which persists within wild populations of mammals, where mortality patterns of some groups remain largely understudied, such as small rodents and bats. It also reflects that much of the evidence in AnAge is linked to longevity records alone, which gives little support to conclude whether mortality does or does not increase with adult ages. Yet, based on maximum longevity records, AnAge provides the largest coverage across species for comparative studies on longevity.

Regarding the latter mechanism, evidence of negligible and negative senescence is likely to decline, as demonstrated by the massive efforts on revealing senescence in the wild [35, 68]. Further, senescence may also be revealed as new data across environmental conditions accumulate for longer periods of study, as exemplified by painted turtles [70]. A new comparative study across Testudines in captivity, led by da Silva and Colchero [71], finds that the actual senescence pattern of the desert tortoise classified as negative senescence by Jones et al. becomes negligible (for males) and

[1] AnAge information as of 11 March 2021.

positive senescence (for females) when enough data – here in a protected environment – is available [29]. At the same time, the study reveals new cases of negative senescence for two other clades of Testudines, confirming that both mechanisms of increasing as well as decreasing evidence will play a role as new data is emerging. In the same issue of *Science*, Reinke and colleagues report phylogenetically widespread cases of negligible senescence across 77 species of wild tetrapod ectotherms [72]. To put these results into perspective, Austad and Finch highlight that the famous Galápagos tortoise Harriet was reported to die of a heart attack at the age of 170 years, and the oldest living Aldabra tortoise Jonathan (estimated to be 160–190 years old) is now blind, has lost his sense of smell and needs to be fed by hand [73]. These typical signs of ageing suggest that patterns of actuarial senescence may mask patterns of physiological changes. In general, any absence of evidence for senescence should be carefully considered. Data quality and type differs largely across studies. To detect senescence, longer and more detailed data will be required. In any case, species with no evidence for senescence invite a deeper look.

In which clades would we expect to find non-senescence in the future? Current evidence for non-senescence from comparative studies based on age patterns of mortality covers algae, plants and invertebrates. Among the vertebrates, it includes reptiles and amphibians. These ectothermic clades seem to hold great potential for future evidence on non-senescent demographic patterns. Negligible senescence has recently been reported for the meadow viper *Vipera ursinii* under beneficial environmental conditions [54]. Negative senescence has been reported for three tropical snakes [74].

In general, however, it should not be expected that all or most ectotherms escape senescence. Reviewing the evidence in reptiles, Hoekstra and colleagues report 'ample evidence of increasing mortality with advancing age' [36, p. 38]. Thus, reptiles and presumably also amphibians, appear to be promising classes to study shifts and turning points for when senescence evolves, or not. They are particularly interesting because, in contrast to most plants, these classes separate the germ line from the soma, which, as elaborated in Section 5.3.1, should theoretically promote the evolution of senescence [75].

Notably, even in mammals and birds, evidence of non-senescence is emerging. The barn owl (*Tyto alba*) [76] and the naked mole-rat [51, 52, 53] apparently escape actuarial senescence, presumably as a consequence of a much reduced environmental risk of death by flight or by living underground [10, 39, 77].

For future evidence, short-lived species could become particularly valuable. Lifespans of short-lived species can be fully observed within timespans that enable continuous research support. Their typically small body size facilitates handling in the laboratory. Especially within classes that are probable to reveal non-senescence, such as reptiles and amphibians, but also for clades of invertebrates and plants, short-lived species could become invaluable to study the boundary between senescence and non-senescence. Because – as initially proposed by Finch [2], and later formalized by Baudisch in the pace–shape framework of mortality [78, 79, 80], and recently fertility [81] – short-lived species are candidates for non-senescence just as much as long-lived ones.

5.2.8 Evidence with Care

All evidence can only be as good as the underlying data and measures. It needs to be confirmed – if possible – with sharper and finer-grained data and measures of senescence, especially when it comes to apparent negligible and negative senescence.

Baudisch et al. found no signs of senescence for 93% out of 290 Angiosperm species [82], predominantly among the shorter-lived species, such as *P. lanceolata*, in line with the initial evidence by Roach and Gampe [16]. A more nuanced and interesting picture emerged when Roach extended her studies on *P. lanceolata* by several cohorts to investigate if and how plantago senesces. On the demographic level, mortality may still decline over chronological age, in line with previous findings [16]. Because it is still true that the plants grow with age, and as they grow larger, their mortality falls. Yet, independent of chronological age and adjusted for size, Roach and colleagues discovered signs of physiological decline for individual plants in the years prior to death [83]. Roach and colleagues also found comparatively lower stress resistance of older individuals of the same size [84]. Thus, a deeper look at the data reveals physiological senescence in this plant even though this was not initially visible from the population pattern of mortality over age. Just as observed for tortoises [73], senescence 'under the hood' is possible in plants. Negative actuarial senescence over chronological age may co-occur with positive stage-dependent physiological senescence in the stages prior to death.

By contrast, studies of senescence in other species, such as fish or sea urchins [50, 85, 86], may not present evidence for the absence of actuarial senescence, but may still measure signs of negligible or negative senescence on the physiological level. This calls for careful interpretation when reading the existing empirical evidence and when defining and measuring senescence – or the absence of it – across scales of observation. And ageing holds further ruses to tackle.

Colchero and Schaible demonstrate how changes in mortality over chronological age can be a combination of both age and size effects [87]. A positive (or negative) effect on mortality due to larger size might mask a negative (or positive) effect due to higher age. This could explain the negative actuarial senescence in *P. lanceolata* despite underlying physiological ageing, as observed by Shefferson and Roach [83].

Last but not least, heterogeneity in frailty of individuals implies survival of the strongest and removal of the weakest individuals over the life course [88]. This selection effect reduces the level of mortality in the surviving population and dampens the increase in population mortality over age relative to the individual level. This could lead to apparent slow, negligible or even negative senescence [89]. In short, evidence for or against negligible and negative senescence must be carefully validated. This could be done by applying frailty models [88], accounting for selective disappearance statistically [90]. The most reliable evidence would derive from large sample sizes of longitudinal observations on age-specific mortality or survival for several cohorts in favourable and unfavourable conditions, as, for example, produced over the years by Deborah Roach and her colleagues with her studies on *P. lanceolata*.

Though not all evidence will stand the scrutiny of further investigation, the repeatedly confirmed and reliable evidence for negligible senescence in hydra since 1909 until today makes a strong case that senescence is not inevitable [5, 13, 14, 55, 56].

Whether negative senescence will be corroborated in the future is an open question. It seems clear, however, that the general phenomenon of falling mortality with increasing size is real. Biologists have long known that individuals may gain a survival advantage as an organism grows larger. However, whether this gain is substantial after reproductive maturity and for a significant amount of time as a phase of adult life remains to be verified for existing and future evidence, just as 'substantial' and 'significant' in this context will have to be conceptually grounded. As for plantago, it may well turn out that nature permits 'ageing under the hood'. Nature may permit a slow accumulation of errors and damage on the molecular, (epi)genetic and even physiological level that do not visibly affect mortality patterns, that is, they do not become apparent on the actuarial level. Mortality may fall while physiological functioning may remain unchanged and molecular damage may accumulate. Our future image of non-senescence will need to qualify senescence patterns across scales of observation, and eventually, to explain how changes on one level translate into changes into the levels above.

5.3 Theoretical Underpinnings

The appearance of widespread evidence on negligible senescence posed a challenge for the existing theories of the evolution of ageing [10, 11, 24, 75, 91], which manifested the paradigm that senescence was inevitable [15]. To reconcile the emerging evidence with existing theories, it could have been argued that senescence was just extremely slow in species with no apparent senescence. The initial evidence on negative actuarial senescence in plantago [16], however, revealed a large gap between theory and data [17], and left no loophole to escape from a necessary expansion of the theoretical underpinning.

Chapter 2 describes the theories of senescence. In what follows, this chapter adds a further perspective by focusing on how, since the early 2000s, theoretical developments have been inspired by and researchers have pushed for empirical research on non-senescent age patterns of mortality and fertility.

5.3.1 Why Senescence Evolved

Human mortality rises with adult age. Primate mortality rises with adult age. Probably much of mammalian and bird mortality rises with adult age, they senesce. But why does senescence exist at all? Evolution favours organisms with high survival and high fertility. Should evolution not work against a detrimental trait, such as senescence, that reduces survival and fertility? Should any senescent species not have become extinct [24]?

As argued by Sir Peter Medawar in 1952 [9], and later formalized by William D. Hamilton in 1966 [15], senescence evolves because survival and reproduction affect Darwinian fitness whereby species are at their strongest at the youngest ages. Whether older individuals survive and reproduce, by contrast, has less and less an effect on evolutionary outcomes. This is called the declining force of selection.

Based on a declining force of selection, different theories have proposed different mechanisms that lead to the evolution of senescence, as fully elaborated in Chapter 2. Briefly, either senescence evolved because evolution was not strong enough to purge the negative effects of senescence at higher ages, when the force of selection dwindles (mutation accumulation theory [24]) and/or senescence evolved as a side effect of selection for beneficial effects earlier in life that are coupled with detrimental effects later in life. This early–late life trade-off favours early life over late life traits, as the force of selection declines with age (antagonistic pleiotropy theory [91]). A physiological trade-off is central to the disposable soma theory [75]. Resources for repair of somatic cells could be saved and concentrated solely on maintaining the quality of the germ line until a new generation is established. Saved resources for maintenance of the parental body would better be spent on increasing the number of offspring [75, 92, 93].

Other mechanisms, different from the disposable soma theory but in line with the antagonistic pleiotropy theory, have been proposed to explain the evolution of ageing [94, 95]. No matter what the mechanism, however, all classic theories predict – conditional on the existence of age-specific genes or on the distinction between the soma and the germ line – an increase in mortality with age, and thus predict inevitable senescence. This seems at odds with the empirical evidence on negligible and negative senescence. What is missing in the evolutionary theories of senescence?

Classic theories view the organism along an age trajectory, along which the importance of the survival of the individual dwindles as the age-specific force of selection declines. Chronological age, however, is typically just a proxy for an organism's state. Rather than age, it is size or stage that determine an organism's survival and reproduction [96].

5.3.2 Did Senescence Always Evolve?

The case of *P. lanceolata* established size and growth as two key missing elements in the evolutionary theories of senescence [16]. Including these elements, evolutionary demographic models made the case for negative senescence [17], and demonstrated that a diversity of senescence patterns, ranging from positive to negative senescence as well as mixed forms, could be evolutionary optimal life history strategies [97].

A central ingredient in these models was the balance between damage and repair, which was a model outcome to be optimized. The models allowed this balance to be positive, zero or negative, which is important to not a priori constrain the results to either senescence or non-senescence [98]. A further central component was the trade-off between maintenance and growth on the one side and reproduction on the other. Growth increased the size or 'vitality' of the organism [17, 97], which improved survival and increased reproductive potential, but required higher maintenance costs. Growth in these models could be indeterminate, that is, continue throughout adult ages, if parallel allocation in survival and reproduction after maturity were optimal strategies.

Vaupel and colleagues hypothesized that negative senescence characterized species that mature at a size or stage that is relatively small compared to maximum attainable size or stage [17]. Larger size or stage in these species increases their reproductive capacity. Baudisch found that given the potential of indeterminate growth, negative senescence characterized the life history of species 'that can easily share resources between [survival and reproduction] and that do not gain much by specializing in either one of them' [97, p. 117]. In other words, the models found that strategies of negative senescence are optimal if allocating one more unit of resources to survival would improve survival by less than the previous unit. And allocating one more unit of resources to reproduction would increase the number of offspring by less than the previous unit. It thus suffices to allocate just enough resources to either of these processes – an intermediate share – to achieve the evolutionary optimal level of survival and reproduction. The opposite is true of increasing returns to resource allocation. Here, allocating a further unit into one process results in a larger gain than the previous unit.

Decreasing returns to allocation in survival maybe typical for 'organisms that are made of simple, repeated structure that can easily be discarded and regenerated when damaged' [97, p. 119]. This 'throw away and grow new' strategy, Baudisch argues, is an efficient and cheap way of repair and growth, especially for organisms where damaged or lost parts are likely not to interfere with the functioning of the rest of body [97, p. 118]. The degree of interdependence of body parts could hence be a strong indicator of whether senescence evolves and whether the decay of an organism happens somewhat randomly and gradually or follows a predictable downward spiral of death [99]. This line of reasoning suggests that species with a modular body plan are prime candidates to escape senescence, as has been hypothesized by Finch [2], and is supported by recent evidence [33].

Decreasing returns to allocation in reproduction may be typical for 'organisms that are capable of vegetative propagation, where growth can be considered an investment in reproduction' [97, p. 117]. Here, newly grown cells initially contribute to the survival of the focal individuals. Once parted from the focal body to form a new individual during vegetative and other modes of asexual reproduction, the same investment in survival now changes interpretation. By contributing to a new individual, the newly grown biomass becomes effectively an allocation to reproduction. This observation undermines a central adaptive premise of why senescence evolves: the trade-off between survival and reproduction. It disappears once survival (via growth of the focal individual) is turned into reproduction (shedding of parts of the adult body, which now become independent offspring). It also implies that there is no clear distinction between the soma and the germ line, in line with the disposable soma theory [75].

Baudisch highlights that for species capable of both asexual as well as sexual reproduction, the alternative modes of reproduction may be associated respectively with decreasing versus increasing returns to allocation in reproduction [97]. Asexual reproduction may not require (much) additional costs, if growth is turned into a means of reproduction; sexual reproduction may entail much higher costs, requiring switching on a distinct reproductive machinery. The former would favour non-senescence, such as observed in asexually reproducing hydra [14]. The latter would favour senescence, such as observed in sexually reproducing hydra [57]. A switch in returns to allocation from decreasing to increasing could explain the switch between types of senescence [97].

The models by Vaupel, Baudisch and colleagues specify a range of components and are solved using dynamic optimization [17, 97], which may raise the concern that their findings are presupposed by the various model specifications. This concern can be addressed with much simpler, less involved models, which do not specify size, growth or age-specific mortality and fertility. Simple models have demonstrated that it can be worth sacrificing a higher level of reproduction for a comparatively larger gain in lifespan [100]. Crudely speaking, the same fitness can be reached either by reproducing at a high level over a short time, or reproducing at a low level, but over a long time. In line with this argument, a model of 'safe niches' favours the evolution of negligible senescence [100], as for example might be the case for hydra. A small fraction of the population could sometimes become lucky and end up in a safe environment, which offers survival prospects that would be extraordinary compared to the normally brief lifespan of a hydra in the wild. This would make the sacrifice in reproduction worth it. For a simple early–late life trade-off within mortality, and between mortality and fertility, Wensink et al. demonstrate that the trade-off parameters that capture the costs and benefits of senescence typically have a range of values that render a senescent life history strategy suboptimal [101].

Another mechanism has been shown to favour non-senescence in trees. A density-dependent argument favours runaway selection for ever-lower mortality and ever-longer lifespans. By outliving their neighbours, trees would ensure that their own seeds would take advantage of the opening in the forest arising from the death of the surrounding individuals [102].

5.3.3 Reconciling Traditional Theories with Empirical and Theoretical Findings of Non-senescence

The models discussed demonstrate that different mechanisms may allow for the evolution of negligible and negative senescence. Trade-offs appear to be major determinants of whether senescence evolves or not [67, 97, 98, 101]. Two of the classic theories (antagonistic pleiotropy theory and disposable soma theory) include trade-offs of early life benefits versus late life costs, but by construction, these trade-offs do not favour the evolution of non-senescence [98]. Given a declining force of selection, an early–late life trade-off favours younger over older ages, at least as long as the onset of the late life costs is not too early, or the increase of mortality with age not too steep [101]. A trade-off between maintenance of soma versus increasing reproduction favours reproduction in species with a clear separation of the germ line. Many theoretical (and empirical) cases for non-senescence may indeed capture species that do not clearly separate the germ line from the soma. But for those who exhibit negligible or negative senescence despite a separation of soma and germline, one will have to consider alternative and potentially linked trade-offs that could override the maintenance–reproduction trade-off proposed by the disposable soma theory [75].

Age-specific mutation accumulation cannot explain the evolution of non-senescent life history strategies. Evidence for non-senescent strategies proves that age-specific mutation accumulation, at least in non-senescent species, plays a negligible role. Including the effect of non-adaptive mutations in an optimization model, Dańko and colleagues found that [103], relative to the adaptive process of optimizing the age at maturity given a trade-off between growth and reproduction, mutation accumulation over age did not substantially alter mortality and fertility patterns.

How do these theoretical results square with Hamilton's paradigm of inevitable senescence? The answer is straightforward. Though Hamilton explicitly mentioned the important role of trade-offs between survival and reproduction, he did not account for these trade-offs in his approach, that is, his formulas did not include possible links within or between mortality and fertility patterns, either on the genetic, physiological or actuarial level. Using Lotka's age-specific framework of stable populations [104], Hamilton calculated the changes in fitness (measured as the population growth rate) with respect to age-specific survival and reproduction to quantify the importance of different ages for evolutionary fitness. He proved that these fitness sensitivities or fitness gradients – called the force of selection – must decline with age [15].

Using a similar approach, Baudisch contrasted Hamilton's result with conditions (proportional rather than additive mutational effects on mortality) under which, surprisingly, the force of selection may increase [105]. This finding shook the paradigm of Hamilton's proof, but Baudisch's result was static. In a dynamic framework, a given force of selection, whether declining or rising, would imply that mutations accumulate accordingly, and thus change the patterns of mortality and fertility. This would change the force of selection and, eventually, should violate the condition for an increase in the force of selection with age. Eventually, the force of selection is expected to decline at a population genetic equilibrium. Proving this intuition, Giaimo and Traulsen recently developed a fully dynamic model without presupposing specific mutational effects and let the force of selection freely evolve [106]. They found that in the absence of trade-offs for populations at genetic equilibrium, ageing is the only possible stable strategy. Hence, if trade-offs are excluded, an increasing force of selection in a Hamiltonian setting must necessarily be transitory.

Caswell and Salguero-Gómez translated the Hamiltonian setting from Lotka's age-specific framework to a stage-specific population projection matrix framework [107]. This allowed them to account for growth and size as two key elements that were missing from the classic theories. The authors showed that within a life-cycle stage, the age-specific force of selection can increase with age for empirically observed data. During later life stages, the duration of this increase could even exceed the lifespan of the organism, which would work against the evolution of senescence. Based on data from matrix population models for animals and plants [60, 61], Roper and colleagues discuss why and when selection pressure may rise with age in nature, emphasizing the role of growth forms, social structure and trade-offs for shaping reproductive value and stage structure in a population [108].

Taken together, these results demonstrate that a purely age-specific framework is not sufficient to explain why species age, and some may not. They highlight size, structure (physical or social) and ecological niches as important drivers in the evolution of senescence. They confirm that whether senescence evolves, or not, crucially depends on the nature of the life history trade-offs. Selection gradients are mere weights to the costs and benefits that must be balanced in a trade-off. Only from selection gradients, together with knowledge on how costs and benefits are linked and change over the life course, is it sensible to predict whether senescence may evolve or not [67, 98, 101, 108, 109, 110].

Summing up, biodemographers have demonstrated with different models and approaches that accounting for size or stage and a diversity of trade-offs may help going beyond the classic theories for why senescence evolves, especially in organisms with indeterminate growth, modular growth forms and regenerative capacities. Table 5.1 provides an overview over the mechanisms proposed by theoretical approaches that potentially promote an escape from senescence. However, the overview in this table can only be preliminary. A proper theoretical typology of senescence is still lacking.

Table 5.1 Proposed mechanisms promoting an escape from senescence

Specific Types of Trade-Off
Decreasing returns to resource allocation in survival and reproduction [97]
Genetic, physiological, behavioural, ecological, demographic or other types of constraints of the organisms that imply that the costs of senescence are too high [101]
Improvement in Reproductive Value
Indeterminate growth, associated with gains in survival and/or reproductive capacity [17, 109]
An indeterminate potential for continued repair and regeneration [17, 97]
Physiological and Social Structure
No distinction between germ line and soma [75, 92, 93, 112]
Modularity, which favours cheap repair strategies, such as 'throw away and grow new' [10, 97]
Absence of a survival/reproduction trade-off via vegetative and other modes of asexual reproduction [46, 101, 105]
Social structure and division of labour [95]
Ecological Niches
Longevity competition as competition for space to establish the next generation [93]
Existence of 'safe niches', which might be especially relevant for typically short-lived species [100]

5.4 Conclusion

Since Aristotle, the question has been 'Why do we age?' Over recent decades, biodemographers have shifted the question towards 'Why do we age, while some species seem not to age?' Research undertaken to answer this question still has much to uncover. Whenever a species with negligible or even negative senescence is detected, the identification of the underlying physiological/genetic mechanisms becomes an important challenge with possible important biomedical implications, such as in reptiles [36]. The united efforts of empiricists and theoreticians will provide better data, methods, measures, models and theory. These will identify specific mechanisms, key principles and major constraints that determine how growth, mortality and fertility change with age in what kinds of species. This work will provide the necessary building blocks to create a typology of senescence across the tree of life. This typology will be important in order to understand the principal constraints on patterns of senescence and to apply this knowledge to make better population projections in the face of accelerated environmental and global changes.

References

1. Pearl, R. 1928. *The Rate of Living*. Knopf.
2. Finch, C.E. 1990. *Longevity, Senescence, and the Genome*. University of Chicago Press.
3. Graunt J., Benjamin B. 1964. John Graunt's 'Observations'. *J. Inst. Actuar.* **90**, 1–61 (doi:10.1017/S002026810001564X).
4. Pearl, R., Miner, J.R. 1935. Experimental studies on the duration of life. XIV. The comparative mortality of certain lower organisms. *Q. Rev. Biol.* **10**, 60–79 (doi:10.1086/394476).
5. Hase, A. 1909. Ueber die deutsche Suesswasser-Polypen *Hydra Fusca L., Hydra grisea L.*, und *Hydra viridis L.* Arch. f. Rass.- u Gesellsch. *Biol.*, **6**, 721–753.
6. Szabó, I., Szabó, M. 1929. Lebensdauer, Wach stum und Altern, studiert bei der Nackt schneckenart *Agriolimax agrestis L. Biologia Generalis Bd.* **5**, 95–118.
7. Bidder, G.P. 1932. Senescence. *BMJ* **115**, 5831.
8. Calow, P. 1978. Bidder's hypothesis revisited. *GER* **24**, 448–458 (doi:10.1159/000212285).
9. Medawar, P.B. 1981. *The Uniqueness of the Individual*, second revised edition. Dover Publications.
10. Finch, C.E. 1990. *Longevity, Senescence, and the Genome*. University of Chicago Press.
11. Finch, C.E. 1998. Variations in senescence and longevity include the possibility of negligible senescence. *J. Geront.: Ser. A* **53A**, B235–B239 (doi:10.1093/gerona/53A.4.B235).
12. Finch, C.E., Pike, M.C., Witten, M. 1990. Slow mortality rate accelerations during aging in some animals approximate that of humans. *Science* **249**, 902–905.
13. Martínez, D.E. 1998. Mortality patterns suggest lack of senescence in hydra. *Exp. Geront.* **33**, 217–225.
14. Schaible, R., Scheuerlein, A., Dańko, M.J., Gampe, J., Martínez, D.E., Vaupel, J.W. 2015. Constant mortality and fertility over age in hydra. *Proc. Nat. Acad. Sci. USA* **112**, 15701–15706 (doi:10.1073/pnas.1521002112).
15. Hamilton, W.D. 1966. The moulding of senescence by natural selection. *J. Theor. Biol.* **12**, 12–45 (doi:10.1016/0022-5193(66)90184-6).

16. Roach, D.A., Gampe, J. 2004. Age-specific demography in plantago: uncovering age-dependent mortality in a natural population. *Am. Nat.* **164**, 60–69 (doi:10.1086/421301).
17. Vaupel, J.W., Baudisch, A., Dölling, M., Roach, D.A., Gampe, J. 2004. The case for negative senescence. *Theor. Popul. Biol.* **65**, 339–351 (doi:10.1016/j.tpb.2003.12.003).
18. Abrams, P.A. 1993. Does increased mortality favor the evolution of more rapid senescence? *Evolution* **47**, 877–887 (doi:10.1111/j.1558-5646.1993.tb01241.x).
19. Bryant, M.J., Reznick, D., Rowe, A.E.L. 2004. Comparative studies of senescence in natural populations of guppies. *Am. Nat.* **163**, 55–68 (doi:10.1086/380650).
20. Charmantier, A., Perrins, C., McCleery, R.H., Sheldon, B.C. 2006. Quantitative genetics of age at reproduction in wild swans: support for antagonistic pleiotropy models of senescence. *PNAS* **103**, 6587–6592 (doi:10.1073/pnas.0511123103).
21. Hendry, A.P., Morbey, Y.E., Berg, O.K., Wenburg, J.K. 2004. Adaptive variation in senescence: reproductive lifespan in a wild salmon population. *Proc. Biol. Sci.* **271**, 259–266.
22. Williams, P.D., Day, T., Fletcher, Q., Rowe, L. 2006. The shaping of senescence in the wild. *Trends Ecol. Evol.* **21**, 458–463 (doi:10.1016/j.tree.2006.05.008).
23. Williams, P.D., Day, T. 2003. Antagonistic pleiotropy, mortality source interactions, and the evolutionary theory of senescence. *Evolution* **57**, 1478–1488 (doi:10.1111/j.0014-3820.2003.tb00356.x).
24. Medawar, P.B. 1952. *An Unsolved Problem of Biology*. H.K. Lewis & Co. Ltd.
25. Monaghan, P., Charmantier, A., Nussey, D.H., Ricklefs, R.E. 2008. The evolutionary ecology of senescence. *Func. Ecol.* **22**, 371–378 (doi:10.1111/j.1365-2435.2008.01418.x).
26. Nussey, D.H, Coulson, T., Festa-Bianchet, M., Gaillard, J.-M. 2008. Measuring senescence in wild animal populations: towards a longitudinal approach. *Func. Ecol.* **22**, 393–406.
27. Jones, O.R. et al. 2008. Senescence rates are determined by ranking on the fast–slow life-history continuum. *Ecol. Lett.* **11**, 664–673 (doi:10.1111/j.1461-0248.2008.01187.x).
28. Nussey, D.H., Froy, H., Lemaître, J.-F., Gaillard, J.-M., Austad, S.N. 2013. Senescence in natural populations of animals: widespread evidence and its implications for bio-gerontology. *Ageing Res. Rev.* **12**, 214–225.
29. Jones, O.R. et al. 2014. Diversity of ageing across the tree of life. *Nature* **505**, 169–173 (doi:10.1038/nature12789).
30. Barks, P.M., Laird, R.A. 2015. Senescence in duckweed: age-related declines in survival, reproduction and offspring quality. *Func. Ecol.* **29**, 540–548 (doi:10.1111/1365-2435.12359).
31. Barthold, J.A., Loveridge, A.J., Macdonald, D.W., Packer, C., Colchero, F. 2016. Bayesian estimates of male and female African lion mortality for future use in population management. *J. App. Ecol.* **53**, 295–304 (doi:10.1111/1365-2664.12594).
32. Berger, V., Lemaître, J.-F., Dupont, P., Allainé, D., Gaillard, J.-M., Cohas, A. 2016. Age-specific survival in the socially monogamous Alpine marmot (*Marmota marmota*): evidence of senescence. *J. Mammalogy* **97**, 992–1000 (doi:10.1093/jmammal/gyw028).
33. Bernard, C., Compagnoni, A., Salguero-Gómez, R. 2020. Testing Finch's hypothesis: the role of organismal modularity on the escape from actuarial senescence. *Func. Ecol.* **34**, 88–106 (doi:10.1111/1365-2435.13486).
34. Bouwhuis, S., Vedder, O. 2017. Avian escape artists? Patterns, processes and costs of senescence in wild birds. In *The Evolution of Senescence in the Tree of Life* (eds R.P. Shefferson, O.R. Jones, R. Salguero-Gomez), pp. 156–174. Cambridge University Press (doi:10.1017/9781139939867.008).

35. Gaillard, J.-M., Lemaître, J.-F. 2020. An integrative view of senescence in nature. *Func. Ecol.* **34**, 4–16 (doi:10.1111/1365-2435.13506).
36. Hoekstra, L.A., Schwartz, T.S., Sparkman, A.M., Miller, D.A.W., Bronikowski, A.M. 2020. The untapped potential of reptile biodiversity for understanding how and why animals age. *Func. Ecol.* **34**, 38–54 (doi:10.1111/1365-2435.13450).
37. Roach, D.A., Smith, E.F. 2020. Life-history trade-offs and senescence in plants. *Func. Ecol.* **34**, 17–25 (doi:10.1111/1365-2435.13461).
38. Zajitschek, F., Zajitschek, S., Bonduriansky, R. 2020. Senescence in wild insects: key questions and challenges. *Func. Ecol.* **34**, 26–37 (doi:10.1111/1365-2435.13399).
39. Austad, S.N., Fischer, K.E. 1991. Mammalian aging, metabolism, and ecology: evidence from the bats and marsupials. *J. Geront.* **46**, B47–B53 (doi:10.1093/geronj/46.2.B47).
40. Speakman, J.R. 2005. Body size, energy metabolism and lifespan. *J. Exp. Biol.* **208**, 1717–1730 (doi:10.1242/jeb.01556).
41. West, G.B., Brown, J.H. 2005. The origin of allometric scaling laws in biology from genomes to ecosystems: towards a quantitative unifying theory of biological structure and organization. *J. Exp. Biol.* **208**, 1575–1592 (doi:10.1242/jeb.01589).
42. Colchero, F. et al. 2016. The emergence of longevous populations. *Proc. Nat. Acad. Sci. USA* **113**, E7681–E7690 (doi:10.1073/pnas.1612191113).
43. Vaupel, J.W. 2010. Biodemography of human ageing. *Nature* **464**, 536–542 (doi:10.1038/nature08984).
44. Burger, O., Baudisch, A., Vaupel, J.W. 2012. Human mortality improvement in evolutionary context. *PNAS* **109**, 18210–18214 (doi:10.1073/pnas.1215627109).
45. Colchero, F. et al. 2021. The long lives of primates and the 'invariant rate of ageing' hypothesis. *Nat. Commun.* **12**, 3666 (doi:10.1038/s41467-021-23894-3).
46. Stroustrup, N., Anthony, W.E., Nash, Z.M., Gowda, V., Gomez, A., López-Moyado, I.F., Apfeld, J., Fontana, W. 2016. The temporal scaling of *Caenorhabditis elegans* ageing. *Nature* **530**, 103–107 (doi:10.1038/nature16550).
47. Kenyon, C., Chang, J., Gensch, E., Rudner, A., Tabtiang, R. 1993. A *C. elegans* mutant that lives twice as long as wild type. *Nature* **366**, 461–464 (doi:10.1038/366461a0).
48. Mair, W., Goymer, P., Pletcher, S.D., Partridge, L. 2003. Demography of dietary restriction and death in drosophila. *Science* **301**, 1731–1733 (doi:10.1126/science.1086016).
49. Carey, J.R., Liedo, P., Müller, H.-G., Wang, J.-L., Vaupel, J.W. 1998. Dual modes of aging in Mediterranean fruit fly females. *Science* **281**, 996–998 (doi:10.1126/science.281.5379.996).
50. Purchase, C.F., Rooke, A.C., Gaudry, M.J., Treberg, J.R., Mittell, E.A., Morrissey, M.B., Rennie, M.D. 2022. A synthesis of senescence predictions for indeterminate growth, and support from multiple tests in wild lake trout. *Proc. R. Soc. B: Biol. Sci.* **289**, 20212146 (doi:10.1098/rspb.2021.2146).
51. Dammann, P. et al. 2019. Comment on 'Naked mole-rat mortality rates defy Gompertzian laws by not increasing with age'. *eLife* **8**, e45415 (doi:10.7554/eLife.45415).
52. Ruby, J.G., Smith, M., Buffenstein, R. 2018. Naked mole-rat mortality rates defy Gompertzian laws by not increasing with age. *eLife* **7**, e31157 (doi:10.7554/eLife.31157).
53. Ruby, J.G., Smith, M., Buffenstein, R. 2019. Response to comment on 'Naked mole-rat mortality rates defy Gompertzian laws by not increasing with age'. *eLife* **8**, e47047 (doi:10.7554/eLife.47047).

54. Tully, T., Le Galliard, J.F., Baron, J.P. 2020. Micro-geographic shift between negligible and actuarial senescence in a wild snake. *J. Anim. Ecol.* **89**, 2704–2716 (doi:10.1111/1365-2656.13317).
55. Pearl, R. 1928. *The Rate of Living*. Alfred A. Knopf. http://archive.org/details/rateofliving031726mbp.
56. Schaible, R., Ringelhan, F., Kramer, B.H., Scheuerlein, A. 2017. Hydra: evolutionary and biological mechanisms for non-senescence. In *The Evolution of Senescence in the Tree of Life* (eds R.P. Shefferson, O.R. Jones, R. Salguero-Gomez), pp. 238–254. Cambridge University Press (doi:10.1017/9781139939867.012).
57. Sun, S., White, R.R., Fischer, K.E., Zhang, Z., Austad, S.N., Vijg, J. 2020. Inducible aging in *Hydra oligactis* implicates sexual reproduction, loss of stem cells, and genome maintenance as major pathways. *Geroscience* **42**, 1119–1132 (doi:10.1007/s11357-020-00214-z).
58. Conde, D.A. et al. 2019. Data gaps and opportunities for comparative and conservation biology. *PNAS* **116**, 9658–9664 (doi:10.1073/pnas.1816367116).
59. DATLifeDatabase. 2019. DATLife – The Demography Across the Tree of Life – database. www.datlife.org.
60. Salguero-Gómez, R. et al. 2016. COMADRE: a global data base of animal demography. *J. Anim. Ecol.* **85**, 371–384 (doi:10.1111/1365-2656.12482).
61. Salguero-Gómez, R. et al. 2015. The COMPADRE Plant Matrix Database: an open online repository for plant demography. *J. Ecol.* **103**, 202–218 (doi:10.1111/1365-2745.12334).
62. Species360 Zoological Information Management System (ZIMS)(2020) www.Species360.org.
63. HFD (Human Fertility Database). www.humanfertility.org.
64. HMD (Human Mortality Database). www.mortality.org/.
65. Wang, T., Gao, X., Jakovlić, I., Liu, H.-Z. 2017. Life-tables and elasticity analyses of Yangtze River fish species with implications for conservation and management. *Rev. Fish Biol. Fisheries* **27**, 255–266 (doi:10.1007/s11160-016-9464-8).
66. Jones, O.R., Vaupel, J.W. 2017. Senescence is not inevitable. *Biogerontology* **18**, 965–971 (doi:10.1007/s10522-017-9727-3).
67. Baudisch, A., Vaupel, J.W. 2012. Getting to the root of aging. *Science* **338**, 618–619 (doi:10.1126/science.1226467).
68. Nussey, D.H., Coulson, T., Festa-Bianchet, M., Gaillard, J.-M. 2008. Measuring senescence in wild animal populations: towards a longitudinal approach. *Func. Ecol.* **22**, 393–406 (doi:10.1111/j.1365-2435.2008.01408.x).
69. Tacutu, R., Craig, T., Budovsky, A., Wuttke, D., Lehmann, G., Taranukha, D., Costa, J., Fraifeld, V.E., de Magalhães, J.P. 2013. Human ageing genomic resources: Integrated Databases and Tools for the Biology and Genetics of Ageing. *Nucleic Acids Res.* **41**, D1027–D1033 (doi:10.1093/nar/gks1155).
70. Warner, D.A., Miller, D.A.W., Bronikowski, A.M., Janzen, F.J. 2016. Decades of field data reveal that turtles senesce in the wild. *Proc. Nat. Acad. Sci.* **113**, 6502–6507 (doi:10.1073/pnas.1600035113).
71. da Silva, R., Conde, D.A., Baudisch, A., Colchero, F. 2022. Slow and negligible senescence among Testudines challenges evolutionary theories of senescence. *Science* **376**, 1466–1470.
72. Reinke, B.A. et al. 2022. Diverse aging rates in ectothermic tetrapods provide insights for the evolution of aging and longevity. *Science* **376**, 1459–1466 (doi:10.1126/science.abm0151).
73. Austad, S.N., Finch, C.E. 2022. How ubiquitous is aging in vertebrates? *Science* **376**, 1384–1385 (doi:10.1126/science.adc9442).

74. Cayuela, H., Akani, G.C., Hema, E.M., Eniang, E.A., Amadi, N., Ajong, S.N., Dendi, D., Petrozzi, F., Luiselli, L. 2019. Life history and age-dependent mortality processes in tropical reptiles. *Bio. J. Linn. Soc.* **128**, 251–262 (doi:10.1093/biolinnean/blz103).
75. Kirkwood,T.B.L.1977.Evolutionofageing.*Nature***270**,301–304(doi:10.1038/270301a0).
76. Altwegg, R., Schaub, M., Roulin, A. 2007. Age-specific fitness components and their temporal variation in the barn owl. *Am. Nat.* **169**, 47–61 (doi:10.1086/510215).
77. Austad, S.N. 1997. Comparative aging and life histories in mammals. *Exp. Geront.* **32**, 23–38 (doi:10.1016/S0531-5565(96)00059-9).
78. Baudisch, A. 2011. The pace and shape of ageing. *Methods Ecol. Evol.* **2**, 375–382 (doi:10.1111/j.2041-210X.2010.00087.x).
79. Wrycza, T., Baudisch, A. 2014. The pace of aging: intrinsic time scales in demography. *DemRes* **30**, 1571–1590 (doi:10.4054/DemRes.2014.30.57).
80. Wrycza, T.F., Missov, T.I., Baudisch, A. 2015. Quantifying the shape of aging. *PLoS ONE* **10**, e0119163 (doi:10.1371/journal.pone.0119163).
81. Baudisch, A., Stott, I. 2019. A pace and shape perspective on fertility. *Methods Ecol. Evol.* **10**, 1941–1951 (doi:10.1111/2041-210X.13289).
82. Baudisch, A., Salguero-Gómez, R., Jones, O.R., Wrycza, T., Mbeau-Ache, C., Franco, M., Colchero, F. 2013. The pace and shape of senescence in angiosperms. *J. Ecol.* **101**, 596–606 (doi:10.1111/1365-2745.12084).
83. Shefferson, R.P., Roach, D.A. 2013. Longitudinal analysis in plantago: strength of selection and reverse-age analysis reveal age-indeterminate senescence. *J. Ecol.* **101**, 577–584 (doi:10.1111/1365-2745.12079).
84. Roach, D.A., Ridley, C.E., Dudycha, J.L. 2009. Longitudinal analysis of plantago: age-by-environment interactions reveal aging. *Ecology* **90**, 1427–1433 (doi:10.1890/08-0981.1).
85. Sauer, D.J., Heidinger, B.J., Kittilson, J.D., Lackmann, A.R., Clark, M.E. 2021. No evidence of physiological declines with age in an extremely long-lived fish. *Sci. Rep.* **11**, 9065 (doi:10.1038/s41598-021-88626-5).
86. Amir, Y., Insler, M., Giller, A., Gutman, D., Atzmon, G. 2020. Senescence and Longevity of Sea Urchins. *Genes* **11**, 573 (doi:10.3390/genes11050573).
87. Colchero, F., Schaible, R. 2014. Mortality as a bivariate function of age and size in indeterminate growers. *Ecosphere* **5**, article 161 (doi:10.1890/ES14-00306.1).
88. Vaupel, J.W., Manton, K.G., Stallard, E. 1979. The impact of heterogeneity in individual frailty on the dynamics of mortality. *Demography* **16**, 439–454 (doi:10.2307/2061224).
89. Vaupel, J.W., Yashin, A.I. 1985. Heterogeneity's ruses: some surprising effects of selection on population dynamics. *Am. Stat.* **39**, 176–185 (doi:10.2307/2683925).
90. Marzolin, G., Charmantier, A., Gimenez, O. 2011. Frailty in state-space models: application to actuarial senescence in the dipper. *Ecology* **92**, 562–567 (doi:10.1890/10-0306.1).
91. Williams, G.C. 1957. Pleiotropy, natural selection, and the evolution of senescence. *Evolution* **11**, 398–411 (doi:10.1111/j.1558-5646.1957.tb02911.x).
92. Kirkwood, T.B.L., Rose, M.R., Harvey, P.H., Partridge, L., Southwood, S.R. 1991. Evolution of senescence: late survival sacrificed for reproduction. *Philo. Trans. R. Soc. Lond. Ser. B: Bio. Sci.* **332**, 15–24 (doi:10.1098/rstb.1991.0028).
93. Kirkwood, T.B.L., Austad, S.N. 2000. Why do we age? *Nature* **408**, 233–238 (doi:10.1038/35041682).
94. Carlsson, H., Ivimey-Cook, E., Duxbury, E.M.L., Edden, N., Sales, K., Maklakov, A.A. 2021. Ageing as 'early-life inertia': disentangling life-history trade-offs along a lifetime of an individual. *Evol. Lett.* **5**, 551–564 (doi:10.1002/evl3.254).

95. Maklakov, A.A., Chapman, T. 2019. Evolution of ageing as a tangle of trade-offs: energy versus function. *Proc. Biol. Sci.* **286**, 20191604 (doi:10.1098/rspb.2019.1604).
96. McNamara, J.M., Houston, A.I. 1996. State-dependent life histories. *Nature* **380**, 215–221 (doi:10.1038/380215a0).
97. Baudisch, A. 2008. An optimization model based on vitality. In *Inevitable Aging?*, pp. 75–122. Springer.
98. Baudisch, A. 2012. Birds do it, bees do it, we do it: contributions of theoretical modelling to understanding the shape of ageing across the tree of life. *GER* **58**, 481–489 (doi:10.1159/000341861).
99. Vural, D.C., Morrison, G., Mahadevan, L. 2013. Increased network interdependency leads to aging. arXiv:1301.6375 *[q-bio.PE]*.
100. Baudisch, A., Vaupel, J. 2010. Senescence vs. sustenance: evolutionary-demographic models of aging. *DemRes* **23**, 655–668 (doi:10.4054/DemRes.2010.23.23).
101. Wensink, M.J., Wrycza, T.F., Baudisch, A. 2014. No senescence despite declining selection pressure: Hamilton's result in broader perspective. *J. Theor. Biol.* **347**, 176–181 (doi:10.1016/j.jtbi.2013.11.016).
102. Seymour, R.M., Doncaster, C.P. 2007. Density dependence triggers runaway selection of reduced senescence. *PLoS Com. Bio.* **3**, 2580–2589 (doi:10.1371/journal.pcbi.0030256).
103. Dańko, M.J., Kozłowski, J., Vaupel, J.W., Baudisch, A. 2012. Mutation accumulation may be a minor force in shaping life history traits. *PLoS ONE* **7**, e34146 (doi:10.1371/journal.pone.0034146).
104. Lotka, A.J. 1924. *Elements of Mathematical Biology*. Dover Publications. http://archive.org/details/in.ernet.dli.2015.458876.
105. Baudisch, A. 2005. Hamilton's indicators of the force of selection. *Proc. Acad. Sci. USA* **102**, 8263–8268 (doi:10.1073/pnas.0502155102).
106. Giaimo, S., Traulsen, A. 2022. The selection force weakens with age because ageing evolves and not vice versa. *Nat. Commun.* **13**, 686 (doi:10.1038/s41467-022-28254-3).
107. Caswell, H., Salguero-Gómez, R. 2013. Age, stage and senescence in plants. *J. Ecol.* **101**, 585–595 (doi:10.1111/1365-2745.12088).
108. Roper, M., Capdevila, P., Salguero-Gómez, R. 2021. Senescence: why and where selection gradients might not decline with age. *Proc. Biol. Sci.* **288**, 20210851 (doi:10.1098/rspb.2021.0851).
109. Baudisch, A. 2008. *Inevitable Aging? Contributions to Evolutionary-Demographic Theory*. Springer-Verlag (doi:10.1007/978-3-540-76656-8).
110. Wensink, M.J., Caswell, H., Baudisch, A. 2017. The rarity of survival to old age does not drive the evolution of senescence. *Evol. Biol.* **44**, 5–10 (doi:10.1007/s11692-016-9385-4).
111. Colchero, F. et al. 2016. The emergence of longevous populations. *Proc. Nat. Acad. Sci.* **113**, E7681–E7690.
112. Kirkwood, T.B.L., Holliday, R., Maynard Smith, J., Holliday, R. 1979. The evolution of ageing and longevity. *Proc. R. Soc. Lond. Ser. B. Bio. Sci.* **205**, 531–546 (doi:10.1098/rspb.1979.0083).
113. Lucas, E., Keller, L. 2017. Explaining extraordinary life spans: the proximate and ultimate causes of differential life span in social insects. In *The Evolution of Senescence in the Tree of Life* (eds R.P. Shefferson, O.R. Jones, R. Salguero-Gomez), pp. 198–219. Cambridge University Press (doi:10.1017/9781139939867.010).

6 The Untapped Potential of Zoo and Aquarium Data for the Comparative Biology of Ageing

Morgane Tidière, Johanna Staerk, Michael J. Adkesson, Dalia A. Conde and Fernando Colchero

6.1 Introduction

Age- and sex-specific information of sufficient quality for many individuals, populations and species are prerequisites for comparative ageing analyses. Longitudinal studies of marked and recognizable individuals provide the most reliable sources of information [1], but obtaining these data from wild populations requires tremendous human and financial resources. This is particularly challenging for long-lived species that can contribute most to our understanding of ageing and potential anti-ageing mechanisms [2, 3]. Gathering sufficient age-specific data to unravel species' survival and fertility patterns, let alone the physiological mechanisms of ageing and morbidity, might require decades of field research.

Although the number of long-term monitoring projects of wild populations continuously increases [4], age-specific parameters crucial for ageing studies are widely lacking. For example, high-quality demographic information (i.e. life tables or matrix projection models) is available for only 1.3% of all known tetrapods, while 54.8% of species do not even have basic measures for survival and fertility [5]. Even among mammals, more than 95% of species lack sufficient demographic data for advanced research. This knowledge gap is even greater among lesser-known groups, with no basic demographic information available for 88% of amphibians and 65% of reptiles [5].

In this context, age-specific information already collected from non-domestic animals living under human care (e.g. in zoos and aquariums, hereafter Z&A) can greatly help bridge this information gap for ageing research. Over the years, Z&A have become increasingly self-reliant, learning how to best manage demographically and genetically viable populations under their care (see Box 6.1). Part of the development of Z&A has been the establishment of systematic record-keeping such

This chapter was made possible by the worldwide information network of zoos and aquariums, which are members of Species360. We warmly thank the record keepers for recording the information in ZIMS, and the sponsors of the Conservation Science Alliance: Copenhagen Zoo, Mandai Wildlife Group and Toronto Zoo. We thank Dr Lucie Bland for professionally editing the manuscript. We also thank Dr Rachel Thompson and Elizabeth Hunt for their help in obtaining up-to-date medical data summaries. Finally, we thank Dr Rachel Thompson and Katelyn Mucha for proofreading the manuscript.

Box 6.1 The establishment of modern Z&A

Since the establishment of the first modern zoo approximately 200 years ago, Z&A have undergone considerable change, transforming from menageries to conservation centres engaged in education, conservation and research [83, 84, 85]. Although recreation and entertainment remain an important mission for Z&A (with >700 million people visiting Z&A annually [86]), zoological institutions have a unique opportunity to educate the public about environmental problems and to connect humans with nature [87]. Even though the term 'zoo' can refer to any permanent exhibition of wild animals to the public, modern reputable Z&A are members of global, regional and/or nationally recognized associations (e.g. Association of Zoos and Aquariums (AZA), European Association of Zoos and Aquaria (EAZA), World Association of Zoos and Aquariums (WAZA)). Such organizations promote collaboration among members and require rigorous accreditation standards related to animal welfare, animal health and record-keeping while contributing to conservation efforts both in Z&A and in the wild (e.g. [88]). Of the current ~10000 Z&A worldwide, more than 1200 are organized under such professional associations, which represent high standards of science and animal care [89].

Many modern zoological institutions, aquariums, marine parks and reserves have collected various information about their animals. Much of it has been pooled in the ZIMS (a global database that facilitates animal and population management within and across Z&A, see Box 6.2). The establishment of systematic record-keeping was one of the important steps made by the Z&A to improve the management and knowledge of their animals. For example, the first studbook (i.e. the curated demographic, pedigree and genetic data of a species, used to facilitate population management across multiple organizations) of an endangered species in captivity was established in 1932 for the European bison (*Bison bonasus*) after its extinction in the wild in 1919 [90, 91].

as the Zoological Information Management System (ZIMS) database (see Box 6.2, Figure 6.1). The utilization of zoo-held animals (i.e. *ex situ* populations) to better understand ageing through comparative analysis has been suggested by numerous researchers (e.g. [6, 7, 8, 9, 10], but see Box 6.3). Indeed, the protected environment provided by Z&A to their animals, along with the age-specific parameters collected for thousands of individuals, can greatly enhance our global understanding of ageing processes across the tree of life.

In this chapter, we define 'Z&A' as science-driven zoos and aquariums involved in improving animal management and welfare as well as in the conservation of species. In Section 6.2, we highlight the potential of Z&A data to test evolutionary theories and reveal evolutionary drivers of ageing at inter- and intra-specific levels. Section 6.3 focuses on the physiological mechanisms of ageing and the potential of the rich Z&A medical data. Finally, Sections 6.4 and 6.5 concludes that Z&A data is not only an important resource for ageing studies but also for improving zoo management, veterinary medicine and species conservation.

Box 6.2 Species360 and the ZIMS database

Species360 (formerly called ISIS, the International Species Information System) is a non-profit organization governed by a global volunteer board of trustees experts and was founded in 1974 to share and maintain standardized records for animals in Z&A [92]. At present, Species360 has a global network that includes over 1 300 zoos, aquariums, universities, research and governmental members, spread across six continents and more than 100 countries, spanning a wide range of geographic and climatic regions. Species360 members contribute their data to ZIMS, the world's most comprehensive database on *ex situ* wildlife, including standardized demographic, husbandry and medical knowledge for more than 22 000 species (www.species360.org). In 2017, Species360 launched its scientific team, and the inception of the Conservation Science Alliance (CSA, https://conservation.species360.org/get-involved/), to exploit this database to support evidence-based discussion to: (1) help Z&A in the management and care of their animals and populations; (2) inform policy decisions for conservation; and (3) improve global knowledge through data sharing. Today, the goal is to expand the CSA to bring together a diverse group of experts to harness the value of ZIMS to advance scientific research. While its main focus is on conservation, animal care and welfare, Species360 acknowledges the importance of basic science as well.

Historical animal data records date back to the mid-1800s for some species, including information on an animal's origins, reproduction, husbandry, morphometrics, housing and veterinary data. ZIMS compiles data from ~10 million animals (individually or group-tracked) that are curated by thousands of wildlife professionals worldwide (Figure 6.2). Each animal can be uniquely identified, and its life history from birth to death is maintained by each institution as a single, real-time record, even if the animal moves between Species360 member institutions. These long retrospective data, combined with the exact animal's date of birth (i.e. allowing any event in the animal's life to be age-specific), make this database a real asset for ageing research (see Box 6.3 for more information about the possibility of accessing such data for science). ZIMS data is divided into different modules that are continuously updated and evolving to follow the members' needs:

- ZIMS for husbandry – Represents more than 220 million records, including animal's sex, birth and death dates (or if unknown, its acquisition and departure date of birth and/or death), birth location (i.e. birth in captivity or collection from the wild), parentage and inter-institution transfers. In addition, behavioural observations, measurements (e.g. size, weight), enrichment, training, feed logs or enclosure information (e.g. size, substrate, number of occupants) might also be recorded.
- ZIMS for aquatics – Includes additional information relevant to the specific needs of managing aquatic animals. For instance, information can be recorded at the group level instead of at the individual level and might include enclosure maintenance data, such as water quality.

- ZIMS for medical – Includes over 48 million records from more than 6000 species, such as anaesthesia, treatment, diagnostic testing, clinical observation, prescription, morbidity and mortality, and reference intervals (i.e. species-specific physiologic reference data) (see Figure 6.1). Almost 8 million records refer to information on drug dosages, administration routes and adverse effects including over 670000 anaesthetic records. Over 4 million records refer to standardized diagnostic test results (including details on the laboratory, testing methodology, patient health status, fasting duration, restraint method and sample collection techniques) and over 3 million records to gross and histopathologic lesions and diagnoses (including clinical, biopsy and necropsy diagnoses). Finally, Z&A have started sharing data on the biobanking of biological samples (e.g. preserved blood and tissue samples) through ZIMS.
- ZIMS for care and welfare – This module streamlines access to crucial care and welfare indicators by enabling ZIMS to track and monitor animal care inputs and outputs systematically. Following WAZA's Five Domains model for welfare, this module allows monitoring behaviour, physical health, environment, nutrition, and the animal's positive and negative experiences.
- ZIMS for studbooks – Studbooks focus on high-profile species that meet institutional needs for conservation and education, and face particular breeding challenges. As a result, at least 1494 studbooks for 956 Z&A species and subspecies allow managers (i.e. 'studbook keepers') to assess the viability of populations and breeding programmes.

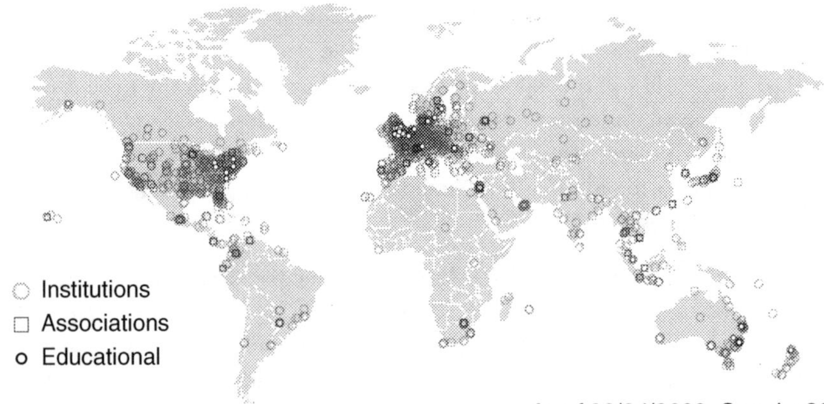

Figure 6.1 Global distribution of Species360 member institutions (incl. Z&A, professional zoological associations and educational facilities (e.g. universities)). All Z&A members of Species360 use ZIMS for their record-keeping. Figure provided by Species360 CSA.

Box 6.3 Z&A data policy

Although the idea of using Z&A data to improve knowledge in the field of ageing was brought up more than 20 years ago (e.g. [6]), only a little progress has been made. The first reason is that ZIMS is a massive database with more than 16000 users generating data streams daily, making data extraction challenging. The database was not initially created to develop scientific research but to provide a depository and summary information for each institution with the aim to gather enough sample sizes to analyse demographic and veterinary data, and improve the lives of animals living in Z&A. Consequently, it took over 40 years for Species360 to accumulate millions of records on >25000 species. Furthermore, extracting the data for research took six years and significant funding from external grants to support scientific studies. Moreover, because of the tremendous amount of information registered daily, data entry errors exist despite the registrars' care. Therefore, one of the first challenges of the CSA after its inception in 2017 was, in collaboration with the Interdisciplinary Centre on Population Dynamics (CPop) at the University of Southern Denmark (SDU), to develop tools to extract and curate the data to develop demographic analyses. Today, husbandry data for vertebrates are extracted, curated and updated once a year. Research requests are also regularly reviewed by the Species360 Research Committee and the Board of Trustees. The next objective is to apply a similar method to data from the ZIMS for the medical module.

The second reason is related to the sensitivity of these data, given that keeping animals in Z&A fuels a sustained debate around animal welfare. While Z&A agree to record information about their animals for legal (traceability) reasons and to improve the management and care of their animals through data sharing across institutions, they do not want to see their improvement and transparency efforts backfire. Therefore, Species360 works to process and analyse these data to provide critical information to its members and the conservation community while maintaining the anonymity of its members. As part of this vision, they support researchers and members around the globe by providing open-source data, supplying select data on request and helping members use data in ZIMS to advance wildlife conservation. To learn more about accessing data through reasonable research requests, visit the Species360 website or read their 'Sharing Data' document (https://conservation.species360.org/wp-content/uploads/2020/08/Species360-Sharing-Data-v3-3_komprimeret.pdf).

6.2 The Potential of Z&A Data to Uncover Evolutionary Drivers of Ageing

The tremendous volume of individual demographic records with exact birth and death dates from thousands of species living in Z&A (Figure 6.2) offers remarkable opportunities to calculate age- and sex-specific demographic rates for many species lacking data from the wild [5]. These data thus provide a unique resource to unravel evolutionary drivers of ageing and longevity through comparative analyses [11].

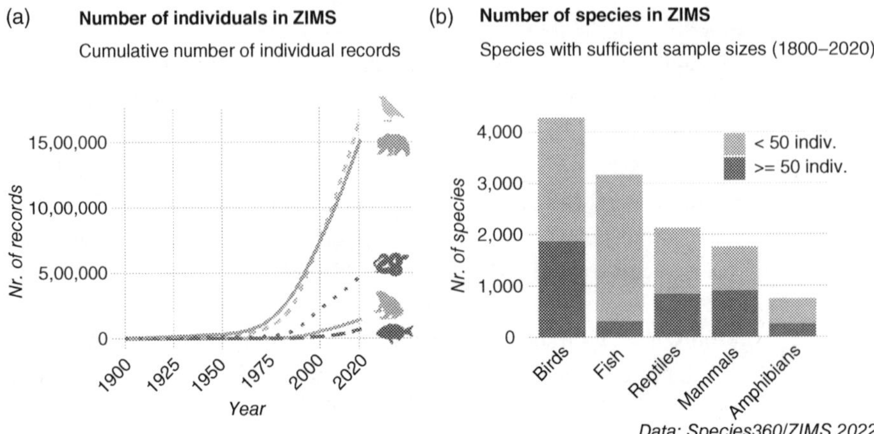

Figure 6.2 (a) Cumulative number of individual animals with demographic records in the husbandry module of ZIMS (Box 6.2) for birds, mammals, reptiles, amphibians and fishes. (b) Number of vertebrate species recorded in ZIMS for husbandry (see Box 6.3) according to the number of individuals available for demographic analyses. Darker bars show the number of species with a sample size of ≥50 individuals, the typical minimum sample size necessary for demographic analyses. Figure provided by Species360 Conservation Science Alliance (CSA).

6.2.1 The Potential of Z&A Data to Test Evolutionary Theories of Ageing

The protected environment in Z&A provides ideal conditions for testing different theories of ageing [11]. It enables, for instance, the disentanglement of the effects of environmental versus genetic or physiological factors on ageing. For example, comparing wild and zoo populations of 28 bird species, Ricklefs demonstrated that environmentally driven mortality rates in zoo populations are, on average, less than 30% of those of wild populations [8]. Studying 59 mammal species, Tidière et al. highlighted that the difference in ageing and longevity between zoo and wild populations is related to the species' pace of life, with fast species (i.e. short longevity and high reproductive output) delaying their ageing more and living longer in Z&A than in the wild, compared to slow species [12]. Also, comparing ageing patterns of wild and zoo-held populations among 13 primate species, Colchero et al. found support for the 'invariant rate of ageing' hypothesis [13], which predicts that the rate of ageing is relatively fixed within species or genera. Their findings stress that, at least in primates, changes in survival between populations of the same species are primarily driven by differences in infant and juvenile survival and age-independent sources of mortality, but that ageing rates change little, irrespective of the environmental context. However, a recent study focusing on the rate of ageing of zoo-held and wild populations of 52 species of Testudines showed that some species might reduce ageing rates in response to improvements in environmental conditions [14]. This last study is an excellent example of how demographic data compiled in ZIMS can be useful for testing evolutionary hypotheses, as the authors have shed light on the unique ageing patterns in Testudines. Their results show that there are species and even entire taxonomic groups with extremely different ageing patterns than those of species commonly used for ageing research. Therefore, if we are to deepen our understanding of ageing, we need to study the demographic and physiological mechanisms of ageing in taxonomic groups that are distant from us.

Evolutionary theories of senescence, such as the antagonistic pleiotropy theory can be tested with Z&A data on numerous species with different life history strategies. The antagonistic pleiotropy theory posits that increased mortality at advanced age results from the expression of alleles that enhance fitness early in life at the expense of survival or reproduction later in life [15] (see also Chapter 2). This trade-off between early life reproductive effort and late life survival is often not seen in animals living under human care. A reason might be the absence of competition for resources, which drastically reduces the pressure exerted on this trade-off (e.g. [16, 17, 18]). However, Z&A populations can still be beneficial to highlight the diversity of ageing patterns between individuals and species across the tree of life. For instance, using Z&A populations of birds and mammals, Ricklefs highlighted the covariation of biological times in zoo-held populations [19], revealing a correlation between rapid early development (i.e. embryo growth rate) and accelerated ageing. Moreover, a recent study using data from studbooks (Box 6.1) highlights an unexpected shorter lifespan for fathers with sons when comparing the reproductive costs of producing and weaning sons versus daughters on parental longevity in tigers (*Panthera tigris*) and ruffed lemurs (*Varecia sp.*) [20]. These examples emphasize the potential of Z&A data to improve our understanding of the early–late life trade-off leading to the progressive decline of individual performances. Importantly, ZIMS includes information on the type and amount of resources (e.g. enclosure size, food, enrichment, number of individuals sharing the enclosure etc.) the animals have access to in Z&A (recorded in the module ZIMS for husbandry, Box 6.2), which could be used as covariables to improve our understanding of the influence that environmental factors have on these processes.

Knowledge about evolutionary drivers mediating reproductive ageing and somatic lifespan remains limited. As found in humans and other mammals, the reproductive function may end before the end of life, resulting in a period of post-reproductive life [21]. Many animals in Z&A show extended lifespans as compared to the wild (e.g. [12] for mammals), allowing for better detection of the species' potential post-reproductive lifespan. Thus, Z&A demographic data could facilitate the testing of several hypotheses, such as the 'grandmother hypothesis', which stipulates that the ability to provide grandparental care alongside an age-related increase in the cost of breeding promotes early cessation of reproduction or post-reproductive survival. Until recently, only three species of toothed whales, in addition to humans, were known to show clear evidence of a large post-reproductive lifespan in the wild [22, 23, 24]. Using published demographic data from wild and zoo populations of mammals, Péron et al. recently identified eight species (among whales, but also primates and elephants) exhibiting post-reproductive lifespan [25]. Their findings show that females of species with grandparental care live 43% longer than males, compared to 12% in other species. The authors explained this pattern as the result of lower baseline mortality rates and delayed onset of actuarial senescence (i.e. the progressive decline in survival probability with age) in females of species with grandparental care compared to females of other species.

In addition, Z&A demographic data can help tackle fundamental challenges in understanding sex-based differences in ageing [26] (see also Chapter 9). First, males and females of the same species are expected to show different ageing and longevity patterns, attributed to differences in trade-offs between reproduction and somatic maintenance influenced by sex-specific pressures of natural and sexual selection [27]. Second, across vertebrates, it has been hypothesized that species with intense male mating competition and high levels of sexual dimorphism in body size exhibit dimorphism in age-specific reproductive success [28]. In these species, males show later ages at first reproduction and earlier reproductive ageing compared with females because they take longer to attain adult body size [28], a state that is only maintained at peak condition for a limited time. In ungulates, research using Z&A data has already shown that species' mating system explains part of the variability in ageing and longevity between sexes and species [29]. Sex chromosomes may also contribute to the sex gap in ageing and longevity [30]. Still, most studies challenging this hypothesis included only birds and mammals, which have sex determination by sex chromosome only (but see [31]). A comparative analysis, including taxa with other sex determination types, could help answer this question. Using the ZIMS database, sex-specific ageing patterns could potentially be defined for hundreds of species of mammals, birds, reptiles, amphibians and fishes (see Figure 6.2). These data could enable scientists to analyse evolutionary factors responsible for differences in ageing and lifespan between sexes and across species, at least in tetrapods. Finally, ZIMS provides a rare opportunity to study sex-specific reproductive ageing, given that reproductive success is monitored for both sexes in Z&A, contrary to most wild populations, where only the females' reproductive success is observable [32].

One of the main advantages of Z&A data for ageing studies is the tremendous diversity of species for which demographic data exists, spanning all classes of vertebrates, including species with a large variability of life history traits, ecology and demographic rates. This diversity allows researchers to unravel factors driving the evolution of ageing and longevity across taxa, highlighting differences between species, populations and individuals. For example, these data could be helpful to study if sociality or regular torpor periods across life are associated with an increased lifespan and slower ageing rates in these species, as repeatedly hypothesized in the ageing literature (e.g. [33, 34, 35]). Comparing demographic data from Z&A populations to their wild counterparts would also help us gain a deeper understanding of environmental factors influencing the variability of ageing and longevity (e.g. [12, 36]).

6.2.2 The Potential of Z&A Data to Challenge Mathematical Models of Ageing

Using longitudinal data from zoo-held populations can help scientists to develop or improve existing mathematical models to describe ageing patterns. Diverse age-specific trajectories of mortality and fertility can be captured using mathematical models (e.g. [37, 38, 39, 40]), which can then be tested on available data using statistical models (e.g. [41]). However, inference from these models is limited by the amount and quality of the data used. With small sample sizes or poor-quality data (e.g. high censoring or low recapture probabilities), statistical inference becomes limited: models may be inaccurate or over-parametrized. Longitudinal data from animals living in Z&A, for which exact birth and death dates are known, enables the development of detailed models describing the variation of demographic rates with age. These models can later be tested on wild populations when sufficient data become available (e.g. [31]).

This two-step approach (models run on data from: (1) Z&A animals and then from (2) wild animals) allows for higher robustness and credibility of results. For instance, using zoo data from 181 species of mammals and birds, Moorad et al. assessed the robustness of estimated metrics of ageing and longevity in a population to sample size [42]. These metrics can then be applied to wild populations, allowing researchers to use robust and comparable metrics, especially as sample sizes for wild populations tend to be limited.

Finally, age-specific fertility is commonly modelled with a simple exponential quadratic function (e.g. [43]), or a segmented regression [44], providing little flexibility for understanding reproductive ageing and how fertility changes with age among populations. Although potentially lower than in the wild (see Section 6.2.3), fertility rates from zoo-held populations can help develop more flexible and accurate models. These models can then be applied to wild populations and improve our knowledge of the change of reproductive rates with age between sexes, populations and species (e.g. [13, 45]). Indeed, as in monitored wild populations, only successful reproductive events are observable in Z&A populations.

6.2.3 The Limitation of Z&A Data

One limitation of Z&A data used for estimating fertility models is that animals living under human care are genetically and demographically managed. For many species, when and whether an individual can reproduce is determined based on population management recommendations, genetic value and availability of an appropriate mate. Because space in Z&A facilities is limited and maintenance of genetic diversity is essential, some individuals will be allowed to reproduce, while permanent or reversible forms of contraception will constrain others. Reproductive management decisions are regularly revisited, and an individual considered for reproduction one year can be removed from the breeding programme the following year. In contrast, an individual under non-surgical contraception can breed later if required. Given the variation in husbandry practices between different facilities, reproductive data from Z&A populations must be used cautiously.

Moreover, animals living under human care may not show their full reproductive potential, with fertility rates often lying in the lower bound of a species' potential. For example, finding suitable breeding pairs often requires moving one or both animals to a different facility, which involves adaptation to new habitats, animal groupings and husbandry practices. Thus, a decrease in observed fertility in Z&A populations can result either from an increase in the proportion of individuals not participating in reproduction (i.e. due to management using contraceptives or keeping sexes separately) or from a decrease in the reproductive success of individuals attempting to reproduce (e.g. elder individuals, or individuals prevented from reproducing with members in their social groups, such as meerkats *Suricata suricatta*). Although the distinction between the two is not always possible, data from managed populations can help describe a species' general age-specific reproductive pattern. For instance, a comparison of litter sizes per age between wild and zoo-held populations of red wolves (*Canis rufus*) revealed that females show a similar age-specific pattern irrespective of environmental conditions. At the same time, males reached their maximum reproductive output at later ages in zoos compared to their free-ranging counterparts [46]. These results hint towards complex age-specific patterns in red wolves that vary with environmental conditions. However, it is also possible to observe a higher reproductive value in zoo-held populations, as many species can have hormonal manipulation to increase frequency and litter size. More importantly, many Z&A benefit from intervention and care of neonates to allow higher offspring survival rates than would be seen in the wild (e.g. giant panda, *Ailuropoda melanoleuca*; polar bear, *Ursus maritimus*; Blanding's turtle, *Emydoidea blandingii*; hellbender, *Cryptobranchus alleganiensis*; Panamanian golden frog, *Atelopus zeteki*; Puerto Rican crested toad, *Peltophryne lemur*). Therefore, even if the average reproductive values of zoo-managed and wild populations differ, these data can still provide valuable information to reconstruct the sex- and species-specific patterns of reproductive rates with age.

6.2.4 The Advantage of Data from Z&A in Ageing Research

Because the accurate understanding of ageing patterns requires long-term monitoring, most studies using population-level metrics focus more on survival than reproductive ageing. This is primarily due to the difficulty of gathering reproductive data (e.g. for arboreal species like the ruffed lemur; elusive and low-density species like the tiger; hidden offspring species like the roe deer *Capreolus capreolus*, or alpine marmot *Marmota marmota*) or reliable filiation in wild populations (without concurrent genetic testing) particularly for the species not providing parental care (e.g. some reptiles or amphibians) [32]. These limitations result in a greater data gap for fertility than survival information [5, 10]. Consequently, very little is known about the impact of ageing on different reproductive traits and the occurrence and determinants of post-reproductive lifespan, even among mammals. For instance, except for a few reports on rhinoceroses, reproductive ageing among non-domestic perissodactyls has seldom been studied. Similarly, except for one semelparous marsupial mouse, no reproductive data related to ageing exist for marsupials ([10]; but see [47] for Tasmanian devil females, *Sarcophilus harrisii*). In this context, the ZIMS demographic data represent an opportunity to improve our knowledge and reduce this data gap. For example, 25% of marsupials (i.e. 82 species) have survival and reproductive data available for analysis in ZIMS. Moreover, 78% of tetrapods (i.e. 3292 species) for which demographic data are available for analysis in ZIMS are not mammalian species. Finally, these data can also enable the assessment of reproductive patterns of species with peculiar reproductive systems according to their biology (e.g. giant panda or white-bellied tree pangolin, *Phataginus tricuspis*), which are challenging to study in their wild habitat.

ZIMS data can also reveal the importance of heritability in ageing patterns and test for parental effects on individual ageing and lifespan (e.g. the Lansing effect [48]). Using Z&A data, Ricklefs and Cadena showed a higher heritability coefficient of age at death in mammals than in birds [49]. Moreover, assigning maternity might be easy in wild populations, but assigning paternity can prove challenging, requiring in most cases genetic paternity tests. As a result, maternal effects on offspring ageing have been identified in many species in the wild (e.g. the effect of maternal age on calf survival in the bottlenose dolphin, *Tursiops aduncus* [50]; the effect of maternal rank on sons' reproductive success in Barbary macaque, *Macaca sylvanus* [51]), while paternal effects remain poorly known or have been primarily tested in human or non-human primate populations [52, 53]. Since parental identity is recorded in ZIMS for most individuals born in Z&A, paternal effects such as the father's age or condition at mating can be derived for many species.

For animals living under human care, most environmental sources of mortality are absent. This context is ideal for testing the different theories of ageing, such as the disposable soma theory, because of the controlled environment provided [11]. The closed environment protects animals from predation, animal care staff provide water and food (mostly eliminating competition for resources) and veterinary care prevents and treats many diseases. Comparing wild and zoo populations of 28 birds, Ricklefs shows that the component of mortality related to ageing does not differ significantly between wild and zoo-held birds [8]. This result supports the hypothesis that ageing-related mortality is more associated with genetic and physiological mechanisms than with the environment [54, 55]. However, it is important to note that even if the rate of ageing does not differ, the onset of ageing might be quite variable and deserves more interest from the research community.

6.3 The Potential of Z&A Data to Unravel Mechanisms of Ageing

Using Z&A data could help us understand how underlying biological mechanisms of ageing relate to and impact longitudinal changes in health trajectories and how these mechanisms differ at intra- and inter-species levels. This would offer a unique opportunity to identify resilience mechanisms, their dynamic changes and their impact on stress responses. For many animals, the temporal depth of data in ZIMS allows evaluating entire life histories from birth to death in the context of disease and physiologic and pathologic changes (Figure 6.3). The ability to access all clinical exam findings, diagnostic test results and ultimately post-mortem diagnoses for an individual over its entire lifespan from a single centralized database rarely exists, even for humans. Moreover, there is a broad consensus about the need to fill research gaps in the characterization of cellular and molecular processes behind the reproductive ageing [10]. In this context, information from Z&A-held animals can significantly contribute to our general knowledge of this crucial phenomenon and provide new approaches to understanding reproductive lifespan.

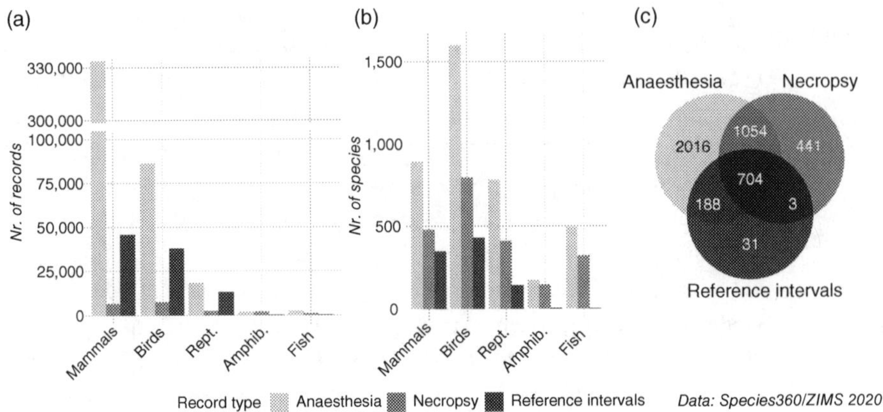

Figure 6.3 Number of records or species available in the ZIMS medical module. (a) Number of records per taxonomic class and record type (note, y-axis break). (b) Number of species per class and record type. (c) Number of species with anaesthesia, necropsy and/or reference interval records. Figure provided by Species360 CSA.

6.3.1 Clinico-Pathological Data

The extent of available medical data in ZIMS allows for robust evaluation of age-related degenerative diseases in animal populations. Clinico-pathologic data can offer insights into the progression of the disease and normal ageing processes. For instance, ZIMS medical data have been recently used to show that the phylogenetic distribution of cancer mortality in mammals is associated with diet: carnivorous mammals, especially mammal-consuming ones, face the highest cancer-related mortality [56]. Moreover, evaluation of specific diseases at the species level is increasingly common today largely due to centralized medical data. The Great Ape Heart Project is one such example of a long-term collaborative project in applied veterinary research across multiple zoos committed to improving the health and life expectancy of apes under human care. Sharing of medical data across facilities has, for example, enabled targeted investigations into cardiac disease and treatment success [57]. In addition to the medical data, ZIMS also collects individual pedigrees. Linking these data could allow the assessment of hereditary traits and genetic predisposition to diseases or cancers. For instance, Comizzoli and Ottinger advocate for using ZIMS to examine the influence of parental age on offspring reproductive health [10]. Finally, the development of a standardized archive for diagnostic images (e.g. radiographs, computerized tomography (CT) scans) is underway (Zoo and Aquarium Radiology Database project) with the intent of providing an interface with ZIMS. This resource will eventually allow new research opportunities and meta-analyses, such as with databases in the human medical field [58]. Such a resource would allow the evaluation of age-related diagnostic findings (e.g. degenerative joint disease, cardiac disease, renal failure).

6.3.2 Physiological Measurements

Physiological measurements, such as body temperature, heart rate and respiratory rate, are a large comparative dataset in ZIMS that remains almost untapped as a source of phylogenetic study. A better characterization of the age-specific pattern of these different parameters could help to understand the physiological changes occurring when organisms age and how these changes differ between individuals, sexes, populations and species. For instance, because of the difficulty of monitoring male reproduction in the wild (as previously mentioned), the physiological process of male reproductive ageing remains poorly understood [10]. The medical module of ZIMS, combined with the husbandry module (see Box 6.2), could provide insightful cues on the variability of physiological parameters across ages for a large diversity of animals, particularly on the physiological process underlying male reproductive ageing. For example, Z&A data allowed researchers to examine the influence of ageing on gamete quality or uterine environment in female cheetahs (*Acinonyx jubatus*) [59, 60]. Using aggregated data in ZIMS, a similar study could focus on the change in male reproductive parameters (e.g. sperm quality, behaviour, hormone dosage etc.) with increasing age over the life course of an individual and at the population level.

6.3.3 Infection Reports

Longitudinal medical data from zoo-held animals could provide new advances in understanding the influence of pathogens on ageing. For example, records on commonly cultured infectious organisms have not been mined for valuable insights. Moreover, morbidity and mortality data reveal the most common causes of death (which ultimately define expected lifespans) and allow for epidemiologic investigation of diseases. Linking the appearance of infection to age, sex and species, we could identify how pathogens influence the deterioration of the immune system with increasing age (i.e. immunosenescence). The collection of information concerning diseases contracted by different species and sexes can provide a broad picture of sex-specific sensitivity to diseases and pathogens and how this affects lifespan and ageing. Indeed, despite excellent veterinary care and protected environments in Z&A, animals living under human care are still susceptible to many diseases. For instance, because animals are hosted in high densities in Z&A, the proximity of other animals and humans can lead to disease outbreaks (e.g. SARS-CoV-2 [61]). These data could help to study the potential influence of pathogens on the origin of sex differences in lifespan and ageing, something rarely investigated so far. Tidière et al. hypothesize that sex-specific immune responses and sex differences in energy allocation strategies driven by sexual selection, leading to differences in exposure and susceptibility to pathogens, should result in an accelerated immunosenescence in males through immune exhaustion [62]. If this hypothesis is correct, sex differences in lifespan, through either direct (host mortality) or indirect (faster immunosenescence) effects, could be modulated according to the pathogenic environment of populations. Using mean adult lifespan in eight carnivores and five primate species from zoo populations, their preliminary analysis supports this hypothesis as the between-sex difference in lifespan increases in favour of females when species-specific pathogen richness increases. Therefore, zoo populations could provide insightful information on the potential effect of chronic and putatively asymptomatic infections on the differential rates of immunosenescence between sexes.

6.3.4 Fluid and Tissue Samples

Z&A routinely maintain an archive of biological samples from their animals, primarily from whole blood, serum and plasma collected during medical exams, and tissue samples collected both ante- and post-mortem. These samples are catalogued and linked to individual medical records, allowing retrospective diagnostic testing. These archives offer easily retrievable samples that can be used as new research projects are identified or as new diagnostic techniques are developed. In many cases, these archives also provide accessible samples that can be utilized for research without the need to restrain or anaesthetize an animal for sample collection. Ultimately, this information also supports the development of technical methods that can be applied to wild populations to improve our knowledge of ageing (see Section 6.4).

6.4 Perspectives and Applications in Veterinary Medicine, Animal Welfare and Conservation Science

Taking advantage of this untapped gold mine of information can lead to significant advances in the scientific field of comparative biology of ageing, allowing the testing of evolutionary theories and exploration of the physiological mechanisms of ageing. For some species, these data may not be available in other resources, making ZIMS a unique source of reference information. Moreover, a real asset of this database for ageing research is that the exact date of birth is recorded for most individuals, which allows for the comparison of differences in age-specific parameters (e.g. survival, reproduction, medical or behavioural parameters) between individuals, sexes, populations and species across tetrapods. These findings can have applications not only to improve our global knowledge of ageing but also to improve veterinary medicine, animal welfare and conservation plans.

6.4.1 Veterinary Medicine

Collectively, this fine-scale data collection allows for complex temporospatial evaluation of disease in professionally managed captive wildlife populations. Increases in mean human life expectancies are directly correlated with advancements in medical science and a population's access to such care (e.g. [63, 64, 65]). Similarly, improvements in anaesthetic safety, diagnostic testing, treatment modalities, nutrition and pharmaceuticals have resulted in positive changes in life expectancy for animals under human care [66]. For instance, preventative veterinary care, such as dental prophylaxis, has been positively correlated with increased lifespan in dogs [67]. Studying the impact of veterinary prevention and treatment on the lifespan of many species is possible with centralized medical data. Current research focuses on measuring the pace of ageing, identifying individuals or species ageing faster, and then testing and developing interventions that could prevent or delay the progression of multi-morbidity and disability with ageing. Also, standardizing other types of medical information, such as antibiotic sensitivity testing, will open new avenues for creating relatively simple resources that can provide vital information to assist veterinarians with diagnosis and treatment. Moreover, ZIMS global medical resources may fill in knowledge gaps for less frequently studied species and provide reference data accessible within the medical record software veterinary teams use to manage the health of the animals in their care. ZIMS also contains lifelong medical records for current and past patients, providing essential historical context for each medical case. Veterinarians can use ZIMS to augment their knowledge and inform their clinical assessments, ultimately enhancing their medical practice.

Animals living under human care can be appropriate model organisms to better understand mechanistic processes and factors influencing health, lifespan and ageing, not solely across the tree of life but also in humans [68]. Many factors have been found to influence health and lifespan in humans, but also in canids, such as genetic, environmental and social factors, as well as sex, previous trauma, stress and lifestyle, including diet and exercise [69, 70, 71]. Similarly, many chronic conditions that commonly occur in human populations (e.g. obesity, arthritis, hypothyroidism and diabetes) and which are associated with comorbidities are also associated with similarly high levels of comorbidity in companion dogs [72]. In this context, ZIMS data can facilitate assessments in animal populations and testing such variables on a larger scale. These data could be the key to developing measures that can stratify risk in younger, healthier populations by detecting deficits earlier in the disease/decline process, which is essential to optimize prevention and assess the effectiveness of intervention in clinical trials and applications. Moreover, the analysis of the ZIMS medical dataset could highlight sex differences in reactions to drugs and subsequent effects on ageing. Such studies could encourage more sex-specific disease treatments in veterinary and human medicine (e.g. [73]).

6.4.2 Animal Welfare

Animals living under human care can experience negative stress [74], which may impact their ageing. There is currently a need to improve the identification of stressors that could affect the well-being, reproduction and survival of animals in Z&A and in the wild. However, the lack of standardized welfare assessment protocols across institutions (for zoo-held or wild populations) prevents researchers and zoo managers from highlighting the general and specific effects of environmental factors on animal welfare. While the ZIMS module ‘Care and Welfare’ has recently been launched (Box 6.2), the information already recorded in the husbandry and medical modules could be of high interest for achieving such analyses by combining medical and demographic data with environmental data (e.g. enclosure size and structure, ambient variables such as light and temperature, visitor interactions etc.).

The negative stressors that animals living in Z&A endure are not so different from the ones experienced by humans. Therefore, identifying the effect of these stressors on ageing in our closely related species, such as non-human primates, could help us understand the effects of negative stressors on human longevity. Given that mortality rates are a commonly used proxy for population welfare [75, 76, 77], a better understanding of the stressors impacting ageing may help improve the overall survival and, thus, population welfare of zoo-held animals. Similar insights may apply to humans, potentially leading to novel directions for research in human ageing.

6.4.3 Conservation Science

Identifying the drivers and mechanisms of ageing of different species and how the environment influences them (i.e. the drivers and mechanisms of ageing) will allow consequent advances in evidence-based conservation programmes. This information could aid in developing technical methods that can be applied to wild populations. For example, toothed whales and dolphins from several institutions have been used to develop a precise DNA method based on the epigenetic ageing clock that provides a mechanism for estimating the age of free-ranging odontocetes from either blood or skin samples [78]. More interestingly, the development of such methods, in combination with archived samples already recorded in ZIMS, might be useful to assess the biological age of the individuals based on the epigenetic clock in addition to the chronological age, a question particularly relevant for the field of ageing. Moreover, understanding sex differences in survival across species can help zoo managers to identify critical ages for reproductive and geriatric care that are sex-specific, thus developing sex- and age-specific management plans for *ex situ* populations of species at risk of extinction.

The information in ZIMS can significantly fill glaring knowledge gaps on life history traits essential for designing species conservation policies, such as the estimation of harvesting quotas or the development of extinction risk assessments to inform conservation priorities. Information such as litter and clutch size, age at first reproduction, reproductive lifespan and life expectancy is the bare minimum for implementing species conservation management plans. In essence, extinction is a demographic process; when mortality rates surpass fertility rates, populations collapse and, thus, species disappear. Therefore, a deep understanding of population dynamics is required to perform population viability analysis to implement species management programmes in Z&A and for populations in the wild. Indeed, developing conservation plans based on population viability analysis but ignoring the ageing stage might be inaccurate [79]. To develop these analyses, information on age- or stage-specific birth and death rates is crucial. However, this information is only available for approximately 1.3% of the extant mammals, birds, reptiles and amphibians [5]. Analyses of mortality and fertility rates from ZIMS data can help fill this gap for up to 15% of these species.

One common criticism of using data from Z&A animals is that results obtained may not be transferable to wild populations because of the protected environment and veterinary care provided in Z&A. However, as anthropogenic activities now impact most wild ecosystems [80], zoo-held populations can be seen as simply occupying a different place on the gradient of anthropogenic environmental effects compared to wild populations. Although wild populations exhibit greater heterogeneity (e.g. [36]), recent studies highlighted that demographic rates related to mortality in some species do not differ drastically between zoo and wild populations [13, 66]. These studies indicate that zoo-held and wild populations of the same species occur along a gradient of environmental conditions, making their demography comparable. Thus, understanding the influence of the environment on species mortality and fertility rates will be essential if we aim to use information from Z&A to support conservation efforts in the wild.

6.5 Conclusion

Conde et al. demonstrated that using ZIMS data could expand the availability of age-specific demographic knowledge by 800%: from less than 700 tetrapod species to more than 4600 species [5]. Moreover, with earlier records dating back to the mid-1800s, these data can be of particular interest for studying the ageing of long-lived animals, which tend to exhibit higher extinction risk [81]. For instance, the recent study from Da Silva et al. using Z&A data represents the first comparative study of ageing in Testudines [14], a grossly underrepresented taxon in the study of ageing across tetrapods. These findings are particularly important for the conservation of this order as 56.3% (out of the 360 currently recognized species of extant and recently extinct Testudines) of all data-sufficient Testudine species are considered 'Threatened' by the International Union for Conservation of Nature (IUCN) Red List [82].

Throughout this chapter, we highlighted how the wealth of information gathered by Z&A about the animals in their care is an untapped resource for the comparative biology of ageing. Using these data, researchers have already provided empirical evidence for different evolutionary hypotheses and mechanisms of ageing. Moreover, from evolutionary theories to mechanistic hypotheses, we emphasized the diversity of timely questions that can still be addressed in the field of ageing using data from Z&A populations and how studies embracing this research avenue will provide valuable information for veterinary medicine, animal welfare and conservation science.

References

1. Nussey, D.H., Coulson, T., Festa-Bianchet, M., Gaillard, J.-M. 2008. Measuring senescence in wild animal populations: towards a longitudinal approach. *Funct. Ecol.* **22**, 393–406 (doi:10.1111/j.1365-2435.2008.01408.x).
2. De Magalhães, J.P. 2006. Species selection in comparative studies of aging and antiaging research. In *Handbook of Models for Human Aging* (ed. P.M. Conn), pp. 9–20. Elsevier Academic Press.
3. De Magalhães, J.P. 2015. The big, the bad and the ugly: extreme animals as inspiration for biomedical research. *EMBO Rep.* **16**, 771–776 (doi:10.15252/embr.201540606).
4. Clutton-Brock, T., Sheldon, B.C. 2010. Individuals and populations: the role of long-term, individual-based studies of animals in ecology and evolutionary biology. *Trends Ecol. Evol.* **25**, 562–573 (doi:10.1016/j.tree.2010.08.002).

5. Conde, D.A. et al. 2019. Data gaps and opportunities for comparative and conservation biology. *Proc. Natl. Acad. Sci.* **116**, 9658–9664 (doi:10.1073/pnas.1816367116).
6. Holmes, D.J., Austad, S.N. 1995. The evolution of avian senescence patterns: implications for understanding primary aging processes. *Am. Zool.* **35**, 307–317 (doi:10.1093/icb/35.4.307).
7. Ricklefs, R.E. 1998. Evolutionary theories of aging: confirmation of a fundamental prediction, with implications for the genetic basis and evolution of life span. *Am. Nat.* **152**, 24–44 (doi:10.1086/286147).
8. Ricklefs, R.E. 2000. Intrinsic aging-related mortality in birds. *J. Avian Biol.* **31**, 103–111 (doi:10.1034/j.1600-048X.2000.210201.x).
9. Ricklefs, R.E., Scheuerlein, A., Cohen, A. 2003. Age-related patterns of fertility in captive populations of birds and mammals. *Exp. Gerontol.* **38**, 741–745 (doi:10.1016/S0531-5565(03)00101-3).
10. Comizzoli, P., Ottinger, M.A. 2021. Understanding reproductive aging in wildlife to improve animal conservation and human reproductive health. *Front. Cell Dev. Biol.* **9**, 680471 (doi:10.3389/fcell.2021.680471).
11. Finch, C.E., Austad, S.N. 2001. History and prospects: symposium on organisms with slow aging. *Exp. Gerontol.* **36**, 593–597 (doi:10.1016/S0531-5565(00)00228-X).
12. Tidière, M., Gaillard, J.-M., Berger, V., Müller, D.W.H., Lackey, L.B., Gimenez, O., Clauss, M., Lemaître, J.-F. 2016. Comparative analyses of longevity and senescence reveal variable survival benefits of living in zoos across mammals. *Sci. Rep.* **6**, 36361 (doi:10.1038/srep36361).
13. Colchero, F. et al. 2021. The long lives of primates and the 'invariant rate of ageing' hypothesis. *Nat. Commun.* **12**, 3666 (doi: 10.1038/s41467-021-23894-3).
14. da Silva, R., Conde, D.A., Baudisch, A., Colchero, F. 2022. Slow and negligible senescence among testudines challenges evolutionary theories of senescence. *Science* **376**, 1466–1470 (doi:10.1126/science.abl7811).
15. Williams, G.C. 1957. Pleiotropy, natural selection, and the evolution of senescence. *Evolution* **11**, 398–411 (doi:10.2307/2406060).
16. Ricklefs, R.E., Cadena, C.D. 2007. Lifespan is unrelated to investment in reproduction in populations of mammals and birds in captivity. *Ecol. Lett.* **10**, 867–875 (doi:10.1111/j.1461-0248.2007.01085.x).
17. Tidière, M., Lemaître, J.-F., Douay, G., Whipple, M., Gaillard, J.-M. 2017. High reproductive effort is associated with decreasing mortality late in life in captive ruffed lemurs. *Am. J. Primatol.* **79**, e22677 (doi:10.1002/ajp.22677).
18. Landes, J., Henry, P.-Y., Hardy, I., Perret, M., Pavard, S. 2019. Female reproduction bears no survival cost in captivity for gray mouse lemurs. *Ecol. Evol.* **9**, 6189–6198 (doi:10.1002/ece3.5124).
19. Ricklefs, R.E. 2006. Embryo development and ageing in birds and mammals. *Proc. R. Soc. Lond. B Biol. Sci.* **273**, 2077–2082 (doi:10.1098/rspb.2006.3544).
20. Tidière, M., Douay, G., Müller, P., Siberchicot, A., Sliwa, A., Whipple, M., Douhard, M. 2021. Lifespan decreases with proportion of sons in males but not females of zoo-housed tigers and lemurs. *J. Evol. Biol.* **34**, 1061–1070 (doi:10.1111/jeb.13793).
21. Lemaître, J.-F., Ronget, V., Gaillard, J.-M. 2020. Female reproductive senescence across mammals: a high diversity of patterns modulated by life history and mating traits. *Mech. Ageing Dev.* **192**, 111377 (doi:10.1016/j.mad.2020.111377).
22. Foote, A.D. 2008. Mortality rate acceleration and post-reproductive lifespan in matrilineal whale species. *Biol. Lett.* **4**, 189–191.

23. Foster, E.A., Franks, D.W., Mazzi, S., Darden, S.K., Balcomb, K.C., Ford, J.K.B., Croft, D.P. 2012. Adaptive prolonged postreproductive life span in killer whales. *Science* **337**, 1313–1313 (doi:10.1126/science.1224198).
24. Photopoulou, T., Ferreira, I.M., Best, P.B., Kasuya, T., Marsh, H. 2017. Evidence for a postreproductive phase in female false killer whales *Pseudorca crassidens*. *Front. Zool.* **14**, 30 (doi:10.1186/s12983-017-0208-y).
25. Péron, G., Bonenfant, C., Lemaitre, J.-F., Ronget, V., Tidiere, M., Gaillard, J.-M. 2019. Does grandparental care select for a longer lifespan in non-human mammals? *Biol. J. Linn. Soc.* **128**, 360–372 (doi:10.1093/biolinnean/blz078).
26. Austad, S.N. 2006. Why women live longer than men: sex differences in longevity. *Gend. Med.* **3**, 79–92 (doi:10.1016/S1550-8579(06)80198-1).
27. Trivers, R.I. 1972. Parental investment and sexual selection. In *Sexual Selection and the Descent of Man, 1871–1971* (ed. B.G. Campbell), pp. 136–208. University of California Los Angeles, Aldine Publishing Company.
28. Andersson, M.B. 1994. *Sexual Selection*. Princeton University Press.
29. Tidière, M., Gaillard, J.-M., Müller, D.W.H., Lackey, L.B., Gimenez, O., Clauss, M., Lemaître, J.-F. 2015. Does sexual selection shape sex differences in longevity and senescence patterns across vertebrates? A review and new insights from captive ruminants. *Evolution* **69**, 3123–3140 (doi:10.1111/evo.12801).
30. Marais, G.A.B., Gaillard, J.-M., Vieira, C., Plotton, I., Sanlaville, D., Gueyffier, F., Lemaître, J.-F. 2018. Sex gap in aging and longevity: can sex chromosomes play a role? *Biol. Sex Differ.* **9**, 33 (doi:10.1186/s13293-018-0181-y).
31. Cayuela, H. et al. 2021. Sex-related differences in aging rate are associated with sex chromosome system in amphibians. *Evolution* **76**, 346–356 (doi:10.1111/evo.14410).
32. Archer, C.R., Paniw, M., Vega-Trejo, R., Sepil, I. 2022. A sex skew in life-history research: the problem of missing males. *Proc. R. Soc. B Biol. Sci.* **289**, 20221117 (doi:10.1098/rspb.2022.1117).
33. Turbill, C., Bieber, C., Ruf, T. 2011. Hibernation is associated with increased survival and the evolution of slow life histories among mammals. *Proc. R. Soc. B Biol. Sci.* **278**, 3355–3363 (doi:10.1098/rspb.2011.0190).
34. Giroud, S., Zahn, S., Criscuolo, F., Chery, I., Blanc, S., Turbill, C., Ruf, T. 2014. Late-born intermittently fasted juvenile garden dormice use torpor to grow and fatten prior to hibernation: consequences for ageing processes. *Proc. R. Soc. B Biol. Sci.* **281**, 20141131 (doi:10.1098/rspb.2014.1131).
35. Gaillard, J., Lemaître, J. 2020. An integrative view of senescence in nature. *Funct. Ecol.* **34**, 4–16 (doi:10.1111/1365-2435.13506).
36. Larson, S.M., Colchero, F., Jones, O.R., Williams, L., Fernandez-Duque, E. 2016. Age and sex-specific mortality of wild and captive populations of a monogamous pair-bonded primate (*Aotus azarae*). *Am. J. Primatol.* **78**, 315–325 (doi:10.1002/ajp.22408).
37. Gompertz, B. 1825. On the nature of the function expressive of the law of human mortality, and on a new mode of determining the value of life contingencies. *Philo. Trans. R. Soc. Lond.* **115**, 513–583.
38. Makeham, W.M. 1866. On the principles to be observed in the construction of mortality tables. *Assur. Mag. J. Inst. Actuar.* **12**, 305–327 (doi:10.1017/S2046165800002823).
39. Pinder III, J.E., Wiener, J.G., Smith, M.H. 1978. The Weibull distribution: a new method of summarizing survivorship data. *Ecology* **59**, 175–179 (doi:10.2307/1936645).

40. Siler, W. 1979. A competing-risk model for animal mortality. *Ecology* **60**, 750–757 (doi:10.2307/1936612).
41. Colchero, F., Jones, O.R., Rebke, M. 2012. BaSTA: an R package for Bayesian estimation of age-specific survival from incomplete mark–recapture/recovery data with covariates. *Methods Ecol. Evol.* **3**, 466–470 (doi:10.1111/j.2041-210X.2012.00186.x).
42. Moorad, J.A., Promislow, D.E.L., Flesness, N., Miller, R.A. 2012. A comparative assessment of univariate longevity measures using zoological animal records. *Aging Cell* **11**, 940–948 (doi:10.1111/j.1474-9726.2012.00861.x).
43. Dugdale, H.L., Pope, L.C., Newman, C., Macdonald, D.W., Burke, T. 2011. Age-specific breeding success in a wild mammalian population: selection, constraint, restraint and senescence. *Mol. Ecol.* **20**, 3261–3274 (doi:10.1111/j.1365-294X.2011.05167.x).
44. Froy, H., Lewis, S., Nussey, D.H., Wood, A.G., Phillips, R.A. 2017. Contrasting drivers of reproductive ageing in albatrosses. *J. Anim. Ecol.* **86**, 1022–1032 (doi:10.1111/1365-2656.12712).
45. Muller, M.N. et al. 2020. Sexual dimorphism in chimpanzee (*Pan troglodytes schweinfurthii*) and human age-specific fertility. *J. Hum. Evol.* **144**, 102795 (doi:10.1016/j.jhevol.2020.102795).
46. Lockyear, K.M., Waddell, W.T., Goodrowe, K.L., MacDonald, S.E. 2009. Retrospective investigation of captive red wolf reproductive success in relation to age and inbreeding. *Zoo Biol.* **28**, 214–229 (doi:10.1002/zoo.20224).
47. Russell, T. et al. 2018. MHC diversity and female age underpin reproductive success in an Australian icon; the Tasmanian devil. *Sci. Rep.* **8**, 4175 (doi:10.1038/s41598-018-20934-9).
48. Monaghan, P., Maklakov, A.A., Metcalfe, N.B. 2020. Intergenerational transfer of ageing: parental age and offspring lifespan. *Trends Ecol. Evol.* **35**, 927–937 (doi:10.1016/j.tree.2020.07.005).
49. Ricklefs, R.E., Cadena, C.D. 2008. Heritability of longevity in captive populations of nondomesticated mammals and birds. *J. Gerontol. A. Biol. Sci. Med. Sci.* **63**, 435–446 (doi:10.1093/gerona/63.5.435).
50. Karniski, C., Krzyszczyk, E., Mann, J. 2018. Senescence impacts reproduction and maternal investment in bottlenose dolphins. *Proc. R. Soc. B Biol. Sci.* **285**, 20181123 (doi:10.1098/rspb.2018.1123).
51. Kuesterl, A.P., Arnemann, J. 1992. Maternal rank affects reproductive success of male Barbary macaques (*Macaca sylvanus*): evidence from DNA fingerprinting. *Behav. Ecol. Sociobiol.* **30**, 337–341.
52. Sartorelli, E.M.P., Mazzucatto, L.F., de Pina-Neto, J.M. 2001. Effect of paternal age on human sperm chromosomes. *Fertil. Steril.* **76**, 1119–1123 (doi:10.1016/S0015-0282(01)02894-1).
53. Huchard, E., Charpentier, M.J., Marshall, H., King, A.J., Knapp, L.A., Cowlishaw, G. 2012. Paternal effects on access to resources in a promiscuous primate society. *Behav. Ecol.* **24**, 229–236 (doi:10.1093/beheco/ars158).
54. Kirkwood, T.B.L., Holliday, R. 1979. The evolution of ageing and longevity. *Proc. R. Soc. B Biol. Sci.* **205**, 97–112 (doi:10.1098/rspb.1979.0083).
55. Partridge, L. 2010. The new biology of ageing. *Philos. Trans. R. Soc. Lond. B Biol. Sci.* **365**, 147–154 (doi:10.1098/rstb.2009.0222).
56. Vincze, O. et al. 2022. Cancer risk across mammals. *Nature* **601**, 263–267 (doi:10.1038/s41586-021-04224-5).

57. Great Ape Heart Project. 2012. The Great Ape Heart Project: a collaboration to understand heart disease, reduce mortality and improve cardiac health in all four great ape taxa, *White Paper*, 1–16. https://greatapeheartproject.files.wordpress.com/2016/07/gahp_whitepaper2012.pdf
58. Johnson, A.E.W., Pollard, T.J., Mark, R., Berkowitz, S.J., Greenbaum, N.R., Lungren, M.P., Deng, C., Mark, R.G., Horng, S. 2019. MIMIC-CXR, a de-identified publicly available database of chest radiographs with free-text reports. *Scientific Data*, **6**, 317 (doi:10.1038/s41597-019-0322-0).
59. Crosier, A.E., Comizzoli, P., Baker, T., Davidson, A., Munson, L., Howard, J., Marker, L.L., Wildt, D.E. 2011. Increasing age influences uterine integrity, but not ovarian function or oocyte quality, in the cheetah (*Acinonyx jubatus*). *Biol. Reprod.* **85**, 243–253 (doi:10.1095/biolreprod.110.089417).
60. Ludwig, C., Dehnhard, M., Pribbenow, S., Silinski-Mehr, S., Hofer, H., Wachter, B. 2019. Asymmetric reproductive aging in cheetah (*Acinonyx jubatus*) females in European zoos. *J. Zoo Aquar. Res.* **7**, 87–93.
61. McAloose, D. et al. 2020. From people to *Panthera*: natural SARS-CoV-2 infection in tigers and lions at the Bronx Zoo. *mBio* **11**, e02220–20 (doi:10.1128/mBio.02220-20).
62. Tidière, M., Badruna, A., Fouchet, D., Gaillard, J.-M., Lemaître, J.-F., Pontier, D. 2020. Pathogens shape sex differences in Mammalian aging. *Trends Parasitol.* **36**, 668–676 (doi:10.1016/j.pt.2020.05.004).
63. Colchero, F. et al. 2016. The emergence of longevous populations. *Proc. Natl. Acad. Sci.* **113**, E7681–E7690 (doi:10.1073/pnas.1612191113).
64. Aburto, J.M., Beltrán-Sánchez, H. 2019. Upsurge of homicides and its impact on life expectancy and life span inequality in Mexico, 2005–2015. *Am. J. Public Health* **109**, 483–489 (doi:10.2105/AJPH.2018.304878).
65. Aburto, J.M., Villavicencio, F., Basellini, U., Kjærgaard, S., Vaupel, J.W. 2020. Dynamics of life expectancy and life span equality. *Proc. Natl. Acad. Sci.* **117**, 5250–5259 (doi:10.1073/pnas.1915884117).
66. Tidière, M. et al. 2023. Survival improvements of marine mammals in zoological institutions mirror historical advances in human longevity. *Proc. Roy. Soc. B.* **290**, 20231895 (doi:10.1098/rspb.2023.1895).
67. Urfer, S.R., Wang, M., Yang, M., Lund, E.M., Lefebvre, S.L. 2019. Risk factors associated with lifespan in pet dogs evaluated in primary care veterinary hospitals. *J. Am. Anim. Hosp. Assoc.* **55**, 130–137 (doi:10.5326/JAAHA-MS-6763).
68. Wallis, L.J., Szabó, D., Erdélyi-Belle, B., Kubinyi, E. 2018. Demographic change across the lifespan of pet dogs and their Impact on health status. *Front. Vet. Sci.* **5**, 200 (doi:10.3389/fvets.2018.00200).
69. Kramer, A.F., Bherer, L., Colcombe, S.J., Dong, W., Greenough, W.T. 2004. Environmental influences on cognitive and brain plasticity during aging. *J. Gerontol. A. Biol. Sci. Med. Sci.* **59**, M940–M957 (doi:10.1093/gerona/59.9.M940).
70. Zimmerman, E., Woolf, S.H. 2014. Understanding the relationship between education and health. *NAM Perspect.*, 1–25 (doi:10.31478/201406a).
71. Szabó, D., Gee, N.R., Miklósi, Á. 2016. Natural or pathologic? Discrepancies in the study of behavioral and cognitive signs in aging family dogs. *J. Vet. Behav.* **11**, 86–98 (doi:10.1016/j.jveb.2015.08.003).
72. Hoffman, J.M., Creevy, K.E., Franks, A., O'Neill, D.G., Promislow, D.E.L. 2018. The companion dog as a model for human aging and mortality. *Aging Cell* **17**, e12737 (doi:10.1111/acel.12737).

73. Klein, S.L., Poland, G.A. 2013. Personalized vaccinology: one size and dose might not fit both sexes. *Vaccine* **31**, 2599–2600 (doi:10.1016/j.vaccine.2013.02.070).
74. Morgan, K.N., Tromborg, C.T. 2007. Sources of stress in captivity. *Appl. Anim. Behav. Sci.* **102**, 262–302 (doi:10.1016/j.applanim.2006.05.032).
75. Broom, D.M. 1991. Animal welfare: concepts and measurement. *J. Anim. Sci.* **69**, 4167–4175 (doi:10.2527/1991.69104167x).
76. Walker, M., Duggan, G., Roulston, N., Van Slack, A., Mason, G. 2012. Negative affective states and their effects on morbidity, mortality and longevity. *Anim. Welf.* **21**, 497–509 (doi:10.7120/09627286.21.4.497).
77. World Health Organization. 2016. World health statistics 2016: monitoring health for the SDGs, sustainable development goals. www.who.int/docs/default-source/gho-documents/world-health-statistic-reports/world-heatlth-statistics-2016.pdf
78. Robeck, T.R. et al. 2021. Multi-species and multi-tissue methylation clocks for age estimation in toothed whales and dolphins. *Commun. Biol.* **4**, 642 (doi:10.1038/s42003-021-02179-x).
79. Robert, A., Chantepie, S., Pavard, S., Sarrazin, F., Teplitsky, C. 2015. Actuarial senescence can increase the risk of extinction of mammal populations. *Ecol. Appl.* **25**, 116–124 (doi:10.1890/14-0221.1).
80. Schipper, J. et al. 2008. The status of the world's land and marine mammals: diversity, threat, and knowledge. *Science* **322**, 225–230 (doi:10.1126/science.1165115).
81. Reynolds, J.D. 2003. Life histories and extinction risk. In *Macroecology: Concepts and Consequences* (eds T.M. Blackburn, K.J. Gaston), pp. 195–217. Blackwell Publishing.
82. Rhodin, A.G.J. et al. 2018. Global conservation status of turtles and tortoises (order Testudines). *Chelonian Conserv. Biol.* **17**, 135–161 (doi:10.2744/CCB-1348.1).
83. Tribe, A., Booth, R. 2003. Assessing the role of zoos in wildlife conservation. *Hum. Dimens. Wildlife* **8**, 65–74 (doi:10.1080/10871200390180163).
84. Fa, J.E., Funk, S.M., O'Connell, D. 2011. *Zoo Conservation Biology*. Cambridge University Press.
85. Grow, S., Lyles, A.M., Greenberg, R., Powell, D.M., Dorsey, C. 2024. Zoos, aquariums, and zoological parks. In *Encyclopedia of Biodiversity (third edition)* (ed. S.M. Scheiner), pp. 474–484. Academic Press (doi:10.1016/B978-0-12-822562-2.00036-0).
86. Gusset, M., Dick, G. 2011. The global reach of zoos and aquariums in visitor numbers and conservation expenditures. *Zoo Biol.* **30**, 566–569 (doi:10.1002/zoo.20369).
87. Fa, J.E., Gusset, M., Flesness, N., Conde, D.A. 2014. Zoos have yet to unveil their full conservation potential. *Anim. Conserv.* **17**, 97–100 (doi:10.1111/acv.12115).
88. Association of Zoos and Aquariums. 2022. Accreditation. www.aza.org/becoming-accredited.
89. Zimmermann, A. 2010. The role of zoos in contributing to in situ conservation. In *Wild Mammals in Captivity: Principles and Techniques for Zoo Management* (eds D.G. Kleiman, K.V. Thompson, C. Kirk Baer), pp. 280–287. University of Chicago Press.
90. Tokarska, M., Pertoldi, C., Kowalczyk, R., Perzanowski, K. 2011. Genetic status of the European bison *Bison bonasus* after extinction in the wild and subsequent recovery: European bison conservation genetics. *Mammal Rev.* **41**, 151–162 (doi:10.1111/j.1365-2907.2010.00178.x).
91. Princée, F.P.G., ed. 2016. *Exploring Studbooks for Wildlife Management and Conservation*. Springer.
92. Seal, U.S., Makey, D.G., Murtfeldt, L.E. 1976. ISIS: an animal census system. *Int. Zoo Yearb.* **16**, 180–184 (doi:10.1111/j.1748-1090.1976.tb00171.x).

7 Perspectives in Comparative Biology of Ageing

Jean-François Lemaître, Jean-Michel Gaillard, Samuel Pavard, François Criscuolo and Fabrice Bertile

7.1 Introduction: Going beyond the Comparative Biology of Lifespan

How long organisms have lived in the past, currently live and could potentially live are biological questions that have fascinated scientists for centuries. The comparative biology of ageing aims to answer these questions by exploring the (large) variance in lifespan observed across species (e.g. [1]) and shaped by evolution. The first comprehensive comparative analysis of lifespan was proposed by Aristotle (350 BCE) [2], who uncovered some key pillars of the biology of lifespan, which, for some, still hold today. Aristotle noticed that, besides the obvious observations that species markedly differ in terms of lifespan (i.e. 'mankind has a longer life than the horse'), there is a strong intraspecific variation in lifespan (i.e. 'some [men] are long-lived, others short-lived'; see also Chapters 9 and 14). Even more importantly, Aristotle was the very first to spot, more than two millennials before Huxley or D'Arcy Thompson [3, 4], that 'it is a general rule that the larger live longer than the smaller'. However, the proposed mechanism based on deleterious influences of increasing dryness and coldness of organisms over time nowadays looks odd at the best, and the prediction that 'males live longer than females' turned to be wrong in several taxonomic groups (e.g. [5] for an analysis across mammals and Chapter 9 for a review of the topic).

Subsequent analyses performed during the last century, mostly in mammals and birds, have identified that a large combination of intertwined factors related to life history, environmental conditions and physiology, do shape inter-specific variation in lifespan. Among them, body size (via allometric constraints [6, 7, 8]), diet (via metabolic pathways [9, 10]), lifestyle (e.g. via mortality risk due to predation [11]), taxonomic position (via phylogenetic inertia [12]) and, more recently, genomic, epigenomic and physiological peculiarities (via modulations of the ageing pathways [13, 14, 15]) have been shown to drive observed variation in lifespan across species.

Currently, species differences in lifespan are mostly understood in the framework of biological (or physiological [16]) time [17], under which all traits with a dimension of time or time frequency should covary isometrically. This means that long-lived organisms develop at a slower pace, reach sexual maturity later and produce offspring at a slower rate than short-lived organisms, which leads organisms to rank along the so-called slow–fast continuum (after [18]) of life histories. As the slow–fast continuum is the major structuring axis of life history variation across species ([19] for a review), a corollary to the existence of this continuum is that any selective pressure or constraint on development or reproductive rate will influence lifespan (see e.g. [20]). Interestingly, for a given position on the slow–fast continuum (or for a given body mass, see Section 7.2.2), lifestyle features also contribute to the species differences in lifespan. Volant birds and flying mammals live longer than their non-volant counterparts [11] and both fossorial and arboreal species live longer than terrestrial species [11, 21].

However, to date, most, if not all, of our current knowledge on lifespan variation relies on the analysis of a single metric: the maximum lifespan. Yet, as repeatedly pointed out, the use of maximum lifespan is flawed, with severe problems, including the poor ability of the maximum value to inform about the distribution of age at death and its strong dependency to sample size (see [22, 23, 24, 25] for detailed information). Other metrics of lifespan, such as life expectancy, adult life expectancy, lifespan 80% (corresponding to the age at which 80% of individuals within a cohort have died) or median age at death (equivalent to lifespan 50%) provide much more reliable descriptors of lifespan variation and should definitely be preferred over maximum lifespan (see Chapters 3 and 4).

The concept of species lifespan by itself, which relies on measuring the duration of the typical life trajectory of organisms, has a limited value and cannot allow a thorough understanding of mortality patterns along the life trajectory. Lifespan is simply the overall outcome of all the mortality risks an organism faces from birth onwards. At the species level, lifespan can thus be largely uncoupled from actuarial senescence (also called survival ageing) [26] that corresponds to the increase in mortality risk with increasing age. In other words, while valuable, results obtained with comparative analyses of lifespan might only provide a limited amount of insights to accurately understand variation in the ageing process across species.

In a wide range of organisms (e.g. vertebrates, arthropods), age is a strong structuring factor of mortality and allows mortality patterns along the life trajectory to be partitioned into successive life stages. Although the exact partition is species-specific, three major life stages are generally defined (Figure 7.1). The first stage, from birth to maturity, corresponds to the developmental stage and is characterized by a high and variable mortality risk, which decreases from birth to maturity (juvenile mortality). The second stage describes the period between the age at first reproduction when mortality risk is very low and the onset of actuarial senescence, defined as the age at which mortality risk starts to increase with age (see Section 7.2.2 for a specific discussion on this parameter). During that prime-age stage, the mortality risk is quite constant and minimal. Lastly, the period beyond the age at the onset of senescence corresponds to the senescence period during which mortality risk exponentially increases with age. To get a full picture of the mortality patterns, one needs to assess both stage-specific mortality risks and the threshold ages defining the transitions between stages. Any metric of lifespan will not allow getting this information. For instance, although lifespan is still often used as a proxy for actuarial senescence, the association between these characteristics is only moderate since variation in both the onset and the rate of actuarial senescence accounts for less than half the variation observed in lifespan across mammalian species [26].

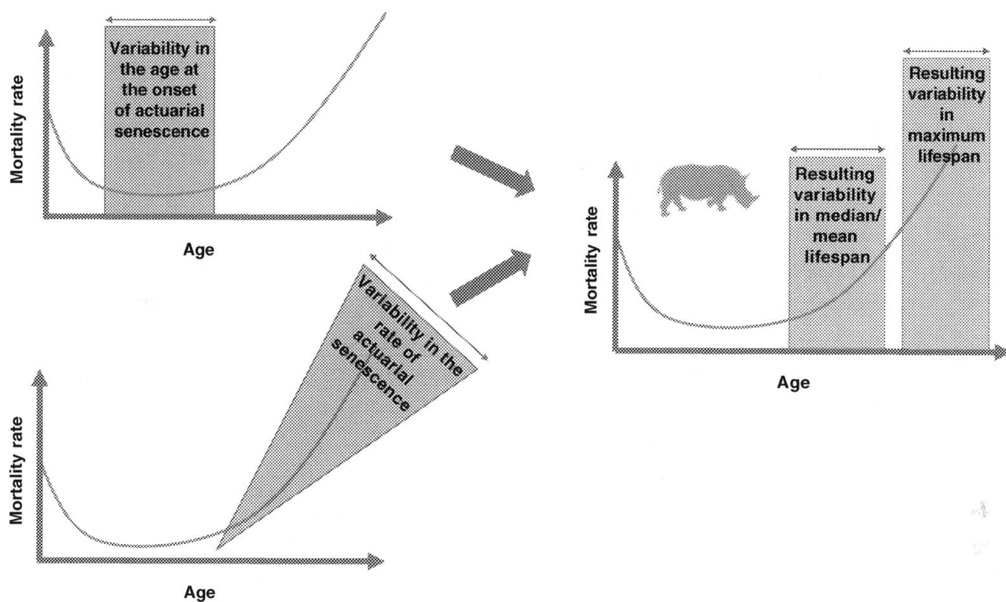

Figure 7.1 Age-specific mortality of a typical mammalian species. The age at the onset of actuarial senescence and the rate of actuarial senescence can both vary a lot, which will ultimately influence lifespan metrics such as the mean lifespan or the maximum lifespan. Future comparative studies of ageing should seek to identify the genetic and physiological determinants of actuarial senescence to understand the biological drivers of the timing and tempo of ageing.

From our current understanding of mortality patterns, lifespan variation across species only informs about the pace of life, with long-lived species living on the slow line and short-lived species living on the fast line. The full assessment of life trajectories requires investigation of the shape of mortality patterns in addition to the pace-of-life component described by the lifespan. Thus, the relative proportion of lifespan spent in developmental, prime-age and senescent stages varies tremendously even after differences in pace of life have been corrected for. A more holistic approach than the focus on lifespan is now necessary. Overall, biogerontological studies aimed at determining whether and how a given molecular pathway influences the age-specific mortality trajectory of a given species through the comparative tool needs to go beyond simple consideration of lifespan and embrace the diversity and complexity of actuarial senescence patterns.

7.2 Actuarial Senescence Patterns across the Tree of Life

7.2.1 Availability of Age-Specific Mortality Data in the Wild

The availability of online databases such as AnAge [27], or Longevity Records [1], which provide maximum lifespan data for a very broad range of species across the tree of life partly explains why actuarial senescence patterns have been much less studied than lifespan through the lens of comparative biology. The other reason that could be invoked is the misbelief that the number of species for which data on age-specific survival (or age-specific mortality risk) allows key senescence parameters (i.e. age at the onset of actuarial senescence, rate of actuarial senescence) to be estimated is too limited. Yet, age-specific data on mortality risk are available for many species, especially vertebrate, and constitute a neglected, but promising, resource for the comparative biology of ageing.

The census of living or dead individuals within wild populations of animals that have emerged in the middle of the twentieth century have allowed the first estimation of age-specific survival probabilities or mortality risk in non-model organisms (e.g. [28, 29] for historical case studies in mammals), although it is worth noticing that for most of these studies the exact chronological age of the individuals was not known so was assessed through alternative approaches (e.g. tooth wear in ungulates [30]). Since then, the emergence of long-term field studies tracking individuals from birth to death [31], as well as the progressive development of statistical methods enabling imperfect detection of individuals under longitudinal monitoring to be accounted for [32], has led to countless publications reporting age-specific mortality risk, mostly in ecological journals, but highly relevant for the comparative biology of ageing. These demographic data are now available for several hundreds of species, displaying contrasting life history strategies and lifestyles [33]. Importantly, they have allowed demonstrating the widespread occurrence of actuarial senescence, in particular among vertebrates (especially birds and mammals) [34], as well as the huge diversity of actuarial senescence patterns across species, notably in terms of both onset and rate of actuarial senescence ([35, 36], see also Chapter 5).

7.2.2 Comparative Analyses of Actuarial Senescence Patterns

The ornithologist and ecologist Robert E. Ricklefs was one of the first to take advantage of the available data on age-specific mortality, in both wild and captive conditions, to perform comparative analyses of actuarial senescence patterns [37]. He was probably the first ecologist to emphasize the untapped biomedical relevance of actuarial senescence metrics of wild populations of animals (in particular the rate of actuarial senescence, coined *rate of ageing* in his publication) and to publish comparative analyses in biogerontological journals such as *Experimental Gerontology* or *Ageing Cell* [38, 39]. The main objective of Ricklefs' first paper on ageing was to identify the best model fitting age-specific mortality data during adulthood in birds [37]. Comparing analyses of actuarial senescence patterns requires the estimation of parameters that are comparable across species, and Ricklefs dedicated several studies, mostly focused on birds, to the accurate estimation of actuarial senescence rates. Ricklefs found that the Weibull model generally provided the best fit to measure age-specific mortality hazard in birds [37, 39]. Subsequent comparative analyses of mortality patterns have revealed that, in mammals, age-specific mortality is best described by a Gompertz model in approximately 80% of the species explored [40], in line with previous observations made across human populations [41]. This consistency, at least in mammals, in the type of models describing mortality patterns gives the opportunity to explore whether, how and why the human rate of actuarial senescence could deviate from that of other mammals [42, 43].

Until very recently, comparative analyses of mortality patterns only focused on the rate of actuarial senescence and have dismissed any other relevant parameters. Based on both verbal and mathematical predictions regarding the evolution of ageing [44, 45], the increase in mortality risk with age was expected to start at the age of first reproduction and models were thus fitted accordingly. However, empirical evidence accumulated across species has revealed that the onset of actuarial senescence is often delayed compared to the age at first reproduction and varies widely between species. A comparative analysis performed by Péron and colleagues in birds and mammals revealed that the age at the onset of actuarial senescence is generally much later than the age at first reproduction [46]. For instance, the mortality risk in female black bears (*Ursus americanus*) starts to increase from 13 years of age (i.e. 8 years after the mean age of first reproduction) onwards. A similar delay of around eight years after the age at first reproduction is also observed in female pronghorn (*Antilocapra americana*), for which an estimate of 10.5 years at the onset of actuarial senescence is given [46] (see also [47] for an illustration of the between-species variation in the onset of actuarial senescence). Interestingly, this diversity in demographic ageing patterns echoes the spate of biogerontological studies highlighting that tissue- or organ-specific dysregulation or failure can start at various ages and with different intensity [48, 49].

It is beyond the scope of this chapter to review the numerous methodological approaches that have been developed to estimate both the onset and the rate of actuarial senescence across species (but see [40, 47] for discussions). However, we will summarize the main ecological factors that have been found to be associated with these parameters, as the corresponding studies can serve as stepping stones for future development in the comparative biology of ageing. Comparative analyses of actuarial senescence patterns have revealed that the onset and rate of actuarial senescence are associated with phylogeny [5], body mass (e.g. [35, 39, 50]) following allometric rules and species position along the slow–fast continuum [35, 51, 52]. These results highlight that slow species generally display a late age at the onset of actuarial senescence and a low rate of actuarial senescence while fast species display the opposite features. Overall, this constitutes an expression of the pattern of covariation among biological times across species, well described in the living world [18, 19], since both age at the onset of actuarial senescence and the rate of actuarial senescence can be expressed in time units (e.g. years and years^{-1}, respectively).

The identification of ecological or biological factors driving inter-specific differences in actuarial senescence patterns thus require researchers to account for the confounding factors structuring life history variation; which include phylogeny, body mass and the slow–fast continuum. However, it is worth noticing that body mass and generation time are tightly allometrically correlated, which prevents both traits being simultaneously included in the same model [53]. It has thus been advocated, when possible, that a correction for generation time over a correction for body mass should be favoured, as the slow–fast continuum constitutes the first axis of variation in life history traits across animals [19]. Such corrections are mandatory to make any relevant biological interpretations (see also Chapter 10 for a specific discussion on primates). For instance, Ricklefs reported a positive association between the rate of actuarial senescence (expressed in years^{-1}) and the embryo growth rate (expressed in days^{-1}) in birds and between the rate of actuarial senescence and the post-natal growth rate (expressed in days^{-1}) in mammals ([54], see also [55]). While these analyses highlight the life history connections between development and senescence [54], whether growth and senescence are functionally related (i.e. there is a direct influence of growth patterns on senescence across species) or are only similarly shaped by the slow–fast continuum of life histories cannot be disentangled. Without explicitly correcting for the confounding influence of the slow–fast continuum in such analyses, it is impossible to conclude that species displaying a fast growth should suffer from earlier and/or steeper rates of actuarial senescence.

Overall, our current knowledge of the drivers of actuarial senescence remains very limited, although exciting results are regularly emerging. For instance, it has been recently documented that among vertebrates, and when accounting for body mass, ectotherms with chemical (venom and skin toxins) or physical (armour and shells) protections display a slower rate of actuarial senescence than ectotherms without such protections [35], in line with early predictions of the evolutionary theory of ageing [44]. Yet, as mentioned, there is absolutely no reason to assume that an ecological or biological factor influencing lifespan will have a similar influence on both onset and rate of actuarial senescence, in terms of directionality and magnitude [26]. This is notably illustrated by comparative analyses seeking to decipher the influence of sexual competition on sex differences in mortality patterns. Indeed, in captive populations of large herbivores, males from polygynous species display an earlier onset of actuarial senescence than males from monogamous species [47]. On the contrary, there is no difference in the rate of actuarial senescence between males from monogamous and polygynous species ([47] see also Chapter 9). This suggests that the physiological costs (e.g. increased telomere loss, impaired immune defence) paid by polygynous males are responsible for the earlier rise in mortality risk observed in these species, but that compensatory mechanisms, yet to be identified, might have evolved to buffer the subsequent increase in age-specific mortality risk.

7.3 The Comparative Biology of Actuarial Senescence in an 'Omic' World

7.3.1 Overview of the Toolkit

Thanks to multiple fast advances in molecular biology and bioinformatics, the comparative study of actuarial senescence can now enter the era of 'omics' approaches. To gain in-depth access to this biological complexity and the mechanisms underlying it, high-throughput approaches for large-scale molecular screening are required. Omics technologies, which have now reached unprecedented performance levels, are emerging at the forefront of next-generation biogerontology. Omics allow now highly detailed comparative analyses based on a systems biology approach to be undertaken [56, 57]. This involves not only considering the many molecules that act in all well-defined pathways of ageing, but also cross-talks within a pathway network, which have yet to be better defined, as well as their age-dependent dynamics [58]. Currently, the accumulation of complementary omics data enables researchers to better understand the evolution of ageing [59]. Omics analyses, combined with the power of modelling, provide an essential tool for testing causal hypotheses regarding the involvement of cellular signalling pathways and metabolic networks in the variability of ageing trajectories, both between species and between people.

Owing to a drastic drop in the cost of DNA (deoxyribonucleic acid) sequencing since around 2014 [60], and the development of next- and third-generation-sequencing technologies [61], the number of sequenced genomes is increasing exponentially, and the target is now to put together a high-quality assembly for each animal species [62]. This widespread availability of genome sequences feeds our understanding of the function of genes through functional genomics technologies, and has allowed age-related single nucleotide polymorphisms to be discovered [56]. Building on these important advances, the omics-based systems biology approach should enable us to understand how gene variation is integrated into a broad and complex set of ageing pathways, and to consider their environmental modulators. In this way, the missing heritability (i.e. the element of heritability that cannot be explained by single genetic variations), which may be partly due to environmental influences and their integration into the biology of organisms, could be better deciphered. It has also opened the way to developing so-called post-genomic sciences (i.e. the various omics that have taken centre stage since the early 2000s [63]). Transcriptomic, proteomic and metabolomic technologies can now be used more or less routinely to follow the flow of information from transcriptional to post-translational stages (i.e. from DNA to proteins), and thus to better understand how cells acquire well-defined functions that ultimately shape the physiology of organisms. In particular, transcriptomics approaches using RNA-seq (ribonucleic acid sequencing) have revolutionized gene expression profiling by providing comprehensiveness and high precision [64], and they offer the essential possibility of analysing epigenetic modifiers such as non-coding RNAs [65]. Mass spectrometry-based proteomics is today powerful enough to analyse the abundance, characterize the structure and evaluate the function of hundreds or thousands of proteins in a single analysis [66, 67]. In addition, the field of metabolomics, which provides quantitative and qualitative information for small molecules (metabolites) that are key to physiology and pathophysiology, has grown rapidly in recent years [68]. The current trend towards systems biology approaches via the integration of multiple omics layers to capture a composite vision of the ageing process is challenging but highly promising [69, 70, 71].

7.3.2 Identifying 'Omics' Candidates for Future Comparative Studies of Actuarial Senescence

Until recently, studies of the mechanistic basis of ageing among species have focused primarily on specific hallmarks (sensu [72]) in an attempt to explain variation in lifespan in the animal kingdom. For instance, several works explored the implication of telomere length (or of its regulation) in explaining disparities in lifespan from unicellular organisms to vertebrates (e.g. [73, 74]). Additional examples of ageing mechanisms that have been proposed include mitochondrial function or the sensitivity of membrane structure to oxidative stress [75]. Overall, a variety of potential biomarkers of ageing (or healthy ageing) have now been identified (e.g. mean arterial pressure, waist circumference, cholesterol levels, telomere attrition, cellular senescence, mitochondrial dysfunction, changes in the composition of the gut microbiome) and the trend towards combining several of them proves promising for future comparative analyses.

Studies with the objective to go a step further, either deeper in physiological, cellular or molecular regulatory pathways, or by using exploratory methodologies, which do not target an *a priori* mechanism, are now emerging. In that context, omics approaches represent a promising tool for the comparative biology of ageing in general, and more specifically for identifying the mechanisms shaping the diversity of actuarial senescence patterns across the tree of life but also between individuals. To date, omics approaches have been used in the field of ageing for different purposes: to estimate chronological age, to discover new biomarkers of mortality risk and/or age-related diseases, to better determine the mechanisms underlying the ageing process. All these approaches can be relevant for the comparative biology of ageing.

7.3.2.1 'Omics' Markers of Chronological Age: Expanding the Current Demographic Dataset

As mentioned, the fine scale estimation of actuarial senescence requires age-specific mortality data. For a long time, scientists have looked for methods to assign the age of individual animals (e.g. tooth wear, number of horn annuli in bovids, size in species with indeterminate growth). However, methods based on phenotypic traits are associated with a relatively high degree of uncertainty. The advent of long-term longitudinal studies where individuals are individually marked from birth (with a degree of uncertainty on age of only a few days) have largely solved the issue of age assignment. However, as the number of longitudinal follow-up of animal species cannot reasonably cover the whole range of the tree of life, and because it remains difficult to assign chronological age in some species (e.g. small rodents), comparative analyses of actuarial senescence patterns are inherently biased towards specific taxa. This is well illustrated by the comparative analyses of actuarial senescence performed so far across mammals, which remain largely biased towards primates, carnivores and ungulates [5]. Therefore, the assessment of chronological age through the use of molecular techniques appears as a relevant and complementary tool to estimate demographic parameters.

The search for molecular predictors of chronological age is not recent. Telomere length was initially proposed as a marker of chronological age ('mitotic clock' hypothesis) [76], but its predictive power rapidly appeared to be limited, with R^2 generally less than 0.6 in populations of animals in the wild (e.g. [77]). Telomeres are now often considered to be a marker of biological age, although this is still subject to lively debate. In mammals, we have recently assisted in making huge progress in the accuracy of chronological age estimation thanks to the development of an epigenetic clock conserved across species (i.e. Universal Mammalian Clock) [78]. The coefficient of determination of the regression linking the so-called epigenetic age and the exact chronological age is generally higher than 0.9 [79]. Yet, even if the percentage of non-explained variation is small, it has been suggested that it could contain meaningful biological information [80].

Additionally, several studies have put forward the relevance of emerging proteomic clocks, notably those involving immune and neuronal pathways that exhibit good associations with ageing phenotypes, mortality, multi-morbidity, healthspan and lifespan (see a review in [57]). Besides the measurements of protein quantitative changes over age, attention has to be paid to the mass spectrometry-based analysis of protein structures. Indeed, N-glycosylation of circulating immunoglobulins has already been reported to help predict chronological age in multiple cohorts [81], and there is no doubt glycomics will provide further ageing clocks based on more glycoprotein characterization in the future. Recently, metabolomic signatures in blood and urine were used for chronological age prediction. However, some studies advocate that metabolomic and epigenetic clocks are only partly synchronized, as they capture changes in different age-related pathways (nutrient signalling or mitochondrial functioning vs genomic stability, respectively). This desynchronization is attributable to the fact that the metabolomics clock is highly sensible to particular pathophysiological features of ageing such as being overweight/obese or having diabetes, hence reflecting a sort of metabolic age [82, 83, 84].

Overall, these discrepancies emphasize that, even if we can expect to see further improvements in the estimation of chronological age by omics techniques over the next few years, future studies should focus on the remaining variance and characterize the molecular and cellular mechanisms that might reflect individual differences in life trajectories, like body condition or health. In this way, omics markers of biological age could become a proxy for the remaining healthspan or lifespan (Figure 7.2).

7.3.2.2 'Omics' Markers of Biological Age: Identifying Mechanisms Associated with a Late Onset or a High Rate of Actuarial Senescence across Species

The identification of markers of biological age that can accurately predict mortality risk in a given species is now a central objective in biogerontology. In that context, 'omics approaches' have rapidly become one of the most efficient tool as they can, notably, track the long-term and delayed effects of environmental conditions such as stress on the diversity of ageing rates among individuals [85]. For example, the measurement of the methylome (levels of DNA methylation) associated with the analysis of the epigenome (i.e. the localization of methylations that identifies the genes thus regulated) in roe deer (*Capreolus capreolus*) made it possible to differentiate between environmental- and sex-related differences in the acceleration of epigenetic age (i.e. ageing independent of chronological age), involving biological mechanisms such as cell cycle control or nervous system homeostasis [79]. Although new generations of epigenetic clocks now appear to be some of the most reliable predictors of biological age (e.g. PhenoAge in humans), a more comprehensive omics analysis should provide additional robust predicting signatures because almost all cellular and biochemical processes are affected during ageing [56, 86, 87]. Transcriptomics in eusocial insects has notably proved to be a groundbreaking tool in showing that DNA transposable elements contribute, as an unexpected source of DNA damage, to differential ageing between social castes [88]. Exploratory omics are also likely to improve our understanding of well-known ageing mechanisms by characterizing all the interacting molecules that modulate such pathways, for example, like TIN2 (TERF1-interacting nuclear factor 2), heat shock and chitinase-like proteins that are involved in immune dysregulation with increasing age in mice [89]. Interestingly, all markers of biological age identified within species might constitute a first step of candidate traits that could contribute to the diversity of actuarial senescence patterns across the tree of life. Thus, moving forward, the acquisition of omics data in relation to actuarial senescence patterns will offer additional opportunities to find new medical targets and to better understand how evolution may have solved ageing challenges in relation to specific environmental conditions.

Available omics data actually point out several cellular or molecular targets that are interesting to link to ageing phenotypes. To cite just a few studies using omics, ageing has already been characterized by drastic transcriptomic changes, notably concerning genes that are key to cellular senescence, the mitochondrial function, the immune and stress responses and DNA repair, but in some cases with important differences according to the tissue, genotype or species considered [56, 90, 91, 92, 93, 94, 95, 96]. Age-related changes in the levels of non-coding RNAs have also been reported, but this type of analysis remains scarce [97, 98, 99]. Further studies are needed to correlate those findings, mainly originating from only a handful of model organisms to the actual actuarial senescence process in an extended set of species. Such works will further identify mechanisms shaping inter-specific differences in the onset and/or rate of actuarial senescence that could likely differ from the mechanism shaping inter-specific differences in lifespan [26, 100].

Few studies have reported changes in protein abundance in different tissues, notably involving aerobic and anaerobic metabolism as a function of age [99, 101], the process of brain ageing [102], or cytokine-cytokine receptor interaction, complement and coagulation cascades and axon guidance [103]. The abundance of a number of plasma proteins has also been associated with biological age within species [104, 105]. As well, structural modification of proteins (e.g. through glycosylation and glycation) and the expression levels of the factors (degradation and repair systems) that ensure protein maintenance that has been associated with age [81, 106]. Proteomics approaches have also highlighted age-dependent liver oxidative costs [107], and a telomere and/or telomerase-related cost of innate immune activation in mice [89]. Today, multi-omics strategies have also begun to help identify the molecular bases of inflammageing, thus offering a promise to better fight it in the future. Hence, the pro-inflammatory state that develops upon ageing appears to be linked to a combination of cellular senescence, mitochondrial dysfunction, impaired mitophagy and DNA damage [108].

Interestingly, integrated omics applied to *in vitro* models can provide interesting information about the molecular mechanisms that underlie the ageing process. A fibroblast cell line (WI-38 lung fibroblasts) has notably been used to show that replicative senescence is characterized by metabolic alterations interpreted as stress responses, in relation to inflammation and changes in several metabolic (like fat and glucose) pathways [109]. Moreover, primary cultures of fibroblasts have provided tests of how cellular ageing is affected by energetics challenges [110], and have pointed out that the alteration of the cell cycle, DNA replication and DNA repair may partly elucidate the mechanism of cellular senescence, ageing and ageing-related diseases [111], as well as differences in patterns of actuarial senescence across species (see [112] for an example of a study comparing lifespans between humans, laboratory mice and a captive population of naked mole rats). Since different types of *in vitro* model systems such as primary cultures can today be developed from small biopsies [113], the *in vitro* approach is open to the omics-based research on ageing across a wide range of animal populations in the wild. Conducting comparative omics studies across a

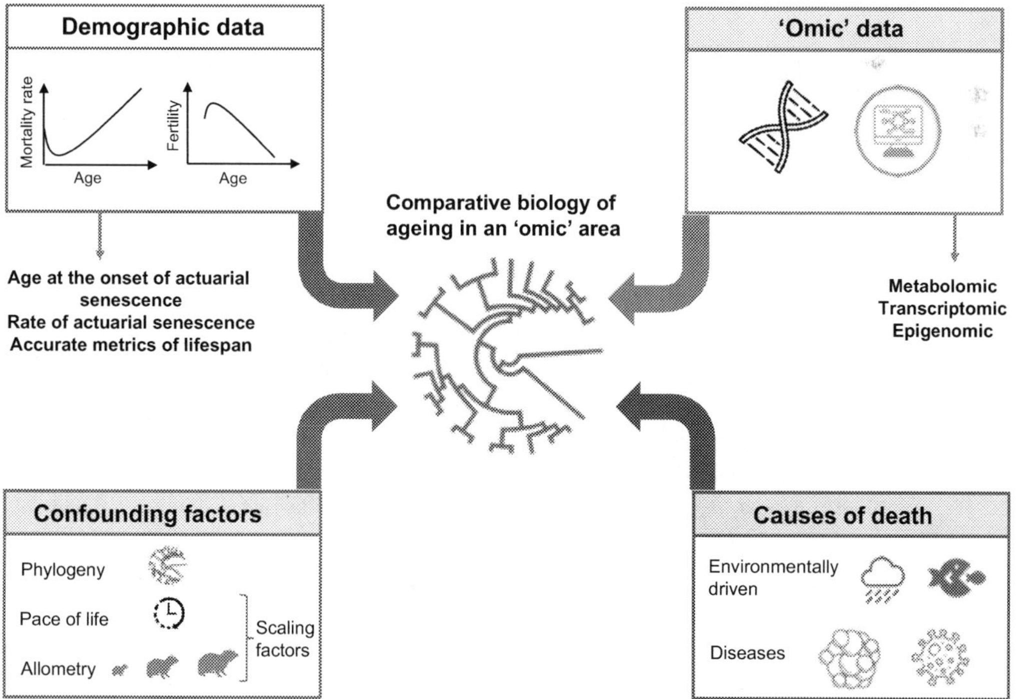

Figure 7.2 Synthetic guideline for performing comparative studies of ageing.

large panel of species, covering a substantial range of the slow–fast continuum, and among individuals displaying contrasted senescence patterns is one of the urgent and particularly exciting objectives in the next future.

7.3.3 From Comparative Omics to Comparative Epidemiology

With the exception of death caused by zoonoses (e.g. [114]), very little is known about the natural causes of death of individuals in wildlife, in particular for old individuals. The conditions and diseases from which these causes arise thus remain a black box. Yet, it is crucial to go beyond aggregate mortality patterns to look at the age-specific causes from which individuals die, in order to measure the weight of the deterioration of the various physiological functions in the ageing phenotypes, across species and environments. Embracing this avenue of research is timely since the dogma that (see [44]), for a given species, all physiological functions and traits should senesce at the same rates is now largely questioned [115, 116, 117].

Species largely differ in their susceptibility to contract, develop and die from diseases. For instance, necropsies performed on dead individuals from zoological gardens have revealed that in mammals cancer is a major cause of death in some species (e.g. in carnivores) while being almost absent in others [118]. Discrepancies between age-specific causes of death across species can largely be explained by eco-evolutionary dynamics that have selected specific genomic characteristics conferring resistance to some specific diseases (e.g. [119] in the context of infectious diseases or [120] in the context of neoplasia). In other words, ageing patterns may to some extent also reflect species-specific evolutionary paths that have balanced differently risks of death from different diseases.

Moreover, it has been recently suggested that species allocate not only resources between growth, reproduction and both germline and somatic maintenance, but also between various costly physiological functions underpinning the whole maintenance function [116]. These various mechanisms (from molecular maintenance, such as genomic stability and DNA repair, to maintenance of physiological functions, such as glucose metabolism, immunity, digestion or cognition) likely result in potential trade-offs in mortality components [116]. One innovative challenge would now be to explore how inter-specific differences in both onset and rate of actuarial senescence emerge from discrepancies in the prevalence and timing of species-specific diseases and, in return, to decipher the mechanisms enabling old individuals from some species to escape some diseases (e.g. [121]). In that context, omics analyses offer a relevant approach to identify the mechanism shaping age-specific causes of death through age-related diseases like cancer, neurodegeneration or diverse tissue-specific pathologies.

So far, comparative genomics has shown that anti-cancer mechanisms, DNA repair machinery and proteostasis [122], or inflammatory response and haemostasis [123], play a central role in modulating lifespan variation among species. Nutrient-sensing pathways are also likely a central node explaining variance in lifespan between species, as many mutations linked to increased or decreased lifespan have been found on various genes along these pathways; from genes in the GH-IGF1 (growth hormone–insulin-like growth factor 1) signalling pathway down to those into mTOR (mammalian target of rapamycin) or FOXO (forkhead box O) pathways [124]. In addition, multiple proteins and metabolites have also been found to be associated with age-related diseases in humans, and omics data, together with functional data related to ageing, keep on accumulating on many cohorts of individuals, thus offering a large set of new pathways that could be targeted in future comparative epidemiology studies [125]. Illustrating this diversity of ageing (or anti-ageing) mechanisms, species traditionally labelled as long-lived (e.g. naked mole rat, *Heterocephalus glaber*; blind mole rat, *Nannospalax ehrenbergi*; little brown bat, *Myotis lucifugus*; African turquoise killifish, *Nothobranchius furzeri*; but see Chapter 4 for a critical reappraisal of the identification of extremely 'long-lived' species) achieved extended lifespan through different molecular pathways because they have to put in place defence mechanisms against acute risk of death due to some specific biological traits (e.g. body size) or environmental conditions [126]. A classic example is cancer where species differ in their ability to maintain genomic integrity, as well as in their exposure to carcinogens (e.g. pollutants), resulting in potential variation in cancer prevalence [118]. Even among species with low prevalence, large African and Asian elephants (*Loxodonta africana*, *Elephas maximus*, respectively) seem to achieve low prevalence through a larger number of copies of the tumour suppressor TP53 (tumour protein p53) gene, while small naked-mole rats maintain low prevalence thanks to many different adaptations, from the abundant production of very high molecular-mass hyaluronan, to the activation of the tumour suppressor protein ARF (ADP-ribosylation factor), and to a more stable epigenome [127].

Overall, it is likely that variance in onset and rate of actuarial senescence are in fact due to underlying variation in species-specific conditions and disease risks arising from common modulation of hallmarks of ageing along broad ecological continua, to which are added taxon-specific adaptations. Deciphering this variation would therefore require the epidemiological profiles of causes-specific mortality to be investigated in more depth. Thankfully, such analyses are now achievable since detailed information on age-specific causes of death for a wide range of species are currently emerging in the literature, in natural and, more predominantly, in captive environments ([118], see also Chapter 6).

Finally, it is worth noticing that accumulated evidence suggests that specific mechanisms protecting against specific disease risk may be linked to drawbacks on other physiological functions. For instance, in the case of cancer, telomere maintenance may increase lifespan of small-bodied organisms while increasing the risk of cancer in larger ones [127, 128]. Also, senescent cells are increasingly seen as a major mechanism preventing emergence of cancer but their accumulation in tissues may favour stem cell exhaustion and inflammageing [129], potentially resulting in evolutionary trade-offs between mortality components and asynchrony of ageing between cancers and other senescence-related causes of death (see [130]). If proved true, life history trade-offs may occur, not only between reproduction and maintenance, but also within maintenance mechanisms and between mortality components, potentially resulting in the evolution of age-specific disease risks.

7.4 Conclusion

In this chapter, we argue that the comparative biology of ageing would substantially benefit from studies bridging advances in comparative demography and recent developments in omics approaches (Figure 7.2). One might question the minimal number of species that it is possible to include in such projects given that pioneer comparative analyses of actuarial senescence patterns were limited to 10–20 species of vertebrates [37, 50]. However, in the case of mammals, the number of species for which parameters such as age at the onset of actuarial senescence or rate of actuarial senescence are known is about 150 in the wild and more than a thousand in captive conditions [5, 52] (see Chapter 6 for a thorough discussion regarding the advantages and disadvantages of captive data in ageing research). This is largely superior to the datasets used in the most insightful comparative analyses performed on mammalian lifespan in the biogerontological field (e.g. [14, 120, 131]). Moreover, the number and the taxonomic range (e.g. [35] for amphibians and reptiles) of species available for such studies is continuously increasing and, thanks to various initiatives (e.g. [33]), age-specific data on mortality are now compiled and accessible in standardized formats. In addition, we are witnessing the development of molecular tools that enable an assessment of chronological age with an increasing level of confidence (e.g. [79, 132]). While demographic datasets for which chronological age is known without error from individual marking at birth should always be preferred over demographic datasets for which chronological age is assessed through molecular tools, these new approaches offer opportunities to include additional species where it is particularly difficult to catch individuals during the juvenile period. This is notably the case for rodents, a mammalian order for which demographic data are sparse in the wild, which impairs the possibility of cross-talking between ageing studies performed on wild and laboratory rodents (e.g. [133, 134]).

Composite epigenetic clocks appear today as one of the most reliable predictors of biological age; however, as nearly all cellular and biochemical processes are affected during ageing, more comprehensive omics analyses are expected to provide additional robust predicting signatures [56, 90, 91, 92]. To cite just a few studies using omics, ageing has already been characterized by drastic transcriptomic changes, notably concerning genes that are key to cellular senescence, the mitochondrial function, the immune and stress responses and DNA repair, but in some cases with important differences according to the tissue, genotype or species considered (e.g. [96]). It will therefore be necessary to refine the interpretation of these data in the future, for instance to better correlate the findings in model organisms with the actual ageing process in humans [135], to establish a link between ageing and age-related diseases like cancer, neurodegeneration or diverse tissue-specific pathologies [8, 136, 137, 138], to understand how transcriptional events vary relative to translational events [139], and to gain insight into the influence of genomic structural variants [140].

Finally, it is worth noticing that data on age-specific reproductive traits, which are mandatory to estimate parameters such as age at the onset of reproductive senescence and rate of reproductive senescence, are also becoming available for a wide range of species (e.g. [141]). In a similar manner to the one described in this chapter for actuarial senescence, we are confident that cross-referencing age-specific reproductive data with data from omics approaches will shed new light on the mechanisms shaping the diversity of reproductive senescence patterns across the tree of life and should ultimately lead to major advances in the study of reproductive health over the life course [36, 141].

References

1. Carey, J., Judge, D. 2000. *Longevity Records: Life Spans of Mammals, Birds, Reptiles, Amphibians and Fish*. University Odense Press.
2. Aristotle. 2001. *On Longevity and Shortness of Life* (trans. G.R.T. Ross). Internet Classics Archive. https://classics.mit.edu/Aristotle/longev_short.html.
3. Huxley, J.S. 1932. *Problems of Relative Growth*. Methuen.
4. Thompson, D.A.W. 1942. *On Growth and Form*. Cambridge University Press.
5. Lemaître, J.-F. et al. 2020. Sex differences in adult lifespan and aging rates of mortality across wild mammals. *Proc. Natl. Acad. Sci.* **117**, 8546–8553.
6. Peters, R.H. 1983. *The Ecological Implications of Body Size*. Cambridge University Press (doi:10.1017/CBO9780511608551).
7. Calder, W.A. 1984. *Size, Function, and Life History*. Courier Corporation.
8. West, G.B., Brown, J.H. 2004. Life's universal scaling laws. *Phys. Today* **57**, 36–43.
9. Hofman, M.A. 1983. Energy metabolism, brain size and longevity in mammals. *Q. Rev. Biol.* **58**, 495–512.
10. Piper, M.D., Skorupa, D., Partridge, L. 2005. Diet, metabolism and lifespan in Drosophila. *Exp. Gerontol.* **40**, 857–862.
11. Healy, K. et al. 2014. Ecology and mode-of-life explain lifespan variation in birds and mammals. *Proc. R. Soc. Lond. B Biol. Sci.* **281**, 20140298.
12. Gaillard, J.-M., Viallefont, A., Loison, A., Festa-Bianchet, M. 2004. Assessing senescence patterns in populations of large mammals. *Anim. Biodivers. Conserv.* **27**, 47–58.
13. Fushan, A.A. et al. 2015. Gene expression defines natural changes in mammalian lifespan. *Aging Cell* **14**, 352–365.
14. Haghani, A. et al. 2023. DNA methylation networks underlying mammalian traits. *Science* **381**, eabq5693 (doi: 10.1126/science.abq5693).
15. Aledo, J.C., Li, Y., de Magalhães, J.P., Ruíz-Camacho, M., Pérez-Claros, J.A. 2011. Mitochondrially encoded methionine is inversely related to longevity in mammals. *Aging Cell* **10**, 198–207.
16. Lindstedt, S.L., Calder III, W.A. 1981. Body size, physiological time, and longevity of homeothermic animals. *Q. Rev. Biol.* **56**, 1–16.
17. West, G.B. 2017. *Scale: The Universal Laws of Growth, Innovation, Sustainability, and the Pace of Life in Organisms, Cities, Economies, and Companies*. Penguin Press.
18. Stearns, S.C. 1983. The influence of size and phylogeny on patterns of covariation among life-history traits in the mammals. *Oikos* **41**, 173–187.
19. Gaillard, J.-M., Lemaître, J.-F., Berger, V., Bonenfant, C., Devillard, S., Douhard, M., Gamelon, M., Plard, F., Lebreton, J.-D. 2016. Life histories, axes of variation. In *Encyclopedia of Evolutionary Biology* (ed. R.M. Kilman), pp. 312–323. Elsevier.

20. Dupoué, A., Blaimont, P., Angelier, F., Ribout, C., Rozen-Rechels, D., Richard, M., Le Galliard, J.F. 2022. Lizards from warm and declining populations are born with extremely short telomeres. *Proc. Natl. Acad. Sci.* **119**, 2201371119.
21. Shattuck, M.R., Williams, S.A. 2010. Arboreality has allowed for the evolution of increased longevity in mammals. *Proc. Natl. Acad. Sci.* **107**, 4635–4639 (doi:10.1073/pnas.0911439107).
22. Krementz, D.G., Sauer, J.R., Nichols, J.D. 1989. Model-based estimates of annual survival rate are preferable to observed maximum lifespan statistics for use in comparative life-history studies. *Oikos* **70**, 203–208.
23. Vaupel, J.W. 2003. Post-Darwinian longevity. *Popul. Dev. Rev.* **29**, 258–269.
24. Moorad, J.A., Promislow, D.E., Flesness, N., Miller, R.A. 2012. A comparative assessment of univariate longevity measures using zoological animal records. *Aging Cell* **11**, 940–948.
25. Ronget, V., Gaillard, J.-M. 2020. Assessing ageing patterns for comparative analyses of mortality curves: going beyond the use of maximum longevity. *Funct. Ecol.* **34**, 65–75.
26. Péron, G., Lemaître, J.-F., Ronget, V., Tidière, M., Gaillard, J.-M. 2019. Variation in actuarial senescence does not reflect life span variation across mammals. *PLoS Biol.* **17**, e3000432.
27. de Magalhaes, J.P., Costa, J. 2009. A database of vertebrate longevity records and their relation to other life-history traits. *J. Evol. Biol.* **22**, 1770–1774.
28. Deevey Jr, E.S. 1947. Life tables for natural populations of animals. *Q. Rev. Biol.* **22**, 283–314.
29. Caughley, G. 1966. Mortality patterns in mammals. *Ecology* **47**, 906–918.
30. Spinage, C.A. 1973. A review of the age determination of mammals by means of teeth, with especial reference to Africa. *Afr. J. Ecol.* **11**, 165–187.
31. Clutton-Brock, T., Sheldon, B.C. 2010. Individuals and populations: the role of long-term, individual-based studies of animals in ecology and evolutionary biology. *Trends Ecol. Evol.* **25**, 562–573.
32. Gimenez, O. et al. 2008. The risk of flawed inference in evolutionary studies when detectability is less than one. *Am. Nat.* **172**, 441–448.
33. Gaillard, J.-M., Ronget, V., Lemaître, J.-F., Bonenfant, C., Péron, G., Capdevila, P., Gamelon, M., Salguero-Gómez, R. 2021. Applying comparative methods to different databases: lessons from demographic analyses across mammal species. In *Demographic Methods across the Tree of Life* (eds R. Salguero-Gómez and M. Gamelon), pp. 299–312. Oxford University Press.
34. Nussey, D.H., Froy, H., Lemaitre, J.-F., Gaillard, J.-M., Austad, S.N. 2013. Senescence in natural populations of animals: widespread evidence and its implications for bio-gerontology. *Ageing Res. Rev.* **12**, 214–225.
35. Reinke, B.A. et al. 2022. Diverse aging rates in ectothermic tetrapods provide insights for the evolution of aging and longevity. *Science* **376**, 1459–1466.
36. Jones, O.R. et al. 2014. Diversity of ageing across the tree of life. *Nature* **505**, 169–173.
37. Ricklefs, R.E. 1998. Evolutionary theories of aging: confirmation of a fundamental prediction, with implications for the genetic basis and evolution of life span. *Am. Nat.* **152**, 24–44.
38. Ricklefs, R.E., Scheuerlein, A. 2001. Comparison of aging-related mortality among birds and mammals. *Exp. Gerontol.* **36**, 845–857.
39. Ricklefs, R.E. 2010. Insights from comparative analyses of aging in birds and mammals. *Aging Cell* **9**, 273–284.

40. Ronget, V., Lemaître, J.-F., Tidière, M., Gaillard, J.-M. 2020. Assessing the diversity of the form of age-specific changes in adult mortality from captive mammalian populations. *Diversity* **12**, 354.
41. Kirkwood, T.B. 2015. Deciphering death: a commentary on Gompertz (1825) 'On the nature of the function expressive of the law of human mortality, and on a new mode of determining the value of life contingencies'. *Philos. Trans. R. Soc. B Biol. Sci.* **370**, 20140379.
42. Bronikowski, A.M. et al. 2011. Aging in the natural world: comparative data reveal similar mortality patterns across primates. *Science* **331**, 1325–1328.
43. Colchero, F. et al. 2021. The long lives of primates and the 'invariant rate of ageing' hypothesis. *Nat. Commun.* **12**, 1–10.
44. Williams, G.C. 1957. Pleiotropy, natural selection, and the evolution of senescence. *Evolution* **11**, 398–411.
45. Hamilton, W.D. 1966. The moulding of senescence by natural selection. *J. Theor. Biol.* **12**, 12–45.
46. Péron, G., Gimenez, O., Charmantier, A., Gaillard, J.-M., Crochet, P.-A. 2010. Age at the onset of senescence in birds and mammals is predicted by early-life performance. *Proc. R. Soc. Lond. B Biol. Sci.* **10**, rspb20100530 (doi:10.1098/rspb.2010.0530).
47. Tidière, M., Gaillard, J.-M., Müller, D.W., Lackey, L.B., Gimenez, O., Clauss, M., Lemaître, J.-F. 2015. Does sexual selection shape sex differences in longevity and senescence patterns across vertebrates? A review and new insights from captive ruminants. *Evolution* **69**, 3123–3140.
48. Lehallier, B. et al. 2019. Undulating changes in human plasma proteome profiles across the lifespan. *Nat. Med.* **25**, 1843–1850.
49. Richardson, R.B., Allan, D.S., Le, Y. 2014. Greater organ involution in highly proliferative tissues associated with the early onset and acceleration of ageing in humans. *Exp. Gerontol.* **55**, 80–91.
50. Ricklefs, R.E. 2000. Intrinsic aging-related mortality in birds. *J. Avian Biol.* **31**, 103–111.
51. Lemaître, J.-F., Gaillard, J.-M. 2013. Polyandry has no detectable mortality cost in female mammals. *PLoS ONE* **8**, e66670 (doi:10.1371/journal.pone.0066670).
52. Ricklefs, R.E. 2010. Life-history connections to rates of aging in terrestrial vertebrates. *Proc. Natl. Acad. Sci.* **107**, 10314–10319.
53. Freckleton, R.P. 2009. The seven deadly sins of comparative analysis. *J. Evol. Biol.* **22**, 1367–1375.
54. Ricklefs, R.E. 2010. Embryo growth rates in birds and mammals. *Funct. Ecol.* **24**, 588–596.
55. Ricklefs, R.E. 2006. Embryo development and ageing in birds and mammals. *Proc. R. Soc. B Biol. Sci.* **273**, 2077–2082.
56. Zierer, J., Menni, C., Kastenmuller, G., Spector, T.D. 2015. Integration of 'omics' data in aging research: from biomarkers to systems biology. *Aging Cell* **14**, 933–944.
57. Rutledge, J., Oh, H., Wyss-Coray, T. 2022. Measuring biological age using omics data. *Nat. Rev. Genet.* **23**, 715–727.
58. Hoffman, J.M., Lyu, Y., Pletcher, S.D., Promislow, D.E. 2017. Proteomics and metabolomics in ageing research: from biomarkers to systems biology. *Essays Biochem.* **61**, 379–388.
59. Kirkwood, T.B.L. 2011. Systems biology of ageing and longevity. *Philos. Trans. R. Soc. B* **366**, 64–70.

60. McGuire, A.L. 2020. The road ahead in genetics and genomics. *Nat Rev. Genet.* **21**, 582–596.
61. Wang, B., Kumar, V., Olson, A., Ware, D. 2019. Reviving the transcriptome studies: an insight into the emergence of single-molecule transcriptome sequencing. *Front. Genet.* **10**, 384.
62. Hotaling, S., Kelley, J.L., Frandsen, P.B. 2021. Toward a genome sequence for every animal: where are we now? *Proc. Natl. Acad. Sci. USA* **118**, 2109019118.
63. Joyce, A.R., Palsson, B.O. 2006. The model organism as a system: integrating 'omics' data sets. *Nat. Rev. Mol. Cell Biol.* **7**, 198–210.
64. Wang, Z., Gerstein, M., Snyder, M. 2009. RNA-seq: a revolutionary tool for transciptomics. *Nat. Rev. Genet.* **10**, 57–63.
65. Micheel, J., Safrastyan, A., Wollny, D. 2021. Advances in non-coding RNA sequencing. *Non-Coding RNA* **7**, 70.
66. Messner, C.B., Demichev, V., Wang, Z., Hartl, J., Kustatscher, G., Mülleder, M., Ralser, M. 2023. Mass spectrometry-based high-throughput proteomics and its role in biomedical studies and systems biology. *Proteomics* **23**, e2200013.
67. Aebersold, R., Mann, M. 2016. Mass-spectrometric exploration of proteome structure and function. *Nature* **537**, 347–355.
68. Wishart, D.S. 2019. Metabolomics for investigating physiological and pathophysiological processes. *Physiol. Rev.* **99**, 1819–1875.
69. Subramanian, I., Verma, S., Kumar, S., Jere, A., Anamika, K. 2020. Multi-omics data integration, interpretation, and its application. *Bioinform. Biol. Insights* **14**, 1–24.
70. Mani, D.R., Krug, K., Zhang, B., Satpathy, S., Clauser, K.R., Ding, L., Ellis, M., Gillette, M.A., Carr, S.A. 2022. Cancer proteogenomics: current impact and future prospects. *Nat. Rev. Cancer* **22**, 298–313.
71. Krassowski, M., Das, V., Sahu, S.K., Misra, B.B. 2020. State of the field in multi-omics research: from computational needs to data mining and sharing. *Front. Genet.* **11**, 610798.
72. López-Otín, C., Blasco, M.A., Partridge, L., Serrano, M., Kroemer, G. 2023. Hallmarks of aging: an expanding universe. *Cell* **186**, 243–278 (doi:10.1016/j.cell.2022.11.001).
73. Gomes, N.M. et al. 2011. Comparative biology of mammalian telomeres: hypotheses on ancestral states and the roles of telomeres in longevity determination. *Aging Cell* **10**, 761–768.
74. Seluanov, A., Chen, Z., Hine, C., Sasahara, T.H., Ribeiro, A.A., Catania, K.C., Presgraves, D.C., Gorbunova, V. 2007. Telomerase activity coevolves with body mass not lifespan. *Aging Cell* **6**, 45–52.
75. Hulbert, A.J., Pamplona, R., Buffenstein, R., Buttemer, W.A. 2007. Life and death: metabolic rate, membrane composition, and life span of animals. *Physiol. Rev.* **87**, 1175–1213.
76. Tobler, M., Gómez-Blanco, D., Hegemann, A., Lapa, M., Neto, J.M., Tarka, M., Xiong, Y., Hasselquist, D. 2022. Telomeres in ecology and evolution: a review and classification of hypotheses. *Mol. Ecol.* **31**, 5946–5965.
77. Haussmann, M.F., Vleck, C.M. 2002. Telomere length provides a new technique for aging animals. *Oecologia* **130**, 325–328.
78. Lu, A.T. et al. 2021. Universal DNA methylation age across mammalian tissues. *bioRxiv*. www.biorxiv.org/content/10.1101/2021.01.18.426733v2.
79. Lemaître, J. et al. 2021. DNA methylation as a tool to explore ageing in wild roe deer populations. *Mol. Ecol. Resour.* **22**. 1002–1015.

80. Simpson, D.J., Chandra, T. 2021. Epigenetic age prediction. *Aging Cell* **20**, e13452.
81. Kristic, J. 2014. Glycans are a novel biomarker of chronological and biological ages. *J Gerontol. Biol. Sci. Med. Sci.* **69**, 779–789.
82. Hertel, J. et al. 2016. Measuring biological age via metabonomics: the metabolic age score. *J. Proteome Res.* **15**, 400–410.
83. van den Akker Erik, B. et al. 2020. Metabolic age based on the BBMRI-NL 1H-NMR metabolomics repository as biomarker of age-related disease. *Circ. Genomic Precis. Med.* **13**, 541–547 (doi:10.1161/CIRCGEN.119.002610).
84. Robinson, O. et al. 2020. Determinants of accelerated metabolomic and epigenetic aging in a UK cohort. *Aging Cell* **19**, e13149.
85. Gaillard, Lemaître J.-F. 2020. An integrative view of senescence in nature. *Funct. Ecol.* **34**, 4–16.
86. Jylhava, J., Pedersen, N.L., Hagg, S. 2017. Biological age predictors. *EBioMedicine* **21**, 29–36.
87. Dato, S., Piras, I.S., eds. 2022. Omics of human aging and longevity in the post genome era: from single biomarkers to systems biology approaches. *Front. Genet.* **13**, 913531.
88. Elsner, D., Meusemann, K., Korb, J. 2018. Longevity and transposon defense, the case of termite reproductives. *Proc. Natl. Acad. Sci. USA* **115**, 5504–5509.
89. Criscuolo, F., Sorci, G., Behaim-Delarbre, M., Zahn, S., Faivre, B., Bertile, F. 2018. Age-related response to an acute innate immune challenge in mice: proteomics reveals a telomere maintenance-related cost. *Proc. R. Soc. B* **285**, 20181877.
90. Frenk, S., Houseley, J. 2018. Gene expression hallmarks of cellular ageing. *Biogerontology* **19**, 547–566.
91. Ham, S., Lee, S.V. 2020. Advances in transcriptome analysis of human brain aging. *Exp. Mol. Med.* **52**, 1787–1797.
92. Palmer, D., Fabris, F., Doherty, A., Freitas, A.A., Magalhaes, J.P. 2021. Ageing transcriptome meta-analysis reveals similarities and differences between key mammalian tissues. *Aging* **13**, 3313–3341.
93. Saitou, M., Lizardo, D.Y., Taskent, R.O., Millner, A., Gokcumen, O., Atilla-Gokcumen, G.E. 2018. An evolutionary transcriptomics approach links CD36 to membrane remodeling in replicative senescence. *Mol. Omics* **14**, 237–246.
94. Srivastava, A., Barth, E., Ermolaeva, M.A., Guenther, M., Frahm, C., Marz, M., Witte, O.W. 2020. Tissue-specific gene expression changes are associated with aging in mice. *Genom. Proteom. Bioinform.* **18**, 430–442.
95. Solovev, I., Shaposhnikov, M., Moskalev, A. 2020. Multi-omics approaches to human biological age estimation. *Mech. Ageing Dev.* **185**, 111192.
96. Tyshkovskiy, A. et al. 2023. Distinct longevity mechanisms across and within species and their association with aging. *Cell.* **186**, 2929–2949.e20.
97. Beheshti, A. 2017. A circulating microRNA signature predicts age-based development of lymphoma. *PLoS ONE* **12**, 0170521.
98. Peffers, M.J. 2016. Age-related changes in mesenchymal stem cells identified using a multi-omics approach. *Eur. Cell Mater.* **31**, 136–159.
99. Griffiths, H.R. et al. 2015. Novel ageing-biomarker discovery using data-intensive technologies. *Mech. Ageing Dev.* **151**, 114–121.
100. Lemaître, J.-F., Garratt, M., Gaillard, J.-M. 2020. Going beyond lifespan in comparative biology of aging. *Adv. Geriatr. Med. Res.* **2**. e200011.
101. Gelfi, C. et al. 2006. The human muscle proteome in aging. *J. Proteome Res.* **5**, 1344–1353.

102. Chakrabarti, A., Mukhopadhyay, D. 2012. Brain senescence-omics. *J. Protein Proteomics* **3**, 15–29.
103. Tanaka, T. et al. 2018. Plasma proteomic signature of age in healthy humans. *Aging Cell* **17**, e12799.
104. Enroth, S., Enroth, S.B., Johansson, A., Gyllensten, U. 2015. Protein profiling reveals consequences of lifestyle choices on predicted biological aging. *Sci. Rep.* **5**, 17282.
105. Menni, C. et al. 2015. Circulating proteomic signatures of chronological age. *J. Gerontol. Ser. Biomed. Sci. Med. Sci.* **70**, 809–816.
106. Vanhooren, V., Santos, A.N., Voutetakis, K., Petropoulos, I., Libert, C., Simm, A., Gonos, E.S., Friguet, B. 2015. Protein modification and maintenance systems as biomarkers of ageing. *Mech. Ageing Dev.* **151**, 71–84.
107. Plumel, M.I., Benhaim-Delarbre, M., Rompais, M., Thiersé, D., Sorci, G., van Dorsselaer, A., Criscuolo, F., Bertile, F. 2016. Differential proteomics reveals age-dependent liver oxidative costs of innate immune activation in mice. *J. Proteomics* **135**, 181–190.
108. Walker, K.A., Basisty, N., Wilson III, D.M., Ferrucci, L. 2022. Connecting aging biology and inflammation in the omics era. *J. Clin. Invest.* **132**, e158448.
109. Chan, M. et al. 2022. Novel insights from a multiomics dissection of the Hayflick limit. *eLife* **11**, e70283.
110. Sturm, G. et al. 2022. A multi-omics longitudinal aging dataset in primary human fibroblasts with mitochondrial perturbations. *Sci. Data* **9**, 751.
111. Song, Q. et al. 2022. Integrated multi-omics approach revealed cellular senescence landscape. *Nucleic Acids Res.* **50**, 10947–10963.
112. MacRae, S.L. et al. 2015. DNA repair in species with extreme lifespan differences. *Aging* **7**, 1171.
113. Verma, A., Verma, M., Singh, A. 2020. Animal tissue culture principles and applications. In *Animal Biotechnology*, pp. 269–293. Academic Press.
114. Bermejo, M., Rodríguez-Teijeiro, J.D., Illera, G., Barroso, A., Vilà, C., Walsh, P.D. 2006. Ebola outbreak killed 5000 gorillas. *Science* **314**, 1564–1564.
115. Gaillard, J.-M., Lemaître, J.-F. 2017. The Williams' legacy: a critical reappraisal of his nine predictions about the evolution of senescence. *Evolution* **71**, 2768–2785 (doi:10.1111/evo.13379).
116. Cohen, A.A., Coste, C.F., Li, X.-Y., Bourg, S., Pavard, S. 2020. Are trade-offs really the key drivers of ageing and life span? *Funct. Ecol.* **34**, 153–166.
117. Moorad, J.A., Ravindran, S. 2022. Natural selection and the evolution of asynchronous aging. *Am. Nat.* **199**, 551–563.
118. Vincze, O. et al. 2022. Cancer risk across mammals. *Nature* **601**, 1–5.
119. Gorbunova, V., Seluanov, A., Kennedy, B.K. 2020. The world goes bats: living longer and tolerating viruses. *Cell Metab.* **32**, 31–43.
120. Gorbunova, V., Seluanov, A., Zhang, Z., Gladyshev, V.N., Vijg, J. 2014. Comparative genetics of longevity and cancer: insights from long-lived rodents. *Nat. Rev. Genet.* **15**, 531.
121. Abegglen, L.M. et al. 2015. Potential mechanisms for cancer resistance in elephants and comparative cellular response to DNA damage in humans. *JAMA* **314**, 1850–1860.
122. Tian, X., Seluanov, A., Gorbunova, V. 2017. Molecular mechanisms determining lifespan in short- and long-lived species. *Trends Endocrinol. Metab.* **28**, 722–734 (doi:10.1016/j.tem.2017.07.004).
123. Farré, X. et al. 2021. Comparative analysis of mammal genomes unveils key genomic variability for human life span. *Mol. Biol. Evol.* **38**, 4948–4961.

124. Singh, P.P., Demmitt, B.A., Nath, R.D., Brunet, A. 2019. The genetics of aging: a vertebrate perspective. *Cell* **177**, 200–220.
125. Dato, S., Crocco, P., Migliore, N.R., Lescai, F. 2021. Omics in a digital world: the role of bioinformatics in providing new insights into human aging. *Front Genet.* **12**, 689824.
126. Ma, S., Gladyshev, V.N. 2017. Molecular signatures of longevity: insights from cross-species comparative studies. *Semin Cell Dev Biol* **70**, 190–203.
127. Seluanov, A., Gladyshev, V.N., Vijg, J., Gorbunova, V. 2018. Mechanisms of cancer resistance in long-lived mammals. *Nat. Rev. Cancer* **18**, 433–441.
128. Ujvari, B. et al. 2022. Telomeres, the loop tying cancer to organismal life-histories. *Mol. Ecol.* **31**, 6273–6285.
129. He, S., Sharpless, N.E. 2017. Senescence in health and disease. *Cell* **169**, 1000–1011.
130. Bieuville, M., Tissot, T., Robert, A., Henry, P.-Y., Pavard, S. 2023. Modeling of senescent cell dynamics predicts a late-life decrease in cancer incidence. *Evol. Appl.* **16**, 609–624.
131. Ma, S. et al. 2015. Organization of the mammalian metabolome according to organ function, lineage specialization, and longevity. *Cell Metab.* **22**, 332–343.
132. Prado, N.A. et al. 2021. Epigenetic clock and methylation studies in elephants. *Aging Cell* **20**, e13414.
133. Ruby, J.G., Smith, M., Buffenstein, R. 2018. Naked mole-rat mortality rates defy Gompertzian laws by not increasing with age. *eLife* **7**, e31157.
134. Garratt, M., Erturk, I., Alonzo, R., Zufall, F., Leinders-Zufall, T., Pletcher, S.D., Miller, R.A. 2022. Lifespan extension in female mice by early, transient exposure to adult female olfactory cues. *eLife* **11**, e84060.
135. Zhuang, J. 2019. Comparison of multi-tissue aging between human and mouse. *Sci. Rep.* **9**, 6220.
136. Yang, J. et al. 2015. Synchronized age-related gene expression changes across multiple tissues in human and the link to complex diseases. *Sci. Rep.* **5**, 15145.
137. Aramillo Irizar, P. et al. 2018. Transcriptomic alterations during ageing reflect the shift from cancer to degenerative diseases in the elderly. *Nat. Commun.* **9**, 327.
138. Beheshti, A., Vanderburg, C., McDonald, J.T., Ramkumar, C., Kadungure, T., Zhang, H., Gartenhaus, R.B., Evens, A.M. 2017. A circulating microRNA signature predicts age-based development of lymphoma. *PloS ONE* **12**, e0170521.
139. Caliskan, A., Crouch, S.A.W., Giddins, S., Dandekar, T., Dangwal, S. 2022. Progeria and aging-omics based comparative analysis. *Biomedicines* **10**. 2440.
140. Vialle, R.A., Paiva Lopes, K., Bennett, D.A., Crary, J.F., Raj, T. 2022. Integrating whole-genome sequencing with multi-omic data reveals the impact of structural variants on gene regulation in the human brain. *Nat. Neurosci.* **25**, 504–514.
141. Lemaître, J.-F., Ronget, V., Gaillard, J.-M. 2020. Female reproductive senescence across mammals: a high diversity of patterns modulated by life history and mating traits. *Mech. Ageing Dev.* **192**, 111377.

8 An Integrative Approach to Understanding Variation in the Form, Pattern and Pace of Ageing

Pat Monaghan and Jelle Boonekamp

8.1 Introduction

Age-related deterioration is almost ubiquitous in living organisms, usually manifesting as a progressive decline in performance during adult life. As such, we would expect this deterioration to have negative effects on Darwinian fitness (genetic contribution to the next generation) and there are numerous routes by which ageing might reduce lifetime reproductive output through effects on longevity and reproductive capacity. We use the term ageing here as synonymous with senescence, meaning simply a deterioration in performance with the passage of time. This generally takes place in adults, sometime after sexual maturation [1]. While some researchers draw a distinction between ageing and senescence, for example using the term ageing to mean the passage of time with no deterioration implied, we do not think this is helpful in promoting an integrative approach; such a distinction is not widely used [2], and it could lead to confusion since use of the term 'ageing' to mean age-related deterioration is deeply embedded in the biomedical literature [1]. However, it is very important to distinguish between age and ageing. While obviously related, these are not the same thing. Many traits change with adult age but are not linked to any decline in performance and indeed some traits improve with adult age for various reasons. Body size, for example, can continue to increase with adult age in indeterminate growers. In humans, baldness, greying of hair and wrinkling of skin increase with adult age but are (thankfully) not predictive of mortality risk [3]. In the context of ageing, a distinction therefore needs to be drawn between age in the chronological sense, a defined property of an individual referring to time since birth, and biological age, which refers to the amount of deterioration with adult age that has occurred in the organism. Biological age is therefore predictive of time until death rather than expressing time since birth. Biological age varies among individuals of the same chronological age, reflecting variation in ageing rate. Accordingly, ageing is formally defined here as a progressive, intrinsic, physiological deterioration that takes pace in adult organisms with the passage of time and gives rise to an increase in the instantaneous risk of death.

Age-related deterioration is most obvious in somatic tissues. Contrary to what has previously been assumed, it can occur in germline tissue, albeit to a lesser extent [4].

Often, the increased mortality risk associated with ageing is accompanied by a decline in reproductive performance. It can be difficult to evaluate whether an observed decline in reproductive capability with age is due to ageing of the soma, the germline or to a strategic decline in age-specific reproductive effort. Strategic adjustment of effort might be due to phenotypic differences among individuals in the optimum resolution of the trade-off between reproduction and longevity, with those investing less in reproduction being more likely to live to older ages. Such adjustment can result in an apparent decline in reproductive success with age at the population level independent of, or alongside, within-individual change (see [5] for an example of this in the wild). On the other hand, in many species, ageing-related mortality risk might be accompanied by improved reproductive success due to tactically increased reproductive effort in response to diminished life expectancy [6, 7]. We discuss this further in Section 8.4.1 on frailty. Given that organismal fitness will be maximized by evolution, the pattern and pace of ageing, and its phenotypic manifestation, will be subject to selection pressures that balance the fitness costs and benefits of investment in anti-ageing processes. These selection pressures will be highly specific for different species and populations, depending on the local environmental and ecological conditions. Hence, we see great variation among species and individuals in when, how and at what rate ageing occurs.

Deterioration is clearly bad for the individual organism and there is some controversy over whether or not ageing is inevitable. While it is unlikely to be selected for, it can be selected against [1]. In this chapter, we focus not on whether ageing occurs at all, but rather on the variation that we see within and among species in how organisms 'manage' the ageing process. We are not concerned here with whether ageing is itself adaptive or avoidable. Rather, we are concerned with variation in the form, pattern and pace of ageing and its underlying physiology, all of which can differ among species and individuals. We confine ourselves to animals, and to those animals whose body design is at the unitary end of the unitary–modularity spectrum, where ageing is most obvious [8]. By *form* we mean how and where deterioration occurs in the body; deterioration can vary among different body tissues and organisms will differentially prioritize maintenance and repair according to fitness benefits. Ageing can also vary in terms of its *pace* and in the *pattern* of the mortality distribution [9]. By *pace* we mean the rate at which deterioration and increased risk of death progress; some species age relatively quickly, others slowly or very slowly. Life can be short or long irrespective of the pattern of ageing and we use pace only in the context of the pace of ageing, not the pace of life in life history terms. By *pattern* we mean whether the trajectory of decline in an ageing trait is consistent, or shows ups and downs, for example in different seasons, life stages or environmental circumstances prior to eventual death. The term 'pattern' as used here has a similar meaning to the 'shape' of the mortality distribution with the difference that 'shape' is usually standardized to be time independent. We use a slightly simpler definition, but recognize that time standardization can be important, for example when comparing the shape of ageing for species with differing longevity [10]. Figure 8.1 illustrates variation in the form, pace and pattern of ageing.

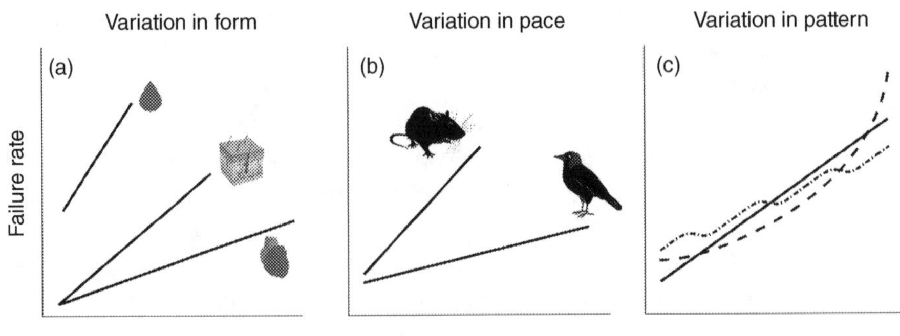

Figure 8.1 Schematic depiction of potential different forms that ageing might take in tissues/species across the life course. On the y-axis is the natural logarithm of instantaneous tissue (a) or organism (b, c) failure rate, effectively ageing rate. On the x-axis is chronological age after maturation. (While some researchers have suggested that ageing begins earlier in life, the changes typical of age-related phenotypic deterioration in adulthood are best considered separately from damage occurring during development, although the latter could influence later life ageing; detailed discussion of this is outside the scope of this chapter.) (a) shows a different pace and baseline level of ageing in different tissues (e.g. heart, skin, blood) within an individual. Red blood cells typically last for a much shorter time than other tissues such as muscle as they are selectively removed from circulation depending on how long they have been present. They are constantly replenished by the haemopoietic stem cells in the bone marrow. The failure rate trajectory depicted here is for red blood cells, but the underlying stem cell population also deteriorates and depletes with age, albeit at a slower pace. Variation across tissues/organs could arise through resource prioritization, where we expect any such differential allocation of resources to yield fitness benefits. (b) shows the failure time trajectories for two different species with the same pattern of ageing, but which differ in the pace of ageing, as shown in the different slopes. (c) shows three failure time trajectories for different species that have roughly the same lifetime average pace of ageing, resulting in similar longevity, but with age-specific periods of accelerated or decelerated failure rate, causing variation in pattern. For example, the short-dashed line may be typical for a hibernating species that could have seasonal intermittent periods, pausing or even reversing ageing during hibernation, while the long-dash line represents a species with catastrophic mortality near end of life.

8.2 The Advantages of an Integrated Approach

Understanding the variation that we see in the form, pattern and pace of ageing within and among species is a challenge that is addressed by scientists from many disciplinary backgrounds, ranging through molecular, cellular and developmental biology, evolutionary ecology, demography, and into psychology and the social and medical sciences. However, while there is some connectivity and exchange of ideas across different disciplinary domains, this remains limited, in part because the questions being asked and the goals pursued are different. Gerontologists and biomedical researchers (which we henceforth refer to collectively as biogerontologists) are interested in understanding how ageing occurs, with the goal of mitigating its effects primarily in humans. To a large extent, these researchers seek out common, conserved, so-called public, ageing processes that can be studied experimentally in 'model organisms' [11, 12]. A very sophisticated toolkit has been developed for work with model organisms that enables common pathways involved in ageing to be identified and manipulated and predictions tested experimentally in the laboratory. The small number of model species used in this kind of research is a somewhat odd selection in taxonomic and life history terms, having been chosen primarily because they are easy to study in a captive, laboratory situation rather than being typical of a particular taxon or lifestyle. They (now mainly the bacterium *Escherichia coli*, brewers yeast *Saccharomyces cerevisiae*, the nematode worm *Caenorhabditis elegans*, the fruitfly *Drosophila melanogaster*, the laboratory mouse *Mus musculus*, the thale cress *Arabidopsis thaliana* and the zebra fish *Danio rerio*) are characterized by short generation times, small bodies, simple social lives, rapid development and a genome that is easy to manipulate; strenuous measures are often put in place to minimize and/or control genetic variation [13]. Some have comparatively unusual traits, such as the fixed cell number and dauer stage in *C. elegans*. The controlled laboratory conditions under which model organisms are usually kept can make it difficult to apply some of the results to the much more variable environment that occurs in the wild (see [14, 15] for further discussion of this). Evolutionary and comparative biologists and ecologists are interested in understanding the nature of differences within and among species. They expect that some species will have evolved particular mechanisms that cause or delay ageing, so-called private ageing processes that might be species or lineage specific [11]. Such biologists (henceforth referred to as evolutionary ecologists) seek to understand why ageing occurs from an evolutionary perspective, asking questions about how the evolutionary history of a trait has influenced its form, how the selection pressures that shape organism life histories might be influenced by the effects of age-related deterioration in performance, why we see within-species heterogeneity in ageing even among individuals with the same genome, and what novel 'private' processes some organisms might have evolved to manage the ageing processes to optimize their life histories. In practice, these researchers are often working with long-lived species in their natural environments and studying ageing is part of a broader study of behaviour, population ecology and life history evolution. Thus, the species for which there are most long-term, individual-based data from wild populations (e.g. ungulates and (sea)birds) have not necessarily been chosen with ageing research or relevant aspects of life history variation in mind. These species too were most probably selected for practical reasons, such as being able to tag individuals to follow them longitudinally. Consequentially, the taxonomic scope of studies of ageing in the wild still tends to be relatively limited, being mainly birds and mammals, although the field is diversifying [16].

Broadening the diversity of life histories in ageing studies is a rich ground for research that would be beneficial to both biogerontologists and evolutionary ecologists. We still know little about ageing in short-lived species in the wild compared with what is known from laboratory studies [17]. Laboratory studies on insects have played a major role in fundamental research on ageing, but the relevance to insects in the wild is unclear [15]. In many social insects the same genome can be associated with a large difference in life histories; in ants for example, there can be a more than twentyfold difference in lifespan between short-lived workers and queens, which potentially challenges some theories on the evolution of ageing [18, 19, 20]. Furthermore, in the honeybee (*Apis mellifera*), ageing even appears to be reversible and depends on the role the individual adopts in the hive [21]. In ectotherms, we again know very little from studies in the wild, with the effect of temperature on ectotherm life histories adding an additional layer of complexity here. In freshwater pearl mussels (*Margaritifera margaritifera*) for example, previously thought to show negligible senescence, a recent study has reported that, in all five populations studied in the wild, there is evidence of an increase in mortality with age, with inter-population differences due to the thermal environment [22]. In the common lizard there is evidence of gradual senescence in reproductive success and survival, followed by an increase in reproductive success in the last year of reproduction before disappearance/death [23]; interestingly, this increase is associated with increased T-cell immunity and resting metabolic rate. These findings show that the form and pattern of ageing can be complex, with different physiological systems showing differing non-linear variation with age, and highlight the added value of an integrative approach. We currently lack a useful framework to fully integrate approaches to the study of ageing, which span such different levels of biological enquiry. We suggest here that the framework provided by Tinbergen's famous 'Four Questions' approach is useful in integrating ageing research. We explore this approach in Section 8.3. We then describe measurements of age-related deterioration, most usually developed for use in model organisms, which can be deployed outside of the standard laboratory setting. Following this, we suggest how this integrative approach could be developed further and highlight future directions.

8.3 Tinbergen's Four Questions Applied to Ageing Research

Niko Tinbergen, the Dutch ethologist who shared the 1973 Nobel Prize in Physiology or Medicine (with ethologists Karl von Frisch and Konrad Lorenz) was a pioneer in the promotion of an integrative approach in biological research, in his case, in the study of animal behaviour. In a landmark paper published in 1963, Tinbergen outlined four questions that could be addressed in the study of any organismal trait [24]:

Question 1: what is the function of the trait?
Question 2: what is its evolutionary origin?
Question 3: how does it develop?
Question 4: what processes cause it to occur?

The first two can be thought of as the 'why' questions, primarily concerned with fitness consequences, and the latter two as 'how' questions, primarily concerned with mechanisms.

8.3.1 The 'Why' Questions

With respect to Question 1, for ageing to have a function in evolutionary terms, we need to identify positive effects on fitness, that is, the benefits need to outweigh the costs for it to be considered adaptive. Considerations of the *function* of ageing are rooted in the links between early life performance and later life ageing. Such links are inherent in the antagonistic pleiotropy and disposable soma theories of ageing in particular (see also Chapter 2), both of which postulate cost–benefit trade-offs (see [2] for a discussion of these). Function, therefore, is best viewed in terms of variation in the fitness outcomes of different ageing profiles. For example, the fitness cost of ageing can be expressed as a percentage decrease in fitness due to ageing, which has been estimated by comparing observed values of lifetime reproductive success with and without natural senescence, that is, the fitness value that would have been achieved had senescence not taken place [25]. This fitness cost of ageing has been estimated to vary substantially among species ranging from only 4% in the great tit to 63% in the mountain goat [25]. The same variation is also observed in the percentage of lifespan variance explained by senescence [26]. The fitness cost of ageing is surprisingly small for some species, probably reflecting low baseline survival probability independent of ageing.

Expressing ageing in terms of a fitness cost is interesting because it might reflect the strength of selection for anti-ageing processes. Traditionally, the 'why' questions have focused on why does ageing evolve, given its apparent negative impact on fitness? We assume the answer to this question is that fundamental environmental and genetic constraints, and the consequences of the decline in the strength of natural selection in older age classes in natural populations, impose a limitation on the evolution of lifespan and fecundity. Physical identification of such constraints remains a major challenge, mostly because of the difficulty in uncovering their manifestation at the phenotypic level [27]. Conceptually, a (genetic) constraint is a multivariate phenomenon where (at least) two fitness traits show a negative (genetic) correlation reflecting the trade-off. However, such negative correlations may be difficult to detect because the underlying traits could also be linked to other performance traits, which is generally well recognized in the life history literature [28].

The inter-specific comparative approach embodied in Question 2 has the potential to provide further insights. The ancestral legacy of a species will not simply have been wiped away through the course of evolution; inbuilt constraints may be present as a consequence of the route taken and the toolkit available to maximize fitness. This might involve lineage-specific 'private' ageing mechanisms. Some of the private mechanisms may have become public through strong evolutionary conservation during speciation events and this could explain why some genetic features of ageing appear common. A well-studied example of a seemingly public mechanism of ageing is the nutrient sensing pathway, with IGF-1 (insulin-like growth factor 1) promoting growth and reproduction, but reducing lifespan, in common laboratory model species and in many mammals. However, the function of IGF-1 in wild vertebrate populations varies among species and appears less universal than suggested from the work on model species [29, 30]. This could be an indication that the seemingly universal function of IGF-1 in laboratory species is a consequence of strong environmental and genetic standardization facilitating (rapid) convergent evolution in laboratory species. More comparative approaches are needed, such as integrating different candidate mechanisms of ageing and also including wild invertebrate species (see [31] for an interesting discussion of this). This approach would shed light on the evolutionary trajectory of private versus public ageing mechanisms across species and laboratory and natural environments.

8.3.2 The 'How' Questions

Researchers in both biogerontology and evolutionary ecology are interested in how ageing occurs. For the former, this might have therapeutic relevance. For the latter, it can provide insights into the environmental factors that influence the form, pattern and pace of ageing, and identify currencies and constraints that can be useful in cost–benefit models of life history evolution. Identifying differences in how organisms age also helps us understand by what means some species have evolved exceptional longevities. Body size explains a large amount of variation in lifespan [32]. But some species have unusual longevity for their body size. Birds for example, have a physiology, lifestyle and morphology that, based on mammalian research, would lead us to expect short lives (e.g. high metabolic rate, high blood glucose, fast growth, high body temperature), yet birds live on average three times as long as mammals of the same size [33, 34]. Similarly, several small mammal species including bats *Chiroptera spp.* and the naked mole rat, *Heterocephalus glaber*, have exceptional longevity [35]. Investigations of such processes are also likely to provide useful insights for gerontologists since this might uncover new ways of avoiding age-related diseases. Tinbergen's Question 3 on how ageing develops interweaves studies of genetic and environmental influences. This can tell us much about early environmental influences, for example, the delayed effects of the tempo of growth on longevity [36]. Early life adversity can accelerate ageing and there is a substantial body of research in this area [37, 38]. Experimental studies on early life effects can tell us much about the mechanisms and trade-offs involved as well as the causal links [39, 40, 41]. Question 4, which is focused on the physiological and molecular processes involved in ageing, is largely the domain of gerontologists, provides insights into where and under what circumstances damage can occur, and whether and at what cost it can be repaired.

Applying Tinbergen's questions to ageing research requires accurate measurements of the phenotypic expression of ageing. Section 8.4 provides a (non-exhaustive) overview of several promising candidate measures of biological age.

8.4 Measuring Biological Age

Assessing the biological age of individuals is important to all ageing researchers, whether for the purposes of understanding the evolution of ageing rates, environmental effects on ageing, how to slow the rate of ageing or how to reduce susceptibility to age-related disease. In humans we know the major age-related pathologies are cancer, diabetes, cardiovascular and neurodegenerative diseases. For studies in the wild (and often also in captivity), we rarely know the cause of death, which is a major gap in our understanding of the effects of age on physiological deterioration and risk of death in the wild (but see Chapter 6 for the specific case of captive populations). One useful approach is estimating age-dependent mortality risk, or actuarial senescence, using Bayesian statistics [42]. This method requires more limited knowledge about birth and death dates, imputing missing data, and it opens up the possibility of estimating the covariation of physiological biomarkers of ageing, for example, telomere dynamics, or life history variables such as reproductive effort, with baseline mortality and the rate of actuarial senescence (see [43] for an example).

Given that ageing is a multifactorial process and that different species, individuals and tissues will deteriorate in different ways, no single marker is likely to accurately reflect biological age. We describe a number of markers, mainly developed by biogerontologists using laboratory species, which are potentially useful in providing measures of variation in the rate of ageing within and among non-laboratory species. We focus here on markers that can be used longitudinally, enabling the tracking of within-individual changes. This is of particular importance in separating age-related change within *populations* due to differential survival of particular phenotypes (for example where individuals whose lower investment in breeding throughout their reproductive life enables them to live longer) from the within-*individual* changes that characterize ageing. Apparent improvements with age in cross-sectional studies can also be due to the poor survival of low-quality individuals to older ages, masking the pattern of within-individual ageing [44]. Such measurements need to be minimally invasive and feasible given the much more limited toolkit available for non-model organisms. Markers based on whole organism performance and/or in fluids or tissues that can be biopsied without causing harm, are likely to be most useful in this context. Measures that can be taken from relatively small blood samples are particularly advantageous [45], and it is sometimes possible to obtain small tissue biopsies or cell samples, for example, of skin, muscle or buccal cells in vertebrates. Urine and faecal samples can also be useful in some cases, as can haemolymph samples in many invertebrates.

From the perspective of human ageing, twelve 'hallmarks of ageing' have been described [46]. These are genomic instability, telomere attrition, epigenetic alterations, loss of proteostasis, deregulated nutrient sensing, mitochondrial dysfunction, cellular senescence, stem cell exhaustion, altered intercellular communication, disabled macroautophagy. Chronic inflammation and dysbiosis. Note that these hallmarks are not independent 'causes' of ageing, but rather are part of a network of interlinked cellular processes whose functionalities are interdependent. For example, epigenetic changes with age are likely to influence all cellular functions via effects on gene expression. Telomere loss can give rise to cellular senescence, which in turn will influence the rate of depletion of stem cell pools. Signalling from the nucleus by master regulators of mitochondrial function and biogenesis is influenced by telomere length [47]. The triggering of the p53 DNA (deoxyribonucleic acid) damage response by short telomeres promotes mitochondrial dysfunction, reduces oxidative defences, and compromises energy-generating capacity [48, 49], which will affect the rate of tissue ageing. Telomeric silencing of genes close to the chromosome ends can be reduced by telomere attrition [50]. For non-model organisms, we generally lack sufficient knowledge of most of the molecular pathways involved, making it very challenging to meaningfully measure some of these hallmarks, let alone relate them to mortality risk. But given that evolutionary ecologists usually want a general index of the amount of deterioration in individuals, there are some very useful options. An important caveat is that most studies do not assess the extent to which a particular maker is predictive of the risk of death. This is crucial because, as mentioned, not all age-related phenotypic changes are due to ageing and the link to increased mortality risk is central to the definition of ageing. Simply showing that a trait changes with age is not a demonstration of ageing. Also, since individuals can differ for reasons other than differences in the rate of ageing (e.g., they might suffer from particular pathologies that are environmentally or genetically acquired), measures of within-individual changes with age are the most valuable. We provide summary information in Section 8.4 on the main markers we think are the most promising measures of biological age with current potential for use in evolutionary ecology at present. While some are generally applicable, in most cases species-specific parameterization is needed based on a sample of individuals whose age and subsequent time of death are known.

8.4.1 Frailty Indices

Frailty is a concept initially used in ageing research to estimate the risk of death for individuals, independent of their chronological age [51]. It is generally interpreted as the age-specific phenotypic state of the individual, with higher frailty being associated with a greater risk of death. The general idea is to use phenotypic and/or behavioural traits that are relatively easy to measure to characterize changes in phenotypic state in the form of a composite trait that is predictive of remaining life expectancy. Conspecifics can differ in frailty throughout their lives, and individuals generally become increasingly frail as they grow old. Calculating a multifactorial index of phenotypic state is something that is commonly used in clinical and biomedical contexts, and various non-invasive, organismal-level indices of frailty have been developed for humans and mice. In humans, this generally involves five clinical characteristics, sometimes classified in a simple binary manner: unintended weight loss of >2 kg in six months; walking speed less than 1.0 m/sec; grip strength <26 kg for men and 18 kg for women; exhaustion in past two weeks; whether or not they engage in physical activity/light sports. Individuals may be classified as 'frail' or 'not frail' if they score negative values on 3/5 measures, but more complex scales have also been developed [52, 53]. These human frailty measures change within individuals with age and show good relationships with mortality risk [54]. While frailty indices have been developed for some other species such as rats and dogs

[52], they are rarely used by evolutionary ecologists. The need to be predictive of mortality risk is obviously an important requirement here and there is a need to distinguish between changes in phenotypic performance that are indicative of ageing from differences between individuals that are due to injury or a pathological condition. Body mass loss, for example, is widespread with age in the wild [55], but individuals can differ in mass loss for reasons that are not related to ageing. Thompson and colleagues examined physical frailty in wild chimps (*Pan troglodytes*) in Uganda using urine creatinine to assess lean body mass as a measure of muscle loss (sarcopaenia) based on 21 years of longitudinal data [56]; while they found a within-individual decline with age, it did not predict mortality and individuals appeared to change their activity patterns to combat their reduced capabilities. Much will depend on the extent to which organisms can survive with reduced muscle mass. Birds, for example, are much less likely to survive if they cannot attain a sufficiently rapid take-off speed or flight performance to enable them to evade predators and so are likely to prioritize maintenance of muscle mass into old age. Cross-sectional studies on long-lived seabirds have found variable evidence of sarcopaenia [57], but within-individual studies are also needed. In an investigation of cross-generational variation in ageing, the age-related decline in the effort that male field crickets (*Gryllus campestris*) in the wild devote to singing in each of 10 years and the actuarial senescence levels in each year were estimated; the authors found that years with a higher actuarial senescence also had high post-peak declines in calling effort [58]. Froy and colleagues found that declining home range size predicts reduced late life survival in two wild ungulates (the red deer *Cervus elaphus* and the Soay sheep *Ovis aries*) based on long-term, within-individual-based studies in the field [59]. Development of frailty indices is probably simplest for traits related to locomotory performance, though maintenance of this may be differentially prioritized depending on species lifestyle. There is also considerable scope for measuring declines in cognitive performance with age in both the lab and the field, but appropriate learning tasks suited to organism life histories need to be used [60].

Measures of reproductive performance have traditionally been studied in the context of (age-specific) fitness variation, but they could potentially contribute to frailty measures. For example, egg and sperm quality change with age in birds and are known to be linked to fitness [61, 62]. However, as mentioned, the use of reproductive performance itself is fraught with difficulties since it can change for many reasons other than ageing, such as strategic changes in investment patterns, the advantages of which will vary among species and individuals. Species can show gradual increases of reproductive performance with age, gradual declines or an abrupt change in the last year of reproduction preceding death. These patterns are very difficult to interpret in the context of phenotypic state and senescence for the following reasons: (1) enhanced reproductive performance with age could reflect an effect of previous breeding experience independently of the physiological state and the amount of reproductive effort or energy spent, or it could reflect adaptive restraint in reproductive effort during early adult life (to maintain high somatic health and survival prospects) that might subsequently be assuaged through the course of adult life [7]; and (2) a reproductive decline could reflect reduced somatic health or reduced reproductive effort or a combination of the two. Reduced reproductive effort with age can be an optimal life history strategy if it efficiently supports somatic repair, increasing survival prospects. The optimal pattern of age-specific reproductive effort therefore depends on the balance between the efficiency with which reproductive effort is turned into reproductive success versus the energy that is required for somatic maintenance and repair to increase survival prospects. It also depends on many ecological factors, such as the condition-dependent risk of disease or predation, because these reduce overall life expectancy, increasing the fitness benefits of early life reproductive success. Whether a single (physiological) variable of reproductive performance can be used as marker of biological age depends on the detailed species-specific knowledge of the underlying life history trade-offs, ecological setting and pattern of optimal age-specific resource allocation. A useful way forward would be to integrate variables of reproductive performance with organismal- and/or physiological-level markers linked to survival, since the pattern of covariation may indicate the relationship between reproductive performance and biological age (e.g. [63]).

8.4.2 Mitochondrial Function

A decline in mitochondrial functionality in the soma with age is considered a key element in the ageing process [46, 64, 65]. In humans this is characterized by a reduced efficiency in oxidative phosphorylation and the generation of ATP (adenosine triphosphate) [66, 67, 68], and results in a loss of muscle strength and neural function [69]. Mitochondrial DNA (mtDNA) has a higher mutation rate than nuclear DNA, in part due to the proximity of mtDNA to the generation of potentially damaging reactive oxygen species (ROS) associated with ATP production and the high rate of mtDNA replication coupled with the absence of many of the DNA repair processes present in the nucleus [70, 71]. Preserving the overall functionality of mitochondria in animals is achieved via multiple mechanisms, including increasing mtDNA copy number [70]. However, the tight regulation of mtDNA copy number, and the rate of fission and fusion of mitochondria, can decrease with age; consequently, reduced age-specific mtDNA copy number is considered a useful biomarker of mitochondrial dysfunction [65, 72]. Measuring and interpreting parameters from mitochondria is difficult and requires specialist knowledge and equipment. Depending on the measure used, it may be necessary to use fresh tissue samples, which can make use in the field very difficult and sometimes impossible, since advanced laboratory facilities are needed and taking such equipment into the field is not for the faint-hearted.[1] In vertebrates with nucleated red blood cells, mitochondria can be extracted from the red cells, making longitudinal studies more feasible [73]. A series of useful papers in this area can be found in *Integrative and Comparative Biology* (volume 58) published in September

[1] See https://journals.biologists.com/jeb/article/224/13/jeb242998/270867/In-the-field-an-interview-with-Wendy-Hood.

2018. It is important to bear in mind that mitochondrial performance is dynamic and can differ among tissues, and that individuals' mitochondrial performance can differ for multiple reasons.

8.4.3 Telomeres

Telomeres are nucleoprotein structures containing a variable number of tandem repeats of a DNA sequence (TTAGGG in vertebrates and TTAGG in most invertebrates). Telomeres cap the ends of the linear chromosomes of eukaryotes, distinguishing true ends from double-stranded chromosomal breaks, and are also involved in other aspects of cell division. Functional telomeres prevent the triggering of a DNA damage response by the chromosome ends, thereby ensuring genome stability [74]. Since the process of DNA replication is incomplete at the end of the lagging DNA strand (the 'end-replication problem'), the sequence loss during cell division is absorbed by the telomere so that the protein-coding DNA sequence is preserved. In addition to the end-replication loss, increased telomere loss can also arise as a consequence of damage to telomeric DNA [75]. Telomeres can be restored by the reverse transcriptase telomerase, or by other routes such recombination-based processes termed Alternative Telomere Lengthening (ALT). Telomere maintenance is essential during development and gametogenesis and is closely regulated, with telomerase and ALT being important at different stages [76, 77]. There is significant variation among species in the pattern of telomere restoration [78]. However, in many species, especially long-lived or large-bodied endotherms, telomere restoration processes are down-regulated in most adult somatic tissues, and are thought to contribute to protection against tumour formation [79, 80, 81]. In the absence of restoration, telomeres therefore become progressively shorter with each round of cell division. Once the telomeres become critically short, the genome becomes unstable, p53-dependent cellular senescence is triggered and the cell enters cell cycle arrest, followed either by apoptosis or an altered, pro-inflammatory secretory profile [82]. Unrestored telomere loss therefore sets a finite limit on the replicative potential of cells in most tissues, and eventually leads to an increase in senescent cells, cell loss and a deterioration in tissue function [74]. Dysfunctional telomeres are associated with a wide range of degenerative diseases, including the main age-related diseases in humans. Telomere length or loss rates have been shown to be predictive of eventual lifespan within and across species, especially in birds [80, 83, 84, 85, 86, 87, 88, 89]. It is important to bear in mind that telomere loss is linked to other ageing processes and that the relationship to mortality risk need not be directly causal but can be the outcome of other processes that deteriorate with age. Telomeres can be measured in multiple cell types relatively non-invasively (e.g. red blood cells in non-mammalian vertebrates, white blood cells in mammals, buccal cells and biopsied tissues) but there can be tissue-specific differences, so it is important that the tissue used is consistent across samples. Most useful is likely to be within-individual loss rate, but this requires repeated measurements across a sufficient time period. Methods for measuring telomere length best suited to evolutionary ecologists can be found in Nussey and colleagues [90].

Not all species show telomere loss, since restoration processes can be active in somatic tissues, and this may occur at different times depending on age and environmental conditions [91, 92, 93]. Further, some species have unusual telomere structures or maintenance methods. This inter-specific variation can be used to test theories about the trade-offs that might be involved in the evolution of telomere maintenance patterns, tumour avoidance and other life history traits [81, 94, 95, 96].

8.4.4 Immune Markers

Major changes in immune function with age have been widely reported in humans, particularly in adaptive immunity, and linked to stem cell exhaustion [46]. Peters et al. carried out a meta-analysis of immunosenescence, investigating the evidence for age-related change in vertebrates, which covered studies in the wild and in wild-derived captive species [97]. In all, they identified 26 immune parameters from 62 studies, covering 44 species, 20 of which were birds, 10 mammals and five reptiles; 35 studies were in the wild and 28 in captive conditions, with no species being studied in both. These studies assessed whether innate or adaptive immunity had declined with age. The authors examined evidence in each vertebrate class separately, but overall, they found no evidence of a significant decline in mammals, birds or reptiles with age, nor any sex-specific effects. However, only five of these studies were longitudinal. When considering some classes of markers separately, they did find some evidence of a decline in adaptive immunity. This paper gives a good appraisal of the range of markers that can be used, but links to mortality risk need to be established. Interestingly, a detailed study on Soay sheep (published too late for inclusion in the meta-analysis just mentioned [97]), showed that a late life decline in an important immune marker of resistance to a prevalent helminth parasite (Immunoglobulin-G binding antigen from Teladorsagia circumcincta, IgGH-Tc) was significantly linked to the number of years before death within individuals; individuals with relatively low levels of this marker compared with the population average were less likely to survive winter [98]. On the other hand, no differences were found in cross-sectional comparisons. While there are a number of caveats associated with this study, it does provide good evidence from the wild of a link between an age-related decline in immune function related to mortality risk. Importantly, the immune marker was a better indicator of time to death than chronological age. This study also emphasizes the importance of looking at within-individual change, and, in the context of immune function, targeting the marker to a prevalent disease.

8.4.5 Oxidative Damage

Oxidative damage occurs when the generation of damaging free radicals, such as ROS, are not sufficiently neutralized by antioxidant defences. Damage can occur in macromolecules such as DNA, lipids and proteins. When unrepaired, damage can accumulate with age, potentially causing reduced tissue functioning. Various measures of antioxidant defences have been used in animals, but the interpretation of differences among individuals is difficult since antioxidant levels vary for many reasons. Various markers of unrepaired oxidative damage are potentially useful indicators of biological age [99], and these have been used by evolutionary ecologists in a wide variety of contexts. With respect to ageing, we might expect animals to manage damage generation to optimize fitness (e.g. by reducing reproductive rate [100]) and/or to have mitigation strategies when the damage risk is high [101]. Currently, it remains unclear to what extent markers of oxidative damage reflect an accumulation with age versus a current transitory state because it is not well known how much oxidative damage progressively accumulates with age, how much damage is being repaired and in general what the individual variability is on both short (e.g. days) and longer (e.g. months/years) timescales [102]. For a marker of oxidative damage to be useful as indicator of biological age, the marker must be linked to mortality risk. One longitudinal study on a wild European shag (*Gulosus aristotelis*) population found that d-ROMs (derivatives-reactive oxygen metabolites; a measure of oxidative damage in blood plasma) increased with age within old individuals and was associated with annual resighting probability, indicative of mortality [103]. The findings of this study further suggest that in old age, oxidative damage may accumulate as a result of reduced antioxidant defences, which could be related to ageing. Other studies using similar methods (e.g. [104]) have found mixed evidence for a relationship between markers of oxidative stress and age or survival prospects, and it is likely that this will vary with species life histories. Short-lived species, for example, may be less 'concerned' about oxidative damage because its accumulation with time may be not sufficient to have any fitness consequences. A study on wild jackdaws (*Corvus monedula*) found that markers of oxidative stress showed differential sensitivity to manipulated environmental conditions; this variation was inversely linked to whether the marker was predictive of survival probability [105]. In this study reduced glutathione was highly predictive of overwintering survival probability, yet unaffected by brood size manipulation, whereas the opposite pattern was found for d-ROMs and TBARS (thiobarbituric acid reactive substances). This could indicate that components of oxidative stress that are important for fitness are prioritized/protected through differential resource allocation, but further integrative work is required to test this hypothesis.

8.4.6 Epigenetic Markers

Epigenetic alterations to DNA affect gene expression without altering the genetic sequence itself and are therefore some of the key factors underpinning genotype-specific phenotypic variation [106]. Epigenetic alterations include DNA methylation (DNAm), and non-coding RNAs (ribonucleic acid) linked to histone and chromatin modification, all of which can regulate gene expression [107, 108]. It has been known for some time, from research in biogerontology, that there is a complex relationship between DNAm and ageing [109]. The early studies reported a decrease in global DNAm (hypomethylation) with age, suggesting increasing dysregulation and genome instability [46]. Later studies reported the surprising finding that, at many loci (particularly CpG sites), methylation actually increases steadily with age. These age-related changes are correlated with the process of ageing [109]. Finding those age-related methylation sites that lead to changes in age-specific gene expression, and what genes are involved, is an exciting area of ageing research and may help uncover the links among chronological age, environmental conditions and ageing. Furthermore, because epigenetic alterations are potentially reversible, it would in principle be possible to develop targeted 'senotherapies', reversing some of the epigenetic ageing that has occurred, to treat, or even cure, ageing-related disease [110, 111].

In humans, much of the contemporary epigenetic ageing research has been focused on the development of epigenetic 'clocks' that can be utilized as age estimator. The first such epigenetic clock was developed by Steve Horvath using an algorithm to predict chronological age based on the age-related variation in DNAm [112]. DNAm is highly tissue specific and most of the observable genomic DNAm variation in a population is likely to be unrelated to ageing. Horvath succeeded in developing a panel of specific genomic sites with age-related changes in DNAm appearing across many types of cell/tissue and across a wide range of mammals. Several other clocks based on different combinations of DNAm sites and predictive algorithms have subsequently been developed; all of these clocks have the same underlying principle in common, each delivering sometimes surprisingly precise age estimation with few loci [109]. In addition to estimating *chronological* age, there is also great interest in utilizing DNAm for estimating *biological* age. The biological age estimation is rooted in the principle that an individual can have a relatively young or old DNAm profile (i.e. DNAm age is older or younger than its chronological age would predict). This DNAm index has been shown to be highly predictive of human age-related morbidity and disease [113].

The great advantage of epigenetic clocks is that they can potentially provide an estimate of the chronological and biological age state of individuals. This powerful approach therefore has tremendous potential for ageing research in non-model wild animals that are difficult to mark and trace through time, such as small invertebrates, diversifying such research into taxa that currently remain understudied. Further, this approach could enable comparisons of laboratory and wild populations of the same species using the same methodology, which could bridge the divide between laboratory and ecological studies. Because genomic sequencing has now become affordable, it is feasible to apply global DNAm methods to wild animal populations. The initial hurdle to overcome is to establish a panel of genomic sites showing age-related variation that is robust across different tissues [111]. Such a project benefits from interdisciplinary collaborations, particularly with computational biologists, to develop species-specific bioinformatic pipelines required for a useful DNAm clock. This has been achieved for a number of different species [109], which demonstrates the potential for ageing research in non-model organisms. A population average age-specific methylation rate needs to be established using known age individuals. The deviation of individuals from the population average age relationship (often termed 'epigenetic age') can then be used an indicator of biological age. However, it is challenging to develop such tool for a novel organism or population, because evaluating whether a marker is a useful indicator of biological age hinges on testing its relationship with prospective survival in a group of known age individuals and hence requires a longitudinal component that may or may not be feasible.

A current limitation of epigenetic research on wild animal populations remains its relatively high per-sample cost, particularly when genome assembly information is lacking, and, as already mentioned, the need for an initial study to calibrate the epigenetic clock. The latter is not a simple task and there are many things to consider, not least how many methylation sites to use and how to separate inaccuracies in the calibration of the clock from differences in individual ageing rates [109]. For the moment, epigenetic studies on wild animal populations remain largely based on mammals and have limited sample sizes ([114, 115, 116], but see [117]), but this is likely to change in the near future. Current initiatives like the Earth BioGenome Project (www.earthbiogenome.org/) are paramount for reducing the costs and will support the extension of epigenetic research to wild animal species. As ageing is a multi-factor phenomenon it will also be important to integrate epigenetic markers with other physiological markers of biological age. For example, a recent study found that rates of telomere attrition were associated with rates of change in DNAm with age and hence that these different physiological markers may in part reflect similar aspects of age-related deterioration [118]. More such studies are required with more accurate DNAm techniques to advance this field.

8.4.7 Other Markers

There are several other markers that have been used in the context of assessing biological age, many of which are based on blood biochemistry. These have largely been developed in a biogerontological context and their use in wild animals remains limited, and they have generally not (yet) been shown to be linked to mortality risk. For example, haematocrit (packed red cell volume) has been reported to decline with age in cross-sectional studies of many vertebrates. It can be measured from blood samples relatively easily provided a suitable centrifuge is available. This was recently examined in a field study of Seychelles warblers (*Acrocephalus sechellensis*), using birds up to 13 years old. This study, which included a longitudinal component, concluded that haematocrit increased up to 1.5 years of age, then declined. Young birds with relatively high haematocrit were more likely to survive but there was no evidence that the decline in haematocrit after 1.5 years was linked to survival or reproductive performance [119]. This study suggests that while haematocrit might provide information on biological state, it is not a useful marker in the context of ageing.

Blood in vertebrates is generally relatively easy to sample and potentially contains much useful information about the biological state [45]. Assessment of biochemical markers from blood is very widely used in studies of ageing in humans and has potential for further development in wild animals. Machines routinely used by veterinarians can produce more than a dozen blood parameters from a single small whole blood sample (e.g. VetScan – www.zoetisus.com/products/diagnostics/instruments/vetscan-vs2). Detailed physiological knowledge is required, however, to interpret these as indicators of physiological state. Birds, for example, show much higher blood glucose levels than mammals; these levels would be associated with diabetes and poor health in mammals but are normal in birds [120]. A recent study in humans examined changes in numerous blood parameters along individual ageing trajectories, which were then combined into a single variable termed the 'Dynamic Organism State Indicator' (DOSI), using a Principal Components Analysis; the DOSI appeared to be indicative of an age-related loss of physiological resilience and linked to the risk of developing degenerative disease [121]. Interestingly, a large-scale study by the MARK-AGE consortium is underway in humans, which aims to examine blood markers in relation to age and mortality in over 3000 humans, and which is likely to yield interesting results. Increased inflammation (inflamm-ageing), which can be measured in different ways, is associated with ageing in studies of model vertebrates but is relatively little used in studies outside of a laboratory setting. A recent review of this area included 12 studies that involved inflammatory markers, but these have not been widely used as yet [97].

Overall, as we have stressed repeatedly, biomarkers of biological age should be linked to mortality risk. This is much more difficult in the wild than it is with organisms in captivity and with humans.

8.5 The Way Forward

What do we gain by an integrative approach to ageing? From the perspective of evolutionary and behavioural ecology, understanding the 'how' questions that are addressed by biogerontological researchers provides us with multiple benefits. Explanations based on differential outcomes of trade-offs feature strongly in understanding the evolution of ageing in a life history context. Importantly, understanding mechanisms and progression of ageing (Tinbergen's Questions 3 and 4 on development and process) helps us to understand the nature, currency and relevance of the fitness costs and benefits involved, and, importantly, the extent to which outcomes are constrained and impacted by environmental conditions and change. While there is a tendency to assume that evolution can get round constraints, it is unlikely that this is always so as evolution does not operate through orchestrated design but through the processes of randomly generated mutations combined with selection and genetic drift. Furthermore, the historical legacy embedded in organismal form (Question 2) may mean that some phenotypic outcomes, even if optimal, are simply not possible. For example, through directional selection the ancestral evolution may have purged genetic variation for some traits, or led to extreme canalization, and may also have introduced (negative) genetic correlations among traits, imposing constraints on the subsequent evolutionary trajectory. These constraints limit the capacity for novel and recurrent mutations to have beneficial effects on relative fitness because, through genetic and/or epigenetic linkage, such mutations may affect a whole network of traits, with contrasting effects on performance. Hence, when the environmental conditions are sufficiently stable, genetic correlation may become fixed, which could occur, for example, due to DNAms becoming permanently embedded onto specific loci. Ultimately, similar processes underpin speciation, but on a finer scale they may also underpin intra- and inter-specific variation in the form, pace and pattern of ageing. The extent to which an organism's ageing trajectory reflects (epi)genetic constraints and hence 'suites' of connected traits, is likely to be informative of the ancestral environmental conditions/ stability to which the population/species was exposed.

We clearly need to expand the breadth of species studied in terms of both taxonomic range and life history variation. Advances in biogerontological research can provide us with the means to identify differences among individuals in how fast they age, and the role of environmental conditions early and late in life. For biogerontologists seeking to find ways to slow ageing, or even reverse its harshest effects, recognizing that evolution has already solved this problem for some species might provide useful therapeutic approaches. Thus, the approaches embedded in Tinbergen's Questions 1 and 2 potentially contain many useful insights. Birds, for example, appear to have more resistance to oxidative damage than mammals and maintain higher cellular function as they age; the fine detail of how they do this is best studied in the laboratory and requires cellular and organismal expertise [122]. An integrative approach based on cross-disciplinary partnerships seems to us the way forward.

References

1. Zhao, X., Promislow, D.E.L. 2019. Senescence and ageing. In *Oxford Handbook of Evolutionary Medicine* (eds M. Brüne, W. Schiefenhövel), Chapter 5, pp. 167–221. Oxford University Press (doi:10.1093/oxfordhb/9780198789666.013.5).
2. Monaghan, P., Charmantier, A., Nussey, D.H., Ricklefs, R.E. 2008. The evolutionary ecology of senescence. *Funct. Ecol.* **22**, 371–378.
3. Schnohr, P., Nyboe, J., Lange, P., Jensen, G. 1998. Longevity and gray hair, baldness, facial wrinkles, and arcus senilis in 13,000 men and women: the Copenhagen City Heart Study. *J. Gerontol. Ser. -Biol. Sci. Med. Sci.* **53**, M347–M350.
4. Marasco, V., Boner, W., Griffiths, K., Heidinger, B., Monaghan, P. 2019. Intergenerational effects on offspring telomere length: interactions among maternal age, stress exposure and offspring sex. *Proc. R. Soc. B-Biol. Sci.* **286**, 20191845 (doi:10.1098/rspb.2019.1845)
5. Reid, J.M., Bignal, E.M., Bignal, S., McCracken, D.I., Bogdanova, M.I., Monaghan, P. 2010. Parent age, lifespan and offspring survival: structured variation in life history in a wild population. *J. Anim. Ecol.* **79**, 851–862.
6. Williams, G.C. 1957. Pleiotropy, natural selection, and the evolution of senescence. *Evolution* **11**, 398–411.
7. Boonekamp, J.J., Bauch, C., Verhulst, S. 2020. Experimentally increased brood size accelerates actuarial senescence and increases subsequent reproductive effort in a wild bird population. *J. Anim. Ecol.* **89**, 1395–1407.
8. Bernard, C., Compagnoni, A., Salguero-Gomez, R. 2020. Testing Finch's hypothesis: the role of organismal modularity on the escape from actuarial senescence. *Funct. Ecol.* **34**, 88–106.
9. Baudisch, A. 2011. The pace and shape of ageing. *Methods Ecol. Evol.* **2**, 375–382.
10. Ronget, V., Gaillard, J.-M. 2020. Assessing ageing patterns for comparative analyses of mortality curves: going beyond the use of maximum longevity. *Funct. Ecol.* **34**, 65–75.

11. Partridge, L., Gems, D. 2002. Mechanisms of ageing: public or private? *Nat. Rev. Genet.* **3**, 165–175.
12. Brunet, A. 2020. Old and new models for the study of human ageing. *Nat. Rev. Mol. Cell Biol.* **21**, 491–493.
13. Bolker, J. 2012. There's more to life than rats and flies. *Nature* **491**, 31–33.
14. Briga, M., Verhulst, S. 2015. What can long-lived mutants tell us about mechanisms causing aging and lifespan variation in natural environments? *Exp. Gerontol.* **71**, 21–26.
15. Zajitschek, F., Zajitschek, S., Bonduriansky, R. 2020. Senescence in wild insects: key questions and challenges. *Funct. Ecol.* **34**, 26–37.
16. Gaillard, J.-M., Lemaître, J.-F. 2020. An integrative view of senescence in nature. *Funct. Ecol.* **34**, 4–16.
17. Promislow, D.E., Flatt, T., Bonduriansky, R. 2021. The biology of aging in insects: from Drosophila to other insects and back. *Annu. Rev. Entomol.* **67**, 83–103.
18. Kramer, B.H., van Doorn, G.S., Weissing, F.J., Pen, I. 2016. Lifespan divergence between social insect castes: challenges and opportunities for evolutionary theories of aging. *Curr. Opin. Insect Sci.* **16**, 76–80 (doi:10.1016/j.cois.2016.05.012).
19. Lucas, E.R., Keller, L. 2014. Ageing and somatic maintenance in social insects. *Curr. Opin. Insect Sci.* **5**, 31–36 (doi:10.1016/j.cois.2014.09.009).
20. Kreider, J.J., Pen, I., Kramer, B.H. 2021. Antagonistic pleiotropy and the evolution of extraordinary lifespans in eusocial organisms. *Evol. Lett.* **5**, 178–186 (doi:10.1002/evl3.230).
21. Quigley, T.P., Amdam, G.V. 2021. Social modulation of ageing: mechanisms, ecology, evolution. *Philo. Trans. R. Soc. B-Biol. Sci.* **376**, 20190738 (doi:10.1098/rstb.2019.0738).
22. Hassall, C., Amaro, R., Ondina, P., Outeiro, A., Cordero-Rivera, A., San Miguel, E. 2017. Population-level variation in senescence suggests an important role for temperature in an endangered mollusc. *J. Zool.* **301**, 32–40.
23. Massot, M. 2011. Ageing and fitness correlates determined in a wild population of lizards. *Herpetol. Rev.* **42**, 133–133.
24. Bateson, P., Laland, K.N. 2013. Tinbergen's four questions: an appreciation and an update. *Trends Ecol. Evol.* **28**, 712–718.
25. Bouwhuis, S., Choquet, R., Sheldon, B.C., Verhulst, S. 2012. The forms and fitness cost of senescence: age-specific recapture, survival, reproduction, and reproductive value in a wild bird population. *Am. Nat.* **179**, E15–E27 (doi:10.1086/663194).
26. Peron, G., Lemaître, J.-F., Ronget, V., Tidiere, M., Gaillard, J.-M. 2019. Variation in actuarial senescence does not reflect life span variation across mammals. *PLoS Biol.* **17**, e3000432 (doi:10.1371/journal.pbio.3000432).
27. Pease, C.M., Bull, J.J. 1988. A critique of methods for measuring life-history trade-offs. *J. Evol. Biol.* **1**, 293–303 (doi:10.1046/j.1420-9101.1988.1040293.x).
28. Lee, W.-S., Monaghan, P., Metcalfe, N.B. 2013. Experimental demonstration of the growth rate–lifespan trade-off. *Proc. R. Soc. B* **280**, 20122370.
29. Lodjak, J., Verhulst, S. 2020. Insulin-like growth factor 1 of wild vertebrates in a life-history context. *Mol. Cell. Endocrinol.* **518**,110978 (doi:10.1016/j.mce.2020.110978).
30. Swanson, E.M., Dantzer, B. 2014. Insulin-like growth factor-1 is associated with life-history variation across Mammalia. *Proc. R. Soc. B-Biol. Sci.* **281**, 20132458 (doi:10.1098/rspb.2013.2458).
31. Cohen, A.A. 2018. Aging across the tree of life: the importance of a comparative perspective for the use of animal models in aging. *Biochim. Biophys. Acta-Mol. Basis Dis.* **1864**, 2680–2689.

32. de Magalhaes, J.P., Costa, J. 2009. A database of vertebrate longevity records and their relation to other life-history traits. *J. Evol. Biol.* **22**, 1770–1774.
33. Holmes, D.J., Fluckiger, R., Austad, S.N. 2001. Comparative biology of aging in birds: an update. *Exp. Gerontol.* **36**, 869–883.
34. Munshi-South, J., Wilkinson, G.S. 2010. Bats and birds: exceptional longevity despite high metabolic rates. *Ageing Res. Rev.* **9**, 12–19.
35. Wilkinson, G.S., South, J.M. 2002. Life history, ecology and longevity in bats. *Aging Cell* **1**, 124–131.
36. Metcalfe, N.B., Monaghan, P. 2003. Growth versus lifespan: perspectives from evolutionary ecology. *Exp. Gerontol.* **38**, 935–940.
37. Belsky, J. 2019. Early-life adversity accelerates child and adolescent development. *Curr. Dir. Psychol. Sci.* **28**, 241–246.
38. Sapolsky, R.M. 2000. Stress hormones: good and bad. *Neurobiol. Dis.* **7**, 540–542.
39. Jimeno, B., Briga, M., Verhulst, S., Hau, M. 2017. Effects of developmental conditions on glucocorticoid concentrations in adulthood depend on sex and foraging conditions. *Horm. Behav.* **93**, 175–183 (doi:10.1016/j.yhbeh.2017.05.020).
40. Lind, M.I., Ravindran, S., Sekajova, Z., Carlsson, H., Hinas, A., Maklakov, A.A. 2019. Experimentally reduced insulin/IGF-1 signaling in adulthood extends lifespan of parents and improves Darwinian fitness of their offspring. *Evol. Lett.* **3**, 207–216.
41. Spagopoulou, F. 2020. Transgenerational maternal age effects in nature: lessons learnt from Asian elephants. *J. Anim. Ecol.* **89**, 936–939 (doi:10.1111/1365-2656.13218).
42. Colchero, F., Jones, O.R., Rebke, M. 2012. BaSTA: an R package for Bayesian estimation of age-specific survival from incomplete mark–recapture/recovery data with covariates. *Methods Ecol. Evol.* **3**, 466–470.
43. Boonekamp, J.J., Salomons, M., Bouwhuis, S., Dijkstra, C., Verhulst, S. 2014. Reproductive effort accelerates actuarial senescence in wild birds: an experimental study. *Ecol. Lett.* **17**, 599–605.
44. van de Pol, M., Wright, J. 2009. A simple method for distinguishing within- versus between-subject effects using mixed models. *Anim. Behav.* **77**, 753–758 (doi:10.1016/j.anbehav.2008.11.006).
45. Stier, A., Reichert, S., Criscuolo, F., Bize, P. 2015. Red blood cells open promising avenues for longitudinal studies of ageing in laboratory, non-model and wild animals. *Exp. Gerontol.* **71**, 118–134 (doi:10.1016/j.exger.2015.09.001).
46. Lopez-Otin, C., Blasco, M.A., Partridge, L., Serrano, M., Kroemer, G. 2023. Hallmarks of aging: An expanding universe. *Cell* **186**, 243–278.
47. Zhu, Y., Liu, X., Ding, X., Wang, F., Geng, X. 2019. Telomere and its role in the aging pathways: telomere shortening, cell senescence and mitochondria dysfunction. *Biogerontology* **20**, 1–16 (doi:10.1007/s10522-018-9769-1).
48. Sahin, E. et al. 2011. Telomere dysfunction induces metabolic and mitochondrial compromise. *Nature* **470**, 359–365 (doi:10.1038/nature09787).
49. Sahin, E., DePinho, R.A. 2012. Axis of ageing: telomeres, p53 and mitochondria. *Nat. Rev. Mol. Cell Biol.* **13**, 397–404.
50. Robin, J.D., Ludlow, A.T., Batten, K., Magdinier, F., Stadler, G., Wagner, K.R., Shay, J.W., Wright, W.E. 2014. Telomere position effect: regulation of gene expression with progressive telomere shortening over long distances. *Genes Dev.* **28**, 2464–2476 (doi:10.1101/gad.251041.114).
51. Vaupel, J.W., Manton, K.G., Stallard, E. 1979. Impact of heterogeneity in individual frailty on the dynamics of mortality. *Demography* **16**, 439–454 (doi:10.2307/2061224).

52. Heinze-Milne, S.D., Banga, S., Howlett, S.E. 2019. Frailty assessment in animal models. *Gerontology* **65**, 610–619.
53. Nakazato, Y. et al. 2020. Estimation of homeostatic dysregulation and frailty using biomarker variability: a principal component analysis of hemodialysis patients. *Sci. Rep.* **10**, 10314 (doi:10.1038/s41598-020-66861-6).
54. Theou, O., Brothers, T.D., Pena, F.G., Mitnitski, A., Rockwood K. 2014. Identifying common characteristics of frailty across seven scales. *J. Am. Geriatr. Soc.* **62**, 901–906 (doi:10.1111/jgs.12773).
55. Lemaître, J.-F., Gaillard, J.-M. 2017. Reproductive senescence: new perspectives in the wild: reproductive senescence in the wild. *Biol. Rev.* **92**, 2182–2199 (doi:10.1111/brv.12328).
56. Thompson, M.E., Machanda, Z.P., Fox, S.A., Sabbi, K.H., Otali, E., Thompson González, N., Muller, M.N., Wrangham, R.W. 2020. Evaluating the impact of physical frailty during ageing in wild chimpanzees (*Pan troglodytes schweinfurthii*). *Philo. Trans. R. Soc. B* **375**, 20190607.
57. Brown, K., Jimenez, A.G., Whelan, S., Lalla, K., Hatch, S.A., Elliott, K.H. 2019. Muscle fiber structure in an aging long-lived seabird, the black-legged kittiwake (*Rissa tridactyla*). *J. Morphol.* **280**, 1061–1070.
58. Rodriguez-Munoz, R., Boonekamp, J.J., Liu, X.P., Skicko, I., Haugland Pedersen, S., Fisher, D.N., Hopwood, P., Tregenza, T. 2019. Comparing individual and population measures of senescence across 10 years in a wild insect population. *Evolution* **73**, 293–302 (doi:10.1111/evo.13674).
59. Froy, H. et al. 2018. Declining home range area predicts reduced late-life survival in two wild ungulate populations. *Ecol. Lett.* **21**, 1001–1009.
60. Macphail, E.M., Bolhuis, J.J. 2001. The evolution of intelligence: adaptive specializations versus general process. *Biol. Rev.* **76**, 341–364.
61. Bogdanova, M.I., Nager, R.G., Monaghan, P. 2006. Does parental age affect offspring performance through differences in egg quality? *Funct. Ecol.* **20**, 132–141.
62. Pizzari, T., Dean, R., Pacey, A., Moore, H., Bonsall, M.B. 2008. The evolutionary ecology of pre-and post-meiotic sperm senescence. *Trends Ecol. Evol.* **23**, 131–140.
63. Sudyka, J. 2019. Does reproduction shorten telomeres? Towards integrating individual quality with life-history strategies in telomere biology. *BioEssays* **41**, 1900095.
64. Sun, N., Youle, R.J., Finkel, T. 2016. The mitochondrial basis of aging. *Mol. Cell* **61**, 654–666 (doi:10.1016/j.molcel.2016.01.028).
65. Castellani, C.A., Longchamps, R.J., Sun, J., Guallar, E., Arking, D.E. 2020. Thinking outside the nucleus: mitochondrial DNA copy number in health and disease. *Mitochondrion* **53**, 214–223.
66. Coen, P.M. et al. 2010. ADSkeletal muscle mitochondrial energetics are associated with maximal aerobic capacity and walking speed in older adults. *J. Gerontol. Ser. -Biol. Sci. Med. Sci.* **68**, 447–455.
67. Conley, K.E., Jubrias, S.A., Cress, M.E., Esselman, P. 2013. Exercise efficiency is reduced by mitochondrial uncoupling in the elderly. *Exp. Physiol.* **98**, 768–777.
68. Distefano, G., Standley, R.A., Zhang, X., Carnero, E.A., Yi, F., Cornnell, H.H., Coen, P.M. 2018. Physical activity unveils the relationship between mitochondrial energetics, muscle quality, and physical function in older adults. *J. Cachexia Sarcopenia Muscle* **9**, 279–294.
69. Shpilka, T., Haynes, C.M. 2018. The mitochondrial UPR: mechanisms, physiological functions and implications in ageing. *Nat. Rev. Mol. Cell Biol.* **19**, 109–120 (doi:10.1038/nrm.2017.110).

70. Szklarczyk, R., Nooteboom, M., Osiewacz, H.D. 2014. Control of mitochondrial integrity in ageing and disease. *Philo. Trans. R. Soc. B-Biol. Sci.* **369**, 20130439 (doi:10.1098/rstb.2013.0439).
71. Ladoukakis, E.D., Zouros, E. 2017. Evolution and inheritance of mitochondrial DNA: rules and exceptions. *J. Biol. Res.-Thessalon.* **24** (doi:10.1186/s40709-017-0060-4).
72. Short, K.R., Bigelow, M.L., Kahl, J., Singh, R., Coenen-Schimke, J., Raghavakaimal, S., Nair, K.S. 2005. Decline in skeletal muscle mitochondrial function with aging in humans. *Proc. Natl. Acad. Sci. USA* **102**, 5618–5623 (doi:10.1073/pnas.0501559102).
73. Stier, A. et al. 2013. Avian erythrocytes have functional mitochondria, opening novel perspectives for birds as animal models in the study of ageing. *Front. Zool.* **10** (doi:10.1186/1742-9994-10-33).
74. Aubert, G., Lansdorp, P.M. 2008. Telomeres and aging. *Physiol. Rev.* **88**, 557–579.
75. Monaghan, P., Eisenberg, D.T.A., Harrington, L., Nussey, D. 2018. Understanding diversity in telomere dynamics. *Philo. Trans. R. Soc. Lond. B. Biol. Sci.* **373**, 20160435.
76. Reig-Viader, R., Garcia-Caldes, M., Ruiz-Herrera, A. 2016. Telomere homeostasis in mammalian germ cells: a review. *Chromosoma* **125**, 337–351 (doi:10.1007/s00412-015-0555-4).
77. Keefe, D.L. 2019. Telomeres and genomic instability during early development. *Eur. J. Med. Genet.* **63**, 103638 (doi:10.1016/j.ejmg.2019.03.002).
78. Smith, S., Hoelzl, F., Zahn, S., Criscuolo, F. 2022. Telomerase activity in ecological studies: what are its consequences for individual physiology and is there evidence for effects and trade-offs in wild populations. *Mol. Ecol.* **31**, 6239–6251 (doi:10.1111/mec.16233).
79. Gorbunova, V., Seluanov, A. 2009. Coevolution of telomerase activity and body mass in mammals: from mice to beavers. *Mech. Ageing Dev.* **130**, 3–9.
80. Gomes, N.M.V., Shay, J.W., Wright, W.E. 2010. Telomeres and telomerase. In *The Comparative Biology of Aging* (ed. N.S. Wolf), pp. 227–258. Springer.
81. Gomes, N.M.V. et al. 2011. Comparative biology of mammalian telomeres: hypotheses on ancestral states and the roles of telomeres in longevity determination. *Aging Cell* **10**, 761–768.
82. Artandi, S.E., Attardi, L.D. 2005. Pathways connecting telomeres and p53 in senescence, apoptosis, and cancer. *Biochem. Biophys. Res. Commun.* **331**, 881–890.
83. Heidinger, B.J., Blount, J.D., Boner, W., Griffiths, K., Metcalfe, N.B., Monaghan, P. 2012. Telomere length in early life predicts lifespan. *Proc. Natl. Acad. Sci.* **109**, 1743–1748.
84. Dantzer, B., Fletcher, Q.E. 2015. Telomeres shorten more slowly in slow-aging wild animals than in fast-aging ones. *Exp. Gerontol.* **71**, 38–47.
85. Tricola, G.M. et al. 2018. The rate of telomere loss is related to maximum lifespan in birds. *Philo. Trans. R. Soc. B* **373**, 20160445 (doi:10.1098/rstb.2016.0445).
86. Wilbourn, R.V., Moatt, J.P., Froy, H., Walling, C.A., Nussey, D.H., Boonekamp, J.J. 2018. The relationship between telomere length and mortality risk in non-model vertebrate systems: a meta-analysis. *Philo. Trans. R. Soc. B* **373**, 20160447.
87. Undroiu, I. 2020. On the correlation between telomere shortening rate and lifespan. *Proc. Natl. Acad. Sci.* **117**, 2248–2249 (doi:10.1073/pnas.1920300117).
88. Remot, F., Ronget, V., Froy, H., Rey, B., Gaillard, J.-M., Nussey, D.H., Lemaitre, J.-F. 2022. Decline in telomere length with increasing age across nonhuman vertebrates: A meta-analysis. *Mol. Ecol.* **31**, 5917–5932 (doi: 10.1111/mec.16145)
89. Sheldon, E.L. et al. 2022. Telomere dynamics in the first year of life, but not later in life, predict lifespan in a wild bird. *Mol. Ecol.* **31**, 6008–6017 (doi:10.1111/mec.16296).

90. Nussey, D.H. et al. 2014. Measuring telomere length and telomere dynamics in evolutionary biology and ecology. *Methods Ecol. Evol.* **5**, 299–310.
91. McLennan, D., Armstrong, J.D., Stewart, D.C., McKelvey, S., Boner, W., Monaghan, P., Metcalfe, N.B. 2018. Links between parental life histories of wild salmon and the telomere lengths of their offspring. *Mol. Ecol.* **27**, 804–814 (doi:10.1111/mec.14467).
92. Olsson, M. 2018. Ectothermic telomeres: it's time they came in from the cold. *Phil. Trans. R. Soc. B.* **373**: 20160449.
93. Boonekamp, J.J., Rodríguez-Muñoz, R., Hopwood, P., Zuidersma, E., Mulder, E., Wilson, W., Verhulst, S., Tregenza, T. 2022. Telomere length is highly heritable and independent of growth rate manipulated by temperature in field crickets. *Mol. Ecol.* **31**, 6128–6140 (doi:10.1111/mec.15888).
94. Gorbunova, V., Seluanov, A., Zhang, Z., Gladyshev, V.N., Vijg, J. 2014. Comparative genetics of longevity and cancer: Insights from long-lived rodents. *Nat. Rev. Genet.* **15**, 531.
95. Ingles, E.D., Deakin, J.E. 2016. Telomeres, species differences, and unusual telomeres in vertebrates: presenting challenges and opportunities to understanding telomere dynamics. *Aims Genet.* **3**, 1–24 (doi:10.3934/genet.2016.1.1).
96. Foley, N.M. et al. 2018. Growing old, yet staying young: the role of telomeres in bats' exceptional longevity. *Sci. Adv.* **4**, eaao0926.
97. Peters, A., Delhey, K., Nakagawa, S., Aulsebrook, A., Verhulst, S. 2019. Immunosenescence in wild animals: meta-analysis and outlook. *Ecol. Lett.* **10**, 1709–1722 (doi: 10.1111/ele.13343)
98. Froy, H., Sparks, A.M., Watt, K., Sinclair, R., Bach, F., Pilkington, J.G., Pemberton, J.M., McNeilly, T.N., Nussey, D.H. 2019. Senescence in immunity against helminth parasites predicts adult mortality in a wild mammal. *Science* **365**, 1296–1298.
99. Monaghan, P., Metcalfe, N.B., Torres, R. 2009. Oxidative stress as a mediator of life history trade-offs: mechanisms, measurements and interpretation. *Ecol. Lett.* **12**, 75–92.
100. Lee, W.-S., Monaghan, P., Metcalfe, N.B. 2016. Perturbations in growth trajectory due to early diet affect age-related deterioration in performance. *Funct. Ecol.* **30**, 625–635.
101. Meniri, M. et al. 2022. Untangling the oxidative cost of reproduction: an analysis in wild banded mongooses. *Ecol. Evol.* **12**, e8644 (doi:10.1002/ece3.8644).
102. Speakman, J. et al. 2015. Oxidative stress and life histories: unresolved issues and current needs. *Ecol. Evol.* **5**, 5745–5757. https://onlinelibrary.wiley.com/doi/full/10.1002/ece3.1790.
103. Herborn, K.A., Heidinger, B.J., Boner, W., Noguera, J.C., Adam, A., Daunt, F., Monaghan, P. 2014. Stress exposure in early post-natal life reduces telomere length: an experimental demonstration in a long-lived seabird. *Proc. R. Soc. B Biol. Sci.* **281**, 20133151.
104. Hammers, M., Kingma, S.A., Bebbington, K., van de Crommenacker, J., Spurgin, L.G., Richardson, D.S., Burke, T., Dugdale, H.L., Komdeur, J. 2015. Senescence in the wild: insights from a long-term study on Seychelles warblers. *Exp. Gerontol.* **71**, 69–79.
105. Boonekamp, J.J., Mulder, E., Verhulst, S. 2018. Canalisation in the wild: effects of developmental conditions on physiological traits are inversely linked to their association with fitness. *Ecol. Lett.* **21**, 857–864.
106. Adrian-Kalchhauser, I., Sultan, S.E., Shama, L.N.S., Spence-Jones, H., Tiso, S., Valsecchi, C.I.K., Weissing, F.J. 2020. Understanding 'non-genetic' inheritance: insights from molecular-evolutionary crosstalk. *Trends Ecol. Evol.* **35**, 1078–1089.
107. Bewick, A.J., Vogel, K.J., Moore, A.J., Schmitz, R.J. 2017. Evolution of DNA methylation across insects. *Mol. Biol. Evol.* **34**, 654–665.

108. Reik, W. 2007. Stability and flexibility of epigenetic gene regulation in mammalian development. *Nature* **447**, 425–432.
109. Simpson, D.J., Chandra, T. 2021. Epigenetic age prediction. *Aging Cell* **20**, e13452.
110. Horvath, S., Raj, K. 2018. DNA methylation-based biomarkers and the epigenetic clock theory of ageing. *Nat. Rev. Genet.* **19**, 371–384.
111. Shiels, P.G., Buchanan, S., Selman, C., Stenvinkel, P. 2019. Allostatic load and ageing: linking the microbiome and nutrition with age-related health. *Biochem. Soc. Trans.* **47**, 1165–1172 (doi:10.1042/bst20190110).
112. Horvath, S. 2013. DNA methylation age of human tissues and cell types. *Genome Biol.* **14**, 3156.
113. Lu, A.T. et al. 2019. DNA methylation GrimAge strongly predicts lifespan and healthspan. *Aging-Us* **11**, 303–327 (doi:10.18632/aging.101684).
114. Hu, J., Barrett, R.D.H. 2017. Epigenetics in natural animal populations. *J. Evol. Biol.* **30**, 1612–1632 (doi:10.1111/jeb.13130).
115. Hu, J., Askary, A.M., Thurman, T.J., Spiller, D.A., Palmer, T.M., Pringle, R.M., Barrett, R.D.H. 2019. The epigenetic signature of colonizing new environments in Anolis lizards. *Mol. Biol. Evol.* **36**, 2165–2170 (doi:10.1093/molbev/msz133).
116. Fargeot, L., Loot, G., Prunier, J.G., Rey, O., Veyssiere, C., Blanchet, S. 2021. Patterns of epigenetic diversity in two sympatric fish species: genetic vs. environmental determinants. *Genes* **12**, 107.
117. Wilkinson, G.S. et al. 2021. DNA methylation predicts age and provides insight into exceptional longevity of bats. *Nat. Commun.* **12**, 1–13.
118. Sheldon, E.L., Riccardo, T., Boner, W., Monghan, P., Raveh, S., Schrey, A.W., Griffith, S.C. 2022. Associations between DNA methylation and telomere length during early life: insight from wild zebra finches (*Taeniopygia guttata*). *Mol. Ecol.* **31**, 6261–6272 (doi:10.1111/mec.16187).
119. Brown, T.J., Hammers, M., Taylor, M., Dugdale, H.L., Komdeur, J., Richardson, D.S. 2020 Hematocrit, age, and survival in a wild vertebrate population. *Ecol. Evol.* **11**, 214–226.
120. Holmes, D.J., Harper, J.M. 2018. Birds as models for the biology of aging and aging-related disease: an update. *J Gerontol A Biol Sci Med Sci.* **50**, B59–66 (doi:10.1016/b978-0-12-811353-0.00022-1).
121. Pyrkov, T.V., Avchaciov, K., Tarkhov, A.E., Menshikov, L.I., Gudkov, A.V., Fedichev, P.O. 2021. Longitudinal analysis of blood markers reveals progressive loss of resilience and predicts human lifespan limit. *Nat. Commun.* **12**, 2765 (doi:10.1038/s41467-021-23014-1).
122. Harper, J.M., Holmes, D.J. 2021. New perspectives on Avian models for studies of basic aging processes. *Biomedicines* **9**, 649 (doi: 10.3390/biomedicines9060649).

9 Sex Differences in Lifespan, Ageing and Health in the Living World

What Can We Learn from Evolutionary Biology?

Jean-François Lemaître, Jean-Michel Gaillard, Dominique Pontier, Hugo Cayuela, Cristina Vieira and Gabriel A. Marais

9.1 The Conundrum of Sex Differences in Mortality and Health across Human Populations

In human populations, women survive better than men at each age, which ultimately translates into a longer lifespan [1]. In mainstream medias, this bias in survival prospects towards longer-lived women is often illustrated by the iconic case of Jeanne Calment who died at the age of 122 – the longest- lived human with a known birth date (1875) ever recorded (see Chapter 14). A closer look at country-specific demographic data highlights the robustness of this pattern across time and space [2, 3]. For instance, an analysis of life expectancy at birth across 54 countries revealed that life expectancy is higher in women than men in all countries, even if the magnitude of the sex gap is highly variable (i.e. from 3 years in both Pakistan and Nigeria to 14 years in Russia [4]). Moreover, historical demographic data have revealed that the female advantage in life expectancy is consistent across time periods, at least since the seventeenth century [2, 5], even during periods of strong environmental harshness (e.g. famine, epidemics) or when controlling for childbirth mortality (Chapter 10). These elements highlight the ubiquitous nature of women's survival advantage and its biological foundations, even if studies focusing on Catholic nuns, monks or the Amish have revealed that the magnitude of sex differences in survival can also be modulated by social factors and lifestyle [6]. Gaining a comprehensive understanding of these sex differences in mortality patterns constitutes a research topic related to many scientific fields, such as biomedical sciences, social sciences, demography and evolutionary biology [7, 8, 9, 10].

We are grateful to Victor Ronget, Morgane Tidière, Vérane Berger, Florentin Remot, Louise Cheynel, Solène Cambreling, Fernando Colchero, Alexander Scheuerlein, Andras Liker, Tamas Székély and Mike Garratt for stimulating discussions regarding the evolution of sex differences in ageing. We thank Samuel Pavard, Peter Lenárt and Tim Clutton-Brock for their insightful comments on an earlier version of this chapter. We thank the Agence National de la Recherche (ANR) for the grants AGEX (ANR-15-CE32-0002-01 to J.-F.L.) and LongevitY (ANR-20-CE02-0015 to CV), the Laboratoire d'Excellence (LabEx) ECOFECT 'Eco-Evolutionary Dynamics of Infectious Diseases' (ANR-11-LABX-0048) of Lyon University.

When comparing mortality patterns between sexes (and by extension the underpinning biological factors), it is crucial to bear in mind that sex differences in lifespan do not necessarily translate into sex differences in ageing parameters (i.e. rate of actuarial ageing and age at the onset of actuarial ageing) [11]. Comparative analyses have shown that variation in the ageing parameters and metrics of lifespan is largely uncoupled across mammalian species [12]. For a given population or species, the mean, median or maximal lifespan corresponds to the outcome of the full age-specific mortality trajectory going from birth to the oldest age reached by an individual. In particular, although it is often overlooked, the magnitude of juvenile mortality and of early adult mortality that is commonly designed as baseline mortality strongly influence metrics of lifespan, so that the combined effect of age at the onset of ageing and rate of ageing only accounts for about half the observed variation in lifespan ([11, 12, 13]. So far, the few analyses that have considered the full set of mortality metrics in a sex-specific ageing context suggest that, in humans, sex differences in lifespan are mostly determined by sex differences in early adult mortality, with no apparent differences in actuarial senescence rates [1] (see also [14] for a discussion on the 'invariant rate of ageing' hypothesis across primates). Yet a recent decomposition of mortality patterns across countries worldwide highlighted that sex differences in life expectancy largely result from the higher mortality risk above 60 years of age in males than females, at least since 1950 [15]. Overall, a clear understanding of how the onset and the rate of actuarial ageing vary or not in a sex-specific way is yet to be achieved to ultimately modulate sex differences in lifespan across human populations.

Sex differences in human lifespan are logically underpinned by a higher frequency of deaths in men than in women for most common diseases [9]. These differences appear particularly pronounced for cancers [16], late-onset diseases such as multiple metabolic diseases [17], and various infectious diseases, as recently exemplified by the SARS-CoV-2 epidemic [18, 19]. Surprisingly, and despite their lower mortality and disease risks, women are generally considered to be in poorer health (as assessed by self-rating and several indicators of the amount of medical attention needed) compared to men [2, 20]. Yet the origin of this so-called mortality–morbidity paradox (also sometimes named the 'health–survival' paradox) remains largely unknown. It has been proposed that these differences could arise from a stronger effect of viability selection (sensu [21]) on men than on women over the life course, leading to a set of older males who would be on average 'healthier' than the age-matching set of older women (reviewed in [22, 23]). In addition, this higher physical illness in elderly women compared to elderly men has been proposed to directly result from more frequent chronic pain associated with severe arthritis and other osteoskeletal conditions [22].

To get a comprehensive understanding of sex differences in health, disease risk and mortality patterns across human populations, from both the demographic and biological sides, an evolutionary look at the problem is required. Understanding the sex-specific selective pressures that have shaped the age-specific health and mortality trajectories of both men and women in various ecological contexts (e.g. epidemics, famine) can shed new light on the aetiology of diseases displaying sex-specific dynamics over the life course [9, 24], as well as on the future of the sex-specific differences in mortality patterns in a world facing unprecedented global changes [25]. Within this approach, the first step is to explore the sex differences in mortality patterns across the tree of life, to determine whether the sex-specific mortality pattern in humans is the rule rather than the exception, and is shared by closely related species such as primates or more generally, mammals.

9.2 Sex Differences in Mortality Patterns across the Tree of Life

The study of sex differences in lifespan and ageing has fascinated naturalists and ecologists for many years and a considerable number of empirical studies have sought to quantify sex differences in survival across species [26, 27, 28]. As it is beyond the scope of this chapter to review all these studies, we will mostly take advantage of the spate of comparative analyses on sex differences in mortality patterns to draw the current picture.

So far, most comparative analyses of sex differences in mortality have been performed on vertebrates, with a focus on mammals and birds (see [10, 29] for reviews). This is because most longitudinal monitoring of known-age individuals has been performed on these species [30]. These comparative analyses have used a very diverse range of mortality or survival metrics to assess sex differences in mortality. In mammals, comparisons of overall adult mortality [31], or of adult life expectancy [32], have revealed that, similarly to humans, males display, on average, higher adult mortality than females [31, 32], which ultimately translates to a longer lifespan in females, as observed in both wild and captive populations [33, 34] (e.g. [35, 36] in ruminant populations). Overall, mammalian females outlive males in the wild in 60% of species, and the magnitude of the female survival advantage is higher compared to that observed in human populations (e.g. adult median lifespan is 18–20% longer in mammals vs 2–7% in humans [34]). Strikingly, the picture is much less clear when we focus on ageing metrics per se. Differences in the rate of actuarial senescence are not consistent and are weak between sexes in mammals [34], as observed in human and non-human primates [1].

Outside mammals, the current picture is much less clear. Studies on birds generally indicate that females display higher annual adult mortality rates than males [37, 38], although the magnitude of sex differences remain, on average, quite small [38]. Unfortunately, and despite the huge amount of age-specific data on survival or mortality rates available in birds thanks to the long tradition of individual ringing [39, 40, 41], there is a clear lack of detailed comparative analysis focused on sex differences in age at the onset and rate of actuarial ageing. Therefore, while the idea that male birds live longer (and potentially age slower) than females is widespread in the literature (e.g. [29, 42, 43]), this still needs to be properly quantified. Data on other taxonomic groups are even scarcer, even if our knowledge on sex differences in mortality patterns among vertebrate ectotherms is currently increasing (e.g. [44, 45]). In amphibians, a detailed study of 36 wild populations revealed that there are no consistent

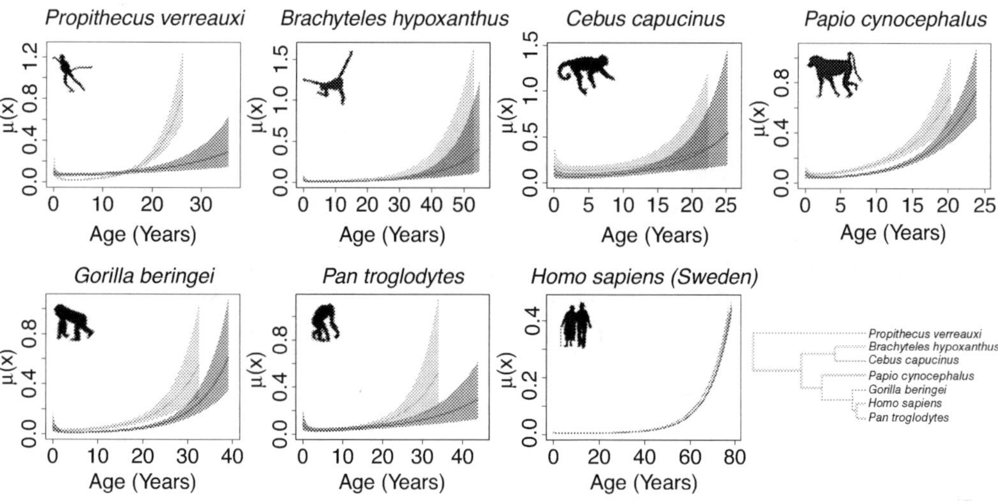

Figure 9.1 Patterns of age-specific changes in hazard mortality rate (i.e. instantaneous mortality rate) for both females (in dark grey) and males (in light grey) of seven wild populations of primates followed longitudinally: sifaka (*Propithecus verreauxi*), muriqui (*Brachyteles hypoxanthus*), capuchin (*Cebus capucinus*), baboon (*Papio cynocephalus*), chimpanzee (*Pan troglodytes*) and gorilla (*Gorilla beringei*). For each species and sex, we fitted a Siler model on age-specific mortality data. Models were performed with the R package Bayesian Survival Trajectory Analysis (BaSTA) [185]). The survival data for the non-human primate populations are extracted from [186] while the data for humans were extracted from the Human Mortality Database (HMD, www.mortality.org/). The phylogeny of the seven species is provided for illustrative purpose.

sex differences in either lifespan or actuarial senescence rates, which appear mainly driven by species differences in the type of sex chromosomes [46]. Besides vertebrates, our knowledge on sex differences in mortality patterns is even more limited. Indeed, although invertebrates constitute the vast majority of extant animal species, the few studies of sex differences in mortality patterns do not allow researchers to draw any firm conclusion about the direction and magnitude of sex differences in lifespan that might occur across invertebrates. A comparative analysis across 35 species of Odonates (i.e. damselflies and dragonflies) revealed that mortality is generally higher in males than females even if multiple exceptions are documented [47]. Moreover, it is important to note that one limitation of most studies performed in invertebrates so far is the use of 'days from marking' as a proxy of chronological age, which can lead to biased estimations of ageing parameters [47]. Sex differences in actuarial senescence patterns of invertebrates are far from consistent (e.g. actuarial senescence documented in male but not in female neriid flies, *Telostylinus angusticollis* [48]; rate of actuarial senescence steeper in females than males in black field crickets, *Teleogryllus commodus* [49]; direction and magnitude of sex differences in actuarial senescence differing among local environmental conditions in the European field crickets, *Gryllus campestris* [50]) (Figure 9.1).

9.3 The Evolutionary Roots of Sex-Specific Mortality

There are three main sets of (non-mutual) hypotheses that have been proposed to explain the evolution of sex differences in mortality patterns [8, 29, 51]. These theories respectively rely on the role played by sex chromosomes, mitochondrial DNA (mtDNA) inheritance and sex differences in life history strategies. At this stage it is worth noticing that, in their original formulation, most of these hypotheses do not make any explicit prediction regarding the direction and the magnitude of the sex differences for each metric of mortality (e.g. average mortality rate, metrics of lifespan or age at the onset and rate of actuarial ageing), even if there are absolutely no biological [29], or theoretical [12], expectations that all the metrics should equally differ between males and females. Some refinements in the theories are thus mandatory to better link the genetic and physiological pathways involved with possible demographic outcomes in terms of sex differences in lifespan, age at the onset of actuarial ageing and/or rate of actuarial ageing.

9.3.1 Sex Chromosome

It has recently been suggested that sex chromosomes influence sex differences in lifespan. Adult sex ratios, which are commonly used as a very rough proxy for sex differences in adult lifespan, have been shown to correlate with sex chromosome types (XY or ZW) in tetrapods (i.e. mammals, birds, reptiles and amphibians). In XY systems, XY males tend to die first, whereas in ZW systems, ZW females tend to die first [52] (see also [53] for similar observations in plants). This was confirmed by using sex-specific data on lifespan and by extending the analysis to fish and insects [54]. More recently, using data on 36 species of amphibians displaying contrasted sex chromosome systems (XY vs ZW species), Cayuela and colleagues found that the heterogametic sex consistently shows a higher ageing rate than the homogametic sex [46]. Two main mechanisms have been proposed to explain these patterns: the unguarded X and the toxic Y (reviewed in [29, 51]).

The unguarded X hypothesis is based on the idea that in the heterogametic sex, with a single X chromosome, recessive deleterious mutations will all be expressed and will have greater effect on lifespan than in the homogametic sex [27]. A similar Z unguarded effect is expected in ZW systems [55]. The toxic Y hypothesis has been proposed more recently [56]. The Y chromosome is generally a chromosome rich in heterochromatin and repeated sequences such as transposable elements (TEs). These TEs, which have the potential to be mobilized in genomes and are potential sources of mutation, are generally silenced by epigenetic mechanisms such as changes in chromatin structure or DNA methylation [57]. When ageing, the epigenetic control of the genome gets weaker, and we can expect a loosening of the control of TE, which will lead to an increase in somatic mutation rates and thus an acceleration of biological ageing, leading to a shorter lifespan and/or accelerated actuarial ageing [29]. Since the heterogametic sex is richer in TE, we expect this sex to have a shorter lifespan. A toxic W is equally possible in ZW systems, and could explain reports of short-lived females compared to long-lived males in such species.

The respective contribution of unguarded X and toxic Y (or equivalent in ZW systems) in determining sex differences in lifespan is currently unknown, although a recent model by Connallon and colleagues suggests that the deleterious effects of the unguarded X on males are too small to explain the observed sex differences in lifespan [58]. A meta-analysis on the sex chromosome size and lifespan in animals indicated that the size of the Y chromosome, but not the size of the X chromosome, correlates best with male survival, which supports the toxic Y [59]. Experimental work has also been done in Drosophila species where increasing the homozygosity of the X chromosome in females reduces female lifespan and equalizes male and female lifespan in line with the prediction of the unguarded X hypothesis [60]. However, these experiments have returned conflicting results, and the latest data does not support the unguarded X [58, 61].

In Drosophila, the available evidence thus supports a strong toxic Y effect. TE activity is stronger in old males than in old females; and manipulating sex chromosome content has clearly shown that the Y chromosome drives reduced lifespan [56, 62, 63]. Whether the toxic Y can explain sex differences in lifespan observed among populations in the wild has now to be investigated. To do that, other avenues of research appear particularly promising, including: (1) testing for the effect of the unguarded X and the toxic Y in animals; (2) exploring the association of sex chromosomes and sex differences in lifespan in plants; and (3) deciphering the molecular mechanisms associated with the unguarded X and the toxic Y hypotheses in a larger range of species.

9.3.2 Mitochondrial DNA Inheritance

In metazoans, the mitochondrial genome has a relatively small size (i.e. a 15–20 kilobase sequence) and only encompasses 37 genes coding for 13 proteins [64]. Despite these limited gene content and products, mtDNA polymorphism may cause important variation in metabolism, health, biological and demographic ageing in both invertebrates and vertebrates [65, 66, 67]. In addition, pathogenic mutations causing oxidative phosphorylation and neurodegenerative disorders have been identified in 30 out of the 37 mtDNA genes, depleting health and lifespan across human populations [68, 69].

Since the mid-2000s, an increasing number of studies have suggested that the accumulation of deleterious mutations in the mitochondrial genome, as well cytonuclear incompatibilities (i.e. detrimental interactions between the nuclear and the mitochondrial genomes), might cause sex-specific variation in lifespan and ageing parameters (reviewed in [70, 71]). Using a population genetic model, Frank and Hurst formulated an evolutionary hypothesis [72] – later coined 'mother's curse' hypothesis by Gemmel et al. [70]. This hypothesis states that the mitochondrial genome contains a high number of mutations with detrimental effect on male health and fitness. Indeed, the mtDNA haplotype of metazoans follows a strict mode of maternal transmission, leading natural selection to be effective at shaping evolutionary changes within the mitochondrial genome through females only. Therefore, mutations that impair male performance but are benign in females are not targeted by purifying selection, resulting in their accumulation within populations. In addition, if mutations have sexually antagonistic effects, positive selection favours alleles that confer a selective advantage on females at the expense of male fitness [73].

Current knowledge about the 'mother's curse' effect on lifespan and ageing metrics is still fragmentary – compared to our understanding of fertility-related mechanisms (reviewed in [64, 71]). In the fruit fly *D. melanogaster*, genetic variance across mtDNA haplotypes, resulting from the accumulation of multiple mutations within the mitochondrial genome, is associated with lifespan and ageing rate variation in males but not females [65]. In addition, the shorter lifespan of males might, to some extent, also be due to detrimental interactions between the nuclear and the mitochondrial genomes [74]. Furthermore, the decrease in male lifespan is tightly linked to bioenergetic mechanisms. The effect of mtDNA haplotypes on the metabolic rate is stronger in males than in females, and there is a male-specific negative correlation between metabolic rate and lifespan in *D. melanogaster* [66]. In humans, the 'mother's curse' effect on health and mortality patterns has been investigated through the study of mtDNA mutations causing Leber hereditary optic neuropathy, a disease with a pronounced male-biased prevalence [75]. A study of the Québec population between 1670 and 1750 found that the T14484C mitochondrial variant causing Leber hereditary optic neuropathy affected the survival of males but not that of females. During the first year of life, the survival of males carrying this mtDNA variant was 20.8% lower than in non-carriers [67]. Yet how much the mother's curse contributes to the sex differences later on to influence lifespan and ageing parameters across species under various environmental conditions is yet to be explored.

9.3.3 Sex Differences in Life History Strategies

In the context of life history evolution, sex differences in mortality patterns constitute a by-product of the evolution of sex-specific life history strategies [76, 77, 78]. More specifically, sex-specific mortality patterns are determined by species-specific intensity of sexual selection that involves competition for gaining mating and/or fertilization opportunities [38, 79], and by the amount of parental care [80, 81]. So far, the role of sexual selection on sex-specific mortality patterns has been mostly studied from the viewpoint of inter-male competition under the hypothesis that sex differences in lifespan and/or ageing parameters have evolved according to the variable intensity of inter-male competition across species [32, 76, 77, 78, 80, 82]. This prediction has generated a spate of comparative analyses that have led to inconsistent findings (reviewed by [83]).

For instance, Clutton-Brock and Isvaran found support for the expectation that sex differences in actuarial ageing rates are larger in polygynous than monogamous species of mammals and birds [79, 80], whereas Lemaître and colleagues did not detect any sex differences in actuarial ageing rates in relation to species-specific mating systems in mammals [34]. As previously pointed out by Tidière and colleagues [83], the wide diversity of traits and metrics used to quantify lifespan and ageing (and also male allocation to sexual competition) likely explains contrasted outcome in different studies. So far, whether the intensity of sexual selection (as measured by mating system and/or sexual size dimorphism) influences sex-specific mortality patterns in wild populations of mammals remains equivocal.

Conversely, the association between the intensity of inter-male competition and sex differences in mortality patterns is much more easily detected in protected environments such as captive populations [35, 36, 83]. In captive environments, the effects of the environmental conditions are strongly buffered and sex differences in mortality patterns are more likely to directly reflect the energetical costs of physiological adaptations to sexual competition (e.g. by-product of the elevated testosterone production associated with the growth and maintenance of secondary sexual traits [78]). This could, notably, explain why males display a shorter lifespan and an earlier age at the onset of actuarial senescence than females in polygynous populations of large herbivores living in zoological gardens [83]. Here, we propose that the effects of inter-male competition are more likely to be detected at both extremes of an environmental continuum (e.g. zoological gardens where resources and veterinary care are provided ad lib and wild populations vulnerable to many pathogens and with low and unpredictable access to resources). Indeed, the mortality cost of inter-male competition could also be exacerbated by increased environmental harshness, when the physiological costs associated with the growth and maintenance of sexual traits markedly escalate.

Deciphering the role of environmental conditions in modulating the relationship between inter-male competition and sex differences in mortality patterns thus constitutes an important challenge. This will require researchers to, first, quantify environmental harshness within a population of interest, which is a far from an easy task because several components, including the pathogenic environment, population density, availability of resources and the climatic conditions can all modulate male allocation to intra-sex competition and its consequences in terms of sex-specific mortality (e.g. [31, 84]). Second, while most studies performed so far have focused on the costs associated with pre-copulatory competition (e.g. growth and maintenance of secondary sexual traits), future studies will have to add the possible costs associated with male allocation to post-copulatory competition (i.e. production of high-quality ejaculates) that are likely far from negligible. Indeed, while there is clear evidence that increased spermatogenesis alter both male's physiological ageing pathways and health (see [85] for a review), the evolutionary consequences for sex differences in mortality remain to be explored. Yet embracing this path could be particularly insightful because intense selection on male's post-copulatory traits is known to set the ground for sexual conflicts, which leads to strong (and sometimes antagonistic) sex-specific impacts on mortality (see [77, 86] for thorough reviews on the complex relationships linking sexual conflict and sex-specific mortality). Third, future studies will also have to consider factors that can modulate female mortality and, through cascading effects, sex differences in mortality patterns. This could include the magnitude of parental care (see Chapter 10), as well as the diversity of female reproductive mode (e.g. oviparity vs viviparity, see [87]), social systems (e.g. level of social competition among females [88]) and reproductive physiology (e.g. hormonal profile, milk nutritional composition).

9.4 Going Beyond Mortality Patterns

9.4.1 Sex Differences in Reproductive Ageing

From an evolutionary biology viewpoint, the process of ageing is traditionally defined as the decline in survival and/or reproductive probabilities with increasing age [90, 91, 92], which originates from the decline in the age-specific contribution to fitness with increasing age throughout the life course ([93, 94], see also Chapter 5). Therefore, a thorough understanding of the evolution of sex differences in ageing needs to embrace the two main components of fitness: fecundity and survival. Yet, as emphasized in the Section 9.3, most studies performed so far have focused on the survival component of fitness and have largely neglected the reproduction component.

While there is now widespread evidence of reproductive ageing (also called 'reproductive senescence') in vertebrate females [95, 96, 97, 98, 99], data on male reproductive success in the wild are much more scarce, since molecular tools are required for paternity assignments [100]. However, recent empirical studies that have started to tackle the topic of sex differences in reproductive ageing have already emphasized the diversity of observed patterns, as illustrated by some recent case studies in birds. Thus, in a monogamous raptor with biparental care, the white-tailed eagle, *Haliaeetus albicilla*, males display more acute reproductive senescence than females [101]. The senescence in breeding probability was also stronger in males than females in another monogamous bird with biparental care, the Nazca boobies (*Sula granti*), but senescence in fledging success was stronger in females than males [102], which highlights the importance of considering a wide range of reproductive traits when studying reproductive ageing [100]. Moreover, it is important to tease apart the effect of both male and female ages when studying the decline in a given reproductive trait [100]. For instance, when comparing the decline in breeding success in three albatross species, Froy and colleagues found that the age of males had a higher impact on the rate of reproductive ageing than the age of females in black-browed albatross (*Thalassarche melanophris*) [103], whereas the reverse occurred in both the wandering albatross (*Diomedea exulans*) and grey-headed albatross (*Thalassarche chrysostoma*).

Thanks to a current increase in the number of studies investigating sex-specific reproductive ageing in a given species (e.g. [104, 105]), it should soon be possible to perform comparative analyses across species to test whether the evolutionary theories of ageing initially proposed to explain the evolution of sex differences in mortality patterns could explain the direction and the magnitude of sex differences in reproductive ageing observed in the wild. For instance, as expected under the 'mother's curse' hypothesis [70], sex-specific deleterious mutations that accumulated on the mtDNA have been reported to be detrimental for men's fertility and associated with severe sperm dysfunction (see [64] for a compilation of studies). Whether such mtDNA mutations leading to impaired male fertility are common across species is largely unknown (but see [106] on the brown hare, *Lepus europaeus*) due to too few studies outside laboratory conditions [64]. Yet, everything else being equal, the bias towards a more pronounced reproductive ageing in males compared to females could be higher in species that have accumulated more mutations on their mtDNA. While this example focuses on the putative role played by the 'Mother's Curse' hypothesis, it is important to note that both sex chromosomes and sex-specific life history strategies could also modulate the direction and the magnitude of sex differences in reproductive ageing patterns [29, 100].

9.4.2 Sex Differences in Body Mass Senescence

In a wide range of species, body mass (or body size) is positively correlated with both survival and reproductive success [107, 108]. Therefore, the decline in body mass with increasing age (i.e. body mass ageing) might constitute an underlying factor of the decrease in reproductive and survival probabilities in late life, as well as a fair indicator of the decline in demographic performance of an individual. This is observed in humans where more pronounced sarcopenia (i.e. the loss of skeletal muscle mass with increasing age [109]) is associated with an increased risk of mortality and multiple adverse health issues (e.g. [110, 111]), but also in grey mouse lemurs (*Microcebus murinus*) where sex differences in seasonal mortality match sex differences in seasonal body mass loss [112]. As a consequence, any sex difference in body mass ageing could ultimately inform possible subsequent differences in actuarial and reproductive ageing patterns between sexes. Some studies performed on populations in the wild (mostly mammals) have investigated sex differences in body mass ageing. So far, these studies have revealed rather contrasted patterns (Table 9.1). However, in several mammalian species, body mass ageing is earlier or steeper in males compared to females (reindeer, *Rangifer tarandus* [113]; Alpine chamois, *Rupicapra rupicapra* [114]; badger, *Meles meles* [115]; Soay sheep, *Ovis aries* [104]; roe deer, *Capreolus capreolus* [116]; see Table 9.1). These sex differences are even more pronounced in Alpine marmots (*Marmota marmota*) for which the decline in body mass in late life is observed in males only [117]. While detrimental effects associated with the Y chromosome and/or unguarded X could potentially contribute to the more pronounced functional decline observed in males (see Section 9.3.2) compared to females, this pattern is generally interpreted as a long-term cost of sexual competition [104, 117]. Body mass trajectories of badgers support this hypothesis, where males facing a high level of intra-sex competition during early life displayed more pronounced body mass ageing than males facing a low intra-sex competition, which can ultimately exacerbate sex differences in body mass ageing [115]. Moreover, environmental conditions likely modulate sex-specific patterns of body mass ageing by amplifying and/or attenuating the decline in body mass with increasing age in males and females independently [114, 116]. Similar to what has been suggested in the context of actuarial ageing [34], sex differences in body mass ageing are likely mediated by complex interactions between environmental conditions and life history strategies acting in sex-specific ways.

While a more pronounced ageing in male than female body mass seems to be relatively common in mammals (Table 9.1), the extent to which sex differences in body mass loss contributes to sex differences in health (e.g. parasite prevalence), lifespan

Table 9.1 Case studies analysing sex differences in body mass ageing across populations of mammals in the wild

Species	Population	Data	Female onset of body mass senescence (years)	Male onset of body mass senescence (years)	Conclusion	References
Alpine marmots (*Marmota marmota*)	Wild	Longitudinal	Ageing not detected	8.00	Ageing detected in males but not in females	[117]
Red squirrel (*Sciurus vulgaris*)	Wild	Longitudinal	Ageing not detected	Ageing not detected	Absence of ageing in both sexes	[187]
Eurasian beaver (*Castor fiber*)	Wild	Longitudinal	Ageing not detected	Ageing not detected	No differences in age-specific body mass patterns in males and females	[118]
Grey mouse lemur (*Microcebus murinus*)	Wild	Longitudinal	Ageing not detected	Ageing not detected	Absence of ageing in both sexes	[188]
Grey mouse lemur (*Microcebus murinus*)	Captive	Longitudinal	4–5	4–5	No differences in ageing rate between males and females	[188]
Badger (*Meles meles*)	Wild	Longitudinal	4–5	4–5	Faster ageing in males than in females[1]	[115]
Meerkat (*Suricata suricatta*)	Wild	Longitudinal	5.56 [5.10–6.13]	5.56 [5.10–6.13]	No sex differences in ageing patterns	[105]
Soay sheep (*Ovis aries*)	Wild	Longitudinal	10.00	6.00	Earlier ageing in males than in females	[104]
Alpine chamois (*Rupicapra rupicapra*)	Wild (Adamello population)	Transversal	Ageing not detected[2]	ca. 8	More pronounced ageing in males	[114]

Alpine chamois (*Rupicapra rupicapra*)	Wild (Presanella population)	Transversal	Ageing not detected	ca. 10	More pronounced ageing in males	[114]
Alpine chamois (*Rupicapra rupicapra*)	Wild (Brenta population)	Transversal	Ageing not detected	Ageing not detected	More pronounced ageing in males	[114]
Roe deer (*Capreolus capreolus*)	Wild (Chizé population)	Longitudinal	12.6 [10.3–14.7]	8.2 [7.6–9.4]	Earlier ageing in males than in females / absence of sex differences in ageing rate	[116]
Roe deer (*Capreolus capreolus*)	Wild (Trois-Fontaines population)	Longitudinal	8 [7–9.8]	6.6 [<6–8.9]	No statistically significant difference in the onset of ageing between males and females / absence of sex differences in ageing rate	[116]
Reindeer (*Rangifer tarandus*)	Wild	Longitudinal	9.00	7.00	Earlier ageing in males than in females[3]	[113]
Asian elephant (*Elephas maximus*)	Semi-captive	Longitudinal	Ageing not detected	Ageing not detected	Absence of ageing in both sexes	[119]

[1] The level of intra-sexual competition during early life increases male body mass ageing and can thus exacerbate the sex differences in body mass ageing.

[2] There is a trend from a decline in body mass with increasing age in females (starting around 4 years of age).

[3] The sex differences in the rate of body mass ageing are not tested.

and actuarial ageing parameters is still unknown. Age-specific body mass trajectories throughout the life course are similar between sexes in beaver, *Castor canadensis* [118], and Asian elephant, *Elephas maximus* [119]; both species where females outlive males [34]. In that context, further studies performed in the wild should now tackle the topic of sex differences in body mass ageing on a finer scale by investigating more directly whether and how body composition changes between sexes over the life course. Thorough investigation of the age-specific decline in muscle mass or efficiency directly involved in competitive abilities among conspecifics and locomotor performance in most animal species is likely to be particularly promising to understand better the link between biological and demographic ageing. Such approaches, based on biopsies, have already been developed in non-model organisms (e.g. measures of myocyte density and collagen content in Weddell seal, *Leptonychotes weddellii* [120], and in both water shrew, *Sorex palustris*, and short-tailed shrew, *Blarina brevicauda* [121]), but are yet to be embraced in the context of male versus female comparison.

9.4.3 The Physiological and Genetic Basis of Sex Differences in Ageing

Multiple mechanisms, such as various physiological pathways, are potentially involved in observed sex differences in lifespan and ageing trajectories in the living world (see Section 9.2). As nicely synthetized by Hägg and Jylhävä [122], the biomedical research performed in this area has repeatedly found that women outperform men in diverse biological functions, which are known to be tightly associated with both health and survival. For instance, for a given chronological age, women show a lower epigenetic age than men [123], which indicates a lower rate of biological ageing [122]. Better performance of women over men has also been documented for other traits and functions such as immunity, telomere length and microbiota composition [17, 122]. Many of the various pathways modulating sex-specific ageing trajectories in humans have also been identified in laboratory organisms [9]. Whether they also contribute to sex differences in lifespan and ageing across wild populations of animals is yet to be established. So far, studies seeking to investigate sex differences in lifespan and ageing parameters in animal populations in the wild from an ecophysiology perspective have mostly focused on immunity, oxidative stress and telomere length. As this latter topic is gaining increasing popularity in ecology [124], we will use it for illustrative purposes.

Telomeres are non-coding repetitive DNA sequences typically located at the extremity of eukaryotic chromosomes involved in the maintenance of the genomic integrity [125]. Basically, telomeres protect coding DNA from both oxidative damages and the incomplete replication of DNA ends [126]. Shorter telomeres have been found to be associated with the occurrence of some diseases (e.g. cardiovascular diseases [127]) and suggested to be associated with lower survival prospects (e.g. [128] in humans; [129] in animal populations) although these effects are far from consistent across studies (e.g. [123] for a counter-example). Therefore, sex-specific patterns of telomere attrition could potentially shape sex-specific patterns of lifespan and ageing parameters [130]. Telomere measurements performed across human populations are in line with this prediction, as telomere length is shorter in men than in women [131]. From a mechanistic viewpoint, if sex differences in telomere length contribute to the evolution of sex differences in lifespan and ageing parameters, we could expect that – everything else being equal – the telomere attrition rate should be consistently faster in males compared to females in mammals, whereas in birds, the telomere attrition rate should be consistently faster in females. From an evolutionary viewpoint, these expectations are also particularly relevant as they can be embedded in both the 'heterogametic sex' and 'life history' frameworks proposed to explain the evolution of ageing (see Section 9.3). For instance, it has been suggested that mammalian males could suffer from both a higher rate of telomere attrition and shorter telomeres because some key telomere maintenance genes (e.g. dyskerin pseudouridine synthase 1, DKC1) are located on the X but not the Y chromosome, and also because the increased level of oxidative stress following the allocation to intra-sex competition might fasten telomere shortening [130, 132]. However, a recent analysis of sex- and age-specific data on telomere length across 52 species of vertebrates did not detect any difference between the sexes in telomere length and thus did not support the expectations [133]. Moreover, this study failed to detect any association between sex differences in telomere length and sex differences in lifespan in vertebrates [133], which is in fact not very surprising. Indeed, the physiology of organisms has been fine-tuned by environmental conditions and the associated risk and cause of death inherent to these conditions [134]. Therefore, the relative importance of the set of mechanisms underlying lifespan is likely to differ according to species-specific habitat, lifestyle or life history features [135, 136]. To further complicate the picture, we can also note that, for a given species, the primary mechanisms underlying lifespan might differ from the primary mechanisms underlying the onset or the rate of actuarial ageing [137], and even the onset or the rate of reproductive ageing. Therefore, even if the ageing mechanisms can be largely conserved across animals [138], the physiological mechanisms underlying sex differences in mortality patterns should be partly determined by the selective pressures acting differently between sexes in a given environment and are thus likely to differ among species.

Identifying the physiological roots of sex differences in lifespan and demographic ageing across species is thus an immense challenge that will be overcome only by bridging conceptual and methodological advances from various scientific fields, especially evolutionary ecology, epidemiology, demography and biogerontology.

9.5 Understanding Sex Differences in Health and Diseases in Light of Evolutionary Biology

The previous sections emphasized that we are only scratching the surface in the study of the evolution of sex differences in ageing across the living world. They highlighted how evolutionary theories constitute a powerful framework to understand the evolution of sex- and age-specific trajectories in terms of survival, fertility and phenotypic traits (e.g. body mass) along the life course, as well as on the genetic and physiological pathways modulating ageing. In this section, we will build on these elements to highlight how evolutionary theories can be particularly relevant in understanding the evolution of sex-specific disorders and health at late ages, with a specific focus on infectious diseases.

9.5.1 Sex-Specific Differences in Infectious Disease Risk in Humans and Other Animals: An Overview

Epidemiological data show that the frequency and outcome of many infectious diseases vary between sexes in both humans and mammalian species [139] (see review in [140, 141, 142], but see [143]). With some exceptions, men are more susceptible to diverse infectious diseases than women – including measles, hepatitis C, hantavirus, human immunodeficiency virus and influenza [144] – and this is consistently the case throughout the lifespan, starting during infancy (see reviews in [145, 146]). The SARS-CoV-2 infection has recently exemplified this sex bias, with a mortality rate for COVID-19 almost double in men compared to women [147]. Notable exceptions to this general rule include pertussis, for which women display higher incidence rates than men in most age classes [148]. The exact mechanisms responsible for sex differences in the incidence and mortality of many infectious diseases are not well understood. There is a growing body of evidence that the immune system responds differently between sexes to pathogens, in most cases to the benefit of females [145]. Overall, females tend to have a more responsive and robust immune system than males. Sex dimorphism in the immune response to infection occurs at several levels, including both the innate and the adaptive arms of the immune system, and both humoral and cellular responses. Contributing physiological reasons for sex bias in the immune response include differences in anatomy, sex hormone levels and X chromosomes. It is also likely that the microbiome contributes to variation in disease phenotypes through its potential vast addition of gene products [146]. For instance, it has been shown that the microbiome can participate in the regulation of reproductive hormonal functions. Reciprocally, the hormonal status of the host can influence microbiome composition, thus contributing to differences in the immune response between females and males [149]. However, the mechanisms of synergy and the pathways involved in the interaction between gut microbial communities, sex steroid hormone levels and the immune response still require further study, especially through the lens of evolutionary biology.

9.5.2 From Evolutionary Theory to Sex Differences in Mortality by Infectious Diseases

Sex steroid hormones (oestrogens, androgens and differential sex hormone receptor-mediated events), beside their multiple roles in reproductive behaviour (e.g. courting, territoriality, aggression, competition) are able to influence the immune system by modulating the activity of immune cells both quantitatively and qualitatively in response to infections [146, 150, 151]. Generally, oestrogens (in higher concentration in females) have an immunoenhancing effect on the immune system, while both progesterone (higher in females) and testosterone (higher in males) have immunosuppressive effects (e.g. [152] for a review). In line with these effects of steroid hormones, the male bias becomes apparent only after sexual maturity, for example, for leishmaniasis or tuberculosis [153, 154], while female vulnerability to disease and mortality rates is altered during their reproductive years [155]. However, this is a simplistic view of the role of hormones on immune system functioning [156, 157]. Protection against pathogens often comes at the cost of collateral damage associated with a powerful immune response, and sex differences in disease outcome are also observed for autoimmune diseases and cancers. While males have a higher incidence of non-reproductive cancer, females are more likely to suffer from immune-mediated diseases such as autoimmune disorders and inflammatory diseases [145, 158]. However, the consequences in terms of sex differences in lifespan are not clear, at least for humans ([159], but see [160]).

The X chromosome is also likely to influence the immune response and sex bias in disease susceptibility. In females one of the two X chromosomes is randomly inactivated in each cell – thus about half of the cells express genes from the maternal X chromosome and half from the paternal X chromosomes –, leading females to have a noticeable physiological immune diversity relative to males, as both gene products from paternal and maternal X chromosomes will be globally available. Additionally, the X chromosome is known to contain the largest number of immune-related genes of the whole human genome, and many X-linked genes are involved in both innate and adaptive immunity [161, 162]. These differences could potentially amplify the ability of females to defend against a diversity of pathogens (e.g. fungi, parasites, viruses, bacteria).

The consequences of all these discrepancies between males and females in terms of sex-specific survival and both actuarial and reproductive ageing are difficult to determine. A better understanding of why males and females respond differently to infectious diseases may give more clues for fathoming out the general regulatory mechanisms of the immune system, and how these could drive earlier or faster actuarial or reproductive ageing between sexes according to ecological conditions (e.g. different exposition levels, variation in resource availability). One of the challenges will be to monitor the dynamic aspects of immune function with age and sex by taking into account the exposition history of individuals to infectious diseases. Promising avenues are opening thanks to recent advances in genomics, to the availability of markers of adaptive and innate functions and to the availability of high-quality datasets on humans and mammals from different populations under variable ecological conditions.

The principle of allocation [163], positing that organisms are under selective pressure to acquire the limited resources available in the environment and have to allocate the obtained energy among competing biological functions – including selection against pathogens, competition for mates and production (and sometimes rearing) of offspring – throughout their life cycle to maximize fitness [164], constitutes a relevant theoretical framework to investigate the evolution of sex differences in health and diseases [165]. Because the host response against infectious diseases is an expensive key related-fitness trait, the extra allocation of resources needed to increase sexual competitiveness inevitably reduces allocation to immune function. Accordingly, trade-offs between reproduction and immune defence have been reported for different species and can manifest at both the physiological (within an individual) and evolutionary (individuals carrying different genotypes within a population) levels [166, 167]. Such a physiological trade-off between reproduction and immunity can explain the much earlier age at menopause of Bangladeshi women living in London, who have had multiple exposures to infectious diseases over their life course, compared to other women living in London [168]. Genetic trade-offs that play out across age and disease susceptibility later in life may arise from trade-offs between reproductive competitiveness increasing with earlier reproduction and higher fertility and defence against diseases. This phenomenon has been termed antagonistic pleiotropy [82], with the pleiotropic functions of the same gene acting antagonistically between early and late life. For instance, the cancer suppressor genes – such as p53 family genes and BRCA1/2 (breast cancer gene 1/2) [172] – which markedly influence female fertility, might be maintained by the trade-off between fertility enhancement early in the adulthood and a risk of cancer later in life [171]. Similar antagonistic pleiotropic effects have been identified in several genes that confer immune protection against pathogens but increase the risk of developing neurological disease later in life (review in [173]).

9.5.3 The Ecology of Host–Pathogen Interactions and Sex Differences in Ageing

Evolutionary trajectories of the parasites themselves can also be important, regardless of whether or not sexes differ in immunocompetence. Males and females represent very different environments (i.e. differences in morphology, behaviour, reproductive tactics, hormonal milieu, immune capacity or genetic architecture) for parasites, and this could oblige the latter to adapt some of their traits differentially between males and females [174, 175]. Adaptation of parasites to male or female hosts may help to explain why sex differences occur in parasite prevalence and disease expression [174, 175]. More specifically, parasites may display sex-differential development and virulence in their hosts because of sex-specific transmission dynamics. Úbeda and Jansen proposed an evolutionary epidemiology approach to understand the progression from human T-cell lymphotropic virus type 1 (HTLV-1) infection to lethal adult T-cell leukaemia [176], which is more likely in Japanese men than women, in contrast to the Caribbean population where the same virus is equally virulent in both sexes. These authors showed that differences in the transmission routes between sexes can lead to a departure in virulence between the sexes. This may explain why in Japan, where the transmission of HTLV-1 occurs through breastfeeding (thus vertically from mothers to offspring) rather than through sexual transmission, natural selection of HTLV-1 favours a slower progression to adult T-cell leukaemia in women than men, and explain the absence of host between-sex differences in virulence of the virus in the Caribbean, where vertical transmission is much less important. This could be interpreted as a sex-specific adaptation of HTLV-1 virulence to maintain women as a route of virus transmission, the reproductive success of hosts and pathogens being tightly interlinked where parasite transmission is vertical [176]. Thus, both within- and between-host processes have to be deciphered to understand how they ultimately translate into the evolutionary trajectory of a pathogen. Finally, a recent investigation suggested that pathogens can retain epigenetic memories of the host sex from which the infection originated across multiple generations, and select for a greater virulence of pathogens inherited from one sex as opposed to the other according to population demographic characteristics (e.g. host sex most commonly encountered) [177]. This may explain the intriguing complex pattern of virulence identified in diseases like measles and polio [178, 179], where infections that are contracted from same-sex individuals are less virulent compared to those contracted by the opposite sex. More research is required to study how sex differences affect the evolution of parasites and the diseases they develop in the host. Moreover, if parasites adapt differentially between sexes, then this is a further argument that both sexes need to be included equally in clinical trials.

Much remains unknown concerning how the evolution of sex-specific life history trajectories interplay with evolutionary trade-offs, threats associated to the presence of numerous infectious diseases and the diversity of immune response within hosts to shape sex differences observed in ageing (see [180]). These links are difficult to infer, partly because interactions with pathogens represent only a small fraction of the activities of the immune system. Recent discoveries suggest that the immune system is likely to play a central role in the physiology and homeostasis regulation of the organism [181], and plays various roles across the adult lifespan, extending far beyond defending the host against pathogens, supporting host–microbiota interactions [1], and biological processes or systems, such as metabolism and the nervous system [183, 184]. The next challenge will be to understand the evolutionary connections among all these systems in males and females as well as their connection to biological and demographic ageing.

9.6 Conclusion

This chapter highlights that, despite the huge societal and biomedical implications associated with sex differences in both health- and age-specific trajectories, our knowledge remains fragmentary. We argue that expanding our knowledge will be possible only by widening the taxonomic range of species studied and by embracing an integrative approach at the crossroad of many scientific fields, such as evolutionary biology, biogerontology and biodemography. More specifically, future research programmes should be conducted to decipher how the strength of natural and sexual selection acting on both males and females in a given environment have shaped, either directly or through antagonistic pleiotropic effects, the biological pathways modulating ageing. As illustrated with the example of infectious diseases, this would undeniably shed a new light on the origin of the sex-specific causes of death, as well as on the age-specific changes in survival and reproductive performance throughout the life course.

References

1. Austad, S.N. 2011. Sex differences in longevity and aging. In *The Handbook of the Biology of Aging* (eds E.J. Masoro and S.N. Austad), pp. 479–496. Academic Press.
2. Austad, S.N. 2006. Why women live longer than men: sex differences in longevity. *Gend. Med.* **3**, 79–92.
3. Zarulli, V., Jones, J.A.B., Oksuzyan, A., Lindahl-Jacobsen, R., Christensen, K., Vaupel, J.W. 2018. Women live longer than men even during severe famines and epidemics. *Proc. Natl. Acad. Sci.* **115**, E832–E840.
4. Rochelle, T.L., Yeung, D.K., Bond, M.H., Li, L.M.W. 2015. Predictors of the gender gap in life expectancy across 54 nations. *Psychol. Health Med.* **20**, 129–138.
5. Bolund, E., Lummaa, V., Smith, K., Hanson, H., Maklakov, A. 2016. Reduced costs of reproduction in females mediate a shift from a male-biased to a female-biased lifespan in humans. *Sci. Rep.* **6**, 24672 (doi:10.1038/srep24672).
6. Luy, M., Wegner-Siegmundt, C. 2015. The impact of smoking on gender differences in life expectancy: more heterogeneous than often stated. *Eur. J. Public Health* **25**, 706–710 (doi:10.1093/eurpub/cku211).
7. Meyer, M.H., Parker, W.M. 2011. Gender, aging, and social policy. In *Handbook of Aging and the Social Sciences* (7th ed.) (ed. L. George), pp. 323–335. Elsevier.
8. Regan, J.C., Partridge, L. 2013. Gender and longevity: why do men die earlier than women? Comparative and experimental evidence. *Best Pract. Res. Clin. Endocrinol. Metab.* **27**, 467–479.
9. Austad, S.N., Fischer, K.E. 2016. Sex differences in lifespan. *Cell Metab.* **23**, 1022–1033.
10. Bronikowski, A.M. et al. 2022. Sex-specific aging in animals: perspective and future directions. *Aging Cell* **21**, e13542.
11. Kowald, A. 2002. Lifespan does not measure ageing. *Biogerontology* **3**, 187–190.
12. Péron, G., Lemaître, J.-F., Ronget, V., Tidière, M., Gaillard, J.-M. 2019. Variation in actuarial senescence does not reflect life span variation across mammals. *PLoS Biol.* **17**, e3000432.
13. Mallard, F., Farina, M., Tully, T. 2015. Within-species variation in long-term trajectories of growth, fecundity and mortality in the *Collembola folsomia* candida. *J. Evol. Biol.* **28**, 2275–2284 (doi:10.1111/jeb.12752).
14. Colchero, F. et al. 2021. The long lives of primates and the 'invariant rate of ageing' hypothesis. *Nat. Commun.* **12**, 1–10.
15. Zarulli, V., Kashnitsky, I., Vaupel, J.W. 2021. Death rates at specific life stages mold the sex gap in life expectancy. *Proc. Natl. Acad. Sci.* **118**, e2010588118.
16. Clocchiatti, A., Cora, E., Zhang, Y., Dotto, G.P. 2016. Sexual dimorphism in cancer. *Nat. Rev. Cancer* **16**, 330.
17. Sampathkumar, N.K. et al. 2020. Widespread sex dimorphism in aging and age-related diseases. *Hum. Genet.* **139**, 333–356.
18. Bhopal, S.S., Bhopal, R. 2020. Sex differential in COVID-19 mortality varies markedly by age. *The Lancet* **396**, 532–533.

19. Promislow, D.E. 2020. A geroscience perspective on COVID-19 mortality. *J. Gerontol. Ser. A* **75**, e30–e33.
20. Oksuzyan, A., Juel, K., Vaupel, J.W., Christensen, K. 2008. Men: good health and high mortality. Sex differences in health and aging. *Aging Clin. Exp. Res.* **20**, 91–102.
21. Fisher, R.A. 1930. *The Genetical Theory of Natural Selection: A Complete Variorum Edition* (new ed.). Oxford University Press.
22. Austad, S.N., Bartke, A. 2016. Sex differences in longevity and in responses to anti-aging interventions: a mini-review. *Gerontology* **62**, 40–46.
23. Wheaton, F.V., Crimmins, E.M. 2016. Female disability disadvantage: a global perspective on sex differences in physical function and disability. *Ageing Soc.* **36**, 1136–1156 (doi:10.1017/S0144686X15000227).
24. Morrow, E.H. 2015. The evolution of sex differences in disease. *Biol. Sex Differ.* **6**, 5.
25. Sage, R.F. 2020. Global change biology: a primer. *Glob. Change Biol.* **26**, 3–30.
26. Darwin, C. 1871. *The Descent of Man*. Penguin Classics.
27. MacArthur, J.W., Baillie, W.H.T. 1932. Sex differences in mortality in Abraxas-type species. *Q. Rev. Biol.* **7**, 313–325.
28. Caughley, G. 1966. Mortality patterns in mammals. *Ecology* **47**, 906–918.
29. Marais, G., Gaillard, J.-M., Vieira, C., Plotton, I., Sanlaville, D., Gueyffier, F., Lemaître, J.-F. 2018. Sex-specific differences in aging and longevity: can sex chromosomes play a role? *Biol. Sex Differ.* **9**, 33.
30. Clutton-Brock, T., Sheldon, B.C. 2010. Individuals and populations: the role of long-term, individual-based studies of animals in ecology and evolutionary biology. *Trends Ecol. Evol.* **25**, 562–573.
31. Toïgo, C., Gaillard, J.-M. 2003. Causes of sex-biased adult survival in ungulates: sexual size dimorphism, mating tactic or environment harshness? *Oikos* **101**, 376–384.
32. Promislow, D.E. 1992. Costs of sexual selection in natural populations of mammals. *Proc. R. Soc. Lond. B Biol. Sci.* **247**, 203–210.
33. Lemaître, J.-F., Gaillard, J.-M. 2013. Male survival patterns do not depend on male allocation to sexual competition in large herbivores. *Behav. Ecol.* **24**, 421–428 (doi:10.1093/beheco/ars179).
34. Lemaître, J.-F. et al. 2020. Sex differences in adult lifespan and aging rates of mortality across wild mammals. *Proc. Natl. Acad. Sci.* **117**, 8546–8553.
35. Müller, D.W., Lackey, L.B., Streich, W.J., Fickel, J., Hatt, J.-M., Clauss, M. 2011. Mating system, feeding type and ex situ conservation effort determine life expectancy in captive ruminants. *Proc. R. Soc. Lond. B Biol. Sci.* **278**, 2076–2080.
36. Bro-Jørgensen, J. 2012. Longevity in bovids is promoted by sociality, but reduced by sexual selection. *PloS ONE* **7**, e45769.
37. Promislow, D.E., Montgomerie, R., Martin, T.E. 1992. Mortality costs of sexual dimorphism in birds. *Proc. R. Soc. Lond. B Biol. Sci.* **250**, 143–150.
38. Liker, A., Székely, T. 2005. Mortality costs of sexual selection and parental care in natural populations of birds. *Evolution* **59**, 890–897 (doi:10.1111/j.0014-3820.2005.tb01762.x).
39. Ricklefs, R.E. 2010. Life-history connections to rates of aging in terrestrial vertebrates. *Proc. Natl. Acad. Sci.* **107**, 10314–10319.
40. Lebreton, J.-D., Devillard, S., Popy, S., Desprez, M., Besnard, A., Gaillard, J.-M. 2012. Towards a vertebrate demographic data bank. *J. Ornithol.* **152**, 617–624.
41. Conde, D.A. et al. 2019. Data gaps and opportunities for comparative and conservation biology. *Proc. Natl. Acad. Sci.* **116**, 9658–9664.

42. Mayr, E. 1939. The sex ratio in wild birds. *Am. Nat.* **73**, 156–179.
43. Székely, T., Liker, A., Freckleton, R.P., Fichtel, C., Kappeler, P.M. 2014. Sex-biased survival predicts adult sex ratio variation in wild birds. *Proc. R. Soc. B Biol. Sci.* **281**, 20140342.
44. Eckhardt, F., Kappeler, P.M., Kraus, C. 2017. Highly variable lifespan in an annual reptile, Labord's chameleon (*Furcifer labordi*). *Sci. Rep.* **7**, 1–5.
45. Tully, T., Le Galliard, J.-F., Baron, J.-P. 2020. Micro-geographic shift between negligible and actuarial senescence in a wild snake. *J. Anim. Ecol.* **89**, 2704–2716.
46. Cayuela, H. et al. 2022. Sex-related differences in aging rate are associated with sex chromosome system in amphibians. *Evolution*. **76**, 346–356.
47. Sherratt, T.N., Hassall, C., Laird, R.A., Thompson, D.J., Cordero-Rivera, A. 2011. A comparative analysis of senescence in adult damselflies and dragonflies (Odonata). *J. Evol. Biol.* **24**, 810–822.
48. Kawasaki, N., Brassil, C.E., Brooks, R.C., Bonduriansky, R. 2008. Environmental effects on the expression of life span and aging: an extreme contrast between wild and captive cohorts of *Telostylinus angusticollis* (Diptera: Neriidae). *Am. Nat.* **172**, 346–357 (doi:10.1086/589519).
49. Zajitschek, F., Bonduriansky, R., Zajitschek, S.R., Brooks, R.C. 2009. Sexual dimorphism in life history: age, survival, and reproduction in male and female field crickets Teleogryllus commodus under seminatural conditions. *Am. Nat.* **173**, 792–802.
50. Rodriguez-Munoz, R., Boonekamp, J.J., Liu, X.P., Skicko, I., Haugland Pedersen, S., Fisher, D.N., Hopwood, P., Tregenza, T. 2019. Comparing individual and population measures of senescence across 10 years in a wild insect population. *Evolution* **73**, 293–302 (doi:10.1111/evo.13674).
51. Maklakov, A.A., Lummaa, V. 2013. Evolution of sex differences in lifespan and aging: causes and constraints. *BioEssays* **35**, 717–724.
52. Pipoly, I., Bókony, V., Kirkpatrick, M., Donald, P.F., Székely, T., Liker, A. 2015. The genetic sex-determination system predicts adult sex ratios in tetrapods. *Nature* **527**, 91–94.
53. Marais, G.A.B., Lemaître, J.-F. 2022. Sex chromosomes, sex ratios and sex gaps in longevity in plants. *Phil. Trans. R. Soc. B.* **377**, 20210219.
54. Xirocostas, Z.A., Everingham, S.E., Moles, A.T. 2020. The sex with the reduced sex chromosome dies earlier: a comparison across the tree of life. *Biol. Lett.* **16**, 20190867.
55. Trivers, R. 1985. *Social Evolution*. Benjamin/Cummings.
56. Brown, E.J., Nguyen, A.H., Bachtrog, D. 2020. The Y chromosome may contribute to sex-specific ageing in Drosophila. *Nat. Ecol. Evol.* **4**, 853–862.
57. Slotkin, R.K., Martienssen, R. 2007. Transposable elements and the epigenetic regulation of the genome. *Nat. Rev. Genet.* **8**, 272–285.
58. Connallon, T., Beasley, I.J., McDonough, Y., Ruzicka, F. 2022. How much does the unguarded X contribute to sex differences in life span? *Evol. Lett.* **6**, 319–329.
59. Sultanova, Z., Downing, P.A., Carazo, P. 2023. Genetic sex determination and sex-specific lifespan across tetrapods. *J. Evol. Biol.* **36**, 480–484.
60. Carazo, P., Green, J., Sepil, I., Pizzari, T., Wigby, S. 2016. Inbreeding removes sex differences in lifespan in a population of *Drosophila melanogaster*. *Biol. Lett.* **12**, 20160337.
61. Brengdahl, M., Kimber, C.M., Maguire-Baxter, J., Malacrinò, A., Friberg, U. 2018. Genetic quality affects the rate of male and female reproductive aging differently in *Drosophila melanogaster*. *Am. Nat.*, **192**, 761–772 (doi:10.1086/700117).

62. Wei, K.H.-C., Gibilisco, L., Bachtrog, D. 2020. Epigenetic conflict on a degenerating Y chromosome increases mutational burden in *Drosophila* males. *Nat. Commun.* **11**, 1–9.
63. Nguyen, A.H., Bachtrog, D. 2021. Toxic Y chromosome: increased repeat expression and age-associated heterochromatin loss in male Drosophila with a young Y chromosome. *PLoS Genet.* **17**, e1009438.
64. Vaught, R.C., Dowling, D.K. 2018. Maternal inheritance of mitochondria: implications for male fertility? *Reproduction* **155**, R159–R168.
65. Camus, M.F., Clancy, D.J., Dowling, D.K. 2012. Mitochondria, maternal inheritance, and male aging. *Curr. Biol.* **22**, 1717–1721 (doi:10.1016/j.cub.2012.07.018).
66. Nagarajan-Radha, V., Aitkenhead, I., Clancy, D.J., Chown, S.L., Dowling, D.K. 2020. Sex-specific effects of mitochondrial haplotype on metabolic rate in *Drosophila melanogaster* support predictions of the mother's curse hypothesis. *Philo. Trans. R. Soc. B* **375**, 20190178.
67. Milot, E., Moreau, C., Gagnon, A., Cohen, A.A., Brais, B., Labuda, D. 2017. Mother's curse neutralizes natural selection against a human genetic disease over three centuries. *Nat. Ecol. Evol.* **1**, 1400–1406 (doi:10.1038/s41559-017-0276-6).
68. Thorburn, D.R. 2004. Mitochondrial disorders: prevalence, myths and advances. *J. Inherit. Metab. Dis.* **27**, 349–362.
69. Schon, E.A., DiMauro, S., Hirano, M. 2012. Human mitochondrial DNA: roles of inherited and somatic mutations. *Nat. Rev. Genet.* **13**, 878–890.
70. Gemmell, N.J., Metcalf, V.J., Allendorf, F.W. 2004. Mother's curse: the effect of mtDNA on individual fitness and population viability. *Trends Ecol. Evol.* **19**, 238–244.
71. Dowling, D.K., Adrian, R.E. 2019. Challenges and prospects for testing the mother's curse hypothesis. *Integr. Comp. Biol.* **59**, 875–889.
72. Frank, S.A., Hurst, L.D. 1996. Mitochondria and male disease. *Nature* **383**, 224 (doi:10.1038/383224a0).
73. Beekman, M., Dowling, D.K., Aanen, D.K. 2014. The costs of being male: are there sex-specific effects of uniparental mitochondrial inheritance? *Philo. Trans. R. Soc. B* **369**, 20130440 (doi:10.1098/rstb.2013.0440).
74. Vaught, R.C., Voigt, S., Dobler, R., Clancy, D.J., Reinhardt, K., Dowling, D.K. 2020. Interactions between cytoplasmic and nuclear genomes confer sex-specific effects on lifespan in *Drosophila melanogaster*. *J. Evol. Biol.* **33**, 694–713.
75. Piotrowska, A., Korwin, M., Bartnik, E., Tońska, K. 2015. Leber hereditary optic neuropathy: historical report in comparison with the current knowledge. *Gene* **555**, 41–49.
76. Vinogradov, A.E. 1998. Male reproductive strategy and decreased longevity. *Acta Biotheor.* **46**, 157–160.
77. Bonduriansky, R., Maklakov, A., Zajitschek, F., Brooks, R. 2008. Sexual selection, sexual conflict and the evolution of ageing and life span. *Funct. Ecol.* **22**, 443–453.
78. Brooks, R.C., Garratt, M.G. 2017. Life history evolution, reproduction, and the origins of sex-dependent aging and longevity. *Ann. N. Y. Acad. Sci.* **1389**, 92–107.
79. Clutton-Brock, T.H., Isvaran, K. 2007. Sex differences in ageing in natural populations of vertebrates. *Proc. R. Soc. Lond. B Biol. Sci.* **274**, 3097–3104.
80. Trivers, R.L. 1972. *Parental Investment and Sexual Selection [W:] B. Campbell (ed.) Sexual Selection and the Descent of Man, 1871–1971*. Aldine.
81. Owens, I.P., Bennett, P.M. 1994. Mortality costs of parental care and sexual dimorphism in birds. *Proc. R. Soc. Lond. B Biol. Sci.* **257**, 1–8.
82. Williams, G.C. 1957. Pleiotropy, natural selection, and the evolution of senescence. *Evolution* **11**, 398–411.

83. Tidière, M., Gaillard, J.-M., Müller, D.W., Lackey, L.B., Gimenez, O., Clauss, M., Lemaître, J.-F. 2015. Does sexual selection shape sex differences in longevity and senescence patterns across vertebrates? A review and new insights from captive ruminants. *Evolution* **69**, 3123–3140.
84. Moore, S.L., Wilson, K. 2002. Parasites as a viability cost of sexual selection in natural populations of mammals. *Science* **297**, 2015–2018.
85. Lemaître, J.-F., Gaillard, J.-M., Ramm, S.A. 2020. The hidden ageing costs of sperm competition. *Ecol. Lett.* **11**, 1573–1588 (doi:10.1111/ele.13593).
86. Promislow, D. 2003. Mate choice, sexual conflict, and evolution of senescence. *Behav. Genet.* **33**, 191–201.
87. Foucart, T., Lourdais, O., DeNardo, D.F., Heulin, B. 2014. Influence of reproductive mode on metabolic costs of reproduction: insight from the bimodal lizard *Zootoca vivipara. J. Exp. Biol.* **217**, 4049–4056.
88. Stockley, P., Bro-Jørgensen, J. 2011. Female competition and its evolutionary consequences in mammals. *Biol. Rev. Camb. Philos. Soc.* **86**, 341–366 (doi:10.1111/j.1469-185X.2010.00149.x).
89. Skibiel, A.L., Downing, L.M., Orr, T.J., Hood, W.R. 2013. The evolution of the nutrient composition of mammalian milks. *J. Anim. Ecol.* **82**, 1254–1264.
90. Monaghan, P., Charmantier, A., Nussey, D.H., Ricklefs, R.E. 2008. The evolutionary ecology of senescence. *Funct. Ecol.* **22**, 371–378.
91. Cohen, A.A. et al. 2020. Lack of consensus on an aging biology paradigm? A global survey reveals an agreement to disagree, and the need for an interdisciplinary framework. *Mech. Ageing Dev.* **191**, 111316.
92. Gaillard, J.-M, Lemaître J.-F. 2020. An integrative view of senescence in nature. *Funct. Ecol.* **34**, 4–16.
93. Medawar, P.B. 1952. *An Unsolved Problem of Biology*. College.
94. Hamilton, W.D. 1966. The moulding of senescence by natural selection. *J. Theor. Biol.* **12**, 12–45.
95. Holmes, D.J., Thomson, S.L., Wu, J., Ottinger, M.A. 2003. Reproductive aging in female birds. *Exp. Gerontol.* **38**, 751–756.
96. Lemaître, J.-F., Ronget, V., Gaillard, J.-M. 2020. Female reproductive senescence across mammals: a high diversity of patterns modulated by life history and mating traits. *Mech. Ageing Dev.* **192**, 111377 (doi:10.1016/j.mad.2020.111377).
97. Vágási, C.I., Vincze, O., Lemaître, J.-F., Pap, P.L., Ronget, V., Gaillard, J.-M. 2021. Is degree of sociality associated with reproductive senescence? A comparative analysis across birds and mammals. *Philo. Trans. R. Soc. B.* **376**, 20190744.
98. Campos, F.A. et al. 2022. Female reproductive aging in seven primate species: patterns and consequences. *Proc. Natl. Acad. Sci.* **119**, e2117669119.
99. Vrtilek, M., Zab, J., Reichard, M. 2023. Evidence for reproductive senescence across ray-finned fishes: a review. *Front. Ecol. Evol.* **10**, 982915.
100. Lemaître, J.-F., Gaillard, J.-M. 2017. Reproductive senescence: new perspectives in the wild. *Biol. Rev.* **92**, 2182–2199 (doi:10.1111/brv.12328).
101. Murgatroyd, M., Roos, S., Evans, R., Sansom, A., Whitfield, D.P., Sexton, D., Reid, R., Grant, J., Amar, A. 2018. Sex-specific patterns of reproductive senescence in a long-lived reintroduced raptor. *J. Anim. Ecol.* **87**, 1587–1599.
102. Tompkins, E.M., Anderson, D.J. 2019. Sex-specific patterns of senescence in Nazca boobies linked to mating system. *J. Anim. Ecol.* **88**, 986–1000.

103. Froy, H., Lewis, S., Nussey, D.H., Wood, A.G., Phillips, R.A. 2017. Contrasting drivers of reproductive ageing in albatrosses. *J. Anim. Ecol.* **86**, 1022–1032.
104. Hayward, A.D., Moorad, J., Regan, C.E., Berenos, C., Pilkington, J.G., Pemberton, J.M., Nussey, D.H. 2015. Asynchrony of senescence among phenotypic traits in a wild mammal population. *Exp. Gerontol.* **71**, 56–68.
105. Thorley, J., Duncan, C., Sharp, S.P., Gaynor, D., Manser, M.B., Clutton-Brock, T. 2020. Sex-independent senescence in a cooperatively breeding mammal. *J. Anim. Ecol.* **89**, 1080–1093.
106. Smith, S., Turbill, C., Suchentrunk, F. 2010. Introducing mother's curse: low male fertility associated with an imported mtDNA haplotype in a captive colony of brown hares. *Mol. Ecol.* **19**, 36–43.
107. Gaillard, J.-M., Festa-Bianchet, M., Delorme, D., Jorgenson, J. 2000. Body mass and individual fitness in female ungulates: bigger is not always better. *Proc. R. Soc. Lond. B Biol. Sci.* **267**, 471–477.
108. Ronget, V., Gaillard, J.-M., Coulson, T., Garratt, M., Gueyffier, F., Lega, J.-C., Lemaître, J.-F. 2018. Causes and consequences of variation in offspring body mass: meta-analyses in birds and mammals. *Biol. Rev.* **93**, 1–27.
109. Cruz-Jentoft, A.J., Sayer, A.A. 2019. Sarcopenia. *The Lancet* **393**, 2636–2646.
110. Landi, F., Liperoti, R., Fusco, D., Mastropaolo, S., Quattrociocchi, D., Proia, A., Tosato, M., Bernabei, R., Onder, G. 2012. Sarcopenia and mortality among older nursing home residents. *J. Am. Med. Dir. Assoc.* **13**, 121–126.
111. Beaudart, C., Zaaria, M., Pasleau, F., Reginster, J.-Y., Bruyère, O. 2017. Health outcomes of sarcopenia: a systematic review and meta-analysis. *PloS ONE* **12**, e0169548.
112. Landes, J., Perret, M., Hardy, I., Camarda, C.G., Henry, P.-Y., Pavard, S. 2017. State transitions: a major mortality risk for seasonal species. *Ecol. Lett.* **20**, 883–891.
113. Reimers, E., Holmengen, N., Mysterud, A. 2005. Life-history variation of wild reindeer (Rangifer tarandus) in the highly productive North Ottadalen region, Norway. *J. Zool.* **265**, 53–62.
114. Mason, T.H., Chirichella, R., Richards, S.A., Stephens, P.A., Willis, S.G., Apollonio, M. 2011. Contrasting life histories in neighbouring populations of a large mammal. *PloS ONE* **6**, e28002.
115. Beirne, C., Delahay, R., Young, A. 2015. Sex differences in senescence: the role of intra-sexual competition in early adulthood. *Proc. R. Soc. B* **282**, 20151086.
116. Douhard, F., Gaillard, J.-M., Pellerin, M., Jacob, L., Lemaître, J.-F. 2017. The cost of growing large: costs of post-weaning growth on body mass senescence in a wild mammal. *Oikos* **126**, 1329–1338 (doi:10.1111/oik.04421).
117. Tafani, M., Cohas, A., Bonenfant, C., Gaillard, J.-M., Lardy, S., Allainé, D. 2013. Sex-specific senescence in body mass of a monogamous and monomorphic mammal: the case of Alpine marmots. *Oecologia* **172**, 427–436.
118. Campbell, R.D., Rosell, F., Newman, C., Macdonald, D.W. 2017. Age-related changes in somatic condition and reproduction in the Eurasian beaver: resource history influences onset of reproductive senescence. *PloS ONE* **12**, e0187484.
119. Lalande, L.D., Lummaa, V., Aung, H.H., Htut, W., Nyein, U.K., Berger, V., Briga, M. 2022. Sex-specific body mass ageing trajectories in adult Asian elephants. *J. Evol. Biol.* **35**, 752–762.
120. Hindle, A.G., Horning, M., Mellish, J.-A.E., Lawler, J.M. 2009. Diving into old age: muscular senescence in a large-bodied, long-lived mammal, the Weddell seal (*Leptonychotes weddellii*). *J. Exp. Biol.* **212**, 790–796 (doi:10.1242/jeb.025387).

121. Hindle, A.G., Lawler, J.M., Campbell, K.L., Horning, M. 2009. Muscle senescence in short-lived wild mammals, the soricine shrews *Blarina brevicauda* and *Sorex palustris*. *J. Exp. Zool. Part Ecol. Genet. Physiol.* **311**, 358–367.
122. Hägg, S., Jylhävä, J. 2021. Sex differences in biological aging with a focus on human studies. *eLife* **10**, e63425.
123. Li, X., Ploner, A., Wang, Y., Magnusson, P.K., Reynolds, C., Finkel, D., Pedersen, N.L., Jylhävä, J., Hägg, S. 2020. Longitudinal trajectories, correlations and mortality associations of nine biological ages across 20-years follow-up. *eLife* **9**, e51507.
124. Monaghan, P., Eisenberg, D.T.A., Harrington, L., Nussey, D. 2018. Understanding diversity in telomere dynamics. *Philo. Trans. R. Soc. Lond. B. Biol. Sci.* **373**, 20160435.
125. Blackburn, E.H. 1991. Structure and function of telomeres. *Nature* **350**, 569.
126. Aubert, G., Lansdorp, P.M. 2008. Telomeres and aging. *Physiol. Rev.* **88**, 557–579.
127. D'Mello, M.J., Ross, S.A., Briel, M., Anand, S.S., Gerstein, H., Paré, G. 2015. Association between shortened leukocyte telomere length and cardiometabolic outcomes: systematic review and meta-analysis. *Circ. Cardiovasc. Genet.* **8**, 82–90.
128. Cawthon, R.M., Smith, K.R., O'Brien, E., Sivatchenko, A., Kerber, R.A. 2003. Association between telomere length in blood and mortality in people aged 60 years or older. *The Lancet* **361**, 393–395.
129. Wilbourn, R.V., Moatt, J.P., Froy, H., Walling, C.A., Nussey, D.H., Boonekamp, J.J. 2018. The relationship between telomere length and mortality risk in non-model vertebrate systems: a meta-analysis. *Philo. Trans. R. Soc. B* **373**, 20160447.
130. Barrett, E.L., Richardson, D.S. 2011. Sex differences in telomeres and lifespan. *Aging Cell* **10**, 913–921.
131. Gardner, M. et al. 2014. Gender and telomere length: systematic review and meta-analysis. *Exp. Gerontol.* **51**, 15–27 (doi:10.1016/j.exger.2013.12.004).
132. Sudyka, J. 2019. Does reproduction shorten telomeres? Towards integrating individual quality with life-history strategies in telomere biology. *BioEssays* **41**, 1900095.
133. Remot, F., Ronget, V., Froy, H., Rey, B., Gaillard, J.-M., Nussey, D.H., Lemaître, J.-F. 2020. No sex differences in adult telomere length across vertebrates: a meta-analysis. *R. Soc. Open Sci.* **7**, 200548.
134. Omotoso, O., Gladyshev, V.N., Zhou, X. 2021. Lifespan extension in long-lived vertebrates rooted in ecological adaptation. *Front. Cell Dev. Biol.* **18**, 704966.
135. Gorbunova, V., Seluanov, A., Zhang, Z., Gladyshev, V.N., Vijg, J. 2014. Comparative genetics of longevity and cancer: insights from long-lived rodents. *Nat. Rev. Genet.* **15**, 531.
136. Vincze, O. et al. 2022. Cancer risk across mammals. *Nature* **601**, 263–267.
137. Lemaître, J.-F., Garratt, M., Gaillard, J.-M. 2020. Going beyond lifespan in comparative biology of aging. *Adv. Geriatr. Med. Res.* **2**, e200011.
138. Flatt, T., Partridge, L. 2018. Horizons in the evolution of aging. *BMC Biol.* **16**, 93 (doi:10.1186/s12915-018-0562-z).
139. Lozano, R. et al. 2012. Global and regional mortality from 235 causes of death for 20 age groups in 1990 and 2010: a systematic analysis for the Global Burden of Disease Study 2010. *The Lancet* **380**, 2095–2128.
140. Zuk, M., McKean, K.A. 1996. Sex differences in parasite infections: patterns and processes. *Int. J. Parasitol.* **26**, 1009–1024.
141. Cevidanes, A., Proboste, T., Chirife, A.D., Millán, J. 2016. Differences in the ectoparasite fauna between micromammals captured in natural and adjacent residential areas are better explained by sex and season than by type of habitat. *Parasitol. Res.* **115**, 2203–2211.

142. Celestino, I. et al. 2018. Differential redox state contributes to sex disparities in the response to influenza virus infection in male and female mice. *Front. Immunol.* **9**, 1747.
143. Kelly, C.D., Stoehr, A.M., Nunn, C., Smyth, K.N., Prokop, Z.M. 2018. Sexual dimorphism in immunity across animals: a meta-analysis. *Ecol. Lett.* **21**, 1885–1894.
144. Klein, S.L. 2000. The effects of hormones on sex differences in infection: from genes to behavior. *Neurosci. Biobehav. Rev.* **24**, 627–638.
145. Klein, S.L., Flanagan, K.L. 2016. Sex differences in immune responses. *Nat. Rev. Immunol.* **16**, 626–638.
146. Gubbels Bupp, M.R., Potluri, T., Fink, A.L., Klein, S.L. 2018. The confluence of sex hormones and aging on immunity. *Front. Immunol.* **9**, 1269.
147. Scully, E.P., Haverfield, J., Ursin, R.L., Tannenbaum, C., Klein, S.L. 2020. Considering how biological sex impacts immune responses and COVID-19 outcomes. *Nat. Rev. Immunol.* **20**, 442–447.
148. Peer, V., Schwartz, N., Green, M.S. 2020. A multi-country, multi-year, meta-analytic evaluation of the sex differences in age-specific pertussis incidence rates. *PLoS ONE* **15**, e0231570.
149. Neuman, H., Debelius, J.W., Knight, R., Koren, O. 2015. Microbial endocrinology: the interplay between the microbiota and the endocrine system. *FEMS Microbiol. Rev.* **39**, 509–521.
150. Fish, E.N. 2008. The X-files in immunity: sex-based differences predispose immune responses. *Nat. Rev. Immunol.* **8**, 737–744.
151. Klein, S.L. 2012. Immune cells have sex and so should journal articles. *Endocrinology* **153**, 2544–2550.
152. Taneja, A., Das, S., Hussain, S.A., Madadin, M., Lobo, S.W., Fatima, H., Menezes, R.G. 2019. Uterine transplant: a risk to life or a chance for life? *Sci. Eng. Ethics* **25**, 635–642.
153. Neyrolles, O., Quintana-Murci, L. 2009. Sexual inequality in tuberculosis. *PLoS Med.* **6**, e1000199.
154. Guerra-Silveira, F., Abad-Franch, F. 2013. Sex bias in infectious disease epidemiology: patterns and processes. *PloS ONE* **8**, e62390.
155. Giefing-Kröll, C., Berger, P., Lepperdinger, G., Grubeck-Loebenstein, B. 2015. How sex and age affect immune responses, susceptibility to infections, and response to vaccination. *Aging Cell* **14**, 309–321.
156. Bereshchenko, O., Bruscoli, S., Riccardi, C. 2018. Glucocorticoids, sex hormones, and immunity. *Front. Immunol.* **9**, 1332.
157. Jacobsen, H., Klein, S.L. 2021. Sex differences in immunity to viral infections. *Front. Immunol.* **31**, 720952.
158. Jaillon, S., Berthenet, K., Garlanda, C. 2019. Sexual dimorphism in innate immunity. *Clin. Rev. Allergy Immunol.* **56**, 308–321.
159. Ansar Ahmed, S., Penhale, W.J., Tala, N. 1985. Sex hormones, immune and autoimmune responses: mechanism of sex hormone action. *Am. J. Pathol.* **121**, 531–559.
160. Guilbault, C. et al. 2002. Influence of gender and interleukin-10 deficiency on the inflammatory response during lung infection with Pseudomonas aeruginosa in mice. *Immunology* **107**, 297–305.
161. Libert, C., Dejager, L., Pinheiro, I. 2010. The X chromosome in immune functions: when a chromosome makes the difference. *Nat. Rev. Immunol.* **10**, 594–604.
162. Schurz, H., Salie, M., Tromp, G., Hoal, E.G., Kinnear, C.J., Möller, M. 2019. The X chromosome and sex-specific effects in infectious disease susceptibility. *Hum. Genomics* **13**, 1–12.

163. Cody, M.L. 1966. A general theory of clutch size. *Evolution* **20**, 174–184.
164. Stearns, S.C. 1992. *The Evolution of Life Histories*. Oxford University Press.
165. Jasienska, G., Bribiescas, R.G., Furberg, A.-S., Helle, S., Núñez-de la Mora, A. 2017. Human reproduction and health: an evolutionary perspective. *The Lancet* **390**, 510–520.
166. Sheldon, B.C., Verhulst, S. 1996. Ecological immunology: costly parasite defences and trade-offs in evolutionary ecology. *Trends Ecol. Evol.* **11**, 317–321.
167. Sandland, G.J., Minchella, D.J. 2003. Costs of immune defense: an enigma wrapped in an environmental cloak? *Trends Parasitol.* **19**, 571–574.
168. Sievert, L.L. 2014. Anthropology and the study of menopause: evolutionary, developmental, and comparative perspectives. *Menopause* **21**, 1151–1159.
169. Leroi, A.M. et al. 2005. What evidence is there for the existence of individual genes with antagonistic pleiotropic effects? *Mech. Ageing Dev.* **126**, 421–429.
170. Boddy, A.M., Kokko, H., Breden, F., Wilkinson, G.S., Aktipis, C.A. 2015. Cancer susceptibility and reproductive trade-offs: a model of the evolution of cancer defences. *Philo. Trans. R. Soc. B* **370**, 20140220.
171. Byars, S.G., Voskarides, K. 2020. Antagonistic pleiotropy in human disease. *J. Mol. Evol.* **88**, 12–25.
172. Hu, W. 2009. The role of p53 gene family in reproduction. *Cold Spring Harb. Perspect. Biol.* **1**, a001073.
173. Provenzano, F., Deleidi, M. 2021. Reassessing neurodegenerative disease: immune protection pathways and antagonistic pleiotropy. *Trends Neurosci.* **44**, 771–780.
174. Duneau, D., Ebert, D. 2012. Host sexual dimorphism and parasite adaptation. *PLoS Biol.* **10**, e1001271.
175. Hall, M.D., Mideo, N. 2018. Linking sex differences to the evolution of infectious disease life-histories. *Philo. Trans. R. Soc. B Biol. Sci.* **373**, 20170431.
176. Úbeda, F., Jansen, V.A. 2016. The evolution of sex-specific virulence in infectious diseases. *Nat. Commun.* **7**, 1–9.
177. McLeod, D.V., Wild, G., Úbeda, F. 2021. Epigenetic memories and the evolution of infectious diseases. *Nat. Commun.* **12**, 1–13.
178. Aaby, P. 1992. Influence of cross-sex transmission on measles mortality in rural Senegal. *The Lancet* **340**, 388–391.
179. Nielsen, N.M., Wohlfahrt, J., Melbye, M., Mølbak, K., Aaby, P. 2002. Does cross-sex transmission increase the severity of polio infection? A study of multiple family cases. *Scand. J. Infect. Dis.* **34**, 273–277.
180. Metcalf, C.J.E., Roth, O., Graham, A.L. 2020. Why leveraging sex differences in immune trade-offs may illuminate the evolution of senescence. *Funct. Ecol.* **34**, 129–140.
181. Eberl, G., Pradeu, T. 2018. Towards a general theory of immunity? *Trends Immunol.* **39**, 261–263.
182. Fagundes, C.T., Amaral, F.A., Teixeira, A.L., Souza, D.G., Teixeira, M.M. 2012. Adapting to environmental stresses: the role of the microbiota in controlling innate immunity and behavioral responses. *Immunol. Rev.* **245**, 250–264.
183. Brestoff, J.R., Artis, D. 2015. Immune regulation of metabolic homeostasis in health and disease. *Cell* **161**, 146–160.
184. Marques, A.H., Bjørke-Monsen, A.-L., Teixeira, A.L., Silverman, M.N. 2015. Maternal stress, nutrition and physical activity: impact on immune function, CNS development and psychopathology. *Brain Res.* **1617**, 28–46.

185. Colchero, F., Jones, O.R., Rebke, M. 2012. BaSTA: an R package for Bayesian estimation of age-specific survival from incomplete mark–recapture/recovery data with covariates. *Methods Ecol. Evol.* **3**, 466–470.

186. Bronikowski, A.M. et al. 2016. Female and male life tables for seven wild primate species. *Sci. Data* **3**, 160006 (doi:10.1038/sdata.2016.6).

187. Descamps, S., Boutin, S., Berteaux, D., Gaillard, J.-M. 2008. Age-specific variation in survival, reproductive success and offspring quality in red squirrels: evidence of senescence. *Oikos* **117**, 1406–1416.

188. Hämäläinen, A., Dammhahn, M., Aujard, F., Eberle, M., Hardy, I., Kappeler, P.M., Perret, M., Schliehe-Diecks, S., Kraus, C. 2014. Senescence or selective disappearance? Age trajectories of body mass in wild and captive populations of a small-bodied primate. *Proc. R. Soc. B Biol. Sci.* **281**, 20140830.

10 Evolution of Human Reproduction, Ageing and Longevity

Samuel Pavard and Michael D. Gurven

10.1 Introduction

A species' longevity, and characteristic age-related changes in reproduction and survival (i.e. ageing), are fundamental components of its evolved life cycle. A growing consensus in evolutionary biology suggests that ageing evolved with other components of the life cycle such as ontogenesis and timing of reproduction, regardless of the evolutionary mechanism underlying ageing (e.g. mutation accumulation, antagonistic pleiotropy, disposable soma, hyperfunction theory, see Chapters 2 and 5). This is why it is important to contemplate the evolution of life cycles across species to understand the different ways in which those species age. Humans are no different. To understand human ageing from an evolutionary perspective, we therefore need to consider our species' phylogenetic history and the evolution of the entire human life cycle.

Furthermore, humans are a species for which the evolution of cognitive abilities – arguably unparalleled in the animal world – has profoundly modified the relationships humans have with each other and with their environment. Technology and culture, and the emergence of complex and diverse social and family systems, have profoundly changed the evolutionary theatre in which humans are born, grow, reproduce and age. In this chapter, we first contemplate how our life cycle is anchored within our phylogenetic history, and highlight its peculiarities. Then, we consider the main theories explaining the joint evolution of ageing with our cognitive and social capabilities.

10.2 Life Goes Slower in Primates

10.2.1 Structure and Length of the Mammalian Life Cycle

Humans are mammals. As such, we are determinate growers (i.e. we grow, mature, then reproduce), are iteroparous (i.e. have repeated bouts of reproduction) and post-natal care is necessary for offspring survival. As mammals, we are also a species

We thank Rebecca Sear, Barry Bogin and Jean-François Lemaître for their insightful comments on an earlier version of this chapter. This study is supported by a grant from the Agence Nationale de la Recherche (ANR-18-CE02-0011, MathKinD).

that displays senescence, that is, where mortality rates rise throughout adulthood, mostly in a Gompertz-like exponential fashion [1]. In other words, as do all mammals, we grow until around the time of sexual maturity, then, if we are lucky to survive throughout adulthood, we can reproduce and raise offspring more than once, grow old and die. Mammals, together with birds, are the only phylogenetic classes where these life history traits are shared by all species. In other vertebrate classes, such as fishes, reptiles and amphibians, species can exhibit determinate or indeterminate growth, non-continuous or decreasing mortality trajectories with age [2], and can exhibit little or no post-natal parental care. Among bony fishes, plants and insects, semelparity (i.e. a single reproductive bout before dying) is also not uncommon.

Beyond this common life history structure, life-cycle length and longevity vary between mammals, as in other classes, according to body mass. This is referred to as the *first-order* (life history) *tactic* [3, p. 198; 4, p. 64], but is most commonly known as the allometric dimension of the slow–fast continuum (see [5] for an extensive review). Because of the allometric relationships between mass and metabolism, body size operates as a scaling factor on the length of the life cycle (e.g. [6] and see [7] for possible mechanistic explanations of this effect): small organisms take less time to grow, reproduce over a shorter period of time and die earlier than large organisms. Because body mass varies widely in mammals, from the bumblebee bat (*Craseonycteris thonglongyai*) of about 2 grams to the blue whale (*Balaenoptera musculus*) of about 150 tons, body mass variation alone explains much life history variation across species (about 50% in [5]; see Figure 10.1).

10.2.2 Primates Are at the Slow End of the Second-Order Slow–Fast Continuum

After controlling for the ubiquitous allometric effect of body mass, life history traits remain correlated across species, reflecting 'a second order (life history) tactic': [4, 6, 8, 9] some species tend to live fast and die young while others live slow and die older. To what extent are features of life history not captured along this eco-evolutionary continuum? When mass and phylogeny are controlled for, about 70% of the variance in life history traits associated with metabolic rates and ecological modes of life are explained by position along the slow–fast axis [10], although the ecological factors shaping this slow–fast continuum have not yet been clearly identified [5]. Finally, the extent to which the position along this continuum relates to patterns of age-specific mortality (e.g. age at onset of actuarial senescence, rate of actuarial senescence) remains a key area of investigation [11].

As for primates, they are the mammals with the slowest life histories: for a given size, primates tend to have longer gestation, slower growth, later age at weaning and sexual maturity, longer intervals between subsequent births and longer lifespans (see [4, 6, 12] and Figure 10.1). In other words, whether primates are very small, such as the pygmy marmoset (*Callithrix pygmaea*, ~100 g), or very large, such as the gorilla (~150 kg), life tends to unfold at a slower pace than other mammals of similar size. Furthermore, most primates' species are monotocous (birthing one offspring at a

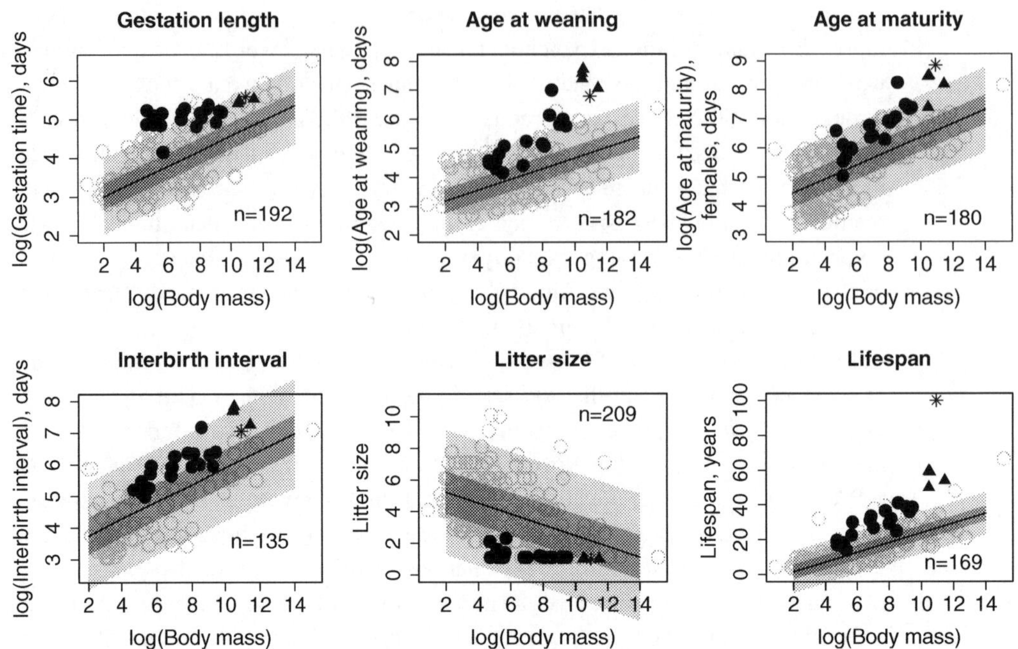

Figure 10.1 Mammalian life history traits, as a function of body mass (excluding cetaceans and chiroptera, for which the relationship between life history traits and body mass differs from that of terrestrial mammals). Grey circles refer to non-primate mammals, dark triangles for great apes (chimpanzee, bonobo, gorilla and orangutan), dark circles for other primates, and a star for humans. Data for humans (i.e. Ache hunter-gatherers), gibbons (*Hylobates lar*), chimpanzees (*Pan troglodytes*), gorillas (here *Gorilla Gorilla beringei*) and orangutans (*Pongo pygmaeus abelii*) are from [27]. Note that ages at weaning and sexual maturity, and interbirth intervals (IBIs), vary substantially among human populations. Data for bonobos (*Pan paniscus*) are from [161] and [162] for weaning time. All other data are from the AnAge database [163]. Values for IBI exclude seasonal species. Lines are linear regressions of the life history traits with respect to body mass for non-primate mammals and the dark and light grey areas encompass 50% and 95% of the predicted values. Life history variation is primarily explained by body size. However, primates' life history traits are always above (or below for litter size) the mean value predicted for non-primates.

time) and rarely give birth to two offspring per reproductive event. Metabolism is known to reflect the energetic cost of living and evolved alongside the pace of life [13]. This is also the case in primates where slower pace of life is linked to a lower metabolism, which explains many differences in life history between primates and other placental mammals. Primates exhibit ~50% lower total daily energy expenditure (TEE) than in other non-primate eutherian mammals of similar mass [14], and removing the effect of mass and metabolism pushes primate life history traits to fall more in line with those of non-primate placental mammals. Lower energetic throughput among primates cannot be explained by differences in physical activity, but instead suggest systemic adaptations affecting cellular metabolism. Of course, low TEE and slow growth rates are phylogenetic trends that obscure some of the variation within the primate clade. For instance, among primates, some subfamilies exhibit similar or faster growth than other mammals of the same size [15, 16].

But 'Why do female primates have such long lifespans and so few babies?' [12]. One dominant hypothesis is that the protracted lifetimes of primates are a direct consequence of the delayed maturation caused by the ontogenetic constraint of developing and maintaining a large functioning brain. Delaying maturation tends to increase the interval between subsequent births, and thus lowers fertility over an extended reproductive life. An extended lifespan is therefore required to achieve a similar lifetime reproductive success.

10.2.3 Do Large Brains Cause Primates to Grow More Slowly?

Primates have large brains relative to their body mass and, among primates, great apes have the largest relative brain size [17]. A large literature has investigated and found correlations between relative brain size and life history traits in mammals. Although relative brain size correlates with basal metabolic rate in mammals [18], the magnitude and significance of this correlation, as well as to what extent it explains life history variation, is still unresolved; for example between marsupial and placental mammals [19], within Carnivora [20, 21], distinguishing monotocous and polytocous mammals [22], or in relation to the level of allocare provided to offspring [23]. With respect to primates, although such correlations have been contested [24], or nuanced [25], recent evidence suggests that larger primate brains are a key factor explaining protracted juvenile period and longer lifespan in these species. Among primates, brain encephalization is correlated with lifespan, group size and home range [26], and with all stages of development except for lactational periods [27]. Across mammals, more generally, encephalization (relative to size) strongly correlates with longevity (also relative to size), with the order of primates exhibiting both larger brains and longer lives [28, 29].

10.2.4 Why Does Having a Large Brain Slow Down the Pace of Life?

An early idea, the *maternal energy hypothesis* [30] postulated that, because large brains are metabolically expensive [31], growing and maintaining a large brain requires a greater investment of energy from mothers. To meet this requirement, primate mothers take more time to transfer this energy by slowing down somatic growth and prolonging developmental stages, including the juvenile period. Expanding upon these ideas about the expensive brain, greater allomaternal care and cooperative breeding have been proposed as social means to bolster limited energy budgets in larger brain species [23]. A second, non-exclusive explanation may be that larger brains with enhanced cognitive capabilities take more time to develop and mature, especially regions associated with higher-order processes, such as the prefrontal cortex [32]. Indeed, neuronal division decreases rapidly after birth, whereas much subsequent brain growth is due to increased neuronal connections and the division of glial cells. This is why neurologists now insist that the brain needs time, as much as energy, to configure itself in contact with post-natal stimuli [33]. Although the generalization of these results across mammalian clades has been debated, a consensus emerges that neonatal encephalization is closely linked to the primate life cycle, and in particular to strategies associated with maternal reproduction [21].

10.2.5 Why Have Primates Evolved Large Brains?

Two principal theories have been proposed to explain why primates have large brains. The first, called the 'foraging brain' or the 'ecological brain' hypothesis, argues that more complex diet and foraging strategies should correlate strongly with brain size. The idea is that improved cognition has jointly evolved with dietary complexity by fostering innovative foraging strategies, planning and navigation over time and space. Jointly, this promotes greater foraging efficiency. Early studies observed that folivorous primates have smaller brains than frugivorous primates [34], while recent work reports convincing relationships between primate brain size and diet, even when adjusting for body mass and phylogeny [35]. In addition to complex diets requiring more complex cognition, another mechanism may link life history traits to diet and foraging strategy: Primates specializing in foraging high-quality foods whose availability varies seasonally or over other time frames, especially in variable environments, will be more sensitive to environmental fluctuations, which in turn may favour low reproductive-effort/large survival tactics (argued in [36]). Lastly, the shift towards a higher-quality diet among large-brained primates entails lower processing costs and smaller guts [37]. This proposed trade-off between digestive tract size and brain size is consistent with primates having lower observed TEE despite having larger brains [14]. However, one recent study showed trade-offs of brain size, not with gut size or other organs, but with mass of fat depots [38], suggesting that encephalization and fat storage both serve to buffer against energetic shortages.

The second hypothesis, called the 'social brain' hypothesis, emphasizes the importance of improved cognition in the context of complex sociality. It has been argued that primates differ from other mammals because they: (1) exhibit stronger feeding competition (which would favour better tactical deception and Machiavellian-type intelligence [39]); (2) transmit behaviour socially (favouring better skill transmission, and 'cultural intelligence' [40, 41]); or (3) solve ecological problems socially as a cooperative process, which would favour more cohesive social bonds and a more social brain [42, 43].

Although fiercely debated [44], these two hypotheses may not be as mutually exclusive as commonly portrayed. First, the prefrontal cortex is implicated in executive function, short-term memory and communication, all potentially mobilized in both foraging and social interactions. Also, much of the social learning observed in primates is of foraging strategies; social, technical and ecological abilities have likely jointly evolved [41]. To the extent that greater dietary complexity is associated with social foraging strategies, larger brains may also help support greater social cognition related to effective cooperation [45, 46]. Second, comparative analyses usually show that brain size and longevity are associated with both group size (as a crude proxy of sociality) and indicators of dietary complexity [26]. Walker and colleagues conclude in favour of 'a mix of social and ecological selective pressures acting at different intensities in particular primate clades' [26, p. 487].

10.2.6 Beyond Brain Size, Does Sociality Increase Primate Longevity?

Another set of theories argue that sociality directly selects for extended lifespan by emphasizing the importance of cooperation and intergenerational transfers as social means of buffering against starvation, predation and other hardships, thereby improving both the fertility and survivorship of kin. Indeed, a major theory for the evolution of sociality is kin selection (see [47] for a review), developed by population geneticists Ronald Fisher and John B.S. Haldane, and adapted for behavioural evolution by William D. Hamilton [48, 49] (but see [50] for a re-evaluation of its generality). Kin selection theory argues that measures of fitness in social species should not only include vertical gene transmission from parents to offspring, but should also include horizontal or oblique transmissions through the increased reproductive success of kin. An individual can increase their *inclusive fitness* by increasing the probability that their siblings transmit their genes to the next generation by helping them in caring for their offspring. To calculate an individual inclusive fitness, one should first weigh the gain of such helping behaviours (i.e. to what extent it is increasing the kin's reproductive success) by the relatedness between the individual and its kin. Then, such gain should be compared to the cost that this behaviour may have in compromising the individual's own reproductive success.

The importance of kin selection for the evolution of longevity and senescence has been extensively reviewed in work by Bourke [51]. The most overt is aid and care of direct descendants: parents (and sometime grandparents) invest time and energy in their (grand)offspring over an extended juvenile period, and during adulthood when they help them attain higher dominance status or find mates. In many cases of intergenerational transfers positive correlations arise between adult survival and younger kin fitness in the form of enhanced survival or fertility [52]. Greater adult survival thus enhances inclusive fitness, opening up the possibility for selection to prolong lifespan. More generally, longer lifespan is expected to emerge in cooperative breeding species when older helpers care for younger individuals; this is not uncommon in nature [53], and is especially common in primates [23]. Sex differences in longevity (see also Chapter 9) may reflect sex differences in parental care. In non-human mammals, grandmaternal care is linked to longer lifespan of females compared to males [54]. In primates, the sex that provides most of the parental care lives longer than the other sex in some primate species [55]. However, while greater longevity among cooperative breeders shows some support, the relationship is inconsistent among birds [56], and in mammals more generally [57].

To conclude, as primates, humans are expected to be long lived. It is also expected, as a fundamental mechanism shaping the life cycle of all primate species, that bigger brain size and configuration delay juvenile developmental stages and increase intergenerational transfers of resources, in turn extending reproductive life and selecting for greater longevity.

10.3 Humans Are Peculiar Primates

The human life cycle appears as an outlier for many traits, including delayed maturation (including childhood, juvenility and adolescent pre-adult phases), and the capacity for a long lifespan. Those are features of 'life in the slow lane', but rapid rates of reproduction in humans are usually features of 'fast' life histories (see Figures 10.1 and 10.2).

10.3.1 Humans Are Long-Lived Primates

Adult life expectancy (i.e. remaining life expectancy at sexual maturity) and maximum lifespan are much larger in humans than in other primates, even accounting for body size (see Figure 10.1). Jeanne Calment (1875–1997), the world's record-holder for longest lifespan died at the canonical age of 122 years, 5 months and 14 days. While the number of supercentenarians is greater now than ever before in human history, 122 is still a difficult record to beat. It's two to three times larger than any longest-living great apes, including those maintained under favourable conditions of captivity (keeping in mind that maximum lifespan may not be the best metric to compare species because it is highly sensitive to sample size).

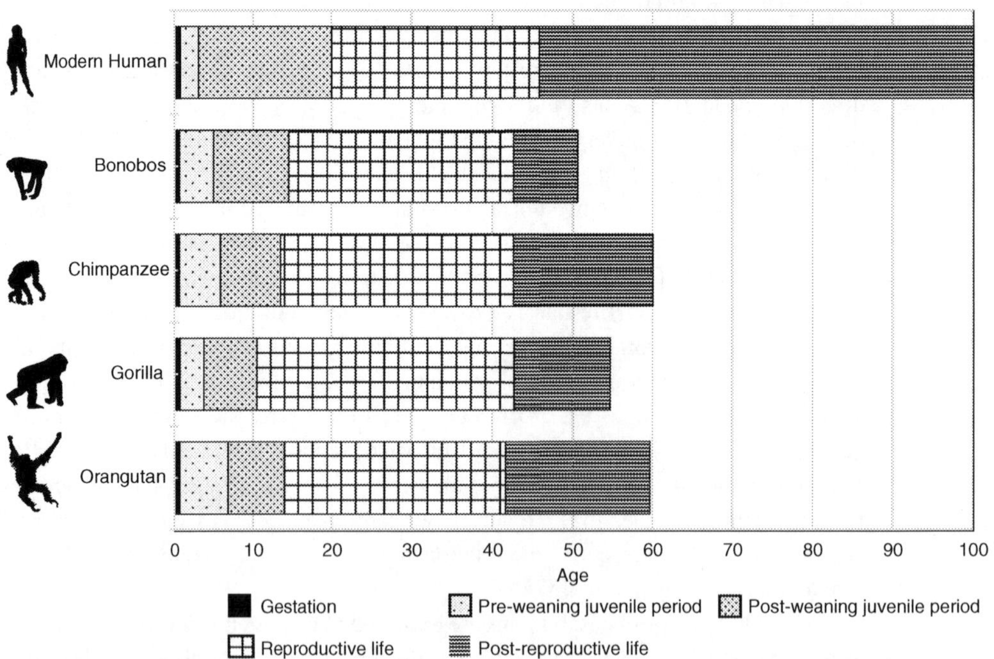

Figure 10.2 Schultz's diagram of the periods of life cycle for orangutan, gorilla, chimpanzee and modern human; template from [164] (p. 149). Data are similar to those in Figure 10.1. Species pictures are from http://phylopic.org/.

A common view in both popular discourse and the biomedical literature is that the kind of longevity illustrated by Jeanne Calment and other supercentenarians has been possible to achieve only very recently in human history, and that seniors emerge as a notable age class only after the demographic transition in the second half of the twentieth century in large income countries (see Chapter 14). We think that this is only partly true. Even in the absence of modern medicine and where life expectancy is low (e.g. below 40 years), a significant proportion of adults live past 50–60 years of age. Among contemporary hunter gatherers with reliable demographic information, those surviving to age 15 have about two chances in three to reach age 45, while those reaching age 45 will live an average of 20 additional years, reaching the age of 65 [58] (see also Chapter 11). If these populations are at all representative of the past, it is likely that seniors were not rare in past human history. After 65, however, mortality rises sharply. It is indeed likely that centenarians were truly exceptional in past human populations (see also [59]).

10.3.2 Human Females Are Fecund Primates

The reproductive window is short for human females. It is first narrowed at the starting end because of late age at sexual maturity. A vast literature has examined the secular declining trends in the age at menarche over the past two centuries (e.g. [60], as well as its genetic and environmental determinants since [61]). In non-contracepting populations of foragers, horticulturalists and pastoralists, age of menarche ranges from 12.6–18.4 [62]. Furthermore, even if age at menarche correlates with age at first reproduction (which occurs four years later, on average [62]), age at first reproduction depends also on cultural factors affecting pair bond formation, marriage and sexual behaviour.

The reproductive period is also narrowed at the other end by the abrupt cessation of reproduction at menopause. The exact timing of this cessation is somewhat variable across populations, due in part to nutritional status, environmental condition and issues surrounding estimation [63]. Age at menopause varies from 43 to 51 years of age, but fertility declines during peri-menopause; births after age 45 are rare in all human populations [64]. Overall, the female reproductive period is about 20 to 35 years in humans (which is about 1.3 to 2 times the length of the juvenile period).

However, despite having a short reproductive period, human females are exceptionally fertile and can bear a large number of children due to short birth spacing [65]. Here, the IBI varies principally because of differences in length of intensive breast-feeding and maternal nutrition affecting lactational amenorrhoea, and cultural prohibitions on post-partum sexual behaviour (e.g. [66, 67]). In non-contracepting hunter-gatherers and horticulturalists, IBI varies between 2.8 and 4.2 years [68]; for example, a median of 2.9 years in Baka pygmies [69]). As a result, high total fertility rates (i.e. the mean number of children born to women that survive until the end of their reproductive life) are possible (6.7 births on average, and as high as 9 births [70]).

10.3.3 In Humans, Females Risk Their Life Giving Birth

In humans, childbirth is complicated, painful and dangerous, as compared to other primates, including chimpanzees [71]. Delivery can be long and energetically demanding, and potential complications, including obstructed labour, haemorrhage, eclampsia and secondary infections, can be fatal for mothers [72]. It is estimated that prior to modern medical care, maternal risk of death was about 1% to 2% per childbirth; for instance maternal mortality was about 1.05% of all live births in eighteenth-century England [73], and was ranging from 2% in 1980s and to 1% in 1990s rural Gambia [74]. Even nowadays, the adult lifetime risk of maternal death has been estimated at about 4.4% in Africa [75], and reducing maternal mortality worldwide is one of the United Nations (UN's) Millennium Development Goals (MDG). Furthermore, in all human populations, giving birth is ritualized, attended by midwives and other helpers, while helping behaviour during delivery has been rarely documented in other primates [76].

10.3.4 To Become a Human Adult Takes Time, Is Perilous and Requires Adults' Care

Comparisons of growth patterns between humans and chimpanzees (reviewed in [77]) shows that, like other primates and mammals, chimpanzees exhibit two main growth phases: an infancy period characterized by a rapid growth up to weaning at about 48 to 60 months, then a juvenile period with lower and constantly declining growth rate up to sexual maturity, after which juveniles are largely responsible for their own care and feeding. By contrast, humans exhibit four developmental periods: an infancy period until weaning at about 36 months, a childhood period (ages 3 to 7) of moderate growth where children are still dependent for care and feeding, a juvenile period where growth is considerably slowed down and where children learn economic and social tasks, and finally adolescence, characterized by a growth spurt, puberty and socio-sexual maturation.

Human infancy is also perilous. Early infant (ages 0–1) and late infant (ages 1–2) mortality is higher among hunter-gatherers than among some chimpanzee populations while, by contrast, mortality rates are lower among hunter-gatherers throughout adulthood [70]. As a rule of thumb, about one in five children die in their first year of life in the absence of modern sanitation and medicine. This is mainly explained by the fact that, although they are larger for the size of the species (around 6% of adult weight versus 3% in other primates [78]) – human children are born premature by a few weeks or a few months, whether body or brain growth is considered [102, 106] compared to newborns of other great apes. In addition, the psychomotor and sensorimotor abilities of human neonates are vastly limited compared to those of neonatal great apes [79, 80]. The human newborn is also susceptible to infectious disease, and infant survival is highly dependent on birth weight and post-natal care. This parental care is comprised of food, protection, affection and education, and is needed for the physiological and behavioural development of offspring [81, 82].

10.3.5 Human Populations Grow Most of the Time

This 'absurd fertility', as referred to by Simone de Beauvoir in *Le Deuxième Sexe* [83, p. 89], coupled with high adult survival, has the important consequence that humans tend to give birth to a large number of babies, including hunter-gatherers. Although survival to adulthood is low among hunter-gatherers (average survival to age 15 is about 0.57 (0.45–0.66) [58]), about $6.7 \times 0.57 = 3.8$ children are expected to survive to adulthood. Thus, most contemporary hunter-gatherers and other *Homo sapiens* with available data are growing at or above 1% per year ([84] and see Chapter 11). This 'forager population paradox' – no population could have grown continuously at such high rates over long evolutionary time – suggests that *Homo sapiens* were equipped with the potential to rapidly colonize new habitats. But most likely, periodic catastrophes limited population growth over evolutionary history. Early existentialists reflecting on what might now be referred to as the evolution of cumulative culture, noticed that such 'species perpetuation [...] with too much abundance' allowed humans to thrive in non-biological activities other than merely perpetuating life [83, p. 89].

10.3.6 A Large Decoupling between Actuarial and Reproductive Senescence

For women, fertility declines appreciably by the late 30s until menopause. This pattern, combined with a long adult lifespan, means that actuarial and reproductive senescence are not synchronized in our species. Consequently, the proportion of post-fertile years lived compared to fertile ones is much larger in humans than in other primates species, even in the case of populations exhibiting low survival like the Trinidad plantation slaves [85]. Long post-fertile lifespans, as routinely documented in all human populations, are rare in mammals and have only been observed in killer and short-finned pilot whales ([86] and see Figure 11.2). Reproduction may still continue at older ages for men, but declines with age as well, offset from women's fertility decline by five to ten years, depending on polygyny, remarriage rates and spousal age gaps [87].

10.3.7 A Massive Overlap of Immature Children to Care For

Having short IBIs coupled with long offspring dependency means that human females often care for several children of different ages at the same time [88]. Assuming a schematic reproductive schedule, a human female aged 35 may care for five children of different ages simultaneously: a newborn, and four additional children aged 12 and under, whereas a chimpanzee female will never have to care and feed more than one offspring at a time (Figure 10.3). This level of compound offspring dependency is, to the best of our knowledge, a rarity within mammals, birds and maybe even across the whole tree of life. Given this potential for high dependency throughout human history, a large number of actors embedded within a familial and social network would have been needed to help provide care to offspring. The necessity and benefits afforded by 'allomaternal care' have been the focus of many studies since the mid-2000s [89, 90]. A broad range of allomaternal caregivers are mobilized during infancy and childhood.

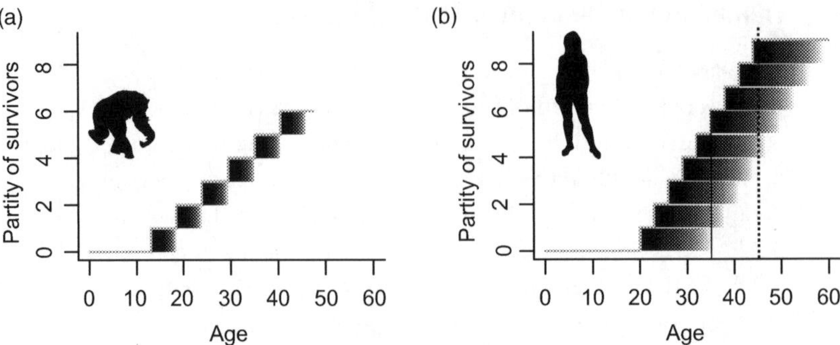

Figure 10.3 Schematics of females' parity with age and post-natal reproductive efforts (represented by grey areas) for chimpanzees and humans. In the absence of juvenile mortality, post-natal reproductive effort can be estimated by the number of still immature offspring to which a female has given birth. (a) female chimpanzees give birth every 5.5 years and care for their offspring up to 4.5 years (the age at weaning). (b) human females give birth every 3 years and care for their offspring for 15 years (taken here as an arbitrary age at independence from parental care).

Ethnographic studies of hunter-gatherers, horticulturalists and agro-pastoralists in sub-Saharan Africa confirm that infants have access to a large and diverse caregiving network, though the size and composition of these networks vary [91].

While motherless children have a higher risk of death than those with living mothers, allomaternal nursing, fostering and kin adoption provide a critical buffer (e.g. [92]; adoption, while observed in the great apes [93, 94], is much less frequent than among humans).

The high prevalence of allomaternal care in humans is often referred to by the ecological concept of *cooperative breeding*. Although genetic kinship is a component of allocare networks in humans, the flexibility in provisioning, learning and care strategies in human societies often relies upon the costly efforts of unrelated individuals [95]. This observation led Barry Bogin and colleagues to prefer the term 'biocultural reproduction' over 'cooperative breeding' to best characterize the human system of reproduction and offspring care [96].

10.4 Along the History of the Human Lineage: The Obstetrical Dilemma and the Birth of Immature Neonates

The complexity and danger of human childbirth has been framed as constituting the *obstetrical dilemma* ([97]; anatomical and fossil evidence reviewed in [98]; extensively discussed in the light of recent advances and alternative hypotheses in [99]). The idea is that, around 4 million years ago, the evolution of bipedalism led to new biomechanical constraints on the pelvis, and changed its form, by narrowing the birth canal. Such anatomical changes would not have originally had major consequences for childbirth. However, 1.5–2 million years ago, brain size increased at a much faster rate than body size [100], leading to a prosaic mechanical problem: How to make an ever-larger head pass through a birth canal relatively constant in size? The dilemma generates different optima for gestational length that would minimize maternal versus infant mortality, in light of encephalization. If the child is born earlier, childbirth is less dangerous for the mother since the child's smaller brain passes easily through the pelvic canal, but, in return, the newborn is more immature at birth, and hence more vulnerable to mortality. If the child is born later, it will be bigger, and hence have a better chance of survival, but the delivery is then more dangerous for the mother. Under the evolutionary pressure of increased brain size, evolution may have led to a compromised outcome favouring neither mother nor infant alone, such that maternal mortality is still relatively high and children are born relatively prematurely (also referred to secondary altriciality).

An alternative view defended by Holly Dunsworth and colleagues is the *energetics of gestational growth* hypothesis ([101]; see also [102]; and [99]). This idea posits that the very fast foetal growth of humans is so energetically costly that the main factor limiting gestation length is maternal metabolic exhaustion [103]. According to this view, labour is timed for when 'foetal energy demands surpass [...] the mother's ability to meet those demands' [101, p. 15214]. This hypothesis is supported by the fact that, across mammals, the invasiveness of the placenta reflects trade-offs between maternal–foetal transfers of nutrients and waste and the necessary protection against invading pathogens. The most invasive placentas are generally observed in species with slower life histories [104]. In humans, the haemochorial placenta facilitates in utero transfers by anchoring directly into the uterine lining by invasive extravillous trophoblasts. Such a strategy helps sustain the especially fast in utero foetal and brain growth of our species [105].

These hypotheses are important because they explain how some degree of excess maternal mortality, and both newborn immaturity and excess infant mortality, may be responses to the increases in brain size that have occurred over the past 2 million years. However, they also raise a puzzling question: How could such a dramatic cost, in terms of both infant and maternal mortality, have evolved, even at the benefit of increased brain size? To answer this, one theory claims that neonate 'secondary altriciality' may be advantageous because it favours later enhanced cognitive and sensorimotor skills by exposing neonates earlier to external stimuli, which in turn may favour adults' survival and reproduction in various environments due to improved sociality and cultural accumulation. This idea is often attributed to Adolf Portmann [106], who coined this period the 'extrauterine spring' [101, p. 15212]. This theory is also consistent with humans showing retained neotenous features over a long period of delayed development, characterized by enhanced cognitive plasticity while immersed in a rich social network (e.g. chapter 10 in [107]; extended among others by [108]). Much earlier, the psychologist Henri Wallon (1879–1962) developed a similar idea about the urge for the human neonate to develop social skills before any others to solicit food and aid and to cope with his native fragility 'his inability to survive without care from others'. Human neonates are thus compelled to develop 'social relationships that go far beyond the relationship with the physical world' [109, p. 48].

Interestingly, several lines of evidence have recently shown that the social skills of human neonates are not only acquired from being born into an intensely social environment. For instance, foetuses train for facial expressions even before delivery (which will later become the expression of their emotions; [110]). The cortical areas of neonates (1–5 days old) also respond to social stimuli (like a human face) but not to dynamic stimuli (like a potentially dangerous moving arm) [111].

10.5 Along the History of the Human Lineage: Tools, Diet, Intelligence and Ontogeny

The use of tools was long thought to have emerged less than 2 MYA; their makers were named as the first of their kind: *Homo habilis* ('able or handy man'). Since the mid-2000s, however, Oldowan assemblages and cut marks have been dated as old as ~2.5 MYA [112, 113]. Even older stone artefacts predating these by over 700 KYA exhibit some technological diversity clearly different from the single-purpose stone tools used by non-human primates discovered in West Turkana, Kenya [114]. This sets the dawn of technology at about 3.3 MYA. The use of tools is therefore not the sole characteristic of the first *Homo* but may have been as old as *Australopithecus africanus*. Parallel to this awareness, genetic and archaeological findings converge on the timing of human–chimpanzee divergence between 4 to 7 MYA ([115, 116], but see [117]).

According to these new understandings of our evolutionary history, it has become increasingly clear that there was more than enough time for our biology, our cognition and our social capabilities to have jointly evolved in response to the use of tools (see Figure 10.4). Rather than viewing changing cognitive, technical, linguistic and social capacities as the sole outcomes of our biological evolution, cultural niche construction and gene-culture coevolution are now thought to have jointly evolved over at least 3 million years [118].

The recognition of tool use earlier in ancient hominids suggests earlier changes in diet towards a greater consumption of more energy-dense foods, for example by crushing tubers, leguminous plants and cereals. In combination with other forms of food processing, as posited by the 'cooking' hypothesis of Wrangham and colleagues [119], changes in diet made possible through technological and cultural innovation are fundamental to explain the greater daily metabolism in the *Homo* lineage. This greater metabolism is critical to explain, because it helps afford the large human large brain and peculiar life history ([100, 120]). A recent study comparing great ape and human foraging energetics shows that hunter-gatherers are as efficient as other apes, when assessed as the ratio of energy gained to energy expended. However, there is a large difference in the rates of energy capture per unit time: hunter-gatherers capture food energy at a rate over three times higher than apes; plant domestication increases the energy capture rate up to three times more than pure foraging [121]. Such changes may also have been fundamental for the evolution of human linguistic and social capabilities. First, more efficient energy capture may have helped divert time from foraging towards socialization and the kind of knowledge transmission [122], using

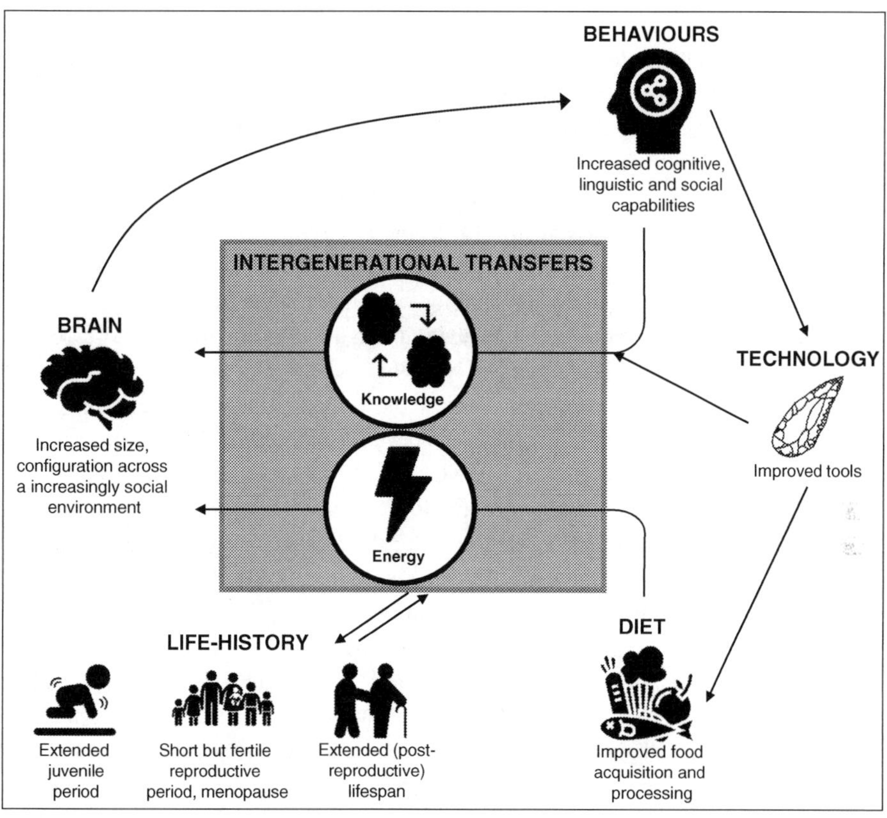

Figure 10.4 Conceptual figure explaining a revised version of the tools, diet and ontogeny theory. Pictograms are from https://thenounproject.com.

complex operatory chains, which would have been required for effective tool building [123]. Furthermore, the transmission of knowledge rarely occurs at random, but role models, teaching and learning strategies are based on ties embedded in complex social networks (e.g. based on gender, kinship, social status and other traits; see [124]).

The greater efficiency of human foraging strategies, enabled by technological innovations, and a social system favouring knowledge and resource transfers, supports the *embodied capital theory of the human adaptive complex* (ECT) [125], posited originally by Hillard Kaplan and colleagues. ECT proposes that our life history and intelligence have jointly evolved due to a 'dietary shift toward high-quality, nutrient-dense, and difficult-to-acquire food resources' [100, p. 156], where intergenerational transfers buffer energy budgets at all ages, but especially in early life during the period of children's brain ontogeny (extensively reviewed by [100]). For instance, the brain's metabolic requirement peaks around age 5, when it consumes about 43% of the body's daily metabolism, and stabilizes at around 20% by sexual maturity when the brain represents only about 2% of body mass [126]. This enormous metabolic cost of growing and maintaining a large functioning brain in our species supports the hypothesis that slowed ontogeny, delayed age at sexual maturity and a net flow of food transfers from adults to children is required to pay this cost (an extension of the maternal energy hypothesis [30] to any allomaternal intergenerational transfers). In turn, ECT posits that these huge costs are compensated by the cognitive benefits, enabling the ability to reap gains from a difficult foraging niche and to navigate the social landscape in a complex foraging ecology.

Complementary to the ECT, the *reserve capacity and biocultural reproduction* hypothesis proposes that the evolution of childhood and adolescence life history stages are prime movers leading to an increase in women's longevity [82, 127, 128]. Energetic surplus and enhanced allomaternal provisioning and care build a 'reserve capacity' that leads to early weaning, shorter IBIs and the insertion of human childhood and adolescence phases into the human ontogenesis, resulting in a more resilient hominin overall. According to this view, this built 'biocultural resilience' at adulthood indirectly leads to the evolution of post-reproductive lifespan, rather than direct fitness effects from intergenerational transfers by grandparents.

These approaches, emphasizing the importance of intergenerational transfers of energy and allomaternal care, are complementary to those emphasizing the role played by social stimuli and intergenerational information transfers that help configure the brain [129]. They emphasize the fact that the relationship between brain size and cognitive function is not straightforward (neither across species, across hominid fossils or among contemporary humans). However, the unique macroscopic organization and post-natal development of the human neocortex require an expanded juvenile period to establish cortical circuitry, thereby providing the 'neural substrate for behavioral and cognitive capacities unique to our species' [129, p. 13]. The extensive volume of intergenerational transfers during the prolonged period of cognitive maturation allows children to acquire not just the technical skills related to subsistence strategies, but the linguistic and social skills necessary for survival and knowledge transmission across generations (as suggested by [130]). Intergenerational transfers may also relate to increased brain size in another way: a larger brain size may have been required to maintain a functioning brain over a longer lifespan made possible by intergenerational transfers (called the *cognitive reserve hypothesis*; [131]).

Finally, we highlight how increased intergenerational transfers are not only fundamental for understanding the evolution of human ontogeny and delayed sexual maturity, but also our rapid pace of reproduction and extended adult lifespan.

10.6 Along the History of the Human Lineage: Intergenerational Transfers and Adult Life History

Since the first explicit theoretical contributions linking intergenerational transfers to the evolution of human lifespan in the mid-2000s [52, 132, 133], a growing body of literature has explored to what extent intergenerational transfers may have jointly evolved with other life history traits, particularly lifespan and reproduction. Most approaches have focused on transfers between age classes at a stable population state (e.g. [52, 132, 134, 135]), while ignoring kinship. Others instead have modelled the demographic effects of transfers, at stable population state, between specific categories of kin, such as the beneficial effect that (grand)parental care has on children's survival (e.g. [136, 137, 138, 139, 140]). The separate approaches arise mainly from the mathematical complexity of modelling both age- and kin-structured populations simultaneously. As a consequence, mathematical models investigating the impact of intergenerational transfers and kin investments on population dynamics and on life history evolution are few, although some major advances have been made ([141, 142, 143, 144]).

These new mathematical models have led to two major breakthroughs (see Box 10.1 for background). First, the potential for intergenerational transfers to impact offspring fitness over a long juvenile period tremendously increases the force of natural selection on adult survival, favouring higher adult life expectancy. This general result is predicted by all theories and models mentioned thus far that invoke the role of intergenerational transfers. Consistent with these models, intergenerational transfers in a kin-structured context have also been proposed as a major force in the negative selection of susceptibility alleles to late-onset diseases in humans, whose onset may occur long after age at menopause and that may explain why most of these alleles are recent and rare ([139]; see also Chapter 16).

Second, intergenerational transfers also tend to favour the emergence of high fertility at the beginning of reproductive life, to the detriment of fertility later in life (see [136, 138]). Consistent with this idea, most non-human primates show modest reproductive senescence prior to death, whereas marked reproductive senescence occurs well before death in all human populations [145]. Reproductive senescence in this case results from a decrease of the strength of natural selection on older age fertility without necessarily favouring a genetically programmed cessation of reproduction (i.e. like menopause). However, early proposals about the emergence of menopause have also relied on the role of intergenerational transfers (e.g. 'the mother hypothesis' [146] and the 'grandmother hypothesis' [133]). To explain the emergence of menopause and its maintenance requires an additional mechanism by which late continuing reproduction must sufficiently compromise (grand)maternal fitness (e.g. [146, 147, 148, 149, 150, 151, 152]). This condition may be met in humans due to increased maternal mortality with age coupled with high and extended offspring dependency (but see [153, 154]), but alternative mechanisms reflecting intergenerational competition between younger and older females have also been suggested (e.g. [155, 156]). The development and testing of models proposed to explain the biology and evolution of menopause and post-reproductive lifespan, both within species and comparatively is still under investigation [151, 157].

In any case, increased lifespan combined with reproductive senescence and menopause leads to a large decoupling between reproductive and actuarial senescence. In these circumstances, the emergence of long post-reproductive life is one of the most peculiar features of our evolved human life history, observed, to date, only in a couple of non-primate species [158]. Some transfers, however, may also occur within generations, as from older to younger siblings [159], which would be expected to mitigate fitness effects of post-reproductive transfers. Formal modelling of both intragenerational and intergenerational transfers together is still mostly lacking.

Box 10.1 The Effects of Intergenerational Transfers on the Evolution of Lifespan and Reproduction: A Primer

Let us assume age trajectories of survival and fertility from age at sexual maturity to maximum lifespan in a species where there are no intergenerational transfers (see Figure 10.5). Here, offspring survival and fertility would be independent of any post-natal transfers of food, information or other resources from their parents or others. In this case, survival and fertility decline with age at the same pace, with little or no post-reproductive period (sixth prediction of [160]). The graphics on the left of Figure 10.5(a) and (b) depict this ancestral state. Note that survival and fertility trajectories appear to overlap with each other for pedagogical purposes, but reflect different scales.

Given the effects of child fitness on the inclusive fitness of adult parents, we now consider that adults can alter the fitness of their children (and other kin) through feeding, teaching, care and any other form of 'transfer'. Let us

(a) Joint evolution of intergenerational transfers and lifespan

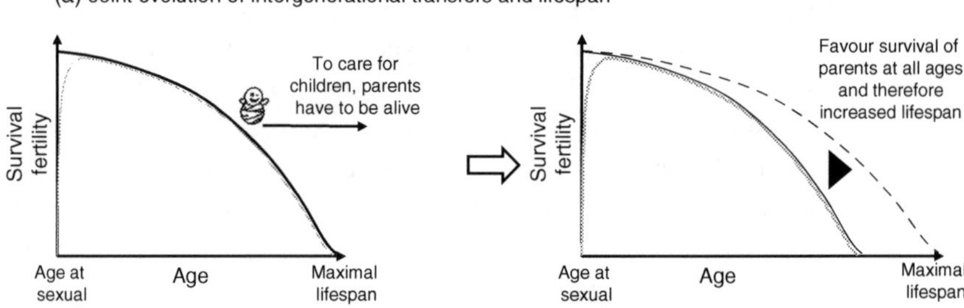

(b) Joint evolution of intergenerational transfers and fertility age trajectory

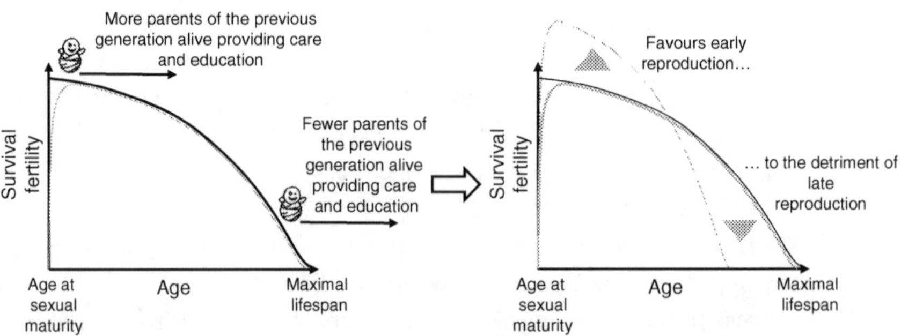

Figure 10.5 Decoupling of survival and reproductive trajectories resulting from a longer period of increased dependency of children on caregiving by relatives.

first examine how lifespan shifts in response to increases in intergenerational transfers (Figure 10.5(a)). The simplest case is of parental care. Because parents have to be alive to care for their children, offspring fitness will be tied to adult survival. Consequently, the force of selection on adult survival depends not only on parents' individual future reproductive value (i.e. on potential future reproduction), but also on the fitness impacts of transfers to offspring (i.e. past reproduction). Alleles that positively (or negatively) affect survival are therefore selected for (or against) at all adult ages where parents are still caring for children. Such selection leads to an increase in adult lifespan.

Next, we focus on the effects of transfers on reproduction (Figure 10.5(b)). As described, offspring fitness often depends on the care received by individuals of the previous generation. To simplify, we can focus on just offspring survival and (grand)parental care. If parents' survival declines with age due to actuarial senescence, then children born to young parents will have more grandparents alive, and their parents will be less likely to die during their infancy or childhood, compared to children born to older parents. As a consequence, offspring born to young parents will receive more care and investment than children born to old parents. This higher investment should improve survivorship among the former compared to the latter. Since children born earlier in life contribute more to parental fitness than those born later in life, any allele increasing early fertility will be favoured over those increasing late reproduction.

10.7 Conclusion

Humans are mammals. As mammals, we grow determinately, we are iteroparous, care intensely for our babies. And we senesce. As primates, we are big brained and long-lived. However, as humans, we are very peculiar primates. Humans sit as outliers on many of the classical ecological continua of life history theory.

Yet, a large literature typically describes humans as the slowest primate species along the slow–fast continuum. This caricature is partially wrong. Primates, relative to their size, exhibit greater longevity than other mammals, and humans are exceptionally long-lived primates. Delayed maturation and extended juvenility are also exacerbated in humans. However, reproduction is undoubtedly fast. Although concentrated over a short reproductive window for females, IBIs are short and the number of children that a woman can give birth to over her short reproductive life is large, well over levels of replacement fertility even when taking high early life mortality into account. Delayed maturation and short IBI leads to a massive overlap of immature children to care for, not by mothers alone, but by a network of multigenerational caregivers. Such overlap of immature children to care for is a rarity across mammals and unique to humans among other primates. A short reproductive life and extended lifespan also lead to a large decoupling between reproductive and actuarial senescence and the existence of a long post-reproductive life. The mix of 'slow' and 'fast' life history characteristics makes humans deviate from both the slow–fast continuum (mostly empirical based on the statistical covariance between life history traits across species) and the earlier *r*/*K* continuum (mostly theoretical inferring evolutionary covariance between life history traits from population dynamic strategies). If humans exhibit mostly characteristics from a *K* species (long gestation periods, slow maturation and extended parental care), evidence for foraging societies show that population abundance is likely not stable at the carrying capacity *K* in the short term. Rather, even hunter-gatherers often show fast population growth of about 1% per year, stabilized in the long term by stochastic catastrophes.

Our understanding of the evolution of this human life cycle along the hominid lineage since divergence with a chimpanzee-like ancestor has improved due to a plethora of theoretical and empirical studies from palaeoanthropology, physiology, neurosciences, biodemography and human behavioural ecology – along two main axes of research. The first aims at understanding the evolution of traits related to the birth of immature children in our species. The second aims at understanding the evolution of the physiological, psychological and social development of human children. Together, they direct our attention to the joint evolution of our brain (its increase in size and in the time required for its configuration), obstetric traits (including the gestation length, the complexity and the dangerousness of human delivery, the immaturity and frailty of neonates), our life history (the emergence of the menopause and extended post-reproductive lifespan, but also increased maternal and neonatal mortality, delayed maturation and fast reproduction), and our biocultural practices (socially structured intergenerational transfers of food and other resources, and of information and other immaterial goods and services, providing sustenance, protection and support, education and affection [82]).

Modelling the correlation in fitness components between individuals of multiple generations resulting from intergenerational transfers can fundamentally alter the classical 'force of selection' believed to impact the evolution of ageing and lifespan. Such amendments tying life history evolution to sociality have proved invaluable so far for improving our understanding of the joint evolution of our extended longevity, peculiar fertility schedule and our cognitive and social capabilities. However, these models are still in their childhood (better than infancy!). In humans, kinship can reflect real or putative family ties, where social kinship is one important factor structuring behaviour (e.g. with whom to forage, marry, raise children, form a coalition, migrate, seek advice etc.). Future directions will benefit from modelling transfers of material and immaterial resources within a social network, consisting of both biological kin and unrelated individuals. Therefore, to fully apprehend the joint evolution of life history traits with our cognitive, linguistic and social capabilities, it will be required to understand how ecology, demography and both biological and social kinship set the stage for the evolution of cumulative culture to occur, and how biological evolution and cultural dynamics have shaped human history.

References

1. Ronget, V., Lemaître, J.-F., Tidière, M., Gaillard, J.-M. 2020. Assessing the diversity of the form of age-specific changes in adult mortality from captive mammalian populations. *Diversity* **12**, 1–12.
2. Jones, O.R. et al. 2014. Diversity of ageing across the tree of life. *Nature* **505**, 169–173.
3. Western, D. 1979. Size, life-history and ecology in mammals. *Afr. J. Ecol.* **17**, 185–204.
4. Gaillard, J.M., Pontier, D., Allaine, D., Lebreton, J.D., Trouvilliez, J., Clobert, J. 1989. An analysis of demographic tactics in birds and mammals. *Oikos* **56**, 59–76.
5. Gaillard, J.-M., Lemaître, J., Berger, V., Bonenfant, C., Devillard, S., Douhard, M., Gamelon, M., Plard, F., Lebreton, J.-D. 2016. Life histories, axes of variation in. In *Encyclopedia of Evolutionary Biology* (ed. R.M. Kilman), pp. 312–323. Academic Press.
6. Sibly, R.M., Brown, J.H. 2007. Effects of body size and lifestyle on evolution of mammal life histories. *Proc. Natl. Acad. Sci. USA* **104**, 17707–17712.
7. Pavard, S., Coste, C.F.D. 2020. Evolution of the human life cycle. In *Encyclopedia of Biomedical Gerontology* (ed. S.I.S. Rattan), pp. 46–56. Academic Press.
8. Harvey, P.H., Zammuto, R.M. 1985. Patterns of mortality and age at first reproduction in natural populations of mammals. *Nature* **315**, 319–320.
9. Promislow, D.E.L., Harvey, P.H. 1990. Living fast and dying young: a comparative-analysis of life-history variation among mammals. *J. Zool.* **220**, 417–437.
10. Healy, K., Ezard, T.H.G., Jones, O.R., Salguero-Gómez, R., Buckley, Y.M. 2019. Animal life history is shaped by the pace of life and the distribution of age-specific mortality and reproduction. *Nat. Ecol. Evol.* **3**, 1217–1224.
11. Péron, G., Lemaître, J.-F., Ronget, V., Tidière, M., Gaillard, J.-M. 2019. Variation in actuarial senescence does not reflect life span variation across mammals. *PLoS Biol.* **17**, e3000432 (doi:10.1371/journal.pbio.3000432).
12. Charnov, E.L., Berrigan, D. 1993. Why do female Primates have such long lifespans and so few babies? Or life in the slow lane. *Evol. Anthropol. Issues News Rev.* **1**, 191–194.
13. Auer, S.K., Dick, C.A., Metcalfe, N.B., Reznick, D.N. 2018. Metabolic rate evolves rapidly and in parallel with the pace of life history. *Nat. Commun.* **9**, 14 (doi:10.1038/s41467-017-02514-z).
14. Pontzer, H. et al. 2014. Primate energy expenditure and life history. *Proc. Natl. Acad. Sci. USA* **111**, 1433–1437.
15. Harvey, P.H., Clutton-Brock, T.H. 1985. Life history variation in Primates. *Evolution* **39**, 559–581.
16. Vinicius, L., Mumby, H.S. 2013. Comparative analysis of animal growth: a primate continuum revealed by a new dimensionless growth rate coefficient. *Evolution* **67**, 1485–1492.
17. Rilling, J.K., Insel, T.R. 1999. The primate neocortex in comparative perspective using magnetic resonance imaging. *J. Hum. Evol.* **37**, 191–223.
18. Isler, K., van Schaik, C.P. 2006. Metabolic costs of brain size evolution. *Biol. Lett.* **2**, 557–560.
19. Weisbecker, V., Goswami, A. 2010. Brain size, life history, and metabolism at the marsupial/placental dichotomy. *Proc. Natl. Acad. Sci. USA* **107**, 16216–16221.
20. Finarelli, J.A., Flynn, J.J. 2009. Brain-size evolution and sociality in Carnivora. *Proc. Natl. Acad. Sci. USA* **106**, 9345–9349.
21. Finarelli, J.A. 2010. Does encephalization correlate with life history or metabolic rate in Carnivora? *Biol. Lett.* **6**, 350–353.

22. Isler, K., van Schaik, C.P. 2009. The expensive brain: a framework for explaining evolutionary changes in brain size. *J. Hum. Evol.* **57**, 392–400.
23. Isler, K., van Schaik, C.P. 2012. Allomaternal care, life history and brain size evolution in mammals. *J. Hum. Evol.* **63**, 52–63.
24. Austad, S.N., Fischer, K.E. 1992. Primate longevity: its place in the mammalian scheme. *Am. J. Primatol.* **28**, 251–261.
25. Allman, J., McLaughlin, T., Hakeem, A. 1993. Brain weight and life-span in primate species. *Proc. Natl. Acad. Sci. USA* **90**, 118–122.
26. Walker, R., Burger, O., Wagner, J., Von Rueden, C.R. 2006. Evolution of brain size and juvenile periods in Primates. *J. Hum. Evol.* **51**, 480–489.
27. Barrickman, N.L., Bastian, M.L., Isler, K., van Schaik, C.P. 2008. Life history costs and benefits of encephalization: a comparative test using data from long-term studies of Primates in the wild. *J. Hum. Evol.* **54**, 568–590.
28. Gonzalez-Lagos, C., Sol, D., Reader, S.M. 2010. Large-brained mammals live longer. *J. Evol. Biol.* **23**, 1064–1074.
29. Sacher, G.A. 1982. The role of brain maturation in the evolution of the Primates. In *Primate Brain Evolution: Methods and Concepts* (eds E. Armstrong, D. Falk), pp. 97–112. Springer US (doi:10.1007/978-1-4684-4148-2_8).
30. Martin, R. 1996. Scaling of the mammalian brain: the maternal energy hypothesis. *Physiology* **11**, 149–156.
31. Mink, J.W., Blumenschine, R.J., Adams, D.B. 1981. Ratio of central nervous system to body metabolism in vertebrates: its constancy and functional basis. *Am. J. Physiol.* **241**, R203–R212.
32. Casey, B.J., Galvan, A., Hare, T.A. 2005. Changes in cerebral functional organization during cognitive development. *Curr. Opin. Neurobiol.* **15**, 239–244.
33. Johnson, M.H. 2001. Functional brain development in humans. *Nat. Rev. Neurosci.* **2**, 475.
34. Clutton-Brock, T.H., Harvey, P.H. 1980. Primates, brains and ecology. *J. Zool.* **190**, 309–323.
35. DeCasien, A.R., Williams, S.A., Higham, J.P. 2017. Primate brain size is predicted by diet but not sociality. *Nat. Ecol. Evol.* **1**, 0112.
36. Jones, J.H. 2011. Primates and the evolution of long, slow life histories. *Curr. Biol.* **21**, R708–R717.
37. Aiello, L.C., Wheeler, P. 1995. The expensive-tissue hypothesis: the brain and the digestive system in human and primate evolution. *Curr. Anthropol.* **36**, 199–221.
38. Navarrete, A., van Schaik, C.P., Isler, K. 2011. Energetics and the evolution of human brain size. *Nature* **480**, 91–93 (doi:10.1038/nature10629).
39. Moll, H., Tomasello, M. 2007. Cooperation and human cognition: the Vygotskian intelligence hypothesis. *Philos. Trans. R. Soc. Lond. B. Biol. Sci.* **362**, 639–648.
40. Reader, S.M., Laland, K.N. 2002. Social intelligence, innovation, and enhanced brain size in Primates. *Proc. Natl. Acad. Sci. USA* **99**, 4436–4441.
41. Reader, S.M., Hager, Y., Laland, K.N. 2011. The evolution of primate general and cultural intelligence. *Philos. Trans. R. Soc. Lond. B. Biol. Sci.* **366**, 1017–1027.
42. Dunbar, R.I.M. 1998. The social brain hypothesis. *Evol. Anthropol. Issues News Rev.* **6**, 178–190.
43. Dunbar, R.I. 2009. The social brain hypothesis and its implications for social evolution. *Ann. Hum. Biol.* **36**, 562–572.
44. Dunbar, R.I.M., Shultz, S. 2017. Why are there so many explanations for primate brain evolution? *Philos. Trans. R. Soc. Lond. B. Biol. Sci.* **372**, 20160244.

45. Kaplan, H.S., Gangestad, S.W., Gurven, M., Lancaster, J., Mueller, T., Robson, A. 2007. The evolution of diet, brain and life history among Primates and humans. In *Guts and Brains: An Integrative Approach to the Hominin Record* (ed. W. Roebroek), pp. 47–81. Leiden University Press.
46. Kaplan, H.S., Gurven, M., Lancaster, J. 2007. Brain evolution and the human adaptive complex: an ecological and social theory. In *The Evolution of Mind: Fundamental Questions and Controversies* (eds S.W. Gangestad, J.A. Simpson), pp. 269–279. Guilford Press.
47. Bourke, A.F.G. 2011. The validity and value of inclusive fitness theory. *Proc. R. Soc. Biol. Sci.* **278**, 3313–3320.
48. Hamilton, W.D. 1964. Genetical evolution of social behaviour I. *J. Theor. Biol.* **7**, 1–16.
49. Hamilton, W.D. 1964. Genetical evolution of social behaviour 2. *J. Theor. Biol.* **7**, 17–52.
50. Clutton-Brock, T. 2002. Breeding together: kin selection and mutualism in cooperative vertebrates. *Science* **296**, 69–72.
51. Bourke, A.F.G. 2007. Kin selection and the evolutionary theory of aging. *Annu. Rev. Ecol. Evol. Syst.* **38**, 103–128.
52. Lee, R.D. 2003. Rethinking the evolutionary theory of aging: transfers, not births, shape senescence in social species. *Proc. Natl. Acad. Sci. USA* **100**, 9637–9642.
53. Clutton-Brock, T.H. 1991. *The Evolution of Parental Care*. Princeton University Press.
54. Péron, G., Bonenfant, C., Lemaitre, J.-F., Ronget, V., Tidiere, M., Gaillard, J.-M. 2019. Does grandparental care select for a longer lifespan in non-human mammals? *Biol. J. Linn. Soc.* **128**, 360–372 (doi:10.1093/biolinnean/blz078).
55. Allman, J., Rosin, A., Kumar, R., Hasenstaub, A. 1998. Parenting and survival in anthropoid Primates: caretakers live longer. *Proc. Natl. Acad. Sci. USA* **95**, 6866–6869.
56. Beauchamp, G. 2014. Do avian cooperative breeders live longer? *Proc. R. Soc.* **281,** 20140844.
57. Thorley, J. 2020. The case for extended lifespan in cooperatively breeding mammals: a re-appraisal. *PeerJ.* **8**, e9214–e9214.
58. Gurven, M., Kaplan, H. 2007. Longevity among hunter-gatherers: a cross-cultural examination. *Popul. Dev. Rev.* **33**, 321–365.
59. Wilmoth, J.R. 1995. The earliest centenarians: a statistical analysis. In *Exceptional Longevity: From Prehistory to the Present* (eds B. Jeune, J. Vaupel), pp. 125–169. Odense University Press.
60. Wyshak, G., Frisch, R.E. 1982. Evidence for a secular trend in age of menarche. *N. Engl. J. Med.* **306**, 1033–1035.
61. Zacharias, L., Wurtman, R.J. 1969. Age at menarche: genetic and environmental influences. *N. Engl. J. Med.* **280**, 868–875.
62. Hochberg, Z., Gawlik, A., Walker, R.S. 2011. Evolutionary fitness as a function of pubertal age in 22 subsistence-based traditional societies. *Int. J. Pediatr. Endocrinol.* **2011**, 2–2.
63. Parsaeian, M. et al. 2017. An explanation for variation in age at menopause in developing countries based on the second national integrated micronutrient survey in Iran. *Arch. Iran. Med.* **20**, 361–367 (doi:0172006/AIM.008).
64. Peccei, J.S. 1999. First estimates of heritability in the age of menopause. *Curr. Anthropol.* **40**, 553–558.
65. Mace, R. 2000. Evolutionary ecology of human life history. *Anim. Behav.* **59**, 1–10.
66. Caldwell, P., Caldwell, J.C. 1981. The function of child-spacing in traditional societies and the direction of change. In *Child Spacing in Tropical Africa: Traditions and Change* (eds H.J. Page, R. Lesthaeghe), pp. 73–92. New York Academic.

67. Pimentel, J., Ansari, U., Omer, K., Gidado, Y., Baba, M.C., Andersson, N., Cockcroft, A. 2020. Factors associated with short birth interval in low- and middle-income countries: a systematic review. *BMC Pregnancy Childbirth* **20**, 156.
68. Davison, R.J., Gurven, M.D. 2020. Human uniqueness illustrated by life history diversity among small-scale societies and chimpanzees. *bioRxiv*, 2020.09.02.280602.
69. Rozzi, F.V.R., Koudou, Y., Froment, A., Le Bouc, Y., Botton, J. 2015. Growth pattern from birth to adulthood in African pygmies of known age. *Nat. Commun.* **6**, 7672.
70. Davison, R.J., Gurven, M.D. 2021. Human uniqueness? Life history diversity among small-scale societies and chimpanzees. *PLoS ONE* **16**, e0239170.
71. Elder, J.H., Yerkes, R.M., Cushing, H.W. 1936. Chimpanzee births in captivity: a typical case history and report of sixTEEn births. *Proc. R. Soc. Lond. Ser. B – Biol. Sci.* **120**, 409–421.
72. Say, L., Chou, D., Gemmill, A., Tunçalp, Ö., Moller, A.-B., Daniels, J., Gülmezoglu, A.M., Temmerman, M., Alkema, L. 2014. Global causes of maternal death: a WHO systematic analysis. *Lancet Glob. Health* **2**, e323–e333.
73. Chamberlain, G. 2006. British maternal mortality in the 19th and early 20th centuries. *J. R. Soc. Med.* **99**, 559–563 (doi:10.1258/jrsm.99.11.559).
74. Walraven, G., Telfer, M., Rowley, J., Ronsmans, C. 2000. Maternal mortality in rural Gambia: levels, causes and contributing factors. *Bull. World Health Organ.* **78**, 603–613.
75. Wilmoth, J. 2009. The lifetime risk of maternal mortality: concept and measurement. *Bull. World Health Organ.*, **87**, 256–262.
76. Yao, M., Yin, L., Zhang, L., Liu, L., Qin, D., Pan, W. 2012. Parturitions in wild white-headed langurs (*Trachypithecus leucocephalus*) in the Nongguan Hills, China. *Int. J. Primatol.* **33**, 888–904.
77. Bogin, B. 2015. Human growth and development. In *Basics in Human Evolution* (ed. M.P. Muehlenbein), pp. 285–293. Academic Press.
78. DeSilva, J.M. 2011. A shift toward birthing relatively large infants early in human evolution. *Proc. Natl. Acad. Sci. USA* **108**, 1022–1027.
79. Parker, S.T. 1977. Piaget's sensorimotor period series in an infant macaquea: model for comparing non stereotyped behavior and intelligence in human and nonhuman Primates. In *Primate Biosocial Development* (ed. S. Chevalier-Skolnikoff, F. Poirier), 43–112. Garland.
80. Chevalier-Skolnikoff, S. 1983. Sensorimotor development in orang-utans and other Primates. *J. Hum. Evol.* **12**, 545.
81. Geary, D.C., Flinn, M.V. 2001. Evolution of human parental behavior and the human family. *Parent. Sci. Pract.* **1**, 1–2.
82. Bogin, B. 2021. *Patterns of Human Growth* (3rd ed.). Cambridge University Press.
83. de Beauvoir, S. 1949. *Le Deuxième Sexe*. Gallimard.
84. Gurven, M.D., Davison, R.J. 2019. Periodic catastrophes over human evolutionary history are necessary to explain the forager population paradox. *Proc. Natl. Acad. Sci.* **116**, 12758 (doi:10.1073/pnas.1902406116).
85. Levitis, D.A., Burger, O., Lackey, L.B. 2013. The human post-fertile lifespan in comparative evolutionary context. *Evol Anthr.* **22**, 66–79.
86. Ellis, S., Franks, D.W., Nattrass, S., Cant, M.A., Bradley, D.L., Giles, D., Balcomb, K.C., Croft, D.P. 2018. Postreproductive lifespans are rare in mammals. *Ecol. Evol.* **8**, 2482–2494.
87. Tuljapurkar, S.D., Puleston, C.O., Gurven, M.D. 2007. Why men matter: mating patterns drive evolution of human lifespan. *PLoS ONE* **2**, e785.

88. Gurven, M., Walker, R. 2006. Energetic demand of multiple dependents and the evolution of slow human growth. *Proc. R. Soc. B Biol. Sci.* **273**, 835–841 (doi:10.1098/rspb.2005.3380).
89. Bentley, G.R., Mace, R. 2009. *Substitute Parents: Biological and Social Perspective on Alloparenting across Human Societies*. Berghahn Books.
90. Sear, R., Mace, R. 2008. Who keeps children alive? A review of the effects of kin on child survival. *Evol. Hum. Behav.* **29**, 1–18.
91. Helfrecht, C., Roulette, J.W., Lane, A., Sintayehu, B., Meehan, C.L. 2020. Life history and socioecology of infancy. *Am. J. Phys. Anthropol.* **173**, 619–629.
92. Pavard, S., Gagnon, A., Desjardins, B., Heyer, E. 2005. Mother's death and child survival: the case of early Quebec. *J. Biosoc. Sci.* **37**, 209–227.
93. Morrison, R.E., Eckardt, W., Colchero, F., Vecellio, V., Stoinski, T.S. 2021. Social groups buffer maternal loss in mountain gorillas. *eLife* **10**, e62939.
94. Tokuyama, N., Toda, K., Poiret, M.-L., Iyokango, B., Bakaa, B., Ishizuka, S. 2021. Two wild female bonobos adopted infants from a different social group at Wamba. *Sci. Rep.* **11**, 4967.
95. Hill, K.R. et al. 2011. Co-residence patterns in hunter-gatherer societies show unique human social structure. *Science* **331**, 1286–1289.
96. Bogin, B., Bragg, J., Kuzawa, C. 2014. Humans are not cooperative breeders but practice biocultural reproduction. *Ann. Hum. Biol.* **41**, 368–380.
97. Washburn, S.L. 1960. Tools and human evolution. *Sci. Am.* **203**, 63–75.
98. Wittman, A.B., Wall, L.L. 2007. The evolutionary origins of obstructed labor: bipedalism, encephalization, and the human obstetric dilemma. *Obstet. Gynecol. Surv.* **62**, 739–748.
99. Haeusler, M., Grunstra, N.D.S., Martin, R.D., Krenn, V.A., Fornai, C., Webb, N.M. 2021. The obstetrical dilemma hypothesis: there's life in the old dog yet. *Biol. Rev. Camb. Philos. Soc.* **96**, 2031–2057 (doi:10.1111/brv.12744).
100. Hublin, J.J., Neubauer, S., Gunz, P. 2015. Brain ontogeny and life history in Pleistocene hominins. *Philos. Trans. R. Soc. B-Biol. Sci.* **370**, 20140062
101. Dunsworth, H.M., Warrener, A.G., Deacon, T., Ellison, P.T., Pontzer, H. 2012. Metabolic hypothesis for human altriciality. *Proc. Natl. Acad. Sci. USA* **109**, 15212–15216.
102. Little, B.B. 1989. Gestation length, metabolic-rate, and body and brain weights in Primates: epigenetic effects. *Am. J. Phys. Anthropol.* **80**, 213–218.
103. Butte, N.F., King, J.C. 2005. Energy requirements during pregnancy and lactation. *Public Health Nutr.* **8**, 1010–1027.
104. Garratt, M., Gaillard, J.-M., Brooks, R.C., Lemaître, J.-F. 2013. Diversification of the eutherian placenta is associated with changes in the pace of life. *Proc. Natl. Acad. Sci. USA* **110**, 7760–7765.
105. Zeldovich, V.B., Bakardjiev, A.I. 2012. Host defense and tolerance: unique challenges in the placenta. *PLoS Pathog.* **8**, e1002804.
106. Portmann, A. 1969. *Biologische Fragmente zu einer Lehre vom Menschen [A Zoologist Looks at Humankind]*. Schwabe.
107. Gould, S.J. 1977. *Ontogeny and Phylogeny*. Harvard University Press.
108. Bjorklund, D.F. 1997. The role of immaturity in human development. *Psychol. Bull.* **122**, 153–169.
109. Jalley, E. 1981. *Wallon, lecteur de Freud et Piaget*. Editions Sociales.
110. Reissland, N., Francis, B., Mason, J., Lincoln, K. 2011. Do facial expressions develop before birth? *PLoS ONE* **6**, e24081.

111. Farroni, T., Chiarelli, A.M., Lloyd-Fox, S., Massaccesi, S., Merla, A., Di Gangi, V., Mattarello, T., Faraguna, D., Johnson, M.H. 2013. Infant cortex responds to other humans from shortly after birth. *Sci. Rep.* **3**, 2851.
112. Semaw, S., Renne, P., Harris, J.W.K., Feibel, C.S., Bernor, R.L., Fesseha, N., Mowbray, K. 1997. 2.5-million-year-old stone tools from Gona, Ethiopia. *Nature* **385**, 333–336 (doi:10.1038/385333a0).
113. McPherron, S.P., Alemseged, Z., Marean, C.W., Wynn, J.G., Reed, D., Geraads, D., Bobe, R., Béarat, H.A. 2010. Evidence for stone-tool-assisted consumption of animal tissues before 3.39 million years ago at Dikika, Ethiopia. *Nature* **466**, 857–860 (doi:10.1038/nature09248).
114. Harmand, S. et al. 2015. 3.3-million-year-old stone tools from Lomekwi 3, West Turkana, Kenya. *Nature* **521**, 310–315.
115. Pääbo, S. 2003. The mosaic that is our genome. *Nature* **421**, 409–412 (doi:10.1038/nature01400).
116. Glazko, G.V., Nei, M. 2003. Estimation of divergence times for major lineages of primate species. *Mol. Biol. Evol.* **20**, 424–434 (doi:10.1093/molbev/msg050).
117. Moorjani, P., Amorim, C.E.G., Arndt, P.F., Przeworski, M. 2016. Variation in the molecular clock of Primates. *Proc. Natl. Acad. Sci.* **113**, 10607 (doi:10.1073/pnas.1600374113).
118. Laland, K.N., Odling-Smee, J., Feldman, M.W. 2001. Cultural niche construction and human evolution. *J. Evol. Biol.* **14**, 22–33 (doi:10.1046/j.1420-9101.2001.00262.x).
119. Wrangham, R.W., Jones, J.H., Laden, G., Pilbeam, D., Conklin-Brittain, N. 1999. The raw and the stolen: cooking and the ecology of human origins. *Curr. Anthropol.* **40**, 567–594.
120. Pontzer, H. et al. 2016. Metabolic acceleration and the evolution of human brain size and life history. *Nature* **533**, 390.
121. Kraft Thomas, S. et al. 2021. The energetics of uniquely human subsistence strategies. *Science* **374**, eabf0130 (doi:10.1126/science.abf0130).
122. Fonseca-Azevedo, K., Herculano-Houzel, S. 2012. Metabolic constraint imposes tradeoff between body size and number of brain neurons in human evolution. *Proc. Natl. Acad. Sci.* **109**, 18571–18576.
123. Nowell, A., Davidson, I. 2010. *Stone Tools and the Evolution of Human Cognition*. University Press of Colorado.
124. Terashima, H., Hewlett, B.S., eds. 2016. *Social Learning and Innovation in Contemporary Hunter-Gatherers: Evolutionary and Ethnographic Perspectives*. Springer Japan (doi:10.1007/978-4-431-55997-9).
125. Kaplan, H., Hill, K., Lancaster, J., Hurtado, A.M. 2000. A theory of human life history evolution: diet, intelligence, and longevity. *Evol. Anthropol.* **9**, 156–185.
126. Kuzawa, C.W. et al. 2014. Metabolic costs and evolutionary implications of human brain development. *Proc. Natl. Acad. Sci. USA* **111**, 13010–13015.
127. Larke, A., Crews, D.E. 2006. Parental investment, late reproduction, and increased reserve capacity are associated with longevity in humans. *J. Physiol. Anthropol.* **25**, 119–131 (doi:10.2114/jpa2.25.119).
128. Bogin, B. 2009. Childhood, adolescence, and longevity: a multilevel model of the evolution of reserve capacity in human life history. *Am. J. Hum. Biol.* **21**, 567–577 (doi:10.1002/ajhb.20895).
129. Hrvoj-Mihic, B., Bienvenu, T., Stefanacci, L., Muotri, A., Semendeferi, K. 2013. Evolution, development, and plasticity of the human brain: from molecules to bones. *Front. Hum. Neurosci.* **7**, 707 (doi:10.3389/fnhum.2013.00707).

130. Bogin, B., Smith, B.H. 1996. Evolution of the human life cycle. *Am. J. Hum. Biol.* **8**, 703–716 (doi:10.1002/(SICI)1520-6300(1996)8:6<703::AID-AJHB2>3.0.CO;2-U).
131. Allen, J.S., Bruss, J., Damasio, H. 2005. The aging brain: the cognitive reserve hypothesis and hominid evolution. *Am. J. Hum. Biol.* **17**, 673–689.
132. Kaplan, H.S., Robson, A.J. 2002. The emergence of humans: the coevolution of intelligence and longevity with intergenerational transfers. *Proc. Natl. Acad. Sci. USA* **99**, 10221–10226.
133. Hawkes, K., O'Connell, J.F., Jones, N.G.B., Alvarez, H., Charnov, E.L. 1998. Grandmothering, menopause, and the evolution of human life histories. *Proc. Natl. Acad. Sci. USA* **95**, 1336–1339.
134. Chu, C.Y., Lee, R.D. 2006. The co-evolution of intergenerational transfers and longevity: an optimal life history approach. *Theor. Popul. Biol.* **69**, 193–201.
135. Chu, C.Y.C., Lee, R.D. 2013. On the evolution of intergenerational division of labor, menopause and transfers among adults and offspring. *J. Theor. Biol.* **332**, 171–180.
136. Pavard, S., Koons, D.N., Heyer, E. 2007. The influence of maternal care in shaping human survival and fertility. *Evolution* **61**, 2801–2810.
137. Pavard, S., Sibert, A., Heyer, E. 2007. The effect of maternal care on child survival: a demographic, genetic, and evolutionary perspective. *Evolution* **61**, 1153–1161.
138. Pavard, S., Branger, F. 2012. Effect of maternal and grandmaternal care on population dynamics and human life-history evolution: a matrix projection model. *Theor. Popul. Biol.* **82**, 364–376.
139. Pavard, S., Coste, C.F.D. 2021. Evolutionary demographic models reveal the strength of purifying selection on susceptibility alleles to late-onset diseases. *Nat. Ecol. Evol.* **5**, 392–400 (doi:10.1038/s41559-020-01355-2).
140. Kim, P.S., McQueen, J.S., Coxworth, J.E., Hawkes, K. 2014. Grandmothering drives the evolution of longevity in a probabilistic model. *J. Theor. Biol.* **353**, 84–94 (doi:10.1016/j.jtbi.2014.03.011).
141. Coste, C.F.D., Bienvenu, F., Ronget, V., Ramirez-Loza, J.-P., Cubaynes, S., Pavard, S. 2021. The kinship matrix: inferring the kinship structure of a population from its demography. *Ecol. Lett.* **24**, 2750–2762 (doi:10.1111/ele.13854).
142. Caswell, H. 2019. The formal demography of kinship: a matrix formulation. *Demogr. Res.* **41**, 679–712 (doi:10.4054/DemRes.2019.41.24).
143. Lee, R. 2008. Sociality, selection, and survival: simulated evolution of mortality with intergenerational transfers and food sharing. *Proc. Natl. Acad. Sci.* **105**, 7124 (doi:10.1073/pnas.0710234105).
144. Davison, R., Gurven, M. 2022. The importance of elders: extending Hamilton's force of selection to include intergenerational transfers. *Proc. Natl. Acad. Sci.* **119**, e2200073119 (doi:10.1073/pnas.2200073119).
145. Alberts, S.C. et al. 2013. Reproductive aging patterns in Primates reveal that humans are distinct. *Proc. Natl. Acad. Sci. USA* **110**, 13440–13445.
146. Peccei, J.S. 1995. The origin and evolution of menopause: the altriciality-lifespan hypothesis. *Ethol. Sociobiol.* **16**, 425–449.
147. Peccei, J.S. 2001. Menopause: adaptation or epiphenomenon? *Evol. Anthropol.* **10**, 43–57.
148. Shanley, D.P., Kirkwood, T.B. 2001. Evolution of the human menopause. *Bioessays* **23**, 282–287.
149. Shanley, D.P., Sear, R., Mace, R., Kirkwood, T.B. 2007. Testing evolutionary theories of menopause. *Proc. R. Soc. Biol. Sci.* **274**, 2943–2949.

150. Pavard, S., Metcalf, C.J.E., Heyer, E. 2008. Senescence of reproduction may explain adaptive menopause in humans: a test of the 'mother' hypothesis. *Am. J. Phys. Anthropol.* **136**, 194–203.
151. Thouzeau, V., Raymond, M. 2017. Emergence and maintenance of menopause in humans: a game theory model. *J. Theor. Biol.* **430**, 229–236.
152. Kaplan, H., Gurven, M., Winking, J., Hooper, P.L., Stieglitz, J. 2010. Learning, menopause, and the human adaptive complex. *Ann. N. Y. Acad. Sci.* **1204**, 30–42 (doi:10.1111/j.1749-6632.2010.05528.x).
153. Rogers, A.R. 1993. Why menopause? *Evol. Ecol.* **7**, 406–420 (doi:10.1007/BF01237872).
154. Hill, K., Hurtado, A.M. 1991. The evolution of premature reproductive senescence and menopause in human females. *Hum. Nat.* **2**, 313–350 (doi:10.1007/BF02692196).
155. Cant, M.A., Johnstone, R.A. 2008. Reproductive conflict and the separation of reproductive generations in humans. *Proc. Natl. Acad. Sci.* **105**, 5332 (doi:10.1073/pnas.0711911105).
156. Lahdenperä, M., Gillespie, D.O.S., Lummaa, V., Russell, A.F. 2012. Severe intergenerational reproductive conflict and the evolution of menopause. *Ecol. Lett.* **15**, 1283–1290 (doi:10.1111/j.1461-0248.2012.01851.x).
157. Croft, D.P. et al. 2017. Reproductive conflict and the evolution of menopause in killer whales. *Curr. Biol.* **27**, 298–304 (doi:10.1016/j.cub.2016.12.015).
158. Ellis, S., Franks, D.W., Nattrass, S., Cant, M.A., Bradley, D.L., Giles, D., Balcomb, K.C., Croft, D.P. 2018. Postreproductive lifespans are rare in mammals. *Ecol. Evol.* **8**, 2482–2494.
159. Sear, R. 2008. Kin and child survival in rural Malawi. *Hum. Nat.* **19**, 277 (doi:10.1007/s12110-008-9042-4).
160. Williams, G.C. 1957. Pleiotropy, natural-selection, and the evolution of senescence. *Evolution* **11**, 398–411.
161. Robson, S.L., Wood, B. 2008. Hominin life history: reconstruction and evolution. *J. Anat.* **212**, 394–425.
162. 2023. Bonobo (*Pan paniscus*) fact sheet. San Diego Zoo Wildlife Alliance. http://ielc.libguides.com/sdzg/factsheets/bonobo.
163. de Magalhaes, J.P., Budovsky, A., Lehmann, G., Costa, J., Li, Y., Fraifeld, V., Church, G.M. 2009. The human ageing genomic resources: online databases and tools for biogerontologists. *Aging Cell* **8**, 65–72.
164. Schultz, A.H. 1969. *The Life of Primates*. Universe Books.

11 Lifespan and Mortality in Hunter-Gatherer and Other Subsistence Populations

Michael D. Gurven

11.1 Introduction

Hunter-gatherers with traditional diets and activity regimes and minimal exposure to modern amenities are an important lens for understanding how selection helped shape the evolution of the human life course (see Chapter 10). *Homo sapiens* were hunter-gatherers for most of their 200–300 KYA existence, up until the advent of plant and animal domestication 6–12 KYA. An appreciation of age-specific differences in hunter-gatherer exposures, capacities, lifestyles, physiology and health can therefore provide insight into the selection pressures that have given rise to physiological reaction norms, and ultimately impacted on ageing and evolved lifespan. One starting point for thinking about evolved human lifespan is to examine age-specific demography in small-scale societies. Given that quality demographic data with reasonably accurate age and mortality estimation exist only for a handful of hunter-gatherer populations, small-scale horticulturalists and pastoralists provide an additional source of data on mortality and senescence in non-industrial societies.

This chapter is an attempt to synthesize the best information about mortality in relatively isolated, small-scale subsistence populations. These populations are acculturating at a rapid rate and so future data are unlikely to be forthcoming. For example, 33 of 105 isolated indigenous groups in Brazil became extinct in the first half of the twentieth century [1], and the few remaining are endangered [2]. There are currently very few extant groups of human hunter-gatherers, and probably no large population for which detailed demography on people of all ages will soon be available.

All groups I consider in the sample of *hunter-gatherers* have had minimal or no exposure to modern medicine during the period of study, and minimal to no inclusion of horticulture nor market-derived foods in their diets. Other traits commonly associated with a foraging lifestyle are variable among hunter-gatherer groups, such as mobility, an egalitarian ethic, widespread sharing, minimal storage and other social traits [3]. Forager-horticulturalists have engaged in horticulture for many generations, while some 'acculturated' hunter-gatherers have either recently started horticulture and/or have been exposed to medicines, markets and

Thanks to Kim Hill and Samuel Pavard for their helpful comments.

other modern amenities. I include those as well to examine how alterations in livelihood and lifestyle affect mortality patterns. All studies made efforts to ensure accurate age estimation, demographic data collection was a research objective, data were subject to rigorous procedures to check for errors and ensure no systematic bias in the under-reporting of deaths. Most importantly, survivorship and mortality profiles for these populations are based on actual deaths from prospective or retrospective studies and not on model life table fits to scanty data or census data. They therefore make no assumptions about stable or stationary populations, which can bias estimates of adult mortality. Stable population theory requires that mortality and fertility schedules remain constant over long periods of time, while stationary distributions additionally require zero population growth. If a population is actually growing, and stationarity is instead assumed, the death rate is usually overestimated [see 4].

My general goal is to showcase what is known about survivorship and longevity among hunter-gatherers and other contemporary subsistence populations. Doing so is of paramount importance for those interested in human longevity. But this information can also help situate recent changes in lifeways and other conditions affecting the health and well-being of small-scale indigenous populations. I first provide pertinent ethnographic information about the sample to help contextualize their demographic profiles. I then summarize survivorship profiles, age distributions of adult deaths and causes of death, based largely on verbal autopsies. I summarize changes in mortality over time associated with contact or globalization for groups where diachronic data exist, and I then discuss the utility of using contemporary subsistence populations for representing prehistoric human demography.

11.2 Populations

The ethnographic record of hunter-gatherers includes hundreds of cultures, but only 60 or so groups have been studied. The sample of foraging societies presented here does not adequately cover all geographical areas, and all groups have had various degrees of interaction with neighbouring populations and state intervention. I address the limitations of relying on contemporary subsistence populations later in this chapter. Table 11.1 provides descriptive information on the populations considered here. Only five foraging societies have been explicitly studied using demographic techniques – Hadza of Tanzania [5, 6, 7], Dobe !Kung of Botswana and Namibia [8], Ache of Paraguay [9], Agta of Philippines [10], and the Hiwi of Venezuela [11]. Though these populations are well studied and show many similarities, it should be emphasized how little we really know about the demography of the many hunter-gatherers that once roamed the globe. To these populations, I also include the Northern Territory aborigines of Australia [12, 13], Yanomamö of Venezuela and Brazil [14], Tsimane of Bolivia [15], Gainj of Papua New Guinea [16, 17], and Herero of Botswana [18].

Table 11.1 Descriptive information on subsistence populations

Population		Location	Region	Habitat	Population size	# person-yrs (# deaths)	Study period	Risk period	Data source(s)
Ache	HG	Paraguay	South America	Tropical rainforest	537	16099 (351)	1958–1960	1890–1970	[9]
Agta	HG	Philippines	Oceania	Tropical dry forest	9000	2566 (117)	<1970	1950–1964	[10]
Hadza	HG	Tanzania	East Africa	Savannah woodland	750	6100 (227)	1962–1986	1985–2000	[6, 7]
Hiwi	HG	Venezuela, Columbia	South America	Neotropical savannah	779	4108 (126)	1978–1983	1985–1992	[11]
Ju/'hoansi	HG	Botswana, Namibia	East Africa	Desert	454	4512 (75)	1963–1974	1963–1974	[8]
Aborigine	AHG	Northern Territory	Australia	Desert	17469	69876 (1115)	1987–1989	1958–1960	[12]
Gainj	FH	Papua New Guinea	Oceania	Tropical dry forest	1318	9102 (287)	1985–1992	1970–1978	[17]
Tsimane	FH	Bolivia	South America	Tropical rainforest	16000	47854 (648)	1950s	1950–2000 1950–1989	[15]
Yanomamö	FH	Venezuela, Brazil	South America	Tropical rainforest	~12000	2843 (64)	2002–2003	1930–1996	[26]
Herero	P	Namibia	East Africa	Desert	10–15000	26564 (405)	1987–1989	1909–1966	[18]

Note: (1) At the time of study, there were about 454 Ju/'hoansi living in the study site. Two life tables are used from this time period. One is based on the 94 deaths during the 11-year study period (table 4.4 in [8]) and the other uses the referent study population with a smaller number of deaths during the same time period (table 4.6 in [8]). (2) Population size reflects total estimated census population at the time of study. It was estimated that there were around 12000 Xilixana Yanomamö at the time of study, though census size for the specific villages studied by Early and Peters varied between 96 in 1930 to 361 in 1996 [26]. HG = hunter-gatherer, AHG = acculturated hunter-gatherer, FH = forager-horticulturalist, P = pastoralist.

11.2.1 Dobe Ju/'hoansi (!Kung)

Nancy Howell's Dobe Ju/'hoansi (!Kung) study in the Kalahari Desert of Botswana and Namibia is one of the first and most impressive demographic accounts of a foraging society [8, 19]. At the time of the study, many of the adults had spent most of their lives foraging, although San populations had had interactions with mercantile interests in the nineteenth century and engaged in trade with pastoral and agricultural populations (see [20]). An 'early' sample refers to the time period before the 1950s when the Bantu influence in the Dobe !Kung area was minimal. A 'later' sample refer to the prospective time of study when the lifeways of the !Kung were rapidly changing. Ages were determined through a combination of relative age lists, known ages of children and young adults and checking against the stable age distribution resulting from a 'West' Coale and Demeny model (that was not used to generate the mortality patterns shown here) [21].

11.2.2 Ache

The Ache were mobile tropical forest hunter-gatherers until the 1970s. Kim Hill and A. Magdalena Hurtado separate Ache history into three time periods [9] – a pre-contact 'forest' period of pure foraging with no permanent peaceful interactions with neighbouring groups (before 1970), a 'contact' period (1971–1977) where epidemics had a profound influence on the population and a recent 'reservation' period where they lived as forager-horticulturalists in relatively permanent settlements (1978–1993). During this latter period, the Ache had some exposure to health care. The pre-contact Ache period shows marked population increase, due in part to reduced competition as a direct result of high adult mortality among Paraguayan nationals during the Chaco War with Bolivia in the 1930s. The authors improve on Howell's methods of age estimation by using averaged informant ranking of age, informant estimates of absolute age differences between people and polynomial regression of estimated year of birth on age rank.

11.2.3 Hadza

The Hadza in the eastern rift valley of Tanzania were studied in the mid-1980s by Nicolas G. Blurton Jones and colleagues. Trading with herders and horticulturalists has been sporadic among Hadza since the second half of the twentieth century, and the overall quantity of food coming from horticulturalists varies from 5% to 10% [5]. The Hadza have been exposed to a series of settlement schemes since the mid-1960s, but none of these has been very successful. The 1990s saw a novel form of outsider intervention in the form of further habitat degradation and 'ethno-tourism' [5]. Although some Hadza have spent considerable time living in a settlement with access to maize and other agricultural foods, most have not and continue to forage and rely on wild foods. The population was aged using relative age lists, a group of individuals of known ages and polynomial regression [7]. Two censuses done about 15 years apart, with an accounting of all deaths during the interim, allowed Blurton Jones to construct a life table, and to further show that sporadic access to horticultural foods and other amenities cannot account for the mortality profile.

11.2.4 Hiwi

The Hiwi are neo-tropical savanna foragers of Venezuela also studied by Kim Hill and A. Magdalena Hurtado in the late 1980s [22, 23]. They were contacted in 1959 when cattle ranchers began encroaching into their territory. Although living in semi-permanent settlements, Hiwi continue to engage in violent conflict with other Hiwi groups. At the time of study, almost the entire diet was wild foods, with 68% of calories coming from meat and 27% from roots, fruits and an arboreal legume. The study population contains a total of 781 individuals. Nearby Guahibo-speaking peoples practise agriculture, while at the time of study, the Hiwi inhabited an area poorly suited for agriculture. As among the Hadza, repeated attempts at agriculture by missionaries or government schemes had failed among this group. The life table comes from Hill and colleagues [11].

11.2.5 Agta

The Casiguran Agta of the Philippines are foragers studied by Tom Headland from 1962 to 1986. They live on a peninsula close to mountainous river areas and the ocean. There are 9 000 Agta in eastern Luzon territory, and John Early and Headland's demographic study was focused on the San Ildefonso group of about 200 people [10]. Although the Luzon area is itself very isolated, Agta have maintained trading relationships with lowlander horticulturalists for at least several centuries [24]. The twentieth century introduced schooling, and brief skirmishes during American and Japanese occupation. Age estimation was achieved through reference to known ages of living people and calendars of dated events. As in the Ache study, the Agta demography is divided into a 'forager' period (1950–1965), a transitional period of population decline (1966–1980) and a 'peasant' phase (1981–1993). These latter phases were marked by guerrilla warfare and subjugation by loggers, miners and colonists.

11.2.6 Northern Territory Australian Aborigine

The Northern Territory Australian Aborigine mortality data come from analysis of vital registration from 1958 to 1960 by Frank Lancaster Jones [12, 13]. At this time, few Aborigines in the region were still full-time foragers. There was a significant amount of age-clumping at five-year intervals, and so a smoothing procedure was done on the age distribution of the population. It is likely that infant deaths and more remote-living individuals are under-renumerated, and Frank Lancaster Jones made adjustments to impute missing deaths. I view these data with caution but include them because no other reliable data exist for Australia, apart from a Tiwi sample, culled from the same author.

11.2.7 Yanomamö

The Yanomamö are forager-horticulturalists populations in Amazonian South America. Several different Yanomamö studies have been carried out since the early 1990s. Although often construed as hunter-gatherers, Yanomamö have practised slash-and-burn horticulture of plantains for many generations [14]. They mostly live in small villages of less than 50 people. The effects of the rubber boom and slave trade before the eighteenth century on Yanomamö were minimal [25]. The Yanomamö remained mostly isolated until missionary contact in the late 1950s. The most complete demography comes from John D. Early and John F. Peters [26], based on prospective studies of eight villages in the Parima Highlands of Brazil. Births and deaths have been recorded by missionaries and National Indigenous Peoples Foundation (FUNAI) personnel since 1959. The pre-contact period (1930–1956) pre-dates missionary and other outside influences. The contact period (1957–1960), 'linkage' period (1961–1981) and Brazilian period (1982–1996) saw increased interaction with miners, Brazilian nationals and infectious disease. Ages for Xilixana (Mucajaí) during this period were estimated using a chain of average interbirth intervals for people with at least one sibling of known age and relative age lists in combination with estimated interbirth intervals.

11.2.8 Tsimane

The Tsimane are also forager-horticulturalist populations in Amazonian South America. They inhabit tropical forest areas of the Bolivian lowlands, congregating in small villages near large rivers and small tributaries. There are roughly 17 000 Tsimane living in dispersed settlements in the Beni region. The Tsimane have had sporadic contact with Jesuit missionaries since before the eighteenth century, although were never successfully converted or settled. Evangelical and Catholic missionaries set up missions in the early 1950s, and they later trained some Tsimane to become teachers in the more accessible villages. However, the daily influence of missionaries is minimal. Market integration is increasing, as are interactions with loggers, merchants and colonists. Most Tsimane continue to fish, practise horticulture, hunt and gather for the majority of their subsistence. The demographic sample used here is based on reproductive histories I collected in 12 remote communities during 2002 and 2003 [15]. Changes in mortality are evident over time, and so an 'early' period of 1950–1989 is contrasted with a 'later' period of 1990–2002. Age estimation of older individuals was undertaken using a combination of written records of missionaries and relative age rankings, and by photographic and verbal comparison with individuals of known ages.

11.2.9 Gainj

The Gainj are slash-and-burn horticulturalists of sweet potato, yams and taro in the central highland forests of northern Papua New Guinea. Meat is fairly rare [27]. At the time of the studies by Patricia Johnson and James Wood in 1978–1979 and

1982–1983, there were roughly 1318 Gainj living in 20 communities. Contact was fairly recent, sporadically in 1953 by Australian colonials and more formally in 1963 by the Australian government, and there is genetic and linguistic evidence of the relative isolation of the Gainj [16]. Prior to contact, population growth had been zero for at least four generations [17]. An A2 Hong Kong influenza epidemic reduced the population by 6.5% in 1969–1970, and this probably accounts for the dearth of older people in this population. Data were obtained from government censuses from 1970 to 1977 and include non-Gainj Kalam speakers [17]. Additionally, published mortality estimates were already fitted with a Brass two-parameter logit model.

11.2.10 Herero

The Herero are Bantu-speaking pastoralists studied by Renee Pennington and Henry Harpending in the late 1980s [18]. They are traditionally cattle and goat herders in the Kalahari Desert of the Ngamiland District of north-western Botswana, numbering 10–15000 during the time of the study. They had migrated to this area in the early twentieth century, due to displacements from the Herero-German War. They live in extended family homesteads without running water or electricity, remain endogamous and are now very successful cattle herders. They also raise drought-resistant goats and other livestock. Total fertility rates increased from 2.7 in the first half of the twentieth century to 7 in the 1980s; the lower earlier fertility was likely due to pelvic inflammatory disease stemming from sexually transmitted infections [28].

11.3 Survivorship Patterns

The age-specific probability of survival (l_x) from birth to adulthood shows a modest amount of variation across different populations of human hunter-gatherers and forager horticulturalists (Figure 11.1). The variation is less marked than when comparing with our closest primate relatives, common chimpanzees (*Pan troglodytes*); humans have a higher probability of survival at all stages of life, with the exception of early infancy. Infant survival may be lower in humans than chimpanzees due to birth complications or higher vulnerability of altricial neonates (see Chapter 10). Infant mortality rates among hunter-gatherer populations range from 14% to 40%, with a mean ± standard deviation of 27±7% dying in the first year of life (n = 16) [29]. On average, 57% and 64% of children born survive to age 15 among hunter-gatherers and forager-horticulturalists, respectively. Beyond age 15, the adult mortality rate ranges from 1% to 1.5% per year until about age 40, where it increases exponentially. In spite of the variation, there does appear to be a clear pre-modern human pattern. There is remarkable similarity in age profiles of mortality risk over the lifespan. By age 10, the mortality hazard has declined to 0.01, and then doubles to about 0.02 by age 40, doubling yet again before age 60, and again by age 70. Low mortality therefore persists until about age 40, when mortality accelerations become more evident. Overall, the mortality rate after age 30 doubles every seven to ten years [30, 31].

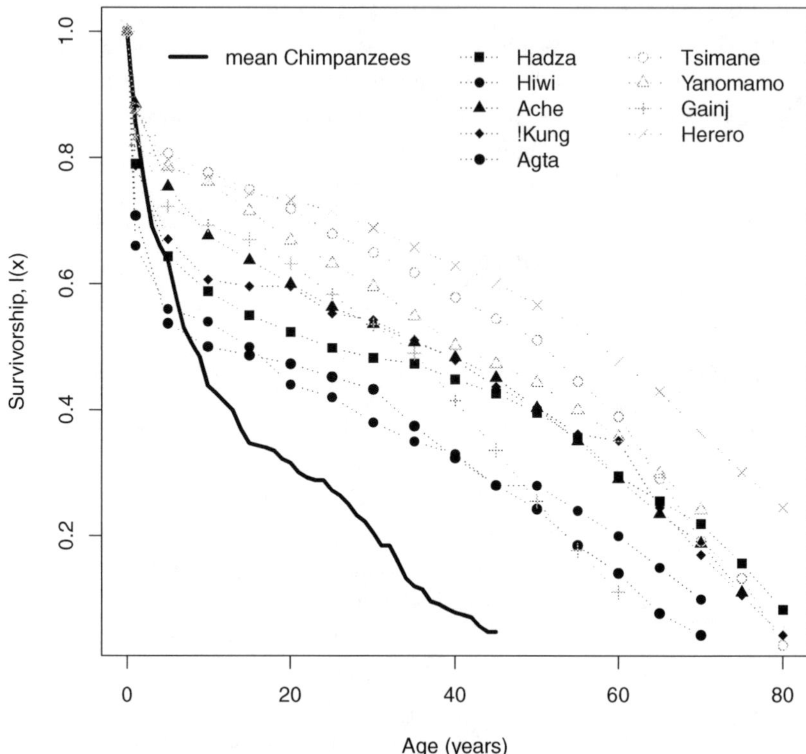

Figure 11.1 Survivorship (l_x) curves for five hunter-gatherer, three horticulturalist and one pastoralist population(s).

While life expectancy at birth ranges from 21 to 43 years of life (average 30 years for hunter-gatherers, 35 years for all groups), women who survive to age 45 can expect to live an additional 20–22 years [5, 30]. While there is substantial variation across groups in life expectancy at early ages, there is significant convergence after about age 30. With the exception of the Hiwi, who show over 10 years less remaining during early ages and over five years less remaining during adulthood, and the Hadza, whose life expectancy at each age is about two years longer than the rest at most adult ages, all other remote groups with minimal medical attention are hardly distinguishable. At age 40, the expected age at death is about 63–66 (i.e. 23–26 additional expected years of life), whereas by age 65, expected age at death is only about 70–76 years of age. By that age, death rates become very high.

These findings support the notion that a substantial post-reproductive lifespan is not just a recent occurrence in human history. To best illustrate this, I use the Post-reproductive Representation (PrR) measure introduced by Daniel Levitis [32]. By considering the proportion of adult years spent in a post-reproductive state, PrR incorporates survival from onset of adulthood to age at menopause, and not just years remaining after menopause. In my sample, PrR varies from 30% to 55%, substantially higher than other mammalian species known to have post-reproductive life [33, 34]

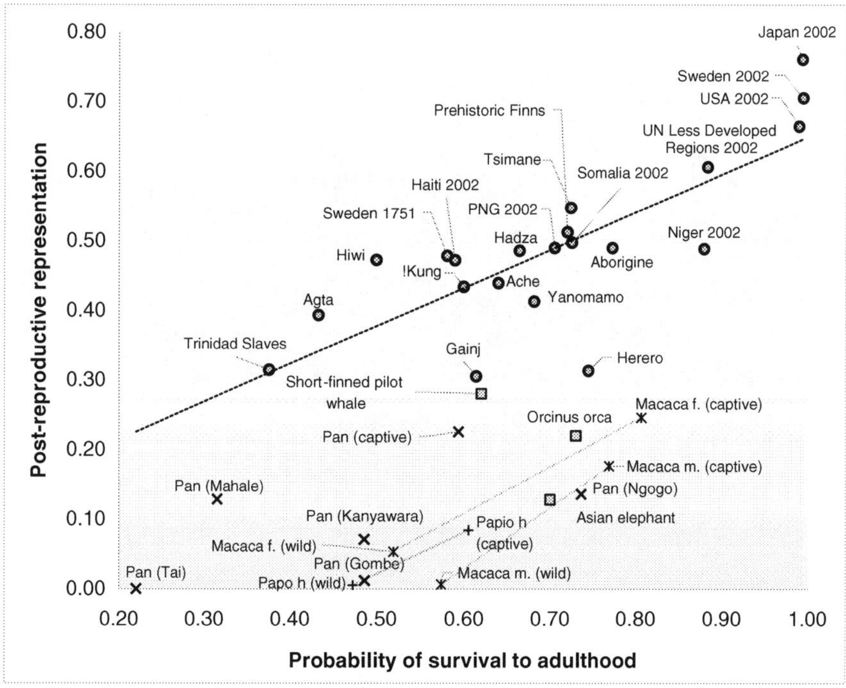

Figure 11.2 Post-reproductive lifespan among humans and other species using PrR, calculated as $l_m/l_b * e_m/e_b$, where l_m is survivorship to post-fertile period (m), e_m is remaining expected lifespan at age m (see [30]). Age b is the age at onset of reproduction in adulthood. By considering the proportion of adult years spent in a post-reproductive state, PrR incorporates adult survival to age at menopause, and not just years remaining after menopause. Human populations are depicted by circles and other primates by crosses and stars. Other species with observed post-reproductive lifespans, including killer whales (*Orcinus orca*), short-finned pilot whales (*Globicephala macrorhynchus*) and Asian elephants (*Elephas maximus*), are depicted by squares.

(Figure 11.2). Most cases of non-trivial PrR in non-humans, however, reflect conditions of captivity – breeding programmes that accelerate reproductive senescence – otherwise the magnitude of post-reproductive life is small [35]. Nonetheless, given that positive PrR is not limited just to humans, adaptive explanations should suggest quantitative rather than qualitative differences between humans and other species.

11.4 A Seven-Decade Human Lifespan?

Following the lead of Väinö Kannisto and Wilhelm Lexis [36, 37], comparing modal ages of adult death and the variance around these modes provide insight into the stability of adult lifespans among populations. The modal age at death may reflect an important stage in physiological decline. The effective length of the human lifespan under traditional conditions seems to be about seven decades. Figure 11.3 shows the

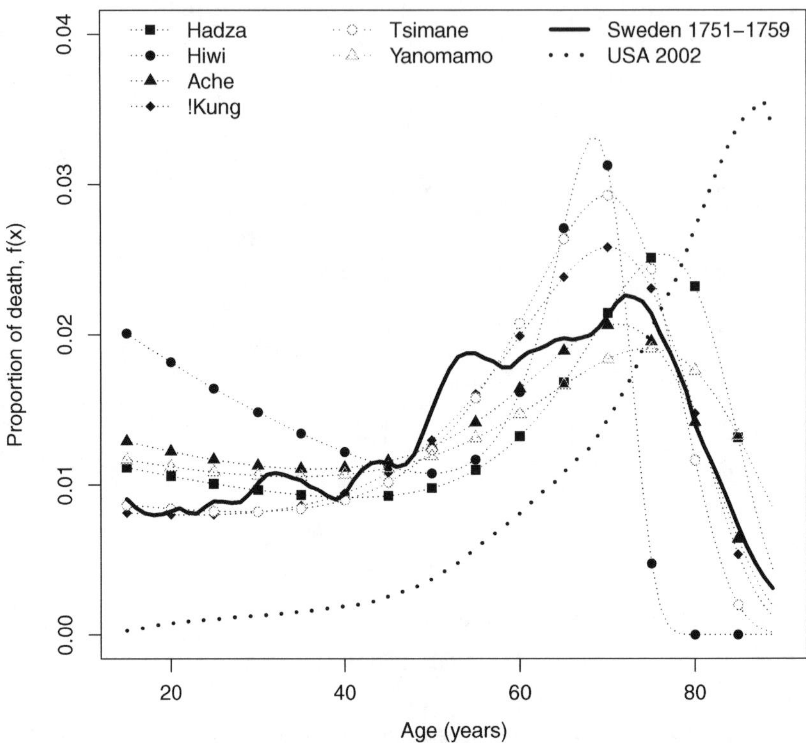

Figure 11.3 Distributions of age at adult death. Subsistence-level populations of hunter-gatherers, horticulturalists and eighteenth-century European agrarian farmers show peak ages of adult death ranging from 65 to 80 years. Industrialized nations like the United States show modal ages of adult death over a decade older than among subsistence populations. Mortality is smoothed using a Siler model for all populations except Sweden and the United States (see [30]). Y-axis is the proportion of all deaths $f(x)$ among those aged 15+.

frequency distribution, $f(x)$, of deaths at age x, conditional upon surviving to age 15, for hunter-gatherers and horticulturalists. Data from eighteenth-century Sweden and the modern United States are shown for comparison. These pre-modern populations show an average modal adult lifespan of about 72 years of age (range: 68–78, standard deviation (SD): 3.3–7.8). There is more variability in the ages of adult death within each of these populations than typically found in modern industrial populations [38], due in part to higher age-independent mortality that may be considered different from the 'normal' course of ageing. The modes are therefore less peaked, accounting for less than 3% of adult deaths. More acculturated hunter-gatherers show the greatest density of deaths after age 55, possibly indicating that as some causes of death (such as violence) are reduced, age-related causes of death become more important, leading to a greater density of death around the mode.

By about seven decades, most people experience sufficient decline that if they do not die from one cause, they are soon to die from another. This is consistent with anecdotal impressions of frailty and work effort among aged foragers. While many individuals remain healthy and vigorous workers through their 60s, few are in good health and capable of significant heavy labour in their 70s, and it is the rare individual who survives to their 80s.

11.5 Causes of Death

Causes of death are difficult to assess without physicians or autopsies, and when causes of deaths are elicited during retrospective interviews. While accidents and homicides may be readily identified, infections and other diseases common in non-industrial societies may be diagnosed with large margin of error. Nonetheless, I assembled causes of death reported for a subset of the sample (n = 3328 deaths), including two additional populations lacking life tables: Aka of Central African Republic and Bakairi of Brazil [39, 40]. Given limitations of the source data, I use general categories to group causes: illness, degenerative disease, accidents or violence.

Overall, the majority (72%) of deaths are due to infections and illness. Respiratory illnesses, such as bronchitis, tuberculosis, pneumonias and other viral infections account for a fifth or more of illness-related deaths. However, most infectious diseases were absent in newly contacted Amazonian groups, consistent with the notion that small, mobile populations cannot support these contagious vectors [41]. Gastrointestinal illnesses account for 5–18% of deaths in traditional human societies. Diarrhoea coupled with malnutrition is and remains one of the most significant causes of infant and early child deaths among forager populations. People living in tropical forest environments are especially vulnerable to helminthic parasites [42], which although not usually lethal, can compromise growth and immune function.

There is a notable lack of data on degenerative disease, but they are probably uncommon. Degenerative disease accounted for about 9% of adult deaths, with the highest representation among (somewhat acculturated) Northern Territory Aborigines. Neoplasms and possible heart disease each accounted for nine of the 49 deaths due to degenerative illness in adults over age 60. Chronic illnesses as causes of death, however, are difficult to identify, since more proximate causes are likely to be mentioned in verbal autopsies. Cases of degenerative deaths largely include perinatal problems early in infancy, late-age cerebrovascular problems and attributions of ‘old age’ in the absence of any obvious symptom or pathology. Heart attacks and strokes appear rarely, and do not account for these old-age deaths [43], which often occur during sleep. Hunter-gatherers and other groups with subsistence lifestyles do not appear to suffer from atherosclerosis or its risk factors [44, 45]. Obesity is rare, hypertension is low, cholesterol and triglyceride levels are low and maximal oxygen uptake (VO_2max) is high [43, 46].

Violent death appears to be a common feature of human societies, accounting for 11% of deaths (Table 11.2). Infanticide is not uncommon; those at greatest risk of being abandoned or killed in humans are the sickly, the unwanted, those of questionable paternity, females, and those viewed as bad omens, such as, sometimes, twins [47]. It is likely, however, that violent deaths among humans decreased with increased

Table 11.2 Reported causes of death in subsistence populations (n = 3 328 deaths)

	Hadza	Yanomamö	Ache forest	Ache settled	!Kung	Tsimane	Aka	Agta	Hiwi	Machiguenga	Northern Territory	Baikari	Gainj	TOTAL	
	%	%	%	%	%	%	%	%	%	%	%	%	%	#	%
a) <15 yrs old			(n = 230)	(n = 84)	(n = 164)	(n = 423)		(n = 112)	(n = 94)	(n = 82)	(n = 74)				
all illness			22.2	65.5	87.8	79.9		95.5	44.8	63.8	67.6			825	65.3
degenerative			8.3	20.2	3.7	10.4			10.4	9.6	24.3			120	9.5
accidents			6.1	3.6		9.7		1.8	15.6	11.7	6.8			102	8.1
violence			63.5	10.7	8.5	7.4		2.7	27.1	2.1	1.4			216	17.1
b) 15–59 yrs old			(n = 125)	(n = 22)	(n = 127)	(n = 192)		(n = 77)	(n = 31)	(n = 19)	(n = 68)				
all illness			28.0	86.4	79.5	74.7		69.5	35.3	33.3	61.8			400	60.5
degenerative			3.2		3.1	16.5		4.9	2.9	14.3	25.0			61	9.2
accidents			23.2	13.6		8.8		4.9	8.8	42.9	0.0			85	12.9
violence			45.6		17.3	12.9		14.6	44.1	0.0	13.2			115	17.4
c) 60+ yrs old			(n = 27)		(n = 52)	(n = 60)				(n = 2)	(n = 33)				
all illness			18.5		51.9	66.1					72.7			95	54.5
degenerative			22.2		40.4	25.4					21.2			49	28.2
accidents			25.9			8.5					6.1			18	10.3
violence			33.3		7.7	1.7				100.0	0.0			12	6.9
d) All ages	(n = 125)	(n = 111)	(n = 382)	(n = 104)	(n = 343)	(n = 690)	(n = 669)	(n = 364)	(n = 139)	(n = 117)	(n = 175)	(n = 65)	(n = 44)		
respiratory		21.6	0.8	31.1		19.9				6.8	28.6	56.9		292	23.7
gastrointestinal		5.4	5.5	13.2		18.2				34.2	17.1	3.1		239	13.8
fever		6.3	8.1	21.7		5.7				1.7	0.0	7.7		107	7.3
other		40.5	9.4	1.9		25.9				14.5	20.6	3.1		317	16.6
all illness	**66.7**	**73.9**	**23.8**	**67.9**	**79.3**	**69.6**	**92.2**	**86.7**	**41.0**	**57.3**	**66.3**	**70.8**	**79.0**	2 333	**70.1**
degenerative	**12.0**	**6.3**	**7.6**	**16.0**	**9.0**	**12.2**	**2.5**	**7.6**	**7.9**	**10.3**	**24.0**	**16.9**	**7.0**	306	**9.2**
accidental	0.8	7.2	13.1	2.8		8.4		2.7	12.9	17.1	4.0	12.3		166	**8.1**
homicide	3.2	4.5	22.0	4.2		7.5				3.4	5.7	0.0		164	6.3
warfare	0.0	8.1	33.5	0.0		0.0				0.0	0.0	0.0		137	5.2
all violence	3.2	12.6	55.5	4.2		7.5		3.0	30.2	3.4	5.7	0.0		354	**12.5**
all violence/ accidental	**4.0**	**19.8**		**7.1**	**11.7**	**15.9**	**5.4**	**5.8**	**43.2**	**20.5**	**9.7**	**12.3**	**14.0**	626	**18.8**
other causes	17.3	0.0	0.0	0.0	0.0	2.2	0.0	0.0	7.9	12.0	0.0	0.0	0.0	62	1.9

state-level intervention and missionary influence in many small-scale groups around the world [48]. The composition of accidental deaths varies across groups, including falls, river drownings, accidental poisonings, snake bites, burns and getting lost. Together, accidental and violent deaths account for 4–43% (average 19%) of all deaths.

Though no comparative data exist on suicide, casual reports suggest that it was rare under more traditional conditions [42], but more common in acculturated groups experiencing rapid loss of territory, livelihood, autonomy and cultural practices [49, 50]. Among Agta and Tsimane, suicides were often young (drunk) men embroiled in conflict involving their spouse or potential mates. However, among Tsimane, suicides were most common among adults age 60+. I estimate a high suicide death rate of about 1900 per 100 000 (though this reflects only four deaths), which is over 90 times higher than US rates. Though high over the entire study period 1950–2002, it was higher from 1950 to 1989 (2 270/100 000) than from 1990 to 2002 (1 579/100 000), consistent with informant reports that suicide among elderly was more common in the past, but is now actively discouraged. Suicide among elderly may also have occurred among Hiwi, but was unheard of by Ache (K. Hill, pers. comm.). Suicide among older Tsimane often followed spousal loss (i.e. widowhood effect), and feelings of being a burden due to physical disability and poor health. The latter are also cited as common reasons for geronticide and neglect in more nomadic hunter-gatherers living in harsh environments (e.g. Ache, Tiwi, Inuit) [51].

Despite expressed fear and the cultural importance of dangerous predators, as represented by mythologies, stories, songs and games, death by predation is rare among extant foragers. Grouping patterns, weapons, warning displays (e.g. fires) and other cultural means of avoiding predators may contribute to the reduced impact of predation on human survivorship [52].

11.6 Diachronic Change: Effects of Contact and Acculturation

One of the best ways to examine the effects of recent changes on traditional small-scale populations is to compare mortality profiles of the same groups at different time periods. These diachronic comparisons can be made for the Yanomamö, Ache, Agta, Hiwi, Ju/'hoansi !Kung, Tsimane and the agro-pastoralist Herero. Mortality profiles thus exist before and after some critical period, be it contact (Yanomamö, Ache), or acculturation and transition to peasant status (Agta and Ju/'hoansi). Figure 11.4 shows the ratio of age-specific mortality hazards from more and less acculturated time periods. Figure 11.4(a) displays populations whose survivorship has improved with acculturation (Ju/'hoansi, Ache, Herero, Hiwi, Tsimane) and Figure 11.4(b) those whose overall condition has worsened (Agta and Yanomamö).

Contact and acculturation had large effects on mortality rates in some groups. Among the Ache, the period of contact brought catastrophic diseases to the population, and about 40% died in the course of less than a decade of contact. Comparing the post-contact with the pre-contact period, there has been a small increase in infant mortality, but mortality at other ages has decreased by a third to a quarter. The effects

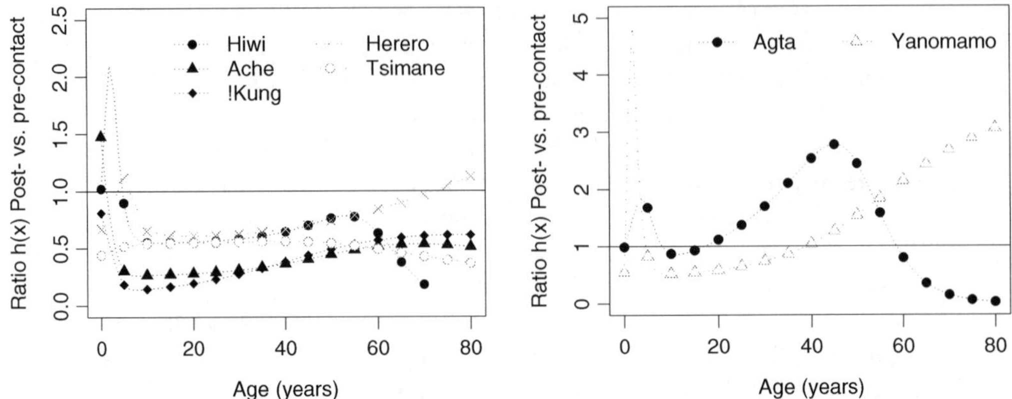

Figure 11.4 Diachronic changes in age-specific mortality patterns within populations. The ratio of estimated mortality hazards from post-contact or acculturated time periods and pre-contact or relatively unacculturated time periods from the same populations. (a) shows the mortality hazard ratio for populations with improvements in survivorship after contact for much of the lifespan, while (b) shows those with higher survivorship before contact and interaction with outsiders.

of improved conditions are greater at younger ages and gradually decay with age. Reduced mortality among settled Ache is largely due to reductions in homicide and forest-related accidents as a result of missionary influence and state intervention [9]. Medical attention also has helped decrease mortality among Ache.

A similar pattern is found for the Ju/'hoansi, with the exception that survival improves at all ages. Although settled !Kung frequently complain about meat scarcity and shifted norms of resource distribution, they also benefit from increased access to milk, protein-rich weaning foods and more predictable diets through greater association with cattle posts and receipt of government rations [53]. It is possible, however, that we have overestimated the effects of acculturation on Ju/'hoansi survivorship due to gaps in the prospective life table created by Howell (see table 4.6 in [8]).[1]

Among the Hiwi, contact has greatly increased infant mortality, but decreases it after infancy by about a half (there are not enough data on the very old to know what is happening after age 60). Among Tsimane, survivorship improves minimally in infancy, but substantially at all other ages [15]. Among the Herero, mortality increases slightly during childhood, but is lower at older ages. The initially higher level of survival among the Herero probably accounts for the smaller effect of acculturation on mortality rates.

Two groups show more deleterious effects of contact. Early child and adult mortality are much higher among acculturated 'peasant' Agta and Xilixana Yanomamö. Infant mortality may be buffered by the protective effects of breastfeeding, and so post-weaning mortality seems to worsen more in acculturated settings among both

[1] Indeed, [8] shows that the estimated life expectancy of acculturated !Kung based on this life table is about 50 years, which is 10 years higher than the national estimate of Botswana during the same time period.

Agta and Yanomamö. Peasant Agta are landless agriculturalists living in more populated and degraded environments with few foraging options and no longer maintain close trading relationships with nearby horticulturalists [10]. John D. Early and Thomas N. Headland suggest that cumulative effects of malnutrition and infectious diseases such as measles have increased child mortality during the peasant phase, and verify this through comparison of post-neonatal and neonatal mortality rates. Additionally, malaria, tuberculosis and other infectious diseases were believed to be largely absent in the forager phase for both Agta and Yanomamö but have been reaching epidemic proportions in recent years. Alcohol abuse and alcohol-induced conflict have also affected the Agta, especially during the peasant phase, though alcohol consumption was also present during the earlier foraging phase. Lower adult mortality among Yanomamö may be due to reduced warfare and homicide in recent years [26]. It is unclear why infectious disease appears to negatively impact adults over age 40 among Yanomamö but reproductive-aged adults among Agta. One possibility is that susceptibility due to differential prior exposure may vary by age groups.

Other studies that have focused on the effects of sedentism on mortality show a general decrease in child mortality, consistent with our description of Ju/'hoansi, Herero and Ache. These include the Ghanzi !Kung [53], Kutchin Athapaskans [54], Turkana pastoralists [55], and Nunamiut [56]. However, a slight increase in mortality was observed among the Adavasi Juang agriculturalists of India [57]. Fertility increases among members of all of these populations, often before a noticeable decline in mortality (see also Chapter 12).

Acculturation in the past several decades is likely to be very different than the transitions to sedentary, agricultural or peasant life in the more remote past. Even when foragers become the new underclass of national society and foraging behaviour becomes rarer, recent post-contact recovery periods are often accompanied by immunization campaigns, public health and sanitation measures that can substantially improve survivorship. It is likely that a worsening of physical health, or at best a lack of improvement, will occur when these benefits are lacking or unavailable.

Among humans, the effects of improved conditions also seem to be greatest during childhood and middle adulthood, tapering off with age (Figure 11.3). Comparing mortality rates between hunter-gatherers and modern Americans, infant mortality is over 30 times greater among hunter-gatherers and early child mortality is over 100 times greater than encountered in the United States. Not until the late teens does the relationship flatten, with over a tenfold difference in mortality. This difference is fivefold by age 50, about fourfold by age 60 and threefold by age 70 (see [30, 58] and Chapter 12).

11.7 Are Contemporary Subsistence Populations Representative of the Past?

Contemporary populations living traditional lifestyles prior to the kinds of recent changes described provide our best estimates of ancestral fertility and mortality.

As stated earlier, what we know about hunter-gatherer demography is skewed by the geography of traditional groups still remaining in the twentieth century. Groups inhabiting coastal and temperate areas are especially under-represented in the ethnographic record. But it is possible that even these twentieth-century post-contact samples may be unrepresentative of the past – or may at best cover a small subset of what past demography may have looked like. Indeed, the Ache were growing during a pioneering period in Paraguayan history after the Chaco War, but they also experienced violent colonial expansion into their core area [9]. The Hadza benefited from colonial pacification of invading herding neighbours but may have also been exposed to significant resource depletion (especially large game) due to increasing outsider settlements in their area [7]. Tribal warfare is likely reduced in most post-contact settings [11], but violent conflict due to expansion of more powerful neighbours is common. On the other hand, environmental conditions may have been more favourable in the distant past than in the present. First, measles, cholera and other infectious diseases requiring large reservoir host populations were unlikely to have proliferated among small groups of nomadic hunter-gatherers, except after contact [41]. Second, contemporary foragers often inhabit marginal environments unsuitable for cultivation and less favourable than those occupied when the planet was filled only with hunter-gatherers [3]. However, the only empirical tests of this claim show minimal differences between forager and non-forager habitat productivity [59, 60].

Nonetheless, most hunter-gatherer and small-scale horticultural populations in the ethnographic record show positive population growth, on average 1%. Such growth could not have represented conditions over long stretches of our species history, resulting in what has been referred to as the 'forager population paradox' (see also Chapter 10). In order to achieve population stationarity (i.e. zero population growth), fertility would have to decline well below that ever recorded in studied natural fertility populations (to a total fertility rate of four births per woman) or survivorship would need to decline below that ever recently reported (to $l_{15} = 0.41$) [30]. While it is possible that current conditions reflected in the demographic data are not representative of the past (i.e. warfare may have been more common, or fertility lower), another possibility is that all groups periodically experience population crashes [9, 61]. Explicit consideration of several scenarios suggests that altered vital rates combined with periodic crashes can feasibly generate long-term zero population growth [62]. These crashes might not be captured in modern surveillance studies because they may happen only once every few hundred years. While evidence suggests that climate varied widely throughout the Pleistocene and into the Holocene epoch [63], the extent to which past foragers typically experienced increasing, declining or zero population growth in past environments is unknown. What we do know, however, is that population growth was likely positive during periods of population expansion well before the advent of farming, particularly during expansions out of Africa to all habitable regions of the world, especially ~45–60 KYA [64]; also, as Page and French highlight, there are other issues that researchers need to treat with caution when making inferences about past populations from the ethnographic present [65].

11.8 Human Lifespan in Broader Perspective

Human lifespan (whether measured as maximal lifespan, life expectancy or modal age at death) is longer than predicted for a typical mammal (or primate) of human body size, but not atypical given the larger than expected brain size of humans [66]. Estimates based on regressions of various primate subfamilies and extant apes suggests a major increase in maximal lifespan between *Homo habilis* (52–56 years) to *Homo erectus* (60–63 years) occurring 1.7–2 million years ago, and further increases in *Homo sapiens* (66–72 years) [67]. Extrapolations for early *Homo sapiens* based on comparative analyses including both brain weights and body sizes among non-human primates similarly suggest a maximum lifespan between 66–78 years [68]. Palaeontological evidence is crude, but suggests that a post-reproductive lifespan emerged anywhere from 150 000 to 1.6 million years ago [69, 70].

Although maximum lifespans are longer than life expectancies (whose average is often lowered due to high infant and child mortality), it is usually reported that Palaeolithic humans had life expectancies of only 15–20 years. This brief lifespan is believed to have persisted over thousands of generations [71, 72], until less than 10 000 years ago when early agriculture presumably caused a slight increase to about 25 years. Gage compiles over 12 reconstructed prehistoric life tables with similar life expectancies to form a composite life table with survivorship to age 50 (l_{50}) of about 2–9% and e_{45} values of about three to seven years [73]. There is a large palaeodemographic literature concerning problematic age estimation in skeletal samples and bias in bone preservation leading to under-representation of older individuals, calling these previous analyses into question [74, 75, 76, 77, 78, 79].

Nonetheless, I point out some observations that further suggest problems with prehistoric life tables. Mortality rates in prehistoric populations are estimated to be lower than those for traditional foragers until about age 2. Estimated mortality rates then increase dramatically for prehistoric populations, so that by age 45 they are over seven times greater than those for traditional foragers. While excessive warfare could explain the shape of one or more of the prehistoric forager mortality profiles, it is improbable that these profiles represent the long-term prehistoric forager mortality profile. Such rapid mortality increase late in life would have severe consequences for our human life history evolution, particularly for senescence in humans. It is noteworthy that more recent prehistoric life tables show greater similarity with those presented here based on ethnographic samples [80].

11.9 Conclusion

Age profiles of mortality among hunter-gatherers and other subsistence populations overlap with those of early Europe up through the mid-nineteenth century [81]. While such similarity across diverse environments is striking, the variability in mortality rates at specific ages is also a good reminder about the dangers of generalizing too much about small-scale societies. The agro-pastoralist population (Herero) and the horticulturalist Tsimane show higher survivorship than hunter-gatherers and other horticulturalists. Globalization and improved access to modern amenities have also altered mortality patterns, mostly for the better, but not always. Despite variability in e_0 (from 21 to 37 years in hunter-gatherers) over early life and adulthood, a modal age of death around seven decades is evident across a wide range of diets and economic livelihoods – whether looking at the Hadza, or late nineteenth-century Finland. Our evolved human lifespan is thus aptly conceived as a population-level distribution of deaths that reflects the broad range of environments and conditions in which our ancestors lived, not only tens of thousands of years ago, but also throughout the Holocene. In these populations, over a third are likely to live to age 50, with an expected 15–20 years remaining. With an average age of first reproduction of ~18 years, up to 40% of hunter-gatherer women could expect to become grandmothers. Given the 5–10 year lag where male fertility trails female fertility in hunter-gatherers [82], many men were also grandfathers, experiencing an effective post-reproductive life.

Life expectancy has more than doubled since the late nineteenth century, and the modal age of death has shifted by over two decades in many countries. It is tempting to suspect from this pattern that a 62-year-old Tsimane woman is biologically 'older' than a 62-year-old Japanese woman. Comparative research that moves beyond mortality rates and actuarial ageing to make desirable inferences about physiological ageing is a nascent enterprise, even more so with longitudinal biomarker collection in non-industrial populations [83]. Further understanding of environmental assaults and energy-limited budgets experienced by hunter-gatherers can provide insight into how optimal allocations to defence and repair over the life course can lead to a characteristic lifespan. A reasonable proposal is that the costs of slowing senescence and preventing mortality, weighed against the benefits of productivity and fitness-enhancing transfers, led to a characteristic human lifespan, with some variance around the central tendency under diverse conditions.

References

1. Ribeiro, D. 1967. Indigenous cultures and languages of Brazil. In *Indians of Brazil in the Twentieth Century* (ed. J.H. Hopper), pp. 77–166. Institute for Cross-Cultural Research.
2. Walker, R.S.; Kesler, D.C.; Hill, K.R. 2016. Are isolated indigenous populations headed toward extinction? *PloS ONE* **11**, e0150987.
3. Kelly, R.L. 2013. *The Lifeways of Hunter-Gatherers: The Foraging Spectrum.* Cambridge University Press.
4. Pennington, R.L. 1996. Causes of early human population growth. *Am. J. Phys. Anthropol.* **99**, 259–274.
5. Blurton Jones, N.B., Hawkes, K., O'Connell, J. 2002. The antiquity of postreproductive life: are there modern impacts on hunter-gatherer postreproductive lifespans? *Hum. Biol.* **14**, 184–205.
6. Blurton Jones, N.G., O'Connell, J.F., Hawkes, K., Kamuzora, C.L., Smith, L.C. 1992. Demography of the Hadza, an increasing and high density population of savanna foragers. *Am. J. Phys. Anthropol.* **89**, 159–181.

7. Blurton Jones, N. 2016. *Demography and Evolutionary Ecology of Hadza Hunter-Gatherers*, vol. 71. Cambridge University Press.
8. Howell, N. 1979. *Demography of the Dobe !Kung*. Academic Press.
9. Hill, K., Hurtado, A.M. 1996. *Ache Life History: The Ecology and Demography of a Foraging People*. Aldine de Gruyter.
10. Early, J.D., Headland, T.N. 1998. *Population Dynamics of a Philippine Rain Forest People: The San Ildefonso Agta*. University Press of Florida.
11. Hill, K., Hurtado, A.M., Walker, R. 2007. High adult mortality among Hiwi hunter-gatherers: implications for human evolution. *J. Hum. Evol.* **52**, 443–454.
12. Lancaster Jones, F. 1965. The demography of the Australian Aborigines. *Int. Soc. Sci. J.* **17**, 232–245.
13. Lancaster Jones, F. 1963. *A Demographic Survey of the Aboriginal Population of the Northern Territory, with Special Reference to Bathurst Island Mission*. Australian Institute of Aboriginal Studies.
14. Chagnon, N. 1968. *Yanomamo: The Fierce People*. Holt, Rinehart & Winston.
15. Gurven, M., Kaplan, H., Supa, A.Z. 2007. Mortality experience of Tsimane Amerindians of Bolivia: regional variation and temporal trends. *Am. J. Hum. Biol.* **19**, 376–398 (doi:10.1002/ajhb.20600).
16. Wood, J.W., Johnson, P.L., Kirk, R.L., McLoughlin, K., Blake, N.M., Matheson, F.A. 1982. The genetic demography of the Gainj of Papua New Guinea. I. Local differentiation of blood group, red cell enzyme, and serum protein allele frequencies. *Am. J. Phys. Anthropol.* **57**, 15–25.
17. Wood, J.W., Smouse, P.E. 1982. A method of analyzing density-dependent vital rates with an application to the Gainj of Papua New Guinea. *Am. J. Phys. Anthropol.* **58**, 403–411.
18. Pennington, R.L., Harpending, H. 1993. *The Structure of an African Pastoralist Community: Demography, History, and Ecology of the Ngamiland Herero*. Clarendon Press.
19. Howell, N. 2010. *Life Histories of the Dobe! Kung: Food, Fatness, and Well-Being over the Life Span*, vol. 4. University of California Press.
20. Solway, J.S., Lee, R.B. 1990. Foragers, genuine or spurious? Situating the Kalahari San in history. *Curr. Anthropol.* **31**, 109–146.
21. Coale, A.J., Demeny, P. 1966. *Regional Model Life Tables and Stable Populations*. Rutgers University Press.
22. Hurtado, A.M., Hill, K. 1990. Seasonality in a foraging society: variation in diet, work effort, fertility, and the sexual division of labor among the Hiwi of Venezuela. *J. Anthropol. Res.* **46**, 293–345.
23. Hurtado, A.M., Hill, K.R. 1987. Early dry season subsistence ecology of the Cuiva foragers of Venezuela. *Hum. Ecol.* **15**, 163–187.
24. Headland, T.N. 1997. Revisionism in ecological anthropology. *Curr. Anthropol.* **38**, 605–630.
25. Ferguson, R.B. 1995. *Yanomami Warfare: A Political History*. School for American Research Press.
26. Early, J.D., Peters, J.F. 2000. *The Xilixana Yanomami of the Amazon: History, Social Structure, and Population Dynamics*. University Press of Florida.
27. Johnson, P.L. 1981. When dying is better than living: female suicide among the Gainj of Papua New Guinea. *Ethnology* **20**, 325–334.
28. Pennington, R., Harpending, H. 1991. Infertility in Herero pastoralists of Southern Africa. *Am. J. Hum. Biol.* **3**, 135–153 (doi:10.1002/ajhb.1310030209).

29. Volk, A.A., Atkinson, J.A. 2013. Infant and child death in the human environment of evolutionary adaptation. *Evol. Hum. Behav.* **34**, 182–192.
30. Gurven, M., Kaplan, H. 2007. Longevity among hunter-gatherers: a cross-cultural comparison. *Popul. Dev. Rev.* **33**, 321–365.
31. Finch, C.E. 1994. *Longevity, Senescence, and the Genome*. University of Chicago Press.
32. Levitis, D.A., Burger, O., Lackey, L.B. 2013. The human post-fertile lifespan in comparative evolutionary context. *Evol. Anthropol. Issues News Rev.* **22**, 66–79.
33. Cohen, A.A. 2004. Female post-reproductive lifespan: a general mammalian trait. *Biol. Rev.* **79**, 733–750.
34. Ellis, S., Franks, D.W., Nattrass, S., Currie, T.E., Cant, M.A., Giles, D., Balcomb, K.C., Croft, D.P. 2018. Analyses of ovarian activity reveal repeated evolution of post-reproductive lifespans in toothed whales. *Sci. Rep.* **8**, 12833.
35. Croft, D.P., Brent, L.J., Franks, D.W., Cant, M.A. 2015. The evolution of prolonged life after reproduction. *Trends Ecol. Evol.* **30**, 407–416.
36. Kannisto, V. 2001. Mode and dispersion of the length of life. *Popul. Engl. Sel.* **13**, 159–171.
37. Lexis, W. 1878. Sur la durée normale de la vie humaine et sur la théorie de la stabilité des rapports statistiques. *Ann. Démographie Int.* **2**, 447–460.
38. Cheung, S.L.K., Robine, J.-M., Jow-Ching, E., Caselli, G. 2005. Three dimensions of the survival curve: horizontalization, verticalization and longevity extension. *Demography* **42**, 243–258.
39. Hewlett, B.S., van de Koppel, J.M.H., van de Koppel, M. 1986. Causes of death among Aka pygmies of the Central African Republic. In *African Pygmies* (ed. L.L. Cavalli Sforza), pp. 45–63. Academic Press.
40. Picchi, D. 1994. Observations about a central Brazilian indigenous population: the Bakairi. *South Am. Indian Stud.* **4**, 37–46.
41. Black, F.L. 1975. Infectious disease in primitive societies. *Science* **187**, 515–518.
42. Dunn, F.L. 1968. Epidemiological factors: health and disease in hunter-gatherers. In *Man the Hunter* (eds R.B. Lee, I. DeVore), pp. 221–228. Aldine.
43. Eaton, S.B., Konner, M.J., Shostak, M. 1988. Stone agers in the fast lane: chronic degenerative diseases in evolutionary perspective. *Am. J. Med.* **84**, 739–749.
44. Eaton, S.B. et al. 1994. Women's reproductive cancers in evolutionary context. *Q. Rev. Biol.* **69**, 353–367.
45. Kaplan, H. et al. 2017. Coronary atherosclerosis in indigenous South American Tsimane: a cross-sectional cohort study. *Lancet* **389**, 1730–1739 (doi:10.1016/s0140-6736(17)30752-3).
46. Raichlen, D.A., Pontzer, H., Harris, J.A., Mabulla, A.Z., Marlowe, F.W., Snodgrass, J., Eick, G., Colette Berbesque, J., Sancilio, A., Wood, B.M. 2017. Physical activity patterns and biomarkers of cardiovascular disease risk in hunter-gatherers. *Am. J. Hum. Biol.* **29**, e22919.
47. Milner, L.S. 2000. *Hardness of Heart/Hardness of Life: The Stain of Human Infanticide*. University Press of America.
48. Keeley, L.H. 1996. *War before Civilization*. Oxford University Press.
49. Long, M., Long, E., Waters, T. 2013. Suicide among the Mla Bri hunter-gatherers of northern Thailand. *J. Siam Soc.* **101**, 155–176.
50. Chachamovich, E., Kirmayer, L.J., Haggarty, J.M., Cargo, M., McCormick, R., Turecki, G. 2015. Suicide among Inuit: results from a large, epidemiologically representative follow-back study in Nunavut. *Can. J. Psychiatry* **60**, 268–275.

51. Glascock, A.P., Feinman, S.L. 1981. Social asset or social burden: treatment of the aged in non-industrial societies. In *Dimensions: Aging, Culture, and Health* (ed. Christine L. Fry), pp. 13–31. Praeger.
52. Wrangham, R.W., Wilson, M.L., Muller, M.N. 2006. Comparative rates of violence in chimpanzees and humans. *Primates* **47**, 14–26.
53. Harpending, H.C., Wandsnider, L.K. 1982. Population structures of Ghanzi and Ngamiland !Kung. *Curr. Dev. Anthropol. Genet.* **2**, 29–50.
54. Roth, E.A. 1981. Sedentism and changing fertility patterns in a northern Athapascan isolate. *J. Hum. Evol.* **10**, 413–425.
55. Brainard, J. 1986. Differential mortality in Turkana agriculturalists and pastoralists. *Am. J. Phys. Anthropol.* **70**, 525–536.
56. Binford, L.R., Chasko, W.J. 1976. Nunamiut demographic history: a provocative case. In *Demographic Anthropology: Quantitative Approaches* (ed. Ezra B.W. Zubrow), pp. 63–143. University of New Mexico Press.
57. Roth, E.A., Ray, A.K. 1985. Demographic patterns of sedentary and nomadic Juang of Orissa. *Hum. Biol.* **57**, 319–325.
58. Burger, O., Baudisch, A., Vaupel, J.W. 2012. Human mortality improvement in evolutionary context. *Proc. Natl. Acad. Sci.* **109**, 18210–18214.
59. Porter, C.C., Marlowe, F.W. 2007. How marginal are forager habitats? *J. Archaeol. Sci.* **34**, 59–68.
60. Cunningham, A.J., Worthington, S., Venkataraman, V.V., Wrangham, R.W. 2019. Do modern hunter-gatherers live in marginal habitats? *J. Archaeol. Sci. Rep.* **25**, 584–599.
61. Boone, J.L. 2002. Subsistence strategies and early human population history: an evolutionary ecological perspective. *World Archaeol.* **34**, 6–25.
62. Gurven, M.D., Davison, R.J. 2019. Periodic catastrophes over human evolutionary history are necessary to explain the forager population paradox. *Proc. Natl. Acad. Sci.* **116**, 12758–12766.
63. Richerson, P.J., Bettinger, R.L., Boyd, R. 2005. Evolution on a restless planet: were environmental variability and environmental change major drivers of human evolution? In *Handbook of Evolution, Vol. 2: The Evolution of Living Systems* (eds F.M. Wuketits, F.J. Ayala), 223–242. Wiley-Blackwell.
64. Henn, B.M., Cavalli-Sforza, L.L., Feldman, M.W. 2012. The great human expansion. *Proc. Natl. Acad. Sci.* **109**, 17758–17764.
65. Page, A.E., French, J.C. 2020. Reconstructing prehistoric demography: what role for extant hunter-gatherers? *Evol. Anthropol. Issues News Rev.* **29**, 332–345.
66. Allman, J., McLaughlin, T., Hakeem, A. 1993. Brain weight and life-span in primate species. *Proc. Natl. Acad. Sci.* **90**, 118–122.
67. Judge, D.S., Carey, J.R. 2000. Postreproductive life predicted by primate patterns. *J. Gerontol. Biol. Sci.* **55A**, B201–B209.
68. Hammer, M., Foley, R. 1996. Longevity, life history and allometry: how long did hominids live? *J. Hum. Evol.* **11**, 61–66.
69. Bogin, B., Smith, B.H. 1996. Evolution of the human life cycle. *Am. J. Hum. Biol.* **8**, 703–716.
70. Caspari, R., Lee, S.-H. 2004. Older age becomes common late in human evolution. *Proc. Natl. Acad. Sci.* **101**, 10895–10900.
71. Cutler, R. 1975. Evolution of human longevity and the genetic complexity governing aging rate. *Proc. Natl. Acad. Sci. USA* **72**, 664–668.

72. Weiss, K.M. 1981. Evolutionary perspectives on human aging. In *Other Ways of Growing Old* (eds P. Amoss, S. Harrell), pp. 25–28. Stanford University Press.
73. Gage, T.B. 1998. The comparative demography of primates: with some comments on the evolution of life histories. *Annu. Rev. Anthropol.* **27**, 197–221.
74. Buikstra, J.E., Konigsberg, L.W. 1985. Paleodemography: critiques and controversies. *Am. Anthropol.* **87**, 316–333.
75. Buikstra, J.E. 1997. Paleodemography: context and promise. In *Integrating Archaeological Demography: Multidisciplinary Approaches to Prehistoric Population* (ed. R.R. Paine), pp. 367–380. Center for Archaeological Investigations.
76. Walker, P.L., Johnson, J.R., Lambert, P.M. 1988. Age and sex biases in the preservation of human skeletal remains. *Am. J. Phys. Anthropol.* **76**, 183–188.
77. Hoppa, R.D., Vaupel, J.W., eds. 2002. *Paleodemography: Age Distributions from Skeletal Samples*. Cambridge University Press.
78. O'Connell, J.F., Hawkes, K., Blurton Jones, N.G. 1999. Grandmothering and the evolution of *Homo erectus*. *J. Hum. Evol.* **36**, 461–485.
79. Kennedy, G.E. 2003. Palaeolithic grandmothers? Life history theory and early Homo. *J. R. Anthropol. Inst.* **9**, 549–572.
80. Konigsberg, L.W., Herrmann, N.P. 2006. The osteological evidence for human longevity in the recent past. In *The Evolution of Human Life History* (eds K. Hawkes, R.R. Paine), pp. 267–306. School of American Research Press.
81. Preston, S.H. 1995. Human mortality throughout history and prehistory. In *The State of Humanity* (ed. Julian L. Simon), pp. 30–36. Blackwell.
82. Tuljapurkar, S., Puleston, C., Gurven, M. 2007. Why men matter: mating pattern drives evolution of post-reproductive lifespan. *PLoS ONE* **2**, e785.
83. Kraft, T.S., Stieglitz, J., Trumble, B.C., Garcia, A.R., Kaplan, H., Gurven, M. 2020. Multi-system physiological dysregulation and ageing in a subsistence population. *Philos. Trans. R. Soc. B* **375**, 20190610.

12 Longevity in Modern Populations

Nadine Ouellette and Julie Choquette

12.1 Introduction

Humans have never lived as long as they do today. The expectation of duration of life at birth for all countries across the world is almost 73 years now [1], and life expectancy values for females in high-income countries (HICs) are reaching or even surpassing 85 years [2]. The contrast with earlier estimates during pre-industrial times is striking: before the mid-eighteenth century, the average length of human life probably never exceeded 25 to 30 years (see [3] and Chapter 11). This remarkable increase in longevity occurred over a relatively short period in human history and it is attributed to a series of relentless battles against epidemics, famines, health insecurity and chronic diseases. The resulting changes in the level and pattern of mortality were spectacular, with equally noticeable shifts in the distribution of deaths by age and cause.

This chapter reviews key concepts explaining these fundamental transformations, including the demographic transition model, which describes a period of paramount changes in modern history. It began as early as the eighteenth century in Europe with the Industrial Revolution and later spread worldwide. This transition is not to be confused with the Neolithic demographic transition that started around 10 000 BC after the onset of the Neolithic revolution, as described in Section 12.2. To shed light on the decline of certain pathologies throughout mortality history in human societies, the notion of epidemiological transitions was proposed and expanded to the conceptual framework of health transition, encompassing the societies' role in health-related responses. With the unexpected decline in old-age mortality (80 years and above) that accelerated around 1970 especially for males when the cardiovascular revolution started, it becomes perhaps more relevant than ever to question whether the increase in human longevity is accompanied by equivalent improvements in health and whether these trends are likely to continue in the future.

12.2 From Hunting and Gathering to Farming

After nearly 2.5 million years during which human beings relied primarily on gathering and hunting, agriculture began to emerge around the tenth millennium BC. This marked the initiation of the Neolithic revolution [4], a period characterized by a significant shift of many populations from nomadic to predominantly sedentary lifestyles.

Table 12.1 World population estimates at various dates since the beginning of the Neolithic revolution

Year	Population size	Doubling time (years)
10000 BC	6 M	–
AD 0	250 M	1 650
1650	500 M	180
1830	1 B	97
1927	2 B	47
1974	4 B	48
2022	8 B	–

Sources: [61, 62, 63, 64].

Scholars believe the shift occurred relatively abruptly, as it took only a few hundred years for humans to cluster and settle themselves into what is now referred to as villages [5]. With the spread of agriculture, the world population began to increase at an unprecedented rate, rising from 6 million on the eve of the Neolithic revolution to 250 million at the beginning of our current era (Table 12.1). The Neolithic demographic transition is presumed to be the driving force of this exceptional spurt in population growth [6].

The causes and mechanisms of the Neolithic demographic transition are intricate and have been a subject of lengthy debates among anthropologists and demographers, among others [chapter 2 in 7]. Initially, there was a belief that population growth during the transition resulted from improved health and survival, as a consequence of the development of agriculture and the domestication of certain animals ensuring better food availability than before [4, 8]. More recently, it was suggested instead that population growth owed to a significant rise in fertility, which was slightly offset by a more modest increase in mortality (see [9] and Chapters 10 and 11).

For the increased fertility hypothesis during the transition, the supporting explanation is based on a reduction in the birth interval, made possible by earlier weaning. With such shorter birth intervals, the resulting rise in total fertility rate is estimated at two live births per woman on average [10]. Among the factors that contributed to the earlier weaning of children, the significantly higher caloric intake from food derived from the cultivation of cereals and pulses, typically consumed by agricultural societies, likely played a role. In contrast, hunter-gatherer populations relied on less calorie-dense foods and faced fluctuations in food availability due to seasonal variations.

Several arguments were put forward to explain the increased mortality hypothesis during the Neolithic demographic transition. Among the most compelling is the fact that in agricultural societies, various conditions were favourable for the development, transmission and survival of parasites and infectious diseases. As facilitating factors, T. Aidan Cockburn mentions [11], for instance, sedentary living (e.g. human waste scattered near the house, mosquitoes, fleas), close contact with animals (e.g. cows, pigs, birds, dogs, rats), certain agricultural practices (e.g. draining swamps, crop fertilization with faeces) and increased population density. Furthermore, though the caloric intake from agricultural foods might have surpassed past levels, Mark N. Cohen emphasizes that it was accompanied by a decrease in nutritional quality and diversity [12]. In particular, the reduced consumption of optimal sources of complete protein, such as meat, and mineral absorption interference in a cereal-based diet, may have heightened susceptibility to infections. While infectious diseases emerged during the Neolithic demographic transition, it was not until the next demographic transition, sometimes labelled as 'contemporary', that substantial gains in the fight against infectious diseases were made.

12.3 The Demographic Transition

Following millennia of minimal progress in human mortality, a profound transformation occurred from the eighteenth century onwards, as life expectancy initiated a decisive and enduring rise during the unfolding of the demographic transition. The transition refers to a gradual shift from the high levels of mortality and fertility, characteristic of ancient agricultural societies, to the low levels typically found in modern industrialized societies. The decline in mortality and fertility during the transition is explained by a combination of economic, social and political factors [13], which are also at the heart of the Industrial Revolution. Warren S. Thompson and Adolphe Landry were the first to discuss the mechanism of transitioning from one demographic regime to the next [14, 15]. About a decade later, Frank W. Notestein introduced the classic formulation of the demographic transition [16], although he did not explicitly refer to it as being a 'transition'. His colleague at the Office of Population Research at Princeton University, however, explicitly referred to the transition [17]. It appears that Notestein was not aware of the works by Landry and Thompson, but he acknowledged having been greatly inspired by the extensive data compilation and demographic processes writings of Alexander M. Carr-Saunders [18].

The theory postulates that the transition is framed by two periods known as pre- and post-transitional, characterized by two quasi-equilibrium regimes in which fertility gains counterbalance mortality losses. Between these two periods lies the transition, which lasted one to two centuries in historical Western societies, and is characterized by a relatively short-lived disequilibrium regime, typically unfolding in the following two stages. First, the decline in mortality rates precedes that of fertility by at least several decades, likely because of improved living conditions and sanitary practices. This results in a significant population growth due to the excess of births over deaths. Second, as awareness of the improved survival grows and social transformations linked to the Industrial Revolution unroll, a decline in fertility rates begins, while the decrease in mortality rates lessens. The intensity of population growth thus slows down gradually. This period of transitional imbalance in two stages is, in fact, the cause of the exceptionally high rate of global population growth during the modern era. Table 12.1 illustrates that it took over 1 500 years for the world population to double in size from year 0. Subsequently, the doubling time dramatically decreased to less than 200 years and continued to drop until stabilizing at around 50 years since the late 1920s.

Although the general principle of demographic transition is fairly universal, substantial discrepancies have been observed among major world regions, particularly regarding the timing, pace and determinants of the transition [19]. The most developed countries in Europe, North America and Australia have already completed their demographic transition, while in other regions the transition is still ongoing. Beyond potential geographical variations, it is worth mentioning that sub-Saharan Africa is the global area that, even now, seems to face the greatest challenge in adhering to the fundamental framework of the transition (see [20] and Chapter 13).

In countries that completed the demographic transition, mortality continued to decline during the post-transitional period instead of stabilizing, thus contradicting the theory's prediction. The rapid increase in life expectancy over more than two centuries is undoubtedly one of the greatest achievements of the human species [21, 22]. The drivers of this longevity revolution are complex. The underlying factors of mortality change are multidimensional, and some of them had to align simultaneously at certain times to sustain the upward trajectory of life expectancy. To gain a clearer understanding, the theories proposed to explain the dynamics of the factors contributing to these advances in survival, beginning with that of Abdel R. Omran [23], and those that followed, proved highly beneficial.

12.4 Omran's Epidemiological Transition and Beyond

In relatively recent human history, the decline of certain pathologies, which significantly altered the epidemiological profile of the population, can explain part of the remarkable increase in longevity recorded since the eighteenth century. The first efforts to theorize about shifts from one dominant pathological structure to another were made by Omran [23]. The epidemiological transition he formulated unfolds in three major epidemiological 'ages' or phases. Initially, before the eighteenth century, 'the age of pestilence and famine' is characterized by high and fluctuating mortality, primarily due to endemic infectious diseases, but also sporadic and unpredictable epidemics (e.g. the plague), famines and wars, which are often highly lethal and significantly increase fluctuations. Next comes 'the age of receding pandemics', simultaneously with the onset of the Industrial Revolution and the rise in life expectancy, which is then made possible by the reduction of epidemics and endemic infectious diseases. Finally, during 'the age of degenerative and man-made diseases', Omran argues that mortality continues to decline and eventually tends to stabilize at a relatively low level [23], as the gains from the decline of certain traditional diseases are increasingly offset by losses caused by the emergence of new degenerative and man-made endemic diseases. As a result, the gains in life expectancy gradually diminish until they reach zero.

It must be understood that at the time Omran formulated his theory, mortality decline had been stalling for a decade in HICs. Some observers were thus speculating that progress against mortality had reached its limit. However, with surprising synchronicity, the cardiovascular revolution began around 1970 in a diverse range of

HICs [24, 25]. The widespread and rapid decline in cardiovascular disease mortality was the result of a combination of therapeutic innovations (e.g. drugs to treat hypertension and hypercholesterolemia, anticoagulants for heart and cerebrovascular diseases) and surgical breakthroughs (e.g. pacemakers, coronary artery bypass surgery), in addition to improvements in health-care systems (e.g. development of emergency medical care services) and behavioural changes (e.g. reduced smoking, diets lower in saturated fats and cholesterol as well as salt, increased physical activity, routine medical appointments) [22, 26, 27]. To account for the cardiovascular revolution, the addition of a fourth phase into Omran's original theory was suggested [23, 28, 29, 30]. However, that was quite unsuccessful because the theory is focused solely on the evolution of pathologies, so is suitable for infectious diseases but not for cardiovascular diseases.

A few years later, Julio Frenk and colleagues introduced a new and broader theory known as the health transition [31], which encompasses not only the evolution of epidemiological characteristics but also society's responses to the population's health status. The first phase of the health transition corresponds to the first epidemiological 'age' of Omran's theory, while the cardiovascular revolution takes centre stage in the second phase. The authors then keep the possibility open for subsequent stages in the future, without, however, characterizing them.

According to Shiro Horiuchi [26], the major shifts in cause-of-death profiles throughout human history can be summarized as five distinct epidemiological transitions. The primary three transitions belong to the past, while the following two are expected in the near future, and each transition has its own unique set of determinants. The first transition concerns the shift from a mortality regime dominated by external causes of death (i.e. accidents), typical of the lifestyle of ancient hunter-gatherer societies, to one governed by infectious diseases in agricultural societies. The factors that drove the rise of infectious diseases have been discussed in Section 12.3. The second epidemiological transition involves a gradual shift from predominantly infectious diseases to mainly degenerative diseases (e.g. cardiovascular diseases, cancers, diabetes, chronic liver and kidney diseases), and it aligns quite well with Omran's epidemiological transition. The determinants of this transition have been the subject of lengthy debates because they were often perceived as one-dimensional, whereas the reality is quite the opposite. To begin with, let's mention medical breakthroughs fighting infectious diseases (e.g. Jenner's smallpox vaccine, Pasteur's discoveries that led to sulphonamides and antibiotics), as well as improvements in food supply (e.g. better agricultural practices, food storage and transportation) aligning with higher standards of living [32, 33]. In addition, public health measures (e.g. improved water quality, sick quarantining) and better personal hygiene practices (e.g. cleanliness) [34, 35] also played a significant role. According to Horiuchi [26], the third epidemiological transition corresponds to the cardiovascular revolution described in the penultimate paragraph.

Looking ahead, Horiuchi proposed a fourth epidemiological transition defined by a decline in cancer mortality. At the time, cancer mortality trends varied substantially by type and between the sexes. Today, we know that in low-mortality countries, cancers caused by infectious agents, such as stomach cancer (*helicobacter pylori*) and cervix cancer (*human papillomavirus*), have indeed declined drastically due to improved infection control [36]. For other major cancers, many of which are closely related to lifestyle-associated factors, such as lung and pancreas (smoking), oesophagus (smoking, alcohol), breast (physical activity, hormonal) and colorectal (physical activity, diet, alcohol, smoking), the authors find that, for the most part, these either declined after peaking, or plateaued, thanks to reduced smoking and considerable improvements in screening and treatment (e.g. surgery, radiotherapy, chemotherapy). Finally, as the fifth epidemiological transition, Horiuchi suggests a potential deceleration in senescence [26], possibly fuelled by advancements in medical technologies addressing degenerative diseases, prolonged healthy lifestyles and revolutionary medical innovations slowing down ageing processes.

12.5 Compression of Mortality and Rectangularization of the Human Survival Curve

The fundamental changes in causes of death described substantially altered the distribution of deaths by age. Instead of primarily occurring at the earliest ages of life, deaths have gradually become concentrated at increasingly advanced ages, resulting in a phenomenon known as 'compression of mortality'. This decrease in lifespan variation indicates that uncertainty in the timing of death greatly diminished over time [37]. Simultaneously, the survival curve shifted towards higher ages and its shape became more rectangular. The biologist Alexander Comfort was the first to introduce the general idea of 'rectangularization of the survival curve' [38], due to a greater concentration of deaths over ages, but the concept of compression was popularized by James Fries [39]. Monitoring changes in the survival curve or in the age distribution of deaths quickly became a topic of great interest among scholars, particularly because of its profound implications at both the individual (e.g. retirement planning, family life participation) and population (e.g. health care and social services planning, income security) levels.

To illustrate, Figure 12.1 presents the gradual shift of deaths in the Swedish female population towards older ages over time, while also concentrating within a narrower age range. Like other developed countries, Sweden experienced substantial reductions in infant and childhood mortality, and to a lesser extent, young adult mortality, resulting in a noticeable relative decline in deaths for these age groups until the late 1950s. The postponement of these deaths to later ages led to a 'global' compression of mortality. Afterwards, the 'global' compression of mortality slowed down considerably, even though there was notable progress in reducing mortality among older age groups [40]. This prompted researchers to distinguish 'old-age' mortality compression, specifically capturing the extent of lifespan inequality among the elderly, from 'global' compression over the entire age range [41].

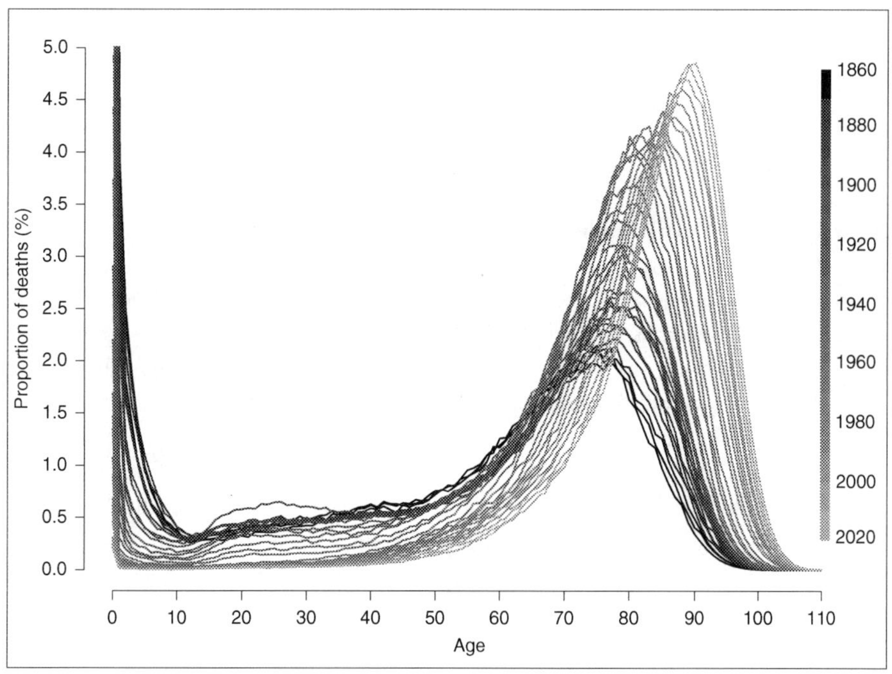

Figure 12.1 Age distribution of life table deaths, females, Sweden, 1860–1864 to 2015–2019. Infant deaths are truncated at 5% for periods 1920–1924 and earlier.
Source: [2].

To describe the evolution of the curves presented in Figure 12.1, it is important to rely not only on indicators summarizing the central tendency of deaths, such as life expectancy at birth, e_0, but also on dispersion indices measuring the variation in age at death [37, 42], like the standard deviation. It is worth noting that e_0 is undoubtedly the most commonly used mortality indicator to track improvements in longevity. However, it is essential to recognize that this measure is not always fully understood. All too often, the gradual slowdown in the rate at which e_0 has increased over the years is cited as the main argument in favour of the existence of an upper limit to the average human lifespan. It is crucial to comprehend that during its early stages, the swift increase in e_0 was predominantly a result of reduced infant mortality. These prevented deaths resulted in a substantial contribution (in person-years lived) to the overall increase in life expectancy. Nowadays, the rise in e_0 is comparatively slow in low-mortality countries because the averted deaths are concentrated among the elderly population (see also Chapter 14). Thus, it is primarily through extending the lives of older individuals, with modest potential contributions in person-years lived gained, that e_0 continues to grow. Sometimes, e_0 is interpreted as the typical (i.e. most frequent) lifespan. But in fact, the most frequent age at death, commonly known as the adult modal age at death, M, acts as a complementary indicator to e_0 by representing the location of old-age death heap in the age distribution of deaths (Figure 12.1). Unlike e_0, M is determined mostly by old-age mortality [43].

Moreover, when compared with life expectancies at advanced ages (e.g. e_{55}, e_{65}, e_{75}), M provides a more accurate representation of the mortality gains recorded since 1950 because it is not affected by any arbitrary selection of an age range for 'old ages'. In HICs, M therefore continues to show a steady increase over time [44].

12.6 Trade-Off between Lifespan and Healthspan

The increase in e_0 is often viewed as a sign of improved population health. While this assumption may seem reasonable under a mortality regime dominated by acute infectious and contagious diseases, as chronic degenerative diseases progressively come to dominate cause-of-death patterns its validity becomes less clear. Mortality is relatively easy to measure since it involves directly counting deaths, but defining and measuring health poses much greater challenges. Even before time-series data on health status became available through population-based health interview surveys, researchers were already speculating about the evolution of the quality of years lived. Parallel to the debate on possible limits to lifespan (see Chapters 14 and 15) emerged another equally fundamental one concerning the evolution of the quality of the additional years lived. In short, three theoretical scenarios were proposed to apprehend the future evolution of morbidity: compression of morbidity, expansion of morbidity and dynamic equilibrium.

Under the morbidity compression scenario, the period of ill health prior to death is assumed to become increasingly shorter [39]. It is based on the principle that the length of human life is fixed, or that future increases will eventually become so slow they will be imperceptible. Fries argued that advancements in medicine, changes in behaviour and improvements in lifestyle would be the primary contributing factors to reduce the relative number of years lived in poor health, resulting in a compression of morbidity [39]. The second scenario suggests just the opposite, that is, gains in mortality will lead to longer periods of illness or disability, resulting in an expansion of morbidity. The argument unfolds as follows: since the decline in mortality was above all a result of the decreasing lethality of chronic diseases rather than a reduction in their prevalence or pace, the development of more severe chronic diseases would ensue, particularly among older individuals [45]. Often referred to as the 'pandemic of mental disorders, chronic diseases, and disabilities' [46, p. 382], this scenario would result in a decrease in the average number of years lived in good health, while the average total number of years lived may not necessarily decline. The third scenario falls between the two previous ones and adopts a moderate view on the future health status of human populations. It suggests that while the decline in mortality leads to an increase in the prevalence of chronic diseases, the severity of these diseases experienced by individuals is generally reduced due to improved treatments, resulting in a certain state of dynamic equilibrium [47].

Given the highly subjective nature of assessing an individual's health status, a functional approach that emphasizes the concept of disability is often preferred due to its minimal susceptibility to reporting bias [48]. The approach is particularly relevant in low-mortality countries, where the increased prevalence of disabilities at older ages may bring functional limitations and activities restrictions, impeding the ability to perform daily chores in an autonomous manner, thus requiring assistance, health-care resources and long-term care planning [49]. Numerous indicators were developed to assess the severity of these disabilities, which fall into two main categories. The first one includes measures such as the Disability-Adjusted Life Years (DALYs) and Quality-Adjusted Life Years (QALYs), designed chiefly to assess the burden of disease and set priorities for health interventions. The second category comprises measures to inform our expectations regarding the debate on the evolution of morbidity, specifically whether it is being compressed, expanded or if there is a dynamic equilibrium. The most widely used 'health expectancies' measures for this purpose include the Disability Free Life Expectancy (DFLE) and Health-Adjusted Life Expectancy (HALE). These offer valuable insights into the distribution of years lived in various health states, taking into account different levels of disability (e.g. mild, moderate, severe). All of the measures enumerated are computed using both mortality and health data.

Cross-country comparisons based on empirical data from health surveys remain difficult despite concerted efforts. In fact, most of the surveys are cross-sectional, rather than longitudinal, and they mainly concern HICs. Still, in countries where health expectancies are routinely computed, they tend to increase proportionally with the rise in life expectancy [50]. In other words, it appears that the hypothesis of dynamic equilibrium is the most suitable scenario to represent the relationship between population health and the increase in longevity. It should be noted, however, that less severe disabilities typically exhibit more variations in levels and trends between countries than the severe ones, in part because definitions used varies, while severe disabilities usually rely on the ability to perform activities of daily living (ADLs) [48]. When health outcomes are contrasted by sex, these usually show that women spend a greater share of years with functional and activities limitations than men, despite living longer (Chapter 9). For instance, throughout a considerable portion of their lives, women tend to have worse self-rated health and experience a higher number of hospitalization episodes compared to men, while also exhibiting lower death rates at each age [51]. The reasons for this 'male–female health–survival paradox' or its mechanisms have spilled a lot of ink, but they have yet to be fully understood [52].

12.7 Conclusion

In retracing history, an abundance of evidence strongly suggests that the remarkable decline in human mortality is, above all else, 'driven by a widespread, perhaps universal, desire for a longer, healthier life' ([53], p. 1 124). The latest major achievement is the significant improvement in old-age mortality in many HICs since the 1950s, which were revealed by groundbreaking studies published about three decades ago [40, 54, 55]. This marked the onset of a new era in longevity, as mortality had never

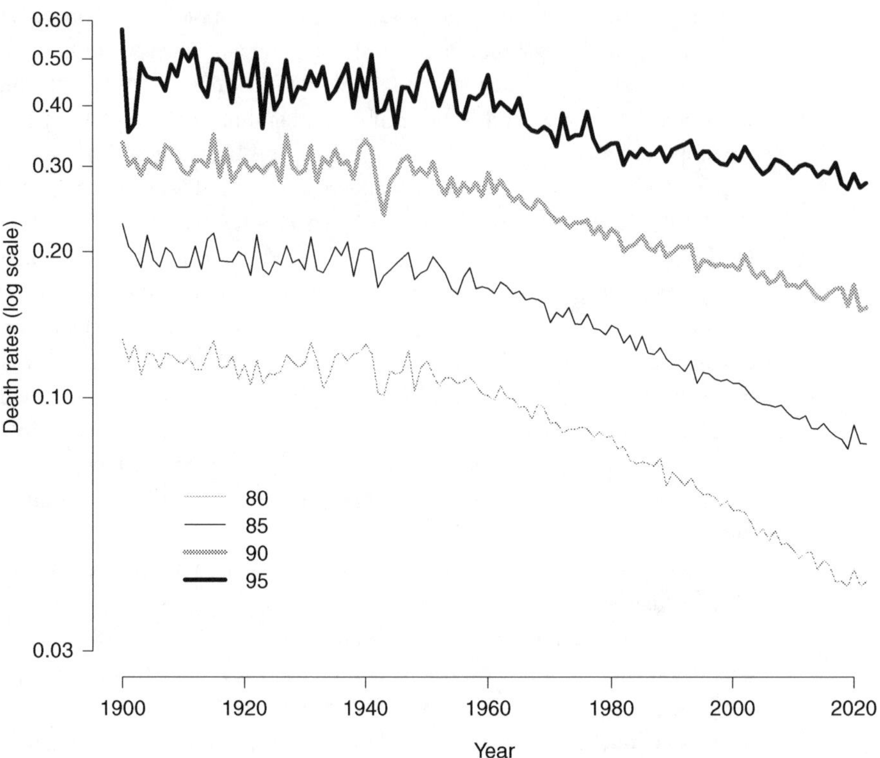

Figure 12.2 Death rates at ages 80, 85, 90 and 95, both sexes, Sweden, 1900 to 2022. Source: [2].

undergone such a decline in the past, except at younger ages. In 2008, Roland Rau and colleagues demonstrated that in countries with lowest mortality, reductions in mortality trends at ages 80 and above persisted without interruption [56]. Moreover, Figure 12.2 illustrates the continuous decline in death rates at ages 80–95 in Sweden up to the present day, which mirrors that in most low-mortality countries. At the same time, some observers are stating that 'old-age survival follows an advancing front, like a travelling wave' ([57], p. 11209). Improvements in old-age mortality not only led to an explosion in numbers of centenarians but also paved the way for the emergence and expansion of new age groups, such as semi-supercentenarians (ages 105–109) and supercentenarians (ages 110 and above).

Although there are reasons to anticipate further decline in death rates at very old ages in low-mortality countries in the coming years, it is crucial to bear in mind that mortality trends have not and may not always be favourable across all other age groups and/or countries. Historic reversals are possible, for instance due to deaths from resurgent infectious diseases. The most recent threatening reversal of this nature was the HIV/AIDS epidemic, which had the most severe impact on southern and eastern African countries. Current treats include, for instance, drug-resistant new strains of previously known pathogens (e.g. tuberculosis), which require close monitoring, along with the potential emergence of new disease outbreaks (e.g. the COVID-19 pandemic). Another important portion of resurgent mortality, both in the past, present and possibly the near future, can be attributed to suicide, homicide, alcoholism and (drug) poisoning. For example, the collapse of the Former Soviet Union (FSU) around 1990 led to a dramatic upsurge in mortality across all these causes of death [58], from which several FSU countries have not yet fully recovered. In the United States, the increase in homicide mortality from the mid-1980s to the mid-1990s, with links to the crack-cocaine epidemic of the mid-1980s, serves as another example of a historical turnaround. Currently, notable threats concerning 'violent' deaths include the escalation of drug-related mortality (e.g. synthetic opioids, stimulants), with a particular impact on the United States and Canada, making them likely the hardest-hit countries thus far [59]. Whether the aforementioned historical setbacks are interpreted as reverse epidemiological transitions [26], or as part of convergence-divergence cycles in mortality [60], they serve as reminders that advancements in survival are not easy to achieve and should never be taken for granted.

References

1. United Nations. 2022. *World Population Prospects 2022: Summary of Results*. United Nations, Department of Economic and Social Affairs, Population Division.
2. Human Mortality Database. 2023. University of California, Berkeley (USA) and Max Planck Institute for Demographic Research (Germany). www.mortality.org.
3. Acsádi, G., Nemeskéri, J. 1970. *History of Human Life Span and Mortality*. Akadémiai Kiadó.
4. Childe, V.G. 1936. *Man Makes Himself*. Watts and Co.
5. Bandy, M.S., Fox, J.R. 2010. Becoming villagers: the evolution of early village societies. In *Becoming Villagers: Comparing Early Village Societies* (eds M.S. Bandy, J.R. Fox), pp. 1–16. University of Arizona Press.
6. Bocquet-Appel, J.-P. 2002. Paleoanthropological traces of a Neolithic demographic transition. *Curr. Anthropol.* **43**, 637–650.
7. Livi-Bacci, M. 2007. *A Concise History of World Population* (4th ed.). Wiley Blackwell.
8. Cohen, M.N. 1984. An introduction to the symposium. In *Paleopathology at the Origins of Agriculture* (eds M.N. Cohen, G.J. Armelagos), pp. 1–11. University Press of Florida.
9. Bocquet-Appel, J.-P. 2008. Explaining the Neolithic demographic transition. In *The Neolithic Demographic Transition and its Consequences* (eds J.-P. Bocquet-Appel, O. Bar-Yosef), pp. 35–55. Springer Netherlands.
10. Bocquet-Appel, J.-P., Dubouloz, J. 2003. Traces paléoanthropologiques et archéologiques d'une transition démographique néolithique en Europe. *Bull. Société Préhistorique Fr.* **100**, 699–714.
11. Cockburn, T.A. 1971. Infectious diseases in ancient populations. *Curr. Anthropol.* **12**, 45–62.
12. Cohen, M.N. 1989. *Health and the Rise of Civilization*. Yale University Press.
13. Kirk, D. 1996. Demographic transition theory. *Popul. Stud.* **50**, 361–387.
14. Thompson, W.S. 1929. Population. *Am. J. Sociol.* **34**, 959–975.

15. Landry, A. 1934. *La Révolution Démographique: Études et Essais sur les Problèmes de la Population*. Librairie du Recueil Sirey.
16. Notestein, F. 1945. Population: the long view. In *Food for the World* (ed. T.W. Schultz), pp. 36–57. University of Chicago Press.
17. Davis, K. 1945. The world demographic transition. *Ann. Am. Acad. Polit. Soc. Sci.* **237**, 1–11.
18. Carr-Saunders, A.M. 1936. *World Population: Past Growth and Present Trends*. Clarendon Press.
19. Chesnais, J.-C. 1986. La Transition Démographique: Étapes, Formes, Implications Économiques. Étude de séries temporelles (1720–1984) relatives à 67 pays. Présentation d'un Cahier de l'INED. *Population* **41**, 1059–1070.
20. Kuate Defo, B. 2014. Demographic, epidemiological, and health transitions: are they relevant to population health patterns in Africa? *Glob. Health Action* **7**, 22443.
21. Oeppen, J., Vaupel, J.W. 2002. Broken limits to life expectancy. *Science* **296**, 1029–1031.
22. Vallin, J., Meslé, F. 2009. The segmented trend line of highest life expectancies. *Popul. Dev. Rev.* **35**, 159–187.
23. Omran, A.R. 1971. The epidemiologic transition: a theory of the epidemiology of population change. *Milbank Mem. Fund Q.* **49**, 509–538.
24. Crimmins, E.M. 1981. The changing pattern of American mortality decline, 1940–77, and its implications for the future. *Popul. Dev. Rev.* **7**, 229–254.
25. Ouellette, N., Barbieri, M., Wilmoth, J.R. 2014. Period-based mortality change: turning points in trends since 1950. *Popul. Dev. Rev.* **40**, 77–106.
26. Horiuchi, S. 1999. Epidemiological transitions in human history. In *Health and Mortality Issues of Global Concern: Proceedings of the Symposium on Health and Mortality, Brussels, 19–22 November 1997* (eds J. Chamie, R.L. Cliquet), pp. 54–71. United Nations and CBGS.
27. Wilmoth, J.R. 2007. Human longevity in historical perspective. In *Physiological Basis of Aging and Geriatrics* (ed. P.S. Timiras), pp. 11–22. CRC Press.
28. Omran, A.R. 1982. Epidemiologic transition. In *International Encyclopedia of Population* (ed. J.A. Ross), pp. 172–183. Free Press.
29. Omran, A.R. 1983. The epidemiologic transition theory: a preliminary update. *J. Trop. Pediatr.* **29**, 305–316.
30. Olshansky, J.S., Ault, B.A. 1986. The fourth stage of the epidemiologic transition: the age of delayed degenerative diseases. *Milbank Q.* **64**, 355–391.
31. Frenk, J., Bobadilla, J.L., Stern, C., Frejka, T., Lozano, R. 1991. Elements for a theory of the health transition. *Health Transit. Rev.* **1**, 21–38.
32. McKeown, T., Record, R.G. 1962. Reasons for the decline of mortality in England and Wales during the nineteenth century. *Popul. Stud.* **16**, 94–122.
33. McKeown, T. 1979. *The Role of Medicine: Dream, Mirage, or Nemesis*. Princeton University Press.
34. Preston, S. 1990. Sources of variation in vital rates: an overview. In *Convergent Issues in Genetics and Demography* (eds J. Adarns, D.A. Lam, A.I. Herrnalin, P.E. Smouse), pp. 335–350. Oxford University Press.
35. Preston, S.H., van de Walle, E. 1978. Urban French mortality in the nineteenth century. *Popul. Stud.* **32**, 275–297.
36. Gersten, O., Barbieri, M. 2021. Evaluation of the cancer transition theory in the US, select European nations, and Japan by investigating mortality of infectious- and noninfectious-related cancers, 1950–2018. *JAMA Netw. Open* **4**, e215322.

37. Wilmoth, J.R., Horiuchi, S. 1999. Rectangularization revised: variability of age at death within human populations. *Demography* **36**, 475–495.
38. Comfort, A. 1954. Biological aspects of senescence. *Biol. Rev.* **29**, 251–366.
39. Fries, J.F. 1980. Aging, natural death, and the compression of morbidity. *N. Engl. J. Med.* **303**, 130–135.
40. Kannisto, V., Lauritsen, J., Thatcher, R.A., Vaupel, J.W. 1994. Reductions in mortality at advanced ages: several decades of evidence from 27 countries. *Popul. Dev. Rev.* **20**, 793–810.
41. Thatcher, A.R., Cheung, S.L.K., Horiuchi, S., Robine, J.-M. 2010. The compression of deaths above the mode. *Demogr. Res.* **22**, 505–538.
42. van Raalte, A.A., Sasson, I., Martikainen, P. 2018. The case for monitoring life-span inequality. *Science* **362**, 1002–1004.
43. Horiuchi, S., Ouellette, N., Cheung, S.L.K., Robine, J.-M. 2013. Modal age at death: lifespan indicator in the era of longevity extension. *Vienna Yearb. Popul. Res.* **11**, 37–69.
44. Ouellette, N., Côté-Gendreau, M., Violo, P. 2022. Modeling contextual determinants of rising adult modal age at death in high-income countries. Paper presented at the Annual Meeting of the Population Association of America, Atlanta, GA, 6–9 April.
45. Gruenberg, E.M. 1977. The failures of success. *Milbank Mem. Fund Q. Health Soc.* **55**, 3–24.
46. Kramer, M. 1980. The rising pandemic of mental disorders and associated chronic diseases and disabilities. *Acta Psychiatr. Scand.* **62**, 382–397.
47. Manton, K.G. 1982. Changing concepts of morbidity and mortality in the elderly population. *Milbank Mem. Fund Q. Health Soc.* **60**, 183–244.
48. Robine, J.-M., Jagger, C., Crimmins, E.M., Saito, Y., Van Oyen, H. 2020. Trends in health expectancies. In *International Handbook of Health Expectancies* (eds C. Jagger, E.M. Crimmins, Y. Saito, R.T.D.C. Yokota, H. Van Oyen, J.-M. Robine), pp. 19–34. Springer.
49. Cambois, E., Duthé, G., Meslé, F. 2023. Global trends in life expectancy and healthy life expectancy. In *Oxford Research Encyclopedia of Global Public Health* (ed. D.V. McQueen), pp. 1–34. Oxford University Press.
50. Jagger, C., Crimmins, E.M., Saito, Y., Yokota, R.T.D.C., Van Oyen, H., Robine, J.-M. 2020. *International Handbook of Health Expectancies*. Springer.
51. Case, A., Paxson, C. 2005. Sex differences in morbidity and mortality. *Demography* **42**, 189–214.
52. Di Lego, V., Di Giulio, P., Luy, M. 2020. Gender differences in healthy and unhealthy life expectancy. In *International Handbook of Health Expectancies* (eds C. Jagger, E.M. Crimmins, Y. Saito, R.T.D.C. Yokota, H. Van Oyen, J.-M. Robine), pp. 151–172. Springer.
53. Wilmoth, J.R. 2000. Demography of longevity: past, present, and future trends. *Exp. Gerontol.* **35**, 1111–1129.
54. Kannisto, V. 1988. On the survival of centenarians and the span of life. *Popul. Stud.* **42**, 389–406.
55. Vaupel, J., Lundström, H. 1994. Longer life expectancy? Evidence from Sweden of reductions in mortality rates at advanced ages. In *Studies in the Economics of Aging* (ed. D.A. Wise), pp. 79–102. University of Chicago Press.
56. Rau, R., Soroko, E., Jasilionis, D., Vaupel, J.W. 2008. Continued reductions in mortality at advanced ages. *Popul. Dev. Rev.* **34**, 747–768.
57. Zuo, W., Jiang, S., Guo, Z., Feldman, M.W., Tuljapurkar, S. 2008. Advancing front of old-age human survival. *Proc. Natl. Acad. Sci.* **115**, 11209–11214.

58. Meslé, F. 2004. Mortality in Central and Eastern Europe: long-term trends and recent upturns. *Demogr. Res.* **2**, 45–70.
59. Humphreys, K. et al. 2022. Responding to the opioid crisis in North America and beyond: recommendations of the Stanford–Lancet Commission. *The Lancet* **399**, 555–604.
60. Vallin, J., Meslé, F. 2004. Convergences and divergences in mortality: a new approach of health transition. *Demogr. Res.* **2**, 12–44.
61. Biraben, J.-N. 1979. Essai sur l'évolution du nombre des hommes. *Population* **34**, 13–25.
62. Biraben, J.-N. 2003. The rising numbers of humankind. *Popul. Soc.* **394**, 1–4.
63. Kaneda, T., Haub, C. 2022. How many people have ever lived on earth? Population Reference Bureau. www.prb.org/articles/how-many-people-have-ever-lived-on-earth.
64. United Nations. 2023. www.un.org/en.

13 Health Transition and Population Ageing

Challenges for the Global South

Géraldine Duthé, Lucie Vanhoutte, Soumaila Ouedraogo and Gilles Pison

13.1 Introduction

In 2022, the world population reached a level of life expectancy at birth (e_0) of almost 72 years on average for both sexes,[1] 25 years more than in the early 1950s (see also Chapter 12). Despite global progress, inequalities in terms of life expectancy at birth are still very high: in 2022, more than 30 years separated the life expectancy at birth estimated for Chad and Lesotho (53 years), the least favoured countries, from the most advanced one, Japan (85 years).

While the demographic transition slowly spread from the eighteenth to the nineteenth century in Europe, North America and Japan, it reached Latin America and parts of Asia and the Middle East only after the First World War, then spread to the rest of the world after the Second World War. The demographic transition was rapid in many regions located in what was called the developing world,[2] with a large decrease in mortality and fertility rates especially in Asia and Latin America. For 2022, e_0 was estimated at 73–74 years in Asia and Latin America, only a few years less than in Europe (Table 13.1).

As Asian and Latin American countries have experienced a rapid demographic transition with gains in life expectancy in addition to a large decrease in fertility, their population are ageing: in 1950, 7% of Asian and 5% of Latin American population were older than 60,[3] the proportion was estimated to be 13% in 2022. This share will grow fast and should reach around a quarter in 2050. The African continent though

[1] In this chapter, we use yearly estimates provided by the last revision (at the time of writing) of the World Population Prospects, published in 2022 by the United Nations (UN), Department of Economic and Social Affairs, Population Division.

[2] Currently, countries are frequently grouped according to the World Bank classification based on their economic level (low- and middle-income countries (LMICs) and high-income countries (HICs)) rather than the UN grouping based on a large development notion. However, this grouping is not constant over time and, in order to have a historical perspective, we also use geographic regions with a focus on Africa, Asia and Latin America, in whose continents are concentrated most of the LMICs. Most of the total human population (85%) live in these three regions – Asia by itself has almost 60% of the total population.

[3] In this chapter, we use 60 years to characterize the beginning of old age. Beyond the facility to use a relatively standardized threshold, it is important to note that this threshold remains rather arbitrary, as, more globally, the concepts of individual ageing and who is considered to be old are culturally dependent [1].

remains characterized by a very young population and despite the fact that the absolute numbers of individuals older than 60 have considerably increased, their share of the population has remained quite stable and was only 5.5% in 2020. Although the expected progress in life expectancy and decrease in fertility should result in population ageing in the future, this proportion will stay lower in Africa for the next decades compared to the other regions (Table 13.1).

This population ageing implies, however, numerous socio-economic and demographic challenges related to the mortality of the elderly among the Global South – that is, the LMICs that are mostly located in Africa, Asia and Latin America[4] – especially in the context of low resources and scarcity of social protection systems. In the framework of the health transition, increase in life expectancy goes along with deaths shifting to older ages and changes in causes of death distribution. These changes are becoming crucial for improving the health of the most numerous populations on earth. In HICs, where almost nine in 10 deaths occur at age 60 or more (Table 13.1), progress in life expectancy depends mostly on old-age mortality trends, whereas mortality remains high at younger ages in countries from the Global South. The international agenda for the development has focused its attention for decades on child, maternal and HIV/AIDS mortality. It is only in the recent Sustainable Development Goals (SDGs), adopted in 2015 by the UN, that attention has also been paid to non-communicable diseases (NCDs) – mostly cardiovascular diseases (CVDs), cancers, chronic respiratory diseases and diabetes – with the need to prevent 'premature' mortality (before age 70 according to the World Health Organization (WHO)). Apart from the African region, the large majority of deaths occur now after age 60 (Table 13.1). However, the health and mortality of (older) adults and their longevity remain poorly known in the Global South despite the dramatic challenges this population has to face. There is indeed a specific issue related to the availability of reliable data and measurement of mortality levels and trends in countries from the Global South. This situation has been called a 'scandal of invisibility' [2]; this is particularly worrying because despite having been a known problem for a long time, little progress has been made. In 2010–2012, six out of 10 worldwide deaths were not properly registered in vital statistics databases [3], especially in Asia and Africa where estimation of the registration coverage is itself challenging [4]. As expected, the figures are even worse for cause-of-death statistics that require a medical diagnosis.

In this chapter, we first present the trends in life expectancy at birth and adult mortality in the Global South according to the estimates available since the 1950s. Second, we propose to contextualize the variety of favourable and unfavourable situations observed, in the framework of the health transition, discussing the estimated burden of the different group of causes of death involved. Third, we focus on the dramatic challenge of the mortality estimation, especially among older adults, which does not allow presenting figures related to longevity. We finally discuss the future challenges of population ageing in the Global South.

[4] The Global South also includes LMICs located in Oceania, however, they only accounted for 0.2% of the world population in 2020.

Table 13.1 Estimates for population, proportion of people aged of 60+, total fertility rate and life expectancy at birth for both sexes, in 1950, 2022 and 2050 by World Bank income groups and regions of the world

	Population in mid-year						Share of people over age 60 years			Total fertility rate (TFR)			Life expectancy at birth (e0)			Share of deaths over age 60 years		
	Thousands			%			%			Children per women			Years			%		
	1950	2022	2050	1950	2022	2050	1950	2022	2050	1950	2022	2050	1950	2022	2050	1950	2022	2050
World Bank income groups																		
High-income countries	687 357	1 250 515	1 281 937	27.5	15.7	13.2	11.6	25.4	34.2	3.01	1.56	1.64	61.5	80.9	85.4	52.1	87.6	94.3
Middle-income countries	1 688 959	5 958 019	7 023 580	67.6	74.7	72.3	6.7	12.6	22.6	5.53	2.16	2.02	43.5	70.8	77.1	23.9	65.6	81.6
Low-income countries	117 442	737 605	1 367 097	4.7	9.2	14.1	5.2	4.9	7.5	6.58	4.54	2.96	31.6	63.0	68.8	13.5	34.1	48.9
Geographic regions																		
Africa	227 549	1 426 736	2 485 136	9.1	17.9	25.6	5.3	5.5	8.7	6.59	4.24	2.87	37.6	62.2	68.3	14.0	35.3	50.1
Asia	1 379 048	4 722 635	5 292 948	55.2	59.2	54.5	6.7	13.7	25.3	5.71	1.94	1.85	42.0	73.2	79.5	23.7	70.2	86.6
Europe	549 722	743 556	703 007	22.0	9.3	7.2	11.8	26.3	35.9	2.70	1.49	1.63	62.8	77.4	83.8	56.4	86.4	94.3
Latin America and the Caribbean	168 336	660 269	749 169	6.7	8.3	7.7	5.2	13.4	25.1	5.80	1.85	1.72	48.6	73.8	80.6	19.3	69.7	87.3
Northern America	162 089	376 871	421 398	6.5	4.7	4.3	12.1	23.7	29.8	2.97	1.64	1.68	68.0	78.7	84.0	62.1	81.6	91.0
Oceania	12 578	45 039	57 834	0.5	0.6	0.6	10.9	17.7	24.1	3.67	2.14	1.96	61.4	79.2	82.1	49.9	76.2	86.3
World	**2 499 322**	**7 975 105**	**9 709 492**	**100.0**	**100.0**	**100.0**	**8.0**	**13.9**	**22.0**	**4.86**	**2.314**	**2.149**	**46.5**	**71.7**	**77.2**	**27.8**	**66.8**	**80.5**

Note: The years 2022 and 2050 are estimates based the UN medium variant.
Source: [28].

13.2 Life Expectancy and Causes of Death in the Global South

13.2.1 Life Expectancy at Birth: Rapid Progress Despite Local and Temporal Disruption

Since the 1950s, regions from the Global South have made rapid progress in e_0 (Figure 13.1). This is especially true in Asia, Latin America and Northern Africa where e_0 is estimated to have been at around 40–50 years in the mid-1950s and has now reached 70 on average and more than 75 years for some (Eastern Asia and Latin America). However, some crises disrupted these trends. This is the case with the large fall in e_0 observed for Eastern Asian during the three-year period 1959–1961: a dramatic consequence of the 'Great Leap Forward' policy of the People's Republic of China (see also Figure 13.2 for an estimation of its impact on adult mortality). More recently, another dramatic decrease occurred in the Caribbean in 2010 due to the earthquake in Haiti that led to a loss of 15 years in Haitian e_0 for this specific year.

More globally, a few regions, such as Central Asian countries, experienced some decline for longer periods in the last decades of the twentieth century. These countries suffered from political and economic crises following the USSR's collapse [5]. They also experienced a health crisis similar to what has been observed in Eastern Europe due to a dramatic rise in alcohol consumption especially among (ethnic Russian) males that led to a rise in NCDs – CVDs in particular – and injuries [6].

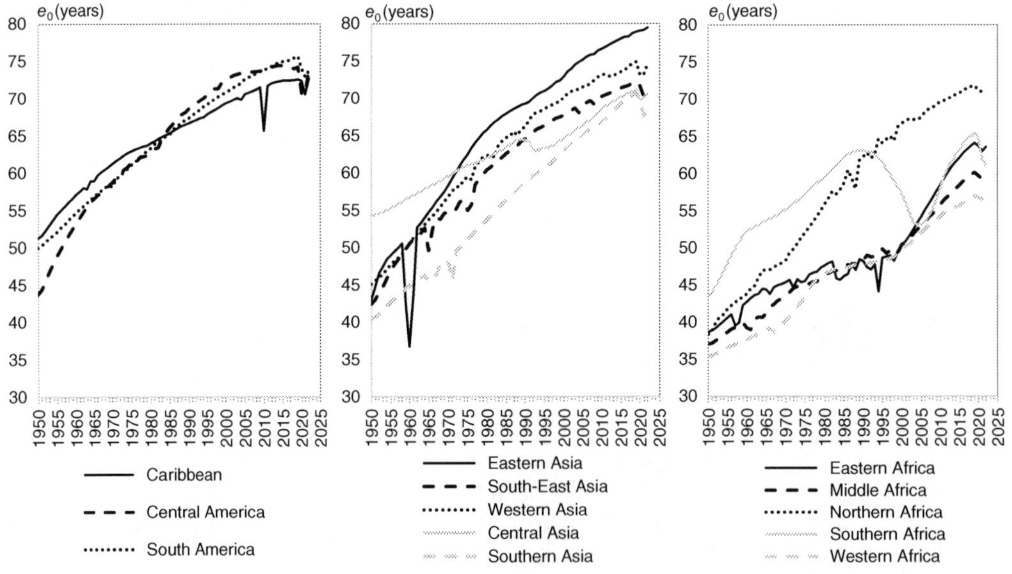

Figure 13.1 Trends in life expectancy at birth, from 1959 to 2022, both sexes, by populated region of the Global South.
Source: [28]. From United Nations World Population Prospects. Copyright © 2022 United Nations. Used with the permission of the United Nations.

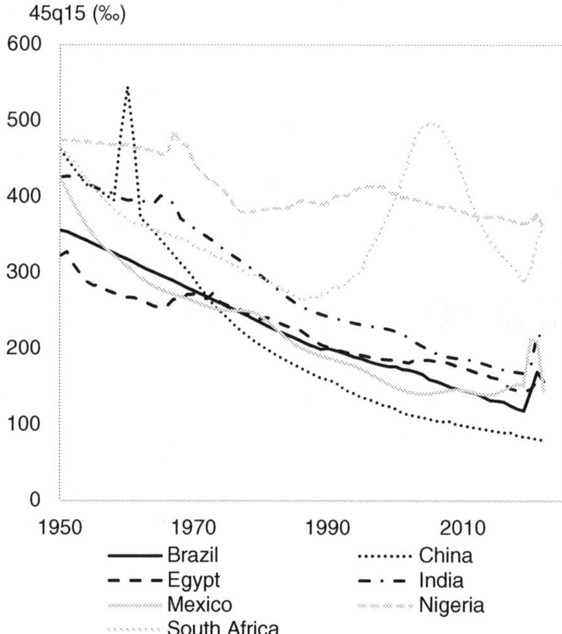

Figure 13.2 Adult mortality trends in a selection of populated southern countries.
Note: 45q15 is the probability a person aged 15 dies before reaching age 60.
Source: [28] From United Nations World Population Prospects. Copyright © 2022 United Nations. Used with the permission of the United Nations.

Nevertheless, it is in sub-Saharan African that the trends were the most unfavourable: HIV/AIDS dramatically affected the southern countries, and the eastern ones to a lesser extent. In South Africa for instance, it is estimated that e_0 dropped from 63 years in the early 1990s to 54 years in 2005 due to a dramatic increase of adult mortality (15–60) during this period (Figure 13.2). However, even in Western and Central Africa, e_0 also stopped progressing during this same period due to different factors, such as malaria re-emerging, economic crises and political conflicts [7]. Indeed, adult mortality increased in Nigeria – the most populated African country – in the same period (Figure 13.2). Despite those declines being responsible for the remaining gap between these least favoured regions and others, progress has resumed in most of them since the early 2000s.

In Central America, after decades of discontinued progress, the beginning of the twenty-first century marked a new break (Figure 13.2). This was mainly due to the rise in NCDs (diabetes notably) and external causes (injuries and violent deaths) and to the decelerating reduction in cardiovascular mortality [8, 9]. Thus, adult mortality has stalled in Mexico.[5]

Finally, 2020 to 2022 were characterized by a global deterioration of the health trends due to the COVID-19 pandemic (Figure 13.1). In many countries from the

[5] And even increased among males.

Global South, data are of poor quality, especially for the oldest population, which is the most affected by COVID-19 mortality. Mortality of the oldest is thus mainly estimated from models (see Section 13.4) and it is challenging to estimate precisely the impact of the COVID-19 pandemic on mortality (either direct, through COVID-19 deaths, or indirect, through deterioration in health systems, economic crises etc.). The technical Advisory Group on COVID-19 Mortality Assessment mandated by WHO and the UN, however, estimate that 8 in 10 of the 15 million excess deaths estimated over the period 2020 and 2021 have occurred in middle-income countries, mostly located in Latin America and South-East Asia [10]. Some countries have been particularly affected, such as Brazil, India and South Africa, as is apparent through their adult mortality figures (Figure 13.2). However, UN projections assume a rapid return to a pre-COVID situation and the world population should hopefully experience a new period of progress, reaching 77 years of e_0 in 2050 (Table 13.1).

13.2.2 Causes of Death: A Recent Picture by Level of Income

Cause-specific mortality is crucial to understand all-cause mortality trends and elaborate targeted health policy and to monitor progress. However, in the absence of a complete and reliable civil registration and vital statistics (CRVS) system at the national level, the available cause-of-deaths statistics, if there are any, are usually the tip of the iceberg. While a cause is known for 90% of the deaths that occur in the American and European regions, this share is less than 10% for the African and South-East Asian regions [11]. In Africa, only six countries provide cause-of-deaths statistics at the national level.[6]

The WHO's Global Health estimates provide the top 10 leading causes of death by level of income,[7,8] comparing 2000 and 2019 [12]. CVDs, Alzheimer and other dementia, respiratory disease (chronic obstructive pulmonary disease) and cancers are the most frequent causes among HICs in 2019. Only one category of infectious diseases – the lower respiratory infections – are in the top 10, at the sixth rank. The upper-middle-income countries converge to this pattern, especially with a dramatic increase of deaths due to ischemic heart diseases since the early 2000s. In the lower-middle- and low-income countries, figures combine NCDs and infectious and parasitic diseases. Neonatal conditions and lower respiratory infections (mostly pneumonia and bronchopneumonia) are, respectively, the first and second leading causes of deaths in low-income countries. Diarrheal diseases, malaria, tuberculosis and HIV/AIDS are also among the leading causes, albeit an important decrease in number of deaths.

[6] Cabo Verde, Egypt, Mauritius, Sao Tome and Principe, Seychelles, South Africa.

[7] www.who.int/data/global-health-estimates. Multiple sources are used to provide these estimates, mainly CRVS when they exist, otherwise estimates are from the work of WHO, the UN and the inter-agency group Global Burden of Disease (GBD). For countries with no reliable and exhaustive statistics, provided numbers are not precise estimates, but this distribution provides a good picture of most important causes of death.

[8] According to the World Bank classification of countries.

Theories have been proposed to explain the epidemiologic changes over the last centuries and imagine its future. Most of them are based on the past experience of Northern countries but trends observed in the Global South, in addition to the regular emergence of new infectious diseases, call for these theories to be questioned, as will be developed in the next section.

13.3 Health Transition and Epidemiological Patterns

13.3.1 Health Transition: A Theoretical Framework for Changes in Lifespan

As previously stated, all regions of the Global South have experienced an increase in life expectancy in the long run. However, progress in child mortality (under 5 years) was largely responsible for the rise in e_0, while progress in adult mortality has been more difficult to establish [7, 13]. Cause-specific mortality is indeed very different between children and adults. Deaths due to infectious and parasitic diseases, perinatal and nutritional conditions represent a large burden on child mortality. Together with maternal mortality, they are generally grouped as 'communicable diseases'. They are also called 'diseases of poverty' because they result from poor socio-economic and health-care conditions. Meanwhile, NCDs and injuries cause most deaths among adults globally. The shift in ages at death thus goes together with an epidemiologic shift, which has some major consequences in terms of public health strategy, with a need to respond to new health issues. The GBD 2019 confirms a gradient where a lower socio-demographic index (SDI) is associated with higher risks stemming from poverty and the environment:[9] malnutrition, air pollution and sanitation [14]. The GBD also finds that the higher the SDI, the more deaths explained by non-communicable conditions, which can be associated with behavioural risks, such as tobacco and high systolic blood pressure [14].

These epidemiological and mortality shifts are best accounted for in the 'health transition' model, which correlates mortality trends with changes in sanitary conditions, and a response to them translated through programmes (nutrition improvements, sanitation, preventive medicine and curative medicine).[10] In addition, the health transition model highlights how the social, economic and political background, at both micro and macro levels, strongly interacts with epidemiology and the implementation of health policies.

In LMICs and the Global South in general, most of the health and social policies have been designed to promote access to primary health care, largely targeting children and women of reproductive ages, and making most progress with infectious diseases

[9] 'Developed by GBD, the Socio-demographic Index (SDI) is a composite indicator of development status. [...] It is the geometric mean of 0 to 1 indices of total fertility rate under the age of 25, mean education for those ages 15 and older, and lag distributed income (LDI) per capita' (Institute for Health Metrics and Evaluation (IHME), from https://ghdx.healthdata.org/record/ihme-data/gbd-2019-socio-demographic-index-sdi-1950-2019).

[10] The model of a 'health transition' was coined by Frenk et al. [15], building upon Omran's 'epidemiological transition' [16, 17] (see also Chapter 12, and a different argument for how the health transition should be understood in [18]).

and health issues related to water, sanitation and hygiene. Nevertheless, 'communicable' causes of death do not fade away as much as we would have expected based on the experience of the high-income settings. Regular phenomena of (re-)emerging infectious diseases are observed for various reasons: opportunistic diseases (i.e. tuberculosis for people living with HIV), drug-resistance development (i.e. malaria) or deterioration of the vaccination coverage (i.e. cholera or measles). The poor living conditions of large shares of southern populations, combined with frailty of health systems, worsen the situation.

Meanwhile, gains in e_0 ineluctably led to a rise in NCDs and external causes because of population ageing. Although urbanization and industrialization globally led to major health benefits for urban dwellers, they also favoured risk factors of NCDs, related both to the environment and to individual behaviours (pollution, injuries, less physical activity, diet saturated with fat, sugar and salt etc.). 'Man-made' diseases resulting from these behaviours (road traffic injuries, occupational diseases, alcohol- and tobacco-related diseases, depression etc.) have been identified, since the 1960s, to be responsible for a large burden of adult mortality in the context of the epidemiologic transition [16]. However, many HICs have implemented health policies in order to prevent and manage such diseases. In the majority of current health systems in LMICs, NCDs are not sufficiently prevented and managed, leading to premature adult deaths. The situation is particularly acute in the context of remaining individual and contextual poverty.

Hence, we have a situation where countries with limited financial resources are increasingly hit by NCDs, while still facing major consequences from infectious diseases and undernutrition. This situation is called the 'double burden of diseases', and it occurs in many countries of the Global South, explaining non-linear health progress in life expectancies [19]. The health transition framework therefore helps to understand persistent divergences between regions [13, 20].

13.3.2 A Global Decrease in Communicable Diseases

As already mentioned, major health progress has been completed since the twentieth century in Latin America, Asia and the Middle East, and even in sub-Saharan Africa, though mortality decline started there much later and at a much slower speed. Some countries in particular have made steady progress during decades, although at various pace, such as Bolivia [13], Ethiopia [21], Thailand [22], and Senegal [23]. The spread of medical interventions, which improved prevention and treatment of infectious diseases (e.g. measles vaccination) and reduced maternal mortality were largely responsible for these health improvements. Controlling or even eradicating some of the deadliest diseases allowed for considerable progress, as illustrated by the case of malaria control in Sri Lanka in the 1940s, which may have caused an increase of up to 12 years in national e_0.[11] Efforts towards reducing new HIV/AIDS infections and

[11] The case of Sri Lanka in the mid-twentieth century also demonstrates the entanglement of all 'communicable' causes of death, managed through prevention of infections, dissemination of treatments and improvement of living conditions [24].

lethality have resulted in a significant global mortality decrease in that regard [25]. The fight was organized around several pillars, from prevention to access to care at local, national and even global level. During the 1990s and 2000s, LMICs struggled to have the right to access generic antiretroviral drugs in order to treat HIV patients who otherwise could not afford them. In Latin America, further major improvements related to the diseases of poverty were made thanks to ambitious public health and social programmes tackling not only health care but also poverty [8].

However, in some countries or regions, especially sub-Saharan Africa, progress in the fight against communicable diseases was limited, leading to a plateau in the levels of life expectancy. This plateau was observed, for example, in Cameroon or Nigeria from the 1990s to the 2000s [23]. These challenges in the fight against diseases are frequently associated with contexts of economic crises, especially in sub-Saharan Africa in the 1980s [7, 23]. Economic crises have major direct impacts, such as famines, but they also weaken health systems, deteriorating vaccination coverage, reproductive health access and so on. They also interact on socio-economic progress such as education. In addition, the HIV/AIDS pandemic caused a dramatic setback in terms of infectious disease control. In sub-Saharan Africa, more extremely than elsewhere, the HIV pandemic strikingly slowed down the rise of health expectancy, not only due to AIDS deaths but also the burden of opportunistic diseases for people living with HIV, principally tuberculosis. Other infectious diseases have contributed to blocking progress, such as the upsurge of malaria in Western Africa during the 1990s and 2000s due to the emergence of drug-resistance strains [26]. Indeed, even if estimates have to be used cautiously, communicable, maternal, neonatal and nutritional diseases would be the leading causes of death in adults from 15 to 49, in 42 of the 47 countries of the WHO African region (source from [27]). Aside from the particularly salient case of sub-Saharan Africa, HIV/AIDS, and more recently the COVID-19 pandemic, have demonstrated that emerging infectious diseases can still have a global impact on life expectancy. COVID-19 has significantly impacted many countries, HICs as well as LMICs, showing that a mortality relapse is always possible. Building on the convergence-divergence model developed by Vallin and Meslé [20], for some countries, this hit may be very temporary, while others may struggle for a longer period of time with the direct and indirect consequences of this outbreak on health.

13.3.3 NCDs: A Fight Yet to Win

Successful achievements against infectious diseases do not completely explain the considerable gains in life expectancy of some countries. In Eastern and South-Eastern Asia, the growth of life expectancy for older adults was steady, and became even stronger in Eastern Asia after 1995 [22]. A few countries thrived, moving from high to low mortality: for example, estimates for e_{60} for South Korea and Chili were 25.7 and 23.3 years old respectively in 2022 [28]. Most advanced countries have made considerable progress in longevity thanks to what has been called the 'cardiovascular revolution', where health systems managed to implement, on the one hand, technological advancements addressing management of CVDs, and, on the other hand, social interventions addressing CVD risk factors (obesity, tobacco, sedentary lifestyles etc.) [20, 22]. As we have seen in Section 13.1, some countries or regions encountered a different situation and reached a plateau due to their inability to take this step.[12] China and Egypt are two countries where progress slowed down [13]. In both countries, CVDs are estimated to be the leading causes of death since the 1990s, and we can assume that those countries successfully achieved an epidemiological shift, but failed to engage with the cardiovascular revolution. Moreover, countries located in Central America, and Mexico in particular, present high and steady rates of deaths due to NCDs and external causes among males [29]. The GBD 2019 estimates that metabolic risk factors are particularly present in Central America (especially diabetes, high body mass index, high systolic blood pressure) [14]. Because these conditions and diseases represent a growing burden on global mortality, for several years now WHO has been encouraging LMICs to tackle NCDs at the primary care level, for instance by implementing their package of essential non-communicable (PEN) disease interventions for primary health care. It is indeed crucial for countries whose health-care system is oriented primarily towards infectious diseases that awareness and diagnoses of NCDs become increasingly integrated into routine health care.

Unfortunately, this is made difficult by limited resources that need to be allocated by considering numerous factors, not only sanitary, but also political and economic. Despite constant advocating for a rise in public expenditure allocated to health, WHO observes that government spending in health remains much larger in HICs than in the rest of the world, and even decreased in low-income countries at the beginning of the twenty-first century. Furthermore, although government spending was evenly allocated between infectious and non-communicable health issues in LMICs, health care in low-income countries is actually mainly financed through external aid and out-of-pocket spending, and these remain primarily oriented towards infectious diseases [25]. Besides, taking on chronic conditions requires high levels of technicity and therefore high investment (in equipment and training). Chronicity can result from communicable diseases (as illustrated by the case of AIDS) as well as from non-communicable risk factors, blurring the lines between 'communicable' and 'non-communicable' diseases, and making health expenditure even more needed at all stages of the health transition.

13.3.4 A Cumulative Burden of Diseases

The cumulative burden of diseases is the result of the addition of health issues that characterize different social groups, highlighting inequality. In addition, this accumulation can also be observed at the individual level, with largely common interactions between infectious diseases and NCDs [30]. This is also true for risk factors, such

[12] This phenomenon is not specific to countries from the Global South, as it has been observed in Eastern Europe (Russia at first) but also more recently in the United States.

as nutrition: some populations currently have to face health issues that are due to undernutrition and overnutrition, mainly in rural areas and urban areas, respectively. In addition, individuals may also experience a nutritional deficiency (anaemia, iron or vitamin A deficiency etc.) while having a cardiometabolic risk factor due to overnutrition (hypertension, hyperglycaemia, diabetes etc.), as has been shown in the context of Ouagadougou, the capital city of Burkina Faso [31].

This situation has major implications for health policies that traditionally split the two groups of diseases. Poverty clearly impedes the ability of societies to face the complexity of the health issues experienced by their populations. In some cases, political and economic crises slow or even reverse transition.

In addition, when factors such as wars, political instability, economic crises, criminality, impact health expenditure, they also directly affect the health status of populations. They keep mortality levels high despite progress made in the prevention and treatment of diseases [8, 20, 32]. For instance, the collapse of the Soviet Union led to recessions in Europe and Asia. In Mongolia, life expectancy after age 65 (e_{65}) dropped so sharply in the 1980s that, despite considerable gains after 1995, e_{65} throughout the period 1980–2010 only rose from 12.5% to 14.0% [22]. In Latin America, violent deaths among adult men contribute to the inequalities between countries of the region [8]. The African continent has also suffered war-related setbacks, sometimes critical (coups in Somalia in 1991, genocide in Rwanda in 1994), which play their part in explaining slow progress on the continent compared with other regions [32]. On the other hand, the case of Columbia shows that when political unrest and violence are controlled, life expectancy can be directly and positively impacted [29].

13.4 Estimating Longevity in the Global South

Demographic and health statistics are essential for planning, monitoring and evaluating health policies and programmes, and for guiding health research [33]. Though the UN produces international statistics on mortality for all countries worldwide, deficiencies in CRVS systems for countries in the Global South make them stand apart from the rest of the world in terms of estimation reliability and precision, especially at old age. This is why, when studying mortality of the elderly and longevity in the Global South, it is of the utmost importance to talk about the available data and methods used to counterbalance its lack.

13.4.1 Data Sources

Mortality data are generally derived from official statistics: headcounts are estimated from enumerations resulting from censuses, and deaths are ideally derived from vital statistics, including a comprehensive system of legal registration of all births and deaths.

In poor-income countries, censuses are often affected by errors and biases such as under- or over-counts and misreporting of ages, affecting sex and age structure estimates, in particular for older age groups [34]. Regarding vital registration, using six criteria,[13] Mikkelsen and colleagues created the vital statistics performance index and established a typology of countries related to their civil registration system performance for the period 2005 to 2012 [3]. In Latin America, most countries globally have a high score, except for a few (Bolivia, Haiti and Honduras). This singularity of Latin America has made it possible to set up a mortality database (LAMBda) following the example of the Human Mortality Database (HMD) devoted to HICs.[14] In contrast, the situation in the Gulf, Asia and Africa, particularly in sub-Saharan Africa, is more alarming and the situation has remained critical since the mid-2000s. Many countries have established vital registration systems, but they are generally incomplete, with very partial geographic coverage and mitigated levels of completeness, especially for deaths. Establishing complete death registration systems is resource-intensive and can take more than a century, as it has been the case in many countries.

This is probably why countries such as India, China and Indonesia, the most populous countries in Asia, have focused on sample-based death registration systems that provide nationally representative mortality statistics while waiting for the deployment of comprehensive systems. In sub-Saharan Africa, the situation is even worse. For the period 2012–2015, official death registers only reported one in three deaths, and in 2016, only four countries in the region met international standards in terms of death registration coverage and certification of cause of death [35]. Because of all these limitations, the study of mortality in these countries relies in general on less conventional data sources. These include censuses and general population-based surveys, particularly household sample surveys, and, increasingly, local population-based demographic and health surveillance data alongside with a few national sample registration systems.

Admittedly, the ‘mortality’ section in censuses captures deaths occurring in the last 12 months in the household, providing the sex and age of each of the deceased. As such, it should allow direct estimation of the mortality of older adults. However, omissions and numerous age errors, including age exaggerations, which are common at older ages, can give the illusion of higher longevity and falsely imply low mortality [36]. Methods exist to address these issues by adjusting the data for incompleteness of death reporting, but they do not always succeed, especially for older people. Moreover, when applied to a single census, and notwithstanding the demographic transitions underway in most LMICs, these methods assume that the population under study is stable. Although other developments have relaxed this unrealistic assumption, renouncing it entailed mobilizing two censuses instead to produce a median estimate only. Given that censuses are not conducted regularly and that intercensal intervals may exceed 10 years, this situation is not suitable for trend analysis. Furthermore, it does not allow for the measurement of mortality in intervals during one-off health crises, such as the COVID-19 pandemic, which was particularly lethal for the elderly.

[13] Completeness of death reporting, quality of death reporting, level of cause-specific detail, internal consistency, quality of age and sex reporting, and data availability or timeliness.

[14] www.ssc.wisc.edu/cdha/latinmortality2/.

Besides censuses, large-scale surveys are used for estimating adult mortality, notably so-called Demographic and Health Surveys (DHS). These surveys collect information related to all the sisters and brothers of the respondents: ages at survey or ages at death and the timing of death, allowing a 'direct' calculation of mortality without having to approximate the duration of exposure to the risk of death. However, they result in implausible estimates beyond age 60 since they are collected from adults under age 50, and prospects for extension to much older ages are limited. Only few national DHS-type surveys allowed a direct measure of mortality at older ages, as they introduced additional questions on deaths at all ages in the last 12 months, as in censuses, but deaths reported beyond age 60 were too few to derive robust conclusions [37]. This is not surprising. Indeed, the variables of interest used to determine the required sample sizes are generally related to maternal and child health rather than adult mortality, especially among the elderly.

In addition to census and survey data, health and demographic surveillance systems (HDSS) have been set up in many countries, mostly in sub-Saharan Africa and Asia. These are locally delimited sites where an initial enumeration of the population is followed by repeated visits during which demographic events are recorded [38]. At each round, information on entries (immigration and births) and exits (emigration and deaths) are collected, as well as other socio-economic and health indicators (unions, pregnancies etc.). In addition, these systems collect information on the circumstances of deaths and produce data on the probable causes of death using the verbal autopsy method [39]. So far, there are more than 45 HDSS sites in LMICs, some of whose data have been made publicly available through the INDEPTH network's iSHARE platform [40].[15] They provide reliable data for direct measurement of mortality that are not available elsewhere, as they are less affected by omissions of deaths, age and date errors than retrospective data. They also allow a better understanding of epidemiological changes over time and a better documentation of the health transition to inform population and health policies [41]. In addition, evidence can be drawn from them for the implementation of a nationally representative sample for vital registration. Although not nationally representative and not intended as a substitute for vital registration systems, an analysis of HDSS data and their comparison with data from national censuses and surveys offers interesting insights for analysing older adult mortality.

13.4.2 Methods for Estimating Mortality in Older Adult Mortality

The limitations of data sources used in countries with poor vital statistics affects the methods to be used for the study of mortality [42]. Table 13.2 summarizes the methods generally used and highlights the limitations associated with each of them. Direct, semi-indirect and indirect estimation methods are all sensitive to age errors, generally restricting their use to the estimation of mortality in adults aged 15 to 50 or 60. Adjustments beyond these ages appear unreliable without evidence to support their plausibility, even by cross-checking various national and local sources or by

[15] www.indepth-ishare.org/index.php/catalog/central.

Table 13.2 Description of the different methods developed to estimate adult mortality, including at older ages

Methods	Description	Observation and limitations
Direct estimation	Applicable when *population data* are collected by age as with *deaths reported during the last 12 months* in censuses and surveys.	Allows for computing mortality rates at all ages. But there are *multiple issues* with such data: • Omissions of deaths due to recall errors and transfers of death out of the year interval of interest for death reporting. • Selection bias due to dissolution of households after a death. • Age heaping and age misstatement, in contexts such as sub-Saharan Africa where being old confers very often a social status within the society, prompting older people to exaggerate their age, as many of them are illiterate and sometimes do not have, or have backdated, birth certificates. • Difficult to be used without adjusting the data for incompleteness of death in censuses. • Sample sizes may be too small to derive robust direct estimates in surveys.
Semi-indirect estimation	Most popular are essentially *death distribution methods* which consist of *estimating the completeness of reported deaths* regarding a population estimate in a single census (when assuming the population to be stable) or using at least two successive censuses (when the stable population assumption is not realistic) using either: the population growth balance equation or the synthetic cohort extinction method.	They allow assessing and adjusting the coverage of one census compared to the next, as well as the completeness of death reporting so as to correct and improve the estimates derived from the raw data. In contrast: • Excludes children in the completeness estimate because of obvious completeness differences with adults. • Provides estimates with large uncertainties at older adult ages, especially beyond age 70, owing to the assumption of a similar level of completeness of deaths at younger and older adult ages.
Indirect estimation	The principle is to convert the proportion of surviving relatives (parents, first spouse of ever married adults, siblings) into conditional probabilities of survival:	• It allows aggregate mortality indices rather than age-specific mortality to be produced. • It is more suitable for estimating mortality from young adulthood onwards than for mortality beyond age 50 or 60.

	The most promising and popular is the *sibling method* that uses information on ages at survey or ages at death and the timing of death, allowing a 'direct' calculation of mortality, without the need to approximate the duration of exposure to the risk of death. It is widely used for producing series of adult mortality estimates.	• It leads to implausible results beyond age 60: probably because it uses survey data such as DHS, which do not cover older people since the data are collected from women aged 15–49, who provide information on siblings of approximately the same age. • Besides the issue of sample size in surveys, it does not account for the year of death and age at death so as to derive direct estimates for old-age mortality from parental survival. • The sibling method may also suffer from large age errors as the ages at death are derived from respondents' reports rather than from death records.
Modelling	This refers to indirect modelling that borrows fragmentary mortality information such as (${}_5q_0$ or both ${}_5q_0$ and ${}_{35}q_{15}/{}_{45}q_{15}$, e_0, etc.) available for an observed population and rely on *empirical mortality experiences* well established elsewhere to *adjust, smooth and predict* the said population's full age pattern of mortality.	Using mortality profiles from countries with an advanced demographic transition would provide standards in terms of ratios between mortality levels at different ages to model mortality age patterns in countries with poor vital statistics. This assumes a universality of mortality patterns that, unfortunately, long-impact epidemics such as HIV have challenged by showing divergent trends between child and adult mortality. As such: • For a ratio between adult and child mortality levels that has no equivalent in the empirical mortality profiles used to calibrate the model, the age pattern of the predicted mortality is likely to be inaccurate unless it is based on a sufficiently diverse universe of mortality profiles that account for these diverging trends. • Due to the lack of credible information on observed mortality beyond age 60, the inferred mortality at these ages from these models are based solely on predictions from observed mortality before age 60 (${}_5q_0$, ${}_{45}q_{15}$): the plausibility of such estimates remains, therefore, questionable.

comparison with model-based estimates from international databases such as those of the UN World Population Prospects. While not well known, a method has recently been developed to estimate the completeness of death registration for better estimating mortality, especially at older ages, based on comparison between the structure of the population at two successive censuses [43]. This method is a step forward in that it is more suitable for high-mortality countries with low completeness of death registration. However, it assumes negligible age errors, an assumption that is difficult to meet in many countries of the Global South. Moreover, it has been little used so far and will have the opportunity to be tested in future studies using data from these countries, which are not homogeneous in terms of mortality level and completeness of deaths. By allowing the reconstruction of complete life tables from limited information, the indirect modelling approach has revolutionized the field. For example, the introduction of singular value decomposition has made it possible to use mortality information before age 60 (5q0, 45q15) from the three previous methods while accounting for additional parameters such as HIV prevalence, and even by extension antiretroviral coverage, to better capture excess HIV mortality [44, 45]. Although this latest avenue for modelling an age-specific schedule of mortality seems highly promising, there are still uncertainties regarding the plausibility of the predicted estimates at older ages. In the absence of specific data or input information to better reflect mortality at these ages, future research should focus on quantifying these uncertainties. The recent revival of interest in Bayesian statistics [46, 47], and their use in mortality modelling [48, 49], could serve this purpose [50].

13.5 Conclusion

In the Global South, progress in life expectancy since the twentieth century has been considerable, with some countries and regions attaining – in a very short period of time – levels observed in Europe and North America. The UN predicts that, by 2070, e_0 could average above 80 years in all geographical regions but sub-Saharan Africa [28]. However, future estimates remain speculative and must be considered with caution. Progress is indeed never fully secured and countries from the Global South will likely face challenges. Transition models tend to give a smoothed picture of these phenomena, which occurred over long periods of time, and which are known through scarce data. Many authors have discussed the divergences between the theoretical framework and reality [18, 20, 32, 51, 52]. The health transition conceptualized by Frenk and colleagues arose from this situation, which was already visible in the 1980s [53]. Indeed, LMIC populations are subject to an enduring situation of 'cumulative burden of diseases'.

On the one hand, new viruses continuously emerge (i.e. HIV/AIDS, Zika, Ebola and most recently COVID-19) with greater or lesser ease of societies in fighting these epidemics on both aspects: (1) progress in medicine; and (2) adequate health policies in terms of prevention and access to care. Furthermore, the eradication of already known infectious and parasitic diseases is never ensured, because of regular phenomena of re-emergence. On the other hand, the inclusion of a target in the SDGs on reducing premature mortality from NCDs illustrates the growing concern about NCDs among LMICs, due primarily to a lack of prevention against behavioural risk factors and to health systems being unable to provide efficient treatments. Progress against NCDs can slow down considerably, as has happened in the Americas. Finally, in addition to what we observe at the population level, individuals may suffer from this cumulative burden, which complicates further the implementation of health policies.

Moreover, WHO has highlighted the dramatic impact on health and mortality from air pollution and climate change; the latter of which has become the 'biggest health threat facing humanity' [54]. Populations from the Global South are more vulnerable to heat, food insecurity, indoor air pollution due to the massive use of solid fuel for cooking and exposure to new arboviruses in rural areas [55]; in addition, they mostly live in countries with fragile economic and public health systems.

Nevertheless, health challenges can be achieved through progress in science, medicine and prevention. In low-mortality countries, the length of life has increased, combatting senescence in addition to CVDs and cancer and leading to progress on longevity [56]. Conditional on socio-economic wealth, the situation is different for populations from the Global South. In 2020, COVID-19 highlighted the inability of many LMICs to estimate and monitor the pandemic's impact on mortality, especially among the elderly (see, for the Indian case, [57]). Reliable and prospective old-age mortality data are still missing for a large share of the world population, while a majority of deaths now occur after age 60.

The question of the quality of the living years becomes more and more acute and there is a need for investigating life quality and well-being for older adults in LMICs. Indicators such as healthy- or disability-free life expectancy [58, 59], as well as aggregate indicators, Disability-Adjusted Life Years (DALYs), have been developed to investigate life quality and to compare different countries and regions of the world [60, 61, 62]. Yet they are based on modelled data, especially among LMICs. Indeed, in many countries, this topic has not yet been properly investigated because public health priorities have long been to the youngest and because data is scarce and often of low quality for older people. Though estimates are neither precise nor always representative, populations from the Global South seem to experience a higher prevalence of disability than populations from HICs [64, 65, 66, 67].[16]

Countries of the Global South also experience specific gender disparities. Gender differences in mortality are usually lower in LMICs than in HICs because of the persistence of gender inequalities as well as the persistence of a relatively high maternal mortality among others. However, all the figures from low-mortality countries show a male–female health–survival paradox in which women live longer than men but with more disability, probably as a result of gender-specific patterns of communicable and chronic diseases (see [68, 69] and Chapter 9). The burden of disability and poor health seems to be also higher among females than males as it has been measured in the

[16] However, there are some exceptions, for instance in the Americas, with the same level of disability-free life expectancy at age 65 between the north and south of the continent [63].

context of Ouagadougou [70]. For instance, it is well established that girls and women are more exposed to household air pollution, leading to many health issues (chronic diseases, limitations etc.) [71, 72]. In addition, the usual age difference between partners implies that most married women survive their spouses and remain widowed for years. Ageing-related functional decline and loss of autonomy require long-term care (assistance in daily activities). It currently relies mostly on informal caregiving provided by close relatives, usually the spouse and children. Family caregiving has economic and health consequences on both the caregiver and the cared-for person. The national welfare systems and the public expenses devoted to formal long-term care are underdeveloped. The increasing burden of dependency requires a reorganization of long-term care at the societal level. Furthermore, 'being old' is associated with a social status that depends on the context and on the individual characteristics, with some positive and/or negative implications on all spheres of life, such as working or receiving a pension, housing and intergenerational relations, which interact with health status.

SDGs want to ensure that progress occurs without leaving anyone behind. Monitoring global health, especially the health of the elderly, depends, crucially, on obtaining precise measurements of trends over time and comparing them across space and social groups.

References

1. Carinnes, M., Alejandria-Gonzalez, P., Subharati, G., Sacco, N.J.C., eds. 2019. *Aging in Global South: Challenges and Opportunities.* Lexington Books.
2. Setel, P.W., Macfarlane, S.B., Szreter, S., Mikkelsen, L., Jha, P., Stout, S., AbouZahr, C. 2007. A scandal of invisibility: making everyone count by counting everyone. *Lancet* **370**, 1569–1577 (doi:10.1016/S0140-6736(07)61307-5).
3. Mikkelsen, L., Phillips, D.E., AbouZahr, C., Setel, P.W., de Savigny, D., Lozano, R., Lopez, A.D. 2015. A global assessment of civil registration and vital statistics systems: monitoring data quality and progress. *Lancet Lond. Engl.* **386**, 1395–1406 (doi:10.1016/S0140-6736(15)60171-4).
4. Karlinsky, A. 2022. International completeness of death registration 2015–2019. *medRxiv* (doi:10.1101/2021.08.12.21261978).
5. Guillot, M., Gavrilova, N., Pudrovska, T. 2011. Understanding the 'Russian mortality paradox' in Central Asia: evidence from Kyrgyzstan. *Demography* **48**, 1081–1104 (doi:10.1007/s13524-011-0036-1).
6. Meslé, F., Vallin, J., Hertrich, V., Andreev, E., Shkolnikov, V. 2003. Causes of death in Russia: assessing trends since the 50s. In *Population of Central and Eastern Europe: Challenges and Opportunities* (eds I.E. Kotowska, J. Józwiak), pp. 389–414. Statistical Publishing Establishment.
7. Masquelier, B., Reniers, G., Pison, G. 2014. Divergences in trends in child and adult mortality in sub-Saharan Africa: survey evidence on the survival of children and siblings. *Popul. Stud.* **68**, 161–177 (doi:10.1080/00324728.2013.856458).
8. Alvarez, J.-A., Aburto, J.M., Canudas-Romo, V. 2020. Latin American convergence and divergence towards the mortality profiles of developed countries. *Popul. Stud.* **74**, 75–92 (doi:10.1080/00324728.2019.1614651).

9. Garcia, J. 2020. Urban bias in Latin American Causes Of Death Patterns. PhD thesis, Ined.
10. World Health Organization. 2022. 14.9 million excess deaths associated with the COVID-19 pandemic in 2020 and 2021. *News*. www.who.int/news/item/05-05-2022-14.9-million-excess-deaths-were-associated-with-the-covid-19-pandemic-in-2020-and-2021.
11. World Health Organization. 2022. WHO launches new mortality database visualization portal. *News*. www.who.int/news/item/13-05-2022-who-launches-new-mortality-database-visualization-portal.
12. World Health Organization. 2020. The top 10 causes of death. WHO Fact Sheets. www.who.int/news-room/fact-sheets/detail/the-top-10-causes-of-death.
13. Tabutin, D., Masquelier, B. 2017. Tendances et inégalités de mortalité de 1990 à 2015 dans les pays à revenu faible et intermédiaire. *Population* **72**, 225 (doi:10.3917/popu.1702.0225).
14. Murray, C.J.L. et al. 2020. Global burden of 87 risk factors in 204 countries and territories, 1990–2019: a systematic analysis for the Global Burden of Disease Study 2019. *The Lancet* **396**, 1223–1249 (doi:10.1016/S0140-6736(20)30752-2).
15. Frenk, J., Bobadilla, J.L., Sepuúlveda, J., Cervantes, M.L. 1989. Health transition in middle-income countries: new challenges for health care. *Health Policy Plan.* **4**, 29–39 (doi:10.1093/heapol/4.1.29).
16. Omran, A.R. 1971. The epidemiologic transition: a theory of the epidemiology of population change. *Milbank Mem. Fund Q.* **49**, 509–538.
17. Caldwell, J., Caldwell, P. 1991. What have we learnt about the cultural, social and behavioural determinants of health? From selected readings to the first health transition workshop. *Health Transit. Rev.* **1**, 3–19.
18. Omran, A.R. 1998. The epidemiologic transition theory revisited thirty years later. *World Health Stat. Q.* **53**, 99–119.
19. Marshall, S.J. 2004. Developing countries face double burden of disease. *Bull. World Health Organ.* **82**, 556–556.
20. Vallin, J., Meslé, F. 2004. Convergences and divergences in mortality: a new approach to health transition. *Demogr. Res.* **10**, 11–44 (doi:10.4054/demres.2004.s2.2).
21. Tabutin, D., Schoumaker, B. 2020. The demography of Sub-Saharan Africa in the 21st century. Transformations since 2000, outlook to 2050. *Population* **75**, 165–286.
22. Gu, D., Gerland, P., Andreev, K.F., Li, N., Spoorenberg, T., Heilig, G. 2013. Old age mortality in Eastern and South-Eastern Asia. *Demogr. Res.* **29**, 999–1038 (doi:10.4054/demres.2013.29.38).
23. Caselli, G., Meslé, F., Vallin, J. 2002. Epidemiologic transition theory exceptions. *Genus* **58**, 9–51.
24. Langford, C.M. 1996. Reasons for the decline in mortality in Sri Lanka immediately after the Second World War: a re-examination of the evidence. *Health Transit. Rev.* **6**, 3–23.
25. World Health Organization. 2021. Global progress report on HIV, viral hepatitis and sexually transmitted infections, 2021; accountability for the global health sector strategies 2016–2021: actions for impact; World Health Organization. Global report. www.who.int/publications/i/item/9789240027077.
26. Trape, J.-F.F. et al. 1998. Impact of chloroquine resistance on malaria mortality. *Comptes Rendus Académie Sci. – Ser. III – Sci. Vie* **321**, 689–697 (doi:10.1016/S0764-4469(98)80009-7).
27. Institute for Health Metrics and Evaluation. 2022. GBD compare – Viz hub – GBD 2019. https://vizhub.healthdata.org.
28. United Nations, Department of Economic and Social Affairs, Population Division. 2022. *World Population Prospects*. Online Edition. https://population.un.org/wpp/.

29. Calazans, J.A., Queiroz, B.L. 2020. The adult mortality profile by cause of death in 10 Latin American countries (2000–2016). *Rev. Panam. Salud PublicaPan Am. J. Public Health* **44**, 1–9 (doi:10.26633/RPSP.2020.1).
30. Remais, J.V., Zeng, G., Li, G., Tian, L., Engelgau, M.M. 2013. Convergence of non-communicable and infectious diseases in low- and middle-income countries. *Int. J. Epidemiol.* **42**, 221–227 (doi:10.1093/ije/dys135).
31. Zeba, A.N., Delisle, H.F., Renier, G., Savadogo, B., Baya, B. 2012. The double burden of malnutrition and cardiometabolic risk widens the gender and socio-economic health gap: a study among adults in Burkina Faso (West Africa). *Public Health Nutr.* **15**, 2210–2219 (doi:10.1017/S1368980012000729).
32. Kuate Defo, B. 2014. Demographic, epidemiological, and health transitions: are they relevant to population health patterns in Africa? *Glob. Health Action* **7**, 22443 (doi:10.3402/gha.v7.22443).
33. Rao, C., Bradshaw, D., Mathers, C.D. 2004. Improving death registration and statistics in developing countries: lessons from sub-Saharan Africa. *South. Afr. J. Demogr.* **9**, 81–99.
34. Pison, G., Ohadike, P.O. 2006. Errors and manipulations in age assessment. In *Human Clocks: The Bio-Cultural Meaning of Age* (eds H. Leridon, N. Mascie-Taylor), pp. 313–336. Peter Lang.
35. Economic Commission for Africa. 2017. Report on the status of civil registration and vital statistics in Africa. Ethiopian Communication Agency.
36. Gavrilov, L.A., Gavrilova, N.S. 2019. Late-life mortality is underestimated because of data errors. *PLoS Biol.* **17**, e3000148 (doi:10.1371/journal.pbio.3000148).
37. Bendavid, E., Seligman, B., Kubo, J. 2011. Comparative analysis of old-age mortality estimations in Africa. *PLoS ONE* **6**, 1–7 (doi:10.1371/journal.pone.0026607).
38. Pison, G. 2005. Population observatories as sources of information on mortality in developing countries. *Demogr. Res.* **13**, 301–334.
39. World Health Organization. 2007. *Verbal Autopsy Standards: Ascertaining and Attributing Cause of Death.* World Health Organization. https://iris.who.int/handle/10665/43764.
40. Sankoh, O.A., Byass, P. 2012. The INDEPTH network: filling vital gaps in global epidemiology. *Int. J. Epidemiol.* **41**, 579–588 (doi:10.1093/ije/dys081).
41. Ye, Y., Wamukoya, M., Ezeh, A., Emina, J.B.O., Sankoh, O. 2012. Health and demographic surveillance systems: a step towards full civil registration and vital statistics system in sub-Sahara Africa? *BMC Public Health* **12**, 741 (doi:10.1186/1471-2458-12-741).
42. Bendavid, E., Seligman, B., Kubo, J. 2011. Comparative analysis of old-age mortality estimations in Africa. *PloS ONE* **6**, e26607.
43. Li, N., Gerland, P. 2019. Evaluating the completeness of death registration at old ages: a new method and its application to developed and developing countries. Technical Paper, United Nations, Department of Economic and Social Affairs, Population Division. www.un.org/en/development/desa/population/publications/pdf/technical/TP2019-5.pdf.
44. Clark, S.J. 2015. A singular value decomposition-based factorization and parsimonious component model of demographic quantities correlated by age: predicting complete demographic age schedules with few parameters. ArXiv Prepr. ArXiv150402057. https://arxiv.org/abs/1504.02057.
45. Sharrow, D.J., Clark, S.J., Raftery, A.E. 2014. Modeling age-specific mortality for countries with generalized HIV epidemics. *PloS ONE* **9**, e96447.
46. Bijak, J., Bryant, J. 2016. Bayesian demography 250 years after Bayes. *Popul. Stud.* **70**, 1–19.

47. Bryant, J., Zhang, J. 2019. *Bayesian Demographic Estimation and Forecasting: Statistics in the Social and Behavioral Sciences*. CRC Press, Taylor & Francis Group.
48. Alexander, M., Zagheni, E., Barbieri, M. 2017. A flexible Bayesian model for estimating subnational mortality. *Demography* **54**, 2025–2041.
49. Schmertmann, C.P., Gonzaga, M.R. 2018. Bayesian estimation of age-specific mortality and life expectancy for small areas with defective vital records. *Demography* **55**, 1363–1388.
50. van Raalte, A.A. 2021. What have we learned about mortality patterns over the past 25 years? *Popul. Stud.* **75**, 105–132 (doi:10.1080/00324728.2021.1967430).
51. Caselli, G. 1995. The key phases of the European health transition. *Pol. Popul. Rev.* **7**, 73–102.
52. Meslé, F., Vallin, J. 2006. The health transition: trends and prospects. In *Demography: Analysis and Synthesis. A Treatise in Population Studies*, vol. 2 (eds G. Caselli, G. Wunsch, J. Vallin), pp. 247–259. Elsevier/Academic Press.
53. Frenk, J., Bobadilla, J.L., Stern, C., Frejka, T., Lozano, R. 1991. Elements for a theory of the health transition. *Health Transit. Rev. Cult. Soc. Behav. Determinants Health* **1**, 21–38.
54. World Health Organization and United Nations. 2021. Climate change and health: Fastfacts. WHO/UN Climate Fast Facts. www.un.org/sites/un2.un.org/files/2021/08/fastfacts-health.pdf.
55. Romanello, M. et al. 2021. The 2021 report of the Lancet countdown on health and climate change: code red for a healthy future. *The Lancet* **398**, 1619–1662 (doi:10.1016/S0140-6736(21)01787-6).
56. Horiuchi, S., Robine, J.-M. 2005. Increasing longevity: causes, trends, and prospects. *Genus* **61**, 11–17.
57. Guilmoto, C.Z. 2022. An alternative estimation of the death toll of the COVID-19 pandemic in India. *PLoS ONE* **17**, e0263187 (doi:10.1371/journal.pone.0263187).
58. Sullivan, D.F. 1971. A single index of mortality and morbidity. *HSMHA Health Rep.* **86**, 347–354 (doi:10.2307/4594169).
59. Robine, J.-M., Romieu, I., Michel, J.-P. 2003. Trends in health expectancies. In *Determining Health Expectancies* (eds J.-M. Robine, C. Jagger, C.D. Mathers, E.M. Crimmins, R.M. Suzman), pp. 75–101. John Wiley & Sons.
60. Murray, C.J., Lopez, A.D., Jamison, D.T. 1994. The global burden of disease in 1990: summary results, sensitivity analysis and future directions. *Bull. World Health Organ.* **72**, 495–509.
61. Murray, C.J.L. et al. 2012. Disability-Adjusted Life Years (DALYs) for 291 diseases and injuries in 21 regions, 1990–2010: a systematic analysis for the Global Burden of Disease Study 2010. *The Lancet* **380**, 2197–2223 (doi:10.1016/S0140-6736(12)61689-4).
62. Forouzanfar, M.H. et al. 2015. Global, regional, and national comparative risk assessment of 79 behavioural, environmental and occupational, and metabolic risks or clusters of risks in 188 countries, 1990–2013: a systematic analysis for the Global Burden of Disease Study 2013. *The Lancet* **386**, 2287–2323 (doi:10.1016/S0140-6736(15)00128-2).
63. Payne, C.F. 2018. Aging in the Americas: disability-free life expectancy among adults aged 65 and older in the United States, Costa Rica, Mexico, and Puerto Rico. *J. Gerontol. Ser. B* **73**, 337–348 (doi:10.1093/geronb/gbv076).
64. Madans, J.H., Loeb, M., Eide, A.H. 2017. Measuring disability and inclusion in relation to the 2030 Agenda on Sustainable Development. *Disability and the Global South* **4**, 1164–1179.

65. Prina, A.M. et al. 2020. Dependence- and disability-free life expectancy across eight low- and middle-income countries: a 10/66 study. *J. Aging Health* **32**, 401–409 (doi:10.1177/0898264319825767).
66. Payne, C.F., Mkandawire, J., Kohler, H.-P. 2013. Disability transitions and health expectancies among adults 45 years and older in Malawi: a cohort-based model. *PLoS Med.* **10**, e1001435 (doi:10.1371/journal.pmed.1001435).
67. Bennett, R., Chepngeno-Langat, G., Evandrou, M., Falkingham, J. 2016. Gender differentials and old age survival in the Nairobi slums, Kenya. *Soc. Sci. Med.* **163**, 107–116 (doi:10.1016/j.socscimed.2016.07.002).
68. Van Oyen, H., Nusselder, W., Jagger, C., Kolip, P., Cambois, E., Robine, J.-M. 2013. Gender differences in healthy life years within the EU: an exploration of the 'health–survival' paradox. *Int. J. Public Health* **58**, 143–155 (doi:10.1007/s00038-012-0361-1).
69. Luy, M., Minagawa, Y. 2014. Gender gaps: life expectancy and proportion of life in poor health. *Health Rep.* **25**, 12–19.
70. Cambois, E., Duthé, G., Soura, A.B., Compaoré, Y. 2019. The patterns of disability in the peripheral neighborhoods of Ouagadougou, Burkina Faso, and the male–female health–survival paradox. *Popul. Dev. Rev.* **45**, 835–863 (doi:10.1111/padr.12294).
71. World Health Organization. 2016. *Burning Opportunity: Clean Household Energy for Health, Sustainable Development, and Wellbeing of Women and Children*. World Health Organization.
72. Haddad, Z., Williams, K.N., Lewis, J.J., Prats, E.V., Adair-Rohani, H. 2021. Expanding data is critical to assessing gendered impacts of household energy use. *BMJ* **375**, n2273 (doi:10.1136/bmj.n2273).

14 Limit of Human Longevity

Historical Perspectives and a New Metric

Carlo Giovanni Camarda and Jean-Marie Robine

Vieillard, lui dit la Mort, je ne t'ai point surpris;
Tu te plains sans raison de mon impatience.
Eh n'as-tu pas cent ans? Trouve-moi dans Paris
Deux mortels aussi vieux, trouve-m'en dix en France.

La mort et le mourant, Jean de La Fontaine, 1677[1]

14.1 Introduction

Since at least the earliest known traces of written history, humans have been consumed by questions on immortality and the comparative longevities of animals, humans and various categories of deities. Such questions vigorously intersect with seeking the longest lifespans and the secrets of eternal youth. Throughout the history of human thought, several scholars have engaged these issues from various perspectives and arrived at diverse conclusions.

We can divide this long intellectual history into two broad periods. Beginning with the *Epic of Gilgamesh* in Mesopotamia, we can clearly distinguish an early history of ideas and speculation, despite the scant data this provides. This marks the beginning of a long period in which the limits of longevity were perceived solely through literary histories, myths and theoretical conjectures based on anecdotal information. Although we can acknowledge that the nineteenth century saw the emergence of a scientific culture [3], it was not until the early twentieth century that scholars of longevity began to test their theories with empirical data. With the birth of an era in which data was more readily available and of higher quality (at least in developed countries), scientists imperatively tested and assessed their hypotheses on human longevity.

The chapter begins by providing an overview of the extensive history of thought on this topic, with a primary emphasis on the perspectives of European thinkers. This is followed by a detailed examination of the empirical trends that have emerged in the twentieth century. We use a new metric called HAPaL30 (Highest Age Providing at Least 30 deaths) to illustrate current trends. This indicator captures the maximum

[1] From [1], translated to English by [2, pp. 187–188] as follows:

> 'Old man'!' Death cries, replying. 'How Dare you complain? Have I not given you A century and more? I vow,
> You would not find me two as old In all of Paris, ten in all of France!'

age at which 30 or more deaths occur in a single year, providing a reliable measure of longevity changes over time while filtering out minor random variations. By analysing the decomposition of this indicator, we can gain insight into the significance of mortality shifts among the elderly population.

14.2 An Historical Perspective on Longevity

14.2.1 The *Epic of Gilgamesh* and the First Greek Poems

The *Epic of Gilgamesh* – Gilgamesh being the king of the city of Uruk in Mesopotamia – is one of the first stories about a quest for immortality. This quest failed because immortality was a prerogative of the gods. Based partially on elements dating from the 2nd millennium BC, the first complete version of this story is known to be part of the Library of Assurbanipal [4].

In Greek mythology, the gods of Olympus enjoyed eternal youth, which may be considered more than 'simple' immortality. An excellent example is given by the story of an unfortunate mere mortal named Tithonos, with whom the goddess Eos was in love. When she asked Zeus to grant her lover immortality, Eos had forgotten to request also that he be granted eternal youth. Consequently, Tithonos experienced an endless ageing process that dried him up to the point that he became like a cicada. This myth can well be viewed as the ancient root of the modern distinction between healthspan and lifespan. Among the texts that have survived from Greek antiquity, the oldest are poems attributed to Hesiod and Homer. These date from the end of the eighth century or the beginning of the seventh century BC. Among those attributed (rightly or wrongly) to Hesiod is a poem comparing the lifespans of men to those of various animals and of the Nymphs, daughters of Jupiter. We read in one of the *Precepts of Chiron*: 'A chattering crow lives out nine generations of aged men, but a stag's life is four times a crow's, and a raven's life makes three stags old, while the phoenix outlives nine ravens, but we, the rich-haired Nymphs, daughters of Zeus the aegis-holder, outlive ten phoenixes' ([5], p. 75). Plutarch (ca. 46–ca. 125 AD) attributes this poem to Hesiod and uses it as a starting point for a long discussion on the units of time used to measure longevity in his essay 'On the Obsolescence of Oracles', dating from the first or second century AD [6].

Aristotle (384–322 BC), to whom we owe the 'History of Animals', dismisses in one sentence the longevity attributed to the deer: 'Fabulous stories are told concerning the longevity of the animal [stag], but the stories have never been verified, and the brevity of the period of gestation and the rapidity of growth in the fawn would not lead one to attribute extreme longevity to this creature' ([7], Book VI, p. 29). Thomas Browne (1605–1682) quoted this sentence in 1646, followed by Georges-Louis Buffon (1707–1788) in 1749, with both men noting the incompatible life history traits that Aristotle reported for this animal.

From our perspective, Aristotle's greatest contribution lies in his quantifying everything that could be counted in animals – the teeth, the age of reproduction, the duration of gestation, the length of life – in order to establish connections between these elements. In other words, Aristotle was among the first to make observation-based assessments and his efforts can be considered as an early precursor to life history trait analysis. What is more, in seeking out the most frequent values while also recording extreme values, he approximated the concept of variability. For instance, although the male horse usually lives about 35 years, Aristotle noted that a horse has been known to live 75 years. He describes similar details for sheep, goats, pigs, men, horses, donkeys and sheep, as well as camels and elephants, as when he wrote, 'Camels live for about thirty years; in some exceptional cases they live much longer, and instances have been known of their living to the age of a hundred. The elephant is said by some to live for about two hundred years; by others, for three hundred' ([7], Book VIII, p. 596). Aristotle obviously could not have observed all these lifespan traits himself and mainly relied on reports of them. Thus, some of these traits may be totally false or at least highly inaccurate, as in the case of bears. In any case, the details gathered by Aristotle were taken up throughout history by Francis Bacon (1561–1626), Thomas Browne, Buffon and William Smellie (1740–1795), who were all satisfied to essentially repeat Aristotle's observations.

It is likely that Aristotle today would have used the words 'average' and 'extreme values', which did not exist in the fourth century BC. Indeed, these concepts would not come into existence until the nineteenth century, when they were pioneered in Adolphe Quetelet's (1796–1874) work on his theory of 'l'homme moyen' (i.e. the average man [8, 9, 10]). On the path of this long scientific journey, at the end of sixteenth century, we find the words of Michel de Montaigne (1533–1592) when he criticizes the confused terminology concerning longevity, which was so often qualified as 'natural' or 'ordinary':

> Let us no longer flatter ourselves with these fine words; we ought rather, peradventure, to call that natural which is general, common, and universal. To die of old age is a death rare, extraordinary, and singular, and, therefore, so much less natural than the others; 'tis the last and extremest sort of dying: and the more remote, the less to be hoped for. It is, indeed, the bourn beyond which we are not to pass, and which the law of nature has set as a limit, not to be exceeded; but it is, withal, a privilege she is rarely seen to give us to last till then. ([11], ch. 57)

This is indeed a major problem underlying studies on human longevity: observations are lacking while everything or nearly everything is only speculation, even in recent times. For example, following the publication of James Fries' theories of 'rectangularization of the survival curve' and 'compression of morbidity' [12], Leonard Hayflick contemplated the possibility of eliminating age-related losses without touching the biological clock itself: 'the result would be a society whose members would live full, physically vigorous, youthful lives until death claimed them at the stroke of midnight on their one- hundredth birthday' ([13], p. 10). Even at the end of the twentieth century, there was still no firm belief that, in the absence of disease, lifespans follow a bell-shaped distribution like other biometric values, despite laboratory observations in genetically identical species living in homogeneous environments [14].

14.2.2 Living for 1000 Years

During the Middle Ages in Europe, religion played a significant role in shaping society, and the belief in the prospect of eternal life after death discouraged questioning about the human experience on Earth, particularly within the context of Christianity: 'Compared thus with timeless eternity, even the few thousand years of the earth's existence was unimportant and the problems of population adjustment a matter of indifference. The life of man, at most a few score years, was also insignificant compared to his future life' ([15], p. 134). Nevertheless, this did not prevent European monks from attributing remarkable longevity to the powerful in their chronicles and from reporting marvellous stories, nor did it prevent the most educated from being passionate about alchemy, the transmutation of metals, and the search for the elixir of long life. Although the numerical data are scant and of poor quality, contemporary discussions on human longevity seem to have been rare. One oft-mentioned exception to the rarity of these discourses is the Franciscan monk Roger Bacon (ca. 1220–ca. 1290), who considered immortality as part of God's original plan for mankind and death a consequence of the original fall from Grace: 'he was commanded not to eat of the fruit of life [...] and hence he was fit for immortality' ([16], p. 624). However, if any fields merit being associated with the term Renaissance, they are demography and biology.

The possibility of immortality was strongly debated in Europe during the Renaissance, as found in the remarkable example of Francis Bacon and his manuscript *Historia Vitæet Mortis*, in which he writes, 'anything that can be repaired gradually, without destroying the original whole, is, like the vestal flame, potentially eternal' ([17], p. 145). Bacon was not the first person seeking the means to prolong life, but his quest distinguished him from the alchemists by laying the foundations of modern science. For instance, the ideas in his posthumously published novel *New Atlantis* notably inspired the formation of the Royal Society of London in 1660.

Some of the Royal Society's more obvious initial objectives were to prolong life, restore youth and slow down the ageing process [18]. Over a century later, Thomas Malthus (1766–1834) placed the question of longevity at the core of the pamphlet he wrote in opposition to the ideas of William Godwin (1756–1836) and the Marquis de Condorcet (1743–1794): 'I own it appears to me that in the train of our present observations, there are no more genuine indications that man will become immortal upon earth than that he will have four eyes and four hands, or that trees will grow horizontally instead of perpendicularly' ([19], p. 233). Ultimately, Malthus reproaches Godwin and Condorcet for having ventured onto God's territory. In his opinion, it is God who has fixed the length of man's life on Earth. Moreover, because God has already granted his followers eternal life, in another world, after death, Malthus believed that pursuing it here on Earth would be nonsense, despite what might be achieved by what Condorcet called 'the progress of the human mind' ([19], p. 142).

Other eighteenth-century authors such as Buffon tried to reconcile biblical texts and the religious beliefs of the time with early observations. For instance, the longevity of the patriarchs was never questioned, although it required great efforts of the imagination to justify it. In fact, Descartes (1596–1650) believed it was possible to prolong human life to a span equalling that of the patriarchs.[2] From today's perspective, it might be difficult to question the ancient texts that were rediscovered during the Renaissance and have been widely recopied since then, even if they contain some legends. It would also be equally difficult to criticize authors such as Montaigne, Bacon, Descartes and Pascal (1601–1665), who formed the foundations of modern science.

For example, in 1749 Buffon accepted the existence of supercentenarians, noting that 'men no doubt there are who have surpassed the usual period of human existence; and not to mention Par, who lived to the age of 144, and Jenkins to that of 165, as recorded in the Philosophical Transactions, we have many instances of the prolongation of life to 110, and even to 120 years' ([20], p. 103). He went so far as to generalize his remarks in 1777:

> there are in animals, as in the human species, some privileged individuals, whose life extends almost to the double of the ordinary term, and I can cite the example of a horse that has lived more than 50 years. [...] In general, this analogy confirms what we know only by a few specific facts, that is, there must be in all species, and consequently in the human species as in that of the horse, some individuals whose life is prolonged to twice the ordinary life, that is, to one 160 years instead of 80. These privileges by the Nature are indeed distributed sporadically over time, and sparse in space; they are jackpots in the universal lottery of life; nevertheless they are sufficient to give even to the most elderly men, the hope of an even higher ages. ([21], pp. 129–130)

Closer to our time, in 1985, the American biologist Roy Walford thought that the chances of increasing the maximum lifespan were good. He wrote, 'Let us not forget that of the three classic dreams of man; the transmutation of metals, such as lead into gold, going to the moon or planets, and increasing the length of life, the first two dreams have already been realized today' [22, p. 85, translated from the French by the authors]. Twenty years later, in 2005, the journal *Science* celebrated its 125th anniversary by compiling a list of 125 unsolved scientific questions. Among the top 25 questions was 'How much can human life span be extended?' [23].

14.2.3 The Slow Rise of Scepticism

James Hart (ca. 1590–1639) was one of the first to express scepticism about the chances of extending the human lifespan: 'And is it not a thing ridiculous, now in these later times, to extend the life of man-kinde to 1000, 900, or, at the least to 600 yeeres?' ([24], p. 7). Criticism became more abundant at the end of the seventeenth century as several authors, such as Descartes and Browne, unequivocally rejected the possibility of becoming immortal. Buffon, writing in 1749, considered it to be a settled matter that death is inevitable:

[2] Once the philosopher Kenelm Digby (1603–1665) asked Descartes about how to prolong the human 'machine'. Descartes replied that he had considered exactly this matter of 'la vie éternelle' and he would not venture to promise that it was possible to render a man immortal, although he was very sure that the length of a man's life could match that of the Patriarchs [18, p. 25].

> The ideas which a few visionaries have formed of perpetuating life by some particular panacea, as that of the transfusion of the blood of one living creature into the body of another, must have died with themselves, did not self-love constantly cherish our credulity, even to the persuasion of some things which are in themselves impossible, and to the doubt of others, of which every day there are demonstrative proofs. ([20], pp. 102–103)

However, for nearly all the seventeenth- and eighteenth-century authors who contemplated the ordinary lifespan, they viewed it as a limit and not a central value. This is then compatible with the existence of octogenarians, nonagenarians and even centenarians, that is, the small number of individuals who reach the broken arches of the Mirzha bridge in Addison's allegory of life and death [25]. Hart notes in 1633 that 'to attaine to 100 is no wonder, having my selfe knowne some of both sexes' ([24, p. 8). Many claimed that the people of the New World lived longer: 'The Brachmans among old Indians, and the Brazilians at the time that country was discovered by the Europeans. Many of these were said then to have lived two hundred, some three hundred years' ([26], pp. 112–113).

Buffon definitively dismisses the latter cases in 1749 when he formulates his general proposition on human longevity: 'If not cut off by accidental diseases, man is found to live to the years of 90 or an 100' ([20, p. 105). However, we have seen that he accepts the existence of men who have lived beyond ordinary terms. In the eighteenth century, Abraham de Moivre (1667–1754) stood out as something of an exception as he expressed scepticism towards claims of individuals with extremely long lifespans: 'As for what is alledged, that by some observations of late years, it appears, that life is carried to 90, 95, and even to 100 years; I am not more moved by it, than by the example of Parr, or Jenkins, the first of which lived 152 years, and the other 167' ([27], p. 263). His conviction regarding the unlikelihood of extreme values was grounded in a practical perspective: 'Another thing was necessary to my calculation, which was, to suppose the extent of life confined to a certain period of time, which I suppose to be at 86' ([27], p. 263). Likewise, in 1740, Nicolas Struyck (1687–1769) also rejected the notion that humans could live to ages above 100 years. In 1753, he clarified his position in response to Johann-Peter Süssmilch's criticisms: 'I have purposely not noted separately the persons of each age above 100 years, because in the [London] register of 1739 there is mention of a person who would have reached the age of 138 years, which I cannot admit without decisive proof' ([28], p. 350). Süssmilch, in the 1741 edition of the *Divine Order*, had not only accepted this age but also emphasized that it was not 'the furthest we can go' before going on to mention the unavoidable cases of Parr and Jenkins ([29], p. 190). However, Süssmilch underlined the exceptional side of these cases: 'It appears from data in London survey that, out of 2 to 300 thousand people, only one was 138 years old. I do not think that one would find its equal in millions, because such an age is a half-miracle' ([29], p. 190). However, Süssmilch's belief was that living up to 200 years was a reasonable possibility, considering that it 'still seems to fit, especially since [it] is only 30 years above Jenkins' age' ([29], p. 167). On the contrary, he claimed that the likelihood of dying at the age of 500 was highly improbable: 'everyone will assign a small degree of acceptance between the tales to the story of the wandering Jew, as well as to that of the American prince Hultazob, who estimates himself 500 years old' ([29], p. 191).

14.2.4 Wilhelm Lexis and the Age-at-Death Distribution

Buffon's proposition continues to be fully endorsed in the nineteenth century, when Pierre Flourens (1794–1867) affirms it using the most recent biological and medical discoveries: 'It is a fact, a law, that is to say, from general experience in this class [mammals], that extraordinary life can be prolonged to double that of ordinary life' ([30], p. 75). When Wilhelm Lexis (1837–1914) presented his dissertation on the normal duration of human life in Paris in 1878, the question was finally specified appropriately. After first crediting 'Quetelet's remarkable research [which] has taught us the interesting fact that individuals belonging to a given nationality are more or less exact copies of a model of given proportions' ([31], p. 450), Lexis goes on to dismiss premature mortality to determine the typical length of time, 'which popular opinion loosely estimates to be seventy to eighty years' ([31], p. 450). He also asserts that 'in any generation sufficiently numerous, a certain group will realize in its lifetime the normal type with deviations in accord with the formula Quetelet calls the binomial law' ([31], p. 450). When applied to the French population, his method shows concentrations around centres of density set at 72.5 years for men and 72 years for women, 'representing the normal lifespan [in France]' ([31], p. 451). Jacques Véron and Jean-Marc Rohrbasser provide a comprehensive presentation of Lexis' approach within its intellectual context [32]. Jacques Bertillon (1851–1922) immediately praised this conceptual and methodological advance, saying that

> [Lexis has] established a maximum of probability of death in those years immediately following birth, and another maximum around 72 to 73 years of age. Indeed, our results are significantly imprecise when we conflate such large differences in probability as we do in calculating the average lifespan [i.e., life expectancy at birth]. In France, this is about 40 years; although this is precisely one of the ages where death occurs most rarely. There is therefore something in the assertion of this average that is in stark contrast to what everyone knows. The actual probability of dying is not to die at 40, but to die in the early years of life, or well beyond 65, 70, and 75 years. It would therefore be useful to consider separately these two age groups: those who have been granted a serious calling for life, and those who have only made a brief appearance on the world stage. ([33], p. 461)

Incidentally, Lexis slightly anticipated Karl Pearson (1857–1936). In his work, *The Chances of Death*, presented in Leeds in January 1895 and published in 1897, Pearson wrote, 'standing in 1875 on the well-known wooden bridge at Luzern, with its pictures of the Dance of Death, it struck me that something might be done to resuscitate the medieval conception of the relation between Death and Chance, and to express it in a more modern scientific form' ([34], pp. 8–11). Karl Pearson must be given credit for introducing the modern term *mode*: 'We may term that occurrence, which happens not necessarily a majority of times, but more frequently than any other the *mode*' ([34], pp. 11–12; emphasis added). From then on, the age-at-death distribution would be described using all the modern statistical terminology. Nevertheless, it seems obvious to everyone from John Graunt (1620–1674) to Karl Pearson that this ordinary lifespan has remained unchanged since King David's time. Scholars mention it only to confirm it.

The only scholar who formulated a clear hypothesis on the possibility that longevity could increase over time was Condorcet in 1795. He first explains his analysis:

> It cannot be doubted that the progress of the sanative art, that the use of more wholesome food and more comfortable habitations, that a mode of life which shall develop the physical powers by exercise, without at the same time impairing them by excess; in fine, that the destruction of the two most active causes of deterioration, penury and wretchedness on the one hand, and enormous wealth on the other, must necessarily tend to prolong the common duration of man's existence, and secure him a more constant health and a more robust constitution. ([35], p. 367)

He then follows up with two sentences that will trigger Malthus' anger:

> Would it even be absurd to suppose this quality of melioration in the human species as susceptible of an indefinite advancement; to suppose that a period must one day arrive when death will be nothing more than the effect either of extraordinary accidents, or of the slow and gradual decay of the vital powers; and that the duration of the middle space, of the interval between the birth of man and this decay, will itself have no assignable limit? Certainly man will not become immortal; but may not the distance between the moment in which he draws his first breath, and the common term when, in the course of nature, without malady, without accident, he finds it impossible any longer to exist, be necessarily protracted? ([35], p. 368)

Notably, Condorcet defines here the two types of indefinite increases that could occur in the average lifespan: (1) a trend 'approaching continually an illimitable extent, it could never possibly arrive at it'; or (2) a law that, 'in the immensity of ages, it may acquire a greater extent than any determinate quantity whatever that may be assigned as its limit' ([35], p. 369). Condorcet finally adds, 'this is precisely the state of the knowledge we have as yet acquired relative to the perfectibility of the [human] species' ([35], p. 369).

14.2.5 The Development of Statistics and the Validation of Ages

The nineteenth century saw the gradual emergence of national statistical systems, as in France, which opened the first ministerial offices dedicated to statistics in 1800. The General Statistics Office was created thirty-three years later and, in 1840, was renamed Statistique Generale de la France (SGF). Other statistical bureaus emerged throughout the developed world: in Britain, Belgium, Denmark and the United States, just to name a few. They covered many aspects of society and population, such as the numbers of births, deaths and registered people, as well as their distributions by age and sex.

Undoubtedly, the Canadian Pierre Joubert's case is the most renowned *cold case*; he apparently having died at the impressive age of 113 in 1814. On one hand, Bowerman [40], who exposed numerous alleged supercentenarian cases from the 1700s and 1800s as fraudulent, deemed Joubert's record to be genuine. On the other hand, it was not until 1990 that some of the mystery surrounding the case was dispelled: Charbonneau revealed that Joubert's wife, [41] who had passed away in 1786, was noted as a widow on her death certificate. In contrast, Wilmoth and colleagues describe a case in which a rigorous validation process was utilized to verify the authenticity of another supercentenarian [42]. In other words, the development of official statistics was accompanied by a decline in the number of counted centenarians, at least until the Second World War [43]. Thoms was surely the first to have investigated systematic errors that affected the declaration and recording of ages at death and, hence, the ages of the so-called centenarians. His work *Human Longevity: Its Facts and Fictions*, published in 1873, marks a real turning point in attitudes regarding the ages of centenarians [36], namely by bringing an end to the credulity that still prevailed among the general public at that time. This work was soon followed by that of Neymark [38], Young [37], and many others. Subsequently, there has been a noticeable proliferation of centenarians, which has been the focus of numerous studies in populations across the world (see [39, 43, 44, 45, 46, 47, 48, 49, 50, 51, 52, 53, 54, 55 56] and Chapter 15, to cite only few of many).

14.2.6 A Revival of Demography

In the twentieth century, Louis I. Dublin and Alfred J. Lotka explored the issue of lifespan limits as part of a broader investigation into life tables [57]. They made a distinction between *mean length of life* and *lifespan*, and, to determine the highest estimate for the former, they developed a hypothetical life table that 'promises an eventual expectation of life at birth of [...] 70 years, the biblical three-score and ten' ([57], p. 196). Concurrently, in the initial chapter of their book, they asserted that 'there is a natural *span* of life, a limit, vaguely defined, it is true, but nevertheless inescapable, beyond which even in the most favorable circumstances, human life cannot extend. What is this extreme limit? The question will never lose its interest. That the limit is beyond the century mark is certain' ([57], p. 3). Furthermore, when examining measures of longevity, they strengthened their stance by suggesting that the potential limit may be unclear but constant over time:

> the life span is the extreme limit set of human life by old age. It is not an exact figure. We only know that very few men outlive one century, so that in rounded numbers we may say that human life span is about hundred years. Now there is no evidence that this life span, this extreme limit, has changed materially within historical times. ([57], pp. 31–32)

Following World War II, two French demographers delved into the topic of lifespan limits. Paul Vincent analyses mortality data from France, Switzerland, the Netherlands and Sweden using his novel approach and comes to the conclusion that 'the benefit of the progress made so far in the fight against mortality [...] decreases with age until it becomes insignificant at the highest ages' ([58], p. 181). He acknowledges that his viewpoint was relevant to the current situation and that it was impossible to observe supercentenarians in the past: 'No human being seems, in the present state of affairs, to be able to live beyond the age of 110, and it is extremely unlikely that a death at this age has ever been observed with certainty' ([58], p. 181). Jean Bourgeois-Pichat attempts to establish a biological limit on the lifespan of humans by contrasting the endogenous and exogenous factors that influence mortality [59]. He discusses possible approaches to reducing mortality in infants and adults, but he believes that 'at very high ages – over 95 for example – […] efforts to fight death are almost in vain' ([59], p. 384). Despite claiming that the biological limit for life expectancy at birth is 77.2 years, Bourgeois-Pichat maintains a possibilist viewpoint and emphasizes that 'this word "limit" should not be misinterpreted: it does not imply a priori any limitation of progress against death' ([59], p. 393).

In the latter part of the twentieth century, a hypothesis gained widespread acceptance among biologists, suggesting that neither the maximum longevity of humans nor the proportion of centenarians would have changed since the most remote periods of history [12, 60, 61, 62, 63]. In their view, human longevity could only be modified by slowing down the intrinsic rate of ageing. In particular, various studies argued that there is a limit to human lifespan, and while it has increased over time, it will eventually plateau due to biological constraints [64, 65, 66, 67]. These researches also noted that extending lifespan could have negative consequences for society, such as overpopulation and strain on resources. Therefore, this line of investigation would give more importance to extending healthy life rather than just extending lifespan, and aim to compress morbidity with only a slight increase in the overall length of life [68, 69].

Conversely, there are demographers who refute these claims and contend that there is no ceiling on human lifespan, which can persist in increasing due to progressions in medicine and technology [70, 71, 72, 73, 74, 75, 76]. Against the existence of a biological limit to lifespan, they argue that 'evolution may have shaped the *age pattern* of human mortality, but current conditions and an individual's life history heavily influence its *level*, even in old age' ([77], p. 793, emphasis in original). In a recent discussion, Milholland and Vijg attempted to reconcile these contrasting views on breaking the biological limits of human lifespan [78]. They acknowledged that current scientific evidence deems it impossible, but also pointed out that past centuries have taught us to never dismiss the possibility, and with ground-breaking research it may be possible to extend the maximum lifespan beyond the current limits.

Looking at empirical data on limits of human longevity, Roger Thatcher was the first to publish an article in 1981 showing a considerable increase in the number of centenarians in England and Wales over the three decades following the end of the Second World War [55]. The research he conducted with Vaïnö Kannisto and James W. Vaupel deserves special mention, as well as the *Monographs on Population Aging* (1994–1998), edited by Bernard Jeune and James W. Vaupel. Particularly noteworthy among these monographs are those by Kannisto, Thatcher and others [79, 80, 81], and the miscellanies on exceptional longevity [82, 83]. Importantly, the work of John Wilmoth and his collaborators showed that maximum age reported at death (MRAD) had increased during the twentieth century in Sweden [75].

Related to this discussion is the continuous debate regarding the search for evidence of a mortality plateau, that is, a phenomenon where the rate of mortality levels off among individuals who have reached an advanced age, such as 105 [84]. We refer to Chapter 15 for a comprehensive exploration of the statistical detection of mortality deceleration and plateaus. If a mortality plateau exists at an advanced age, it suggests that there may not be a fixed upper limit to human lifespan, as mortality rates do not continue to increase indefinitely as individuals reach more advanced ages. This would have significant implications for ageing research and our understanding of the ageing process. It could mean that with continued improvements in health care, lifestyle and technology, people could continue to live longer and healthier lives. Different conclusions have been drawn on this matter as some studies indicate the presence of a mortality plateau [85, 86, 87, 88, 89], while others find weaker statistical support for its existence [90, 91, 92, 93], and some researchers argue for a continual increase in mortality as age advances [94, 95, 96].

To some extent, all this work contributed to the creation of the Human Mortality Database (HMD) [97], which celebrated its 20th anniversary in Paris in 2022. It would also lead a few years later to the creation of the International Database on Longevity (IDL) [98]. These widely available databases serve as highly valued sources of information for scholars. For instance, the HMD currently covers the births, deaths, population sizes and mortality rates for 41 countries, going back as far as 1749 in the case of Sweden. The IDL contains validated individual data from nearly 20 000 people who died at age 105 or older in 13 countries around the world, and it aims to ultimately provide an accurate measure of mortality at very high ages. IDL has been complemented by two monographs on the validation of ages at death and analyses of its data [99, 100].

14.3 Contrasting Theories with Data

14.3.1 A New Metric of Longevity: The HAPaL30

Since the 1980s, demographers have been enriched with a wealth of data, making it possible for many studies to propose challenging theories. At long last, an enormous body of literature can be built on an increasing number of reliable data and international comparisons can be carried out with relatively little effort. However, a literature review of this vast corpus is beyond the scope of this chapter. Most demographic studies have mainly examined changes in average age at death and longevity limits in old age to analyse mortality trends. Fewer studies have looked at changes in mortality at extremely advanced ages considering changes in both mortality and population size. In this chapter, we adopt the latter strategy and suggest a new indicator for analysing mortality at extreme ages. Additionally, we introduce a technique for decomposing this novel metric. Thus, this section has the main purpose of describing longevity from a novel perspective by utilizing the available data from the early twentieth century. This significant availability of data empowers us to critically evaluate theories based on empirical evidence, enabling researchers to either accept or dismiss them.

Throughout history, two distinct measures have frequently been suggested to depict longevity: life expectancy at birth (e_0) and adult modal age at death (M). Life expectancy at birth is defined as a newborn's average length of life if current age-specific mortality rates do not change in a given population. Although this indicator is a measure that applies to a hypothetical cohort, it has likely been the most common longevity indicator since William Farr in 1885 first used it to measure the health of Victorian-era populations [101, 102]. Since e_0 is computed as an average over the whole age range, it is affected by mortality among infants, children and middle-aged adults (see also Chapter 12). Consequently, it cannot be employed to analyse longevity and old-age mortality, especially in economically developed countries. To bypass this issue, we can compute life expectancies conditional on survival at old ages (e.g. 60). However, these measures suffer from an arbitrary selection of starting age.

Kannisto initially promoted the modal age at death (*M*) [103, 104], which an increasing line of research has proposed as an alternative measure of longevity [105, 106, 107, 108, 109, 110, 111, 112]. Solely determined by old-age mortality, this measure can be viewed as the normal age at death for most individuals, and it represents where old-age death is located in an age-at-death distribution. Furthermore, *M* describes lifespan extension more accurately than life expectancy, especially in low-mortality regimes where increased longevity is due primarily to changes in old-age survival. An additional compelling factor that advocates for the use of *M* is its stability with respect to survivorship percentile [113]. This implies that it is less susceptible to selection issues when compared to trend analysis of conditional life expectancies, making it particularly useful for examining socio-economic disparities in mortality at older ages [114]. However, modal age at death is a unique point on the mortality density function and therefore needs to be determined by either parametric or non-parametric techniques. Thus, it is unlike life expectancy because classic decomposition methods cannot be implemented.

Here we change the perspective and employ a new indicator: HAPaL30, which is defined as the highest age at which at least 30 deaths are recorded in any given year. Specifically targeting longevity at the oldest ages, HAPaL30 has been recently proposed by Jean-Marie Robine and François Herrmann (in 2020) to analyse longevity in France [115]. Here, we geographically broaden the overview and Figure 14.1 presents *HAPaL*30 for a set of 11 relatively large populations, for both sexes separately and using available data from at least 1920. In alphabetical order, these countries are Belgium, Denmark, England and Wales, Finland, France, Italy, Netherlands, Norway, Scotland, Sweden and Switzerland. Although we could have included other populations with relatively good statistics, they are either too small (Luxembourg) or the HMD lacks data for them until well after 1900 (Austria, Germany, Greece, Portugal and Spain) [97]. Ultimately, the 11 populations presented in this chapter are properly representative of Western Europe and adequately illustrate longevity changes over the past century. An increase of HAPaL30 for these 11 European countries since at least the Second World War is evident in Figure 14.1. For instance, HAPaL30 for females in France and Finland is equal to, respectively, 100 and 96 in 1950, then rises to 108 in 2018 (France) and 103 in 2019 (Finland).

Here, we must make an important fundamental distinction between e_0, M and our third indicator, HAPaL30. On the one hand, because a life table is used to compute both life expectancy and modal age at death, they are thus age-standardized measures that are unaffected by population structure and temporal changes in cohort size. On the other hand, HAPaL30 is based on empirical death counts and therefore depends on both mortality level and population size. This difference becomes crucial when analysing

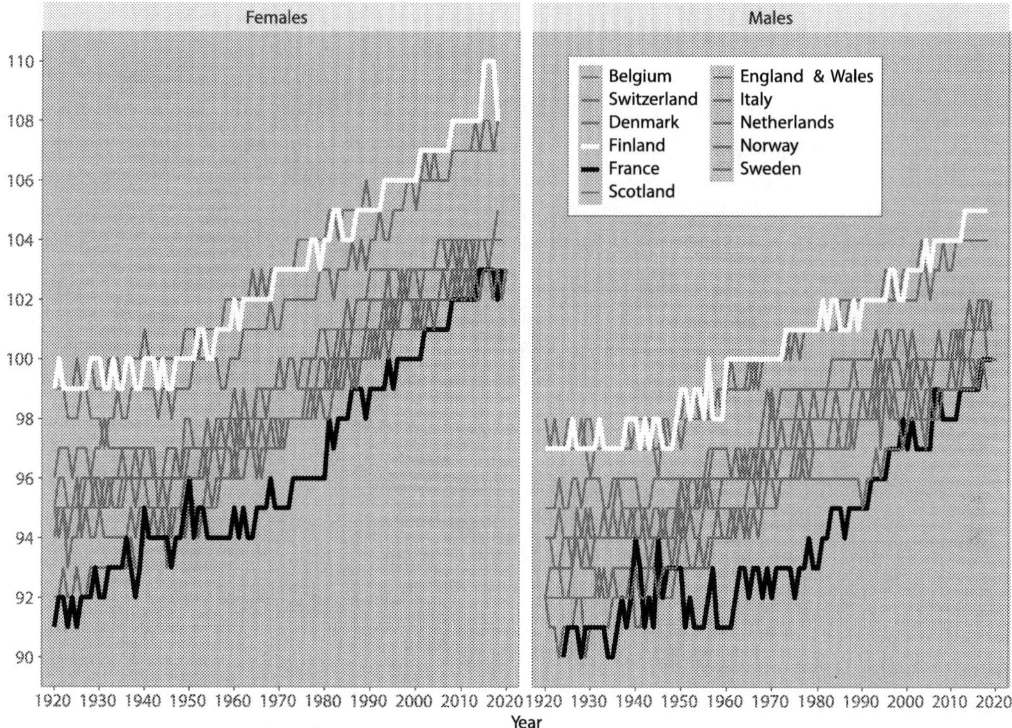

Figure 14.1 HAPaL30 over years for 11 selected populations from 1920, both males and females. Last years depend on data availability (2018–2020).

trends for these indicators and carrying out decomposition methods to reveal changes in the underlying causes of longevity. Furthermore, HAPaL30's computation method shares some similarities with the calculation of the remaining percentage of survival from a life table, as described by Zuo and colleagues [76]. However, implementing a percentile-based approach would require constructing a life table, which at older ages would necessitate modelling observed death rates; something we aimed to avoid.

Among the three indicators, HAPaL30 is thus certainly the most empirical because it does not depend on constructing a life table and its values are determined by the actual numbers of deaths in a given year. It has been developed to overcome issues related to MRAD [116, 117], which is oftentimes employed to depict possible trends in mortality at extreme ages [95, 117, 118, 119, 120, 121, 122, 123, 124, 125]. Whereas MRAD is highly influenced by small random fluctuations and the presence of extreme cases, HAPaL30 is more robust with respect to these issues and it adequately describes temporal changes in longevity at very high ages. Nevertheless, it suffers from some obvious drawbacks. Namely, 30 deaths is an arbitrary choice and, furthermore, the differences in HAPaL30 values clearly depend on population sizes. Consequently, it is advisable to conduct a trend analysis of HAPaL30 in order to directly evaluate its values.

14.3.2 Decomposing HAPaL30

As mentioned previously, HAPaL30 is an empirical indicator that cannot be easily compared among populations and over time because it varies according to population size, cohort size and mortality changes. Nevertheless, we want to know if temporal changes in HAPaL30 are driven by mortality changes at old ages; in other words, we want to find out if variations in longevity trends cause changes to the numbers of deaths at higher ages. Therefore, we present here an approach for decomposing HAPaL30 changes within a population over the last century. First, we isolate the amounts of changes due to mortality variations at the oldest ages and disentangle these from the contributions of cohort size and younger age mortality. This allows us to gauge eventual similarities in the relative changes that occur in HAPaL30 over time among populations.

For our decomposition, we select a threshold of age 90 for two reasons. The first is that HAPaL30 is equal to or above 90 years in all 11 countries from 1920 onwards, even for males in small populations. The second reason is that, after the Second World War, mortality clearly declines until age 90, after which the downward trend is no longer evident, especially above the age of 100 years (trends not shown here).

In practical terms, we are decomposing two components of the HAPaL30 changes: (1) mortality above age 90; and (2) the aggregate effect of mortality below age 90 and varying cohort size over time, which is what we refer to here as the residual effect. The goal is to measure the relevance of mortality at the oldest ages (here 90+) and identify which populations it affects in order to discern patterns that emerge from summary measures of longevity, such as HAPaL30.

Unlike other measures derived from the life table, HAPaL30 is an empirical value that is computed directly from the death counts of age cohorts for each calendar year. Therefore, we cannot decompose this measure assuming an hypothetical cohort [126, 127, 128]. Moreover, HAPaL30 is different to changes in the numbers of centenarians (decomposed, for instance, in [56, 129, 130]) because it can be considered a 'moving target' and its value therefore cannot be attributed to a single cohort. In other words, HAPaL30 clearly depends upon changes in cohort sizes and mortality developments in a rather complex manner. Thus, we decompose the main factors of this indicator by means of the following counterfactual exercise.

First, we assume that mortality above a certain age is constant over time, and we measure the expected variation in HAPaL30 for this theoretical scenario. In this way, we can easily compute the residual increase in HAPaL30, absent any mortality changes at the oldest ages, and thereby indirectly calculate the contribution of mortality changes to the overall increase in HAPaL30.

Although computed for a given year, the value of HAPaL30 is a consequence of the remaining death counts for different cohorts. Hence, we set the mortality condition from a cohort perspective. Figure 14.2 exemplifies the exercise with a Lexis map. Since our analysis begins with 1920, we take deaths from the cohort aged 90 in 1900 (see the first diagonal of dark grey squares in Figure 14.2). Given our data structure, this is the first cohort that could eventually contribute to HAPaL30 in 1920 (see the

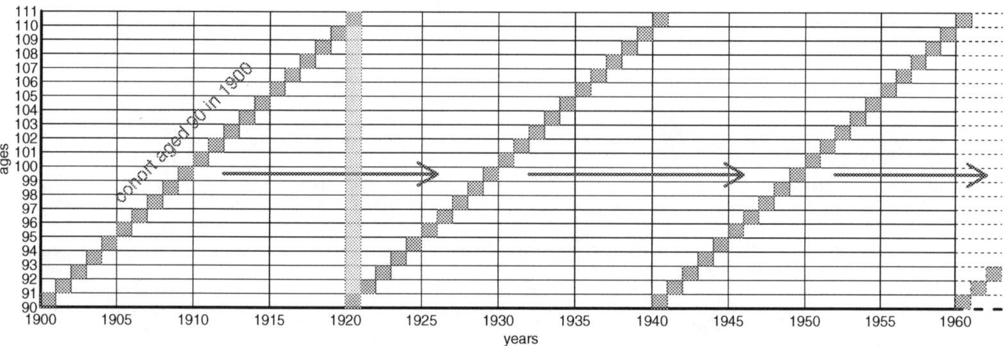

Figure 14.2 Lexis map showing the procedure for creating the counterfactual scenario used to decompose HAPaL30.

light grey bar in Figure 14.2). We then estimate the probability of dying for the cohort aged 90 in 1900 and apply this mortality condition to all cohorts aged 90 afterwards. Before applying this fixed mortality condition to the whole Lexis map, we smooth it with a generalized additive model as a preliminary step for removing the naturally observed random fluctuations in a single cohort at the oldest ages [131].

Once all cohorts aged 90 have been exposed to the same mortality, the counterfactual resulting HAPaL30 is only an outcome of the varying cohort sizes, mortality conditions before age 90 and all interaction effects. To put it differently, we calculate the expected number of deaths that would result from applying a fixed mortality rate to a varying population aged 90 and above. Figure 14.3 presents the outcomes of this exercise for females, and similar outcomes were found for males (not shown here). For instance, the actual increase in HAPaL30 over the past century was nine years for French females, from 99 to 108 years (an average annual gain of 0.09 years). Assuming mortality would not have changed after age 90, HAPaL30 would have increased for that cohort by only four years, up to 103 (average annual increase of 0.04 years). This means that 44% of the overall increase comes from cohort size (residual contribution) and changes in mortality below 90. Consequently, 56% of the overall increase can be attributed to mortality changes above age 90. In other words, the majority of the increase in French female HAPaL30 was due to mortality changes at the oldest ages. Similar divisions in contributions can be found for French males. However, France can be considered an exception among the 11 populations, because it is the only country where mortality changes at the oldest ages are clearly more important than residual effects in describing HAPaL30 changes.

This is evident in Table 14.1, which presents the estimates for both females and males in all 11 countries. Since HAPaL30 depends on population size, the comparison is done by means of the average annual increase. Values are sorted for each sex and by the percentage of contribution from mortality changes above age 90 (i.e. the contribution of changes in longevity). For females, a significantly large contribution of mortality changes above age 90 is evident in Finland, Switzerland, Belgium, England and Wales and, as already mentioned, France. On the other hand, the Netherlands

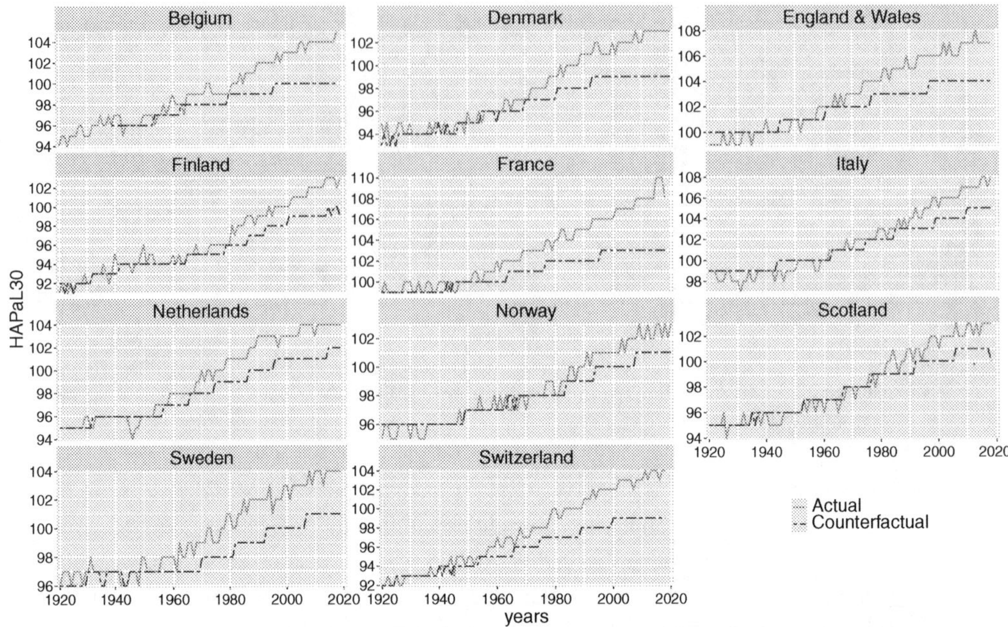

Figure 14.3 Actual and counterfactual HAPaL30 over the years for 11 selected populations from 1920, females. Last years depend on data availability (2018–2020). Note that Belgium counterfactual values cannot be computed for years before 1939: mortality during the First World War is not available in the HMD.

Table 14.1 Average annual increase in HAPaL30 and proportions (in %) of contributions from mortality changes above age 90 and residual effect

			Proportion (in %) of this increase due to:	
	Population	Average annual increase	Mortality change above age 90	Residual effect (mortality below age 90 + varying cohort size)
Females	France	0.09091	56	44
	England & Wales	0.08081	50	50
	Belgium	0.11111	45	55
	Switzerland	0.12121	42	58
	Finland	0.12000	42	58
	Scotland	0.08081	38	62
	Norway	0.07921	38	62
	Sweden	0.08000	38	62
	Italy	0.09091	33	67
	Denmark	0.07921	25	75
	Netherlands	0.09000	22	78

Table 14.1 (cont.)

	Population	Average annual increase	Proportion (in %) of this increase due to: Mortality change above age 90	Residual effect (mortality below age 90 + varying cohort size)
Males	France	0.08081	62	38
	Sweden	0.06000	50	50
	England & Wales	0.07071	43	57
	Finland	0.10000	40	60
	Netherlands	0.08000	38	62
	Switzerland	0.09091	33	67
	Italy	0.07071	29	71
	Norway	0.05941	17	83
	Scotland	0.07071	14	86
	Belgium	0.08081	13	87
	Denmark	0.07921	12	88

Note: Computed from counterfactual scenario and from actual increases over the years for 11 selected populations from 1920. Both males and females. Last years depend on data availability (2018–2020).

and Denmark present a relatively smaller contribution of mortality changes at older ages. For males, increases are relatively smaller, as they show an average annual gain of 0.077 in all 11 populations, compared to 0.093 for females. Moreover, while both sexes in Denmark show no large contribution from mortality changes above age 90, the outcomes for Belgian males are rather different from those of their female counterparts, and they are always due to the old-age contributions to HAPaL30 changes. Finally, it is noteworthy that the contribution of old-age mortality in this decomposition may depend also on the country's initial epidemiological phase.

14.4 Conclusion

In our chapter, we have integrated a range of classic ideas from the history of human thinking about the concept of human longevity limits, with an empirical examination of newly available data that can be used to test longstanding theories. In the absence of statistical information, ancient thinkers were able to imagine almost everything, dreaming of immortality and eternal youth on par with the gods. Above all, they theorized that limits to longevity had been imposed by some god or other transcendental power. Aristotle told us that these limits were specific to each species, and Buffon believed them to be immutable. With the exception of the Marquis de Condorcet in the eighteenth century, it was not until the twentieth century that anyone proposed it possible to expand the limits to human longevity. Buffon did suggest that exceptions to these limits may exist, but he also stated in no uncertain terms that some privileged individuals may possibly extend their life 'almost to the double of the ordinary term' ([21], p. 129).

In all likelihood, given an enduring stagnation in mortality punctuated periodically by cyclical crises, as well as a consistent lack of reliable data, it is not surprising that scholars could hardly believe it was possible to extend the human lifespan. Not even in the first half of the twentieth century did this perception change, despite mortality improvements in Western countries throughout the 1920s and 1930s, when historical limits began to be broken. At the same time, based on all these old texts, one can easily understand not only why statisticians and demographers in the 1970s found it difficult to project past increases in life expectancy into the future, but also why even at the end of the twentieth century so many biologists continued to strongly advocate that it was impossible to have any proportionally more centenarians today than in the Neolithic period [12, 61, 62, 63].

A major break occurred in the history of human longevity around 1950, when a newly established demographic regime extended the limits of human longevity. Traditionally measured by life expectancy at birth, historical longevity is well known in Western countries, at least in recent centuries. The first available statistical series are from the eighteenth century onwards and, although they indicate values that fluctuate considerably from one year to the next, they average around 40 years. These values were again confirmed in the second half of the nineteenth century, a global increase in life expectancy that coincided with the Industrial Revolution. On the eve of the First World War, life expectancy reached about 50 years, then 60 years after the Second World War. Women's life expectancy reached 70 years in 1980 and then 80 years in the 2010s. This suggests that 1950 represents the midpoint of a demographic regime that has approximately doubled historical human longevity.

Wilhelm Lexis (1878) proposed the modal age at death (M) [31], which many have advocated as a better indicator of human longevity [33, 104, 105, 109]. This measure did not vary significantly before the First World War, as it fluctuated between 72 and 75 years of age. A clear increase began in the 1930s and, nowadays, the most frequent age at death is close to 90 years for women. Although life expectancy at birth and M are central indicators of human longevity, the former is modified by changes in mortality at all ages of life, including juvenile ages, while M is influenced only by mortality at old ages.

Our study relies heavily on an alternative and empirical indicator that focuses on the tail of the age-at-death distribution: the oldest age at which 30 deaths still occur in one year. Labelled HAPaL30, this indicator was constructed by trial and error and the number 30 is necessarily a subjective choice that, nevertheless, has no impact on our analysis in this chapter. Our motive for developing HAPaL30 was to find a statistical indicator that would best approximate the age of the oldest person dying each year (MRAD) without being impacted by the natural fluctuations of extreme values. In other words, this indicator aims to reveal trends that could be hidden by annual MRAD fluctuations [75, 115]. HAPaL30 trends in Sweden suggest that longevity experienced a quasi-stagnation after vital statistics first became available in the eighteenth century (not shown here). Additional data from other countries in the nineteenth century validate this pattern until 1950, after which this indicator begins to increase linearly until the most recent years for which data is available. During this time, the indicator increases steadily, showing no sign of acceleration or slowing down. Looking at women in France as an example, it goes from 99 years in 1946 to 109 years in 2016, thus representing an increase of 10 years within 70 calendar years (Figure 14.1).

This first visual inspection may seem to be consistent with the considerable increase in not only the number of centenarians observed since the 1950s, but also in the modal age at death since the 1930s (see references in Sections 14.2.5 and 14.3, respectively). However, mortality changes are not the sole mechanism driving the general increase in the number of survivors at very high ages. This is also particularly true for HAPaL30, which led to the second part of our study, where we decompose the causes behind the increase in HAPaL30. On the one hand, it could be a result of larger numbers of survivors at age 90, whether due to larger birth cohorts, positive migration or drops in mortality before age 90. On the other hand, the cause may be a decrease in mortality at age 90 and beyond. Our choice of age 90 is based on the observed mortality trends: whereas all analysed countries experienced a sharp decline in mortality at age 85 over the period studied, trends beyond this age become increasingly blurry and mortality rates at ages 95 and 100 show no clear pattern (not shown here).

By means of the proposed decomposition, we take a scenario in which mortality at age 90 and beyond remains identical to that observed from 1900 in the cohort born in 1810, then contrast it with the actual observation of mortality trends for each cohort in the 11 Western countries studied. This allowed us to show different outcomes. For example, for females, an average of 61% of the increase in HAPaL30 was due to an increased number of survivors at age 90, while 39% of the increase was indeed due to the mortality improvement beyond age 90. Slightly different outcomes are found for males, with 68% and 32% due to residual and old-age contributions, respectively.

What surprised us most was that the highlighted changes in the history of human longevity date back to the 1950s, which is more than 70 years ago and decades before the current and ongoing debate about what is the 'true' limit to human longevity [85, 94, 119]. We thus envisage the combination of historical scholarship regarding the upper limits of human longevity with new analyses on data at old ages.

In conclusion, the concept of human longevity limits has been shaped by ancient and modern thinkers alike. It was not until the twentieth century that the possibility of extending human life became a topic of discussion among scholars. The midpoint of the demographic regime that doubled historical human longevity is around 1950. The measure of the oldest age at which 30 deaths still occur in one year, known as HAPaL30, suggests that longevity stagnated until 1950, after which it increased linearly without any sign of slowing down. We have finally measured that approximately one third of the increase in longevity can be attributed to changes in mortality rates among the elderly.

References

1. de La Fontaine, J. 1678. *Fables choisies mises en vers*. Denys Thierry and Claude Barbin.
2. Shapiro, N.R. 1900. *The Complete Fables of Jean de La Fontaine*. University of Illinois Press.
3. Gaukroger, S. 2006. *The Emergence of a Scientific Culture: Science and the Shaping of Modernity 1210–1685*. Oxford University Press.
4. Helle, S. 2018. New Gilgamesh fragment: Enkidu's sexual exploits doubled. *World Hist. Encycl.* Last modified November 28, 2018. www.worldhistory.org/article/1286/new-gilgamesh-fragment-enkidus-sexual-exploits-dou/.
5. Hesiod. 1920. *The Homeric Hymns and Homerica*. Willian Heinemann.
6. Plutarch. 1927. *Moralia*, vols. 197–499. Loeb Classical Library (LCL). Harvard University Press.
7. Aristotle. 1910. *A History of Animals*. Clarendon Press.
8. Quetelet, A. 1835. *Sur l'homme et le développement de ses facultés, essai d'une physique sociale*. Bachelier, Imprimeur-Libraire.
9. Quetelet, A. 1848. *Du système social et des lois qui le régissent*. Guillaumin.
10. Quetelet, A. 1871. *Anthropométrie ou mesure des différentes facultés de l'homme*. Muquardt.
11. Montaigne, M. de 1877. *Essays of Michel de Montaigne* (trans. C. Cotton; ed. W.C. Hazlitt). Reeves and Turner 196 Strand.
12. Fries, J.F. 1980. Aging, natural death, and the compression of morbidity. *N. Engl. J. Med.* **303**, 130–135 (doi:10.1056/NEJM198007173030304).
13. Hayflick, L. 1981. Aging: a challenge to science and society. In *Prospects for Human Life Extension by Genetic Manipulation*, vol. 1 (eds D. Danon, N.W. Shock, M. Marois), pp. 162–179. Oxford University Press.
14. Finch, C.E., Kirkwood, T.B.L. 2000. *Chance, Development, and Aging*. Oxford University Press.
15. Russell, J.C. 1958. Late ancient and medieval population. *Trans. Am. Phil. Soc.* **48**, no. 3, 1–152 (doi:10.2307/1005708).
16. Newman, W.R. 1997. An overview of Roger Bacon's alchemy. In *Roger Bacon and the Sciences* (ed. J. Hackett), pp. 317–335. Brill.
17. Rees, G., ed. 1996. *The Oxford Francis Bacon*, vol. 6. Philosophical Studies c.1611–c.1619. Oxford University Press.
18. Haycock, D.B. 2006. *'A Thing Ridiculous'? Chemical Medicines and the Prolongation of Human Life in Seventeenth-Century England*. London School of Economics.
19. Malthus, T.R. 1798. *An Essay on the Principle of Population as It Affects the Future Improvement of Society, with Remarks on the Speculations of Mr. Goodwin, M. Condorcet and Other Writers*. J. Johnson in St Paul's Church-Yard.
20. Barr, J.S., ed. 1797. *Buffon's Natural History*, vol. 4. H.D. Symonds, Paternoster-Row.
21. Buffon, G.-L. 1777. *Histoire naturelle, générale et particulière, avec la description du cabinet du roi*, vol. 12. (ed. B. Frères). Imprimerie Royale.
22. Walford, R.L. 1985. Biologie et sociologie de la prolongation de la vie. *Cah. MURS* **2**, 75–91.
23. Couzin, J. 2005. How much can human life span be extended? *Science* **309**, 5731 (doi:10.1126/science.309.5731.83).
24. Hart, J. 1633. *Klinike, or The Diet of the Diseased*. John Beale for Robert Allot.

25. Addison, J. 1711. *The Vision of Mirza*. The Spectator.
26. Temple, W. 1701. *Miscellanea. The Third Part. Containing I. An Essay on Popular Discontents II. An Essay upon Health and Long Life III. A Defence of the Essay upon Antient and Modern Learning. With Some Other Pieces*. Jonathan Swift.
27. de Moivre, A. 1756. *A Treatise of Annuities on Lives: Preface to the Second Edition in The Doctrine of Chance* (3rd ed.). A. Millar, in the Strand.
28. Vollgraff, J.A. 1912. *Les oeuvres de Nicolaas Struyck*. Algemeene maâtschappij van levensverzekering enlüfrente.
29. Süssmilch, J.-P. 1741. *L'ordre divin dans les changements de l'espèce humaine, démontré par la naissance, la mort et la propagation de celle-ci*. Institut National d'Etudes Démographiques.
30. Flourens, J.P. 1855. *On Human Longevity, and the Amount of Life upon the Globe*. H. Bailliere.
31. Lexis, W. 1878. Sur la durée normale de la vie humaine et sur la théorie de la stabilité des rapports statistiques. *Ann. Démographie Int.* **2**, 447–460.
32. Véron, J., Rohrbasser, J.-M. 2003. Wilhelm Lexis: the normal length of life as an expression of the 'nature of things'. *Population* **58**, 303–322 (doi:10.3917/popu.303.0343).
33. Bertillon, J. 1878. Discussion of Lexis's paper 'Sur la durée normale de la vie humaine et sur la théorie de la stabilité des rapports statistiques'. *Ann. Démographie Int.* **2**, 460–461.
34. Pearson, K. 1897. *The Chances of Death and Other Studies in Evolution*. Cambridge University Press.
35. Condorcet, J.-A.-N. de C. 1795. *Outlines of the Historical View of the Progress of the Human Mind*. J. Jonhson.
36. Thoms, W.J. 1873. *Human Longevity: Its Facts and Its Fictions*. John Murray.
37. Young, T.E. 1899. *On Centenarians and the Duration of the Humane Race*. Charles and Edwin Layton.
38. Neymark, A. 1889. Statistique de la longévité humaine. *J. Société Stat. Paris* **30**, 264–271.
39. Chambre, D. 2021. À La recherche des premiers super-centenaires français (1830–1929). *Gérontologie Société* **43**, 75–108 (doi:10.3917/gs1.166.0075).
40. Bowerman, W.G. 1939. Centenarians. *Trans. Actuar. Soc. Am.* **40**, 360–378.
41. Charbonneau, H. 1990. Pierre Joubert a-t-il vécu 113 ans? *Memoires Société Génealogique Can.-Francaise* **41**, 45–48.
42. Wilmoth, J., Skytthe, A., Friou, D., Jeune, B. 1996. The oldest man ever? A case study of exceptional longevity. *The Gerontologist* **36**, 783–788 (doi:10.1093/geront/36.6.783).
43. Skytthe, A., Jeune, B. 1995. Danish centenarians after 1800. In *Exceptional Longevity: From Prehistory to the Present*. Monographs on Population Aging (eds B. Jeune, J.W. Vaupel), pp. 55–66. Odense University Press.
44. Drefahl, S., Lundström, H., Modig, K., Ahlbom, A. 2012. The era of centenarians: mortality of the oldest old in Sweden. *J. Intern. Med.* **272**, 100–102 (doi:10.1111/j.1365-2796.2012.02518.x).
45. Jeune, B., Skytthe, A. 2001. Centenarians in Denmark in the past and the present. *Popul. Engl. Sel.* **13**, 75–93 (doi:10.2307/3030260).
46. Johnson, S.L., Hogan, H. 2020. Babies no longer: projecting the 100+ population. In *Developments in Demography in the 21st Century* (eds J. Singelmann, D.L. Poston Jr), pp. 95–103. Springer.
47. Kestenbaum, B., Ferguson, R. 2005. Number of centenarians in the United States 01/01/1990, 01/01/2000, and 01/01/2010 based on improved Medicare Data. Paper

presented at Proceedings of the Annual Meeting of the Population Association of America, Philadelphia, 31 March–2 April.

48. Leeson, G.W. 2017. The impact of mortality development on the number of centenarians in England and Wales. *J. Popul. Res.* **34**, 1–15 (doi:10.1007/s12546-016-9178-8).
49. McCormack, J. 2014. The emergence of centenarians and supercentenarians in Australia. *Geriatr. Gerontol. Int.* **4**, S178–S179 (doi:10.1111/j.1447-0594.2004.00192.x).
50. Nepomuceno, M.R., Turra, C.M. 2020. The population of centenarians in Brazil: historical estimates from 1900 to 2000. *Popul. Dev. Rev.* **46**, 813–833 (doi:10.1111/padr.12355).
51. Poon, L.W., Cheung, K.S.L. 2012. Centenarian research in the past two decades. *Asian J. Gerontol. Geriatr.* **7**, 8–13.
52. Robine, J.-M., Caselli, G. 2005. An unprecedented increase in the number of centenarians. *Genus* **61**, 57–82.
53. Robine, J.-M., Cubaynes, S. 2017. Worldwide demography of centenarians. *Mech. Ageing Dev.* **165**, 59–67 (doi:10.1016/j.mad.2017.03.004).
54. Robine, J.-M., Saito, Y., Jagger, C. 2003. The emergence of extremely old people: the case of Japan. *Exp. Gerontol.* **38**, 735–739 (doi:10.1016/S0531-5565(03)00100-1).
55. Thatcher, A.R. 1981. Centenarians. *Popul. Trends* **25**, 11–14.
56. Thatcher, R. 2001. La démographie des centenaires en angleterre et au pays de galle. *Population* **1–2**, 159–180.
57. Dublin, L.I., Lotka, A.J. 1936. *Length of Life: A Study of the Life Table*. The Ronald Press Company.
58. Vincent, P. 1951. La mortalité des vieillards. *Population* **2**, 181–204 (doi:10.2307/1524149).
59. Bourgeois-Pichat, J. 1952. Essai sur la mortalité 'biologique' de l'homme. *Population* **3**, 381–394 (doi:10.2307/1524260).
60. Comfort, A. 1968. *The Conquest of Ageing*. University of Saskatchewan Press.
61. Cutler, R.G. 1985. Biology of aging and longevity. *Gerontol. Biomed. Acta* **1**, 35–61.
62. Hayflick, L. 1996. *How and Why We Age?* Ballantine Books.
63. Walford, R. 1985. *Maximum Life Span*. W.W. Norton & Company.
64. Carnes, B.A., Olshansky, S.J., Grahn, D. 2003. Biological evidence for limits to the duration of life. *Biogerontology* **4**, 31–45 (doi:10.1023/a:1022425317536).
65. Marck, A., Antero, J., Berthelot, G., Saulière, G., Jancovici, J.M., Masson-Delmotte, V., Boeuf, G., Spedding, M., Le Bourg, E., Toussaint, J.F. 2017. Are we reaching the limits of *Homo sapiens*? *Front. Physiology* **8**, 812 (doi:10.3389/fphys.2017.00812).
66. Olshansky, S.J., Carnes, B.A., Désesquelles, A. 2001. Prospect for human longevity. *Science* **292**, 1654 (doi:10.1126/science.291.5508.1491).
67. Olshansky, S.J., Carnes, B.A. 2001. *The Quest for Immortality: Science at the Frontiers of Aging*. Norton.
68. Olshansky, S.J. 2018. From lifespan to healthspan. *J. Am. Med. Assoc.* **320**, E1–E2 (doi:10.1001/jama.2018.12621).
69. Olshansky, S.J., Carnes, B.A. 2019. Inconvenient truths about human longevity. *J. Gerontol. Ser. A* **74**, S7–S12 (doi:10.1093/gerona/glz098).
70. Christensen, K., Doblhammer, G., Rau, R., Vaupel, J.W. 2009. Ageing populations: the challenges ahead. *Lancet* **374**, 1196–1208 (doi:10.1016/S0140-6736(09)61460-4).
71. Oeppen, J., Vaupel, J.W. 2002. Broken limits to life expectancy. *Science* **296**, 1029–1031 (doi:10.1126/science.1069675).
72. Vaupel, J.W. 2010. Biodemography of human ageing. *Nature* **464**, 536–542 (doi:10.1038/nature08984).

73. Vaupel, J.W., Gowan, A.E. 1986. Passage to Methuselah: some demographic consequences of continued progress against mortality. *Am. J. Public Health* **76**, 430–433 (doi:10.2105/ajph.76.4.430).
74. Vaupel, J.W., Villavicencio, F., Bergeron-Boucher, M.-P. 2021. Demographic perspectives on the rise of longevity. *Proc. Natl. Acad. Sci.* **118**, e2019536118 (doi:10.1073/pnas.2019536118).
75. Wilmoth, J.R. 2000. Demography of longevity: past, present, and future trends. *Exp. Gerontol.* **5**, 1111–1129 (doi:10.1016/s0531-5565(00)00194-7).
76. Zuo, W., Jiang, S., Guo, Z., Feldman, M.W., Tuljapurkar, S. 2018. Advancing front of old-age human survival. *Proc. Natl. Acad. Sci.* **115**, 11209–11214 (doi:10.1073/pnas.1812337115).
77. Wilmoth, J.R. 2001. How long can we live? A review essay. *Popul. Dev. Rev.* **27**, 791–800 (doi:10.1111/j.1728-4457.2001.00791.x).
78. Milholland, B., Vijg, J. 2022. Why Gilgamesh failed: the mechanistic basis of the limits to human lifespan. *Nat. Aging* **2**, 878–884 (doi:10.1038/s43587-022-00291-z).
79. Kannisto, V. 1994. *Development of Oldest-Old Mortality, 1950–1990: Evidence from 28 Developed Countries*. Monographs on Population Aging. Odense University Press.
80. Kannisto, V. 1996. *The Advancing Frontier of Survival*. Monographs on Population Aging. Odense University Press.
81. Thatcher, R., Kannisto, V., Vaupel, J.W. 1998. *The Force of Mortality at Ages 80 to 120*. Monographs on Population Aging. Odense University Press.
82. Jeune, B., Vaupel, J.W. 1995. *Exceptional Longevity: From Prehistory to the Present*. Monographs on Population Aging. Odene University Press.
83. Jeune, B., Vaupel, J.W. 1999. *Validation of Exceptional Longevity*. Monographs on Population Aging. Odense University Press.
84. Eisenstein, M. 2022. Does the human lifespan have a limit? *Nature* **601**, 7893 (doi:10.1038/d41586-022-00070-1).
85. Barbi, E., Lagona, F., Marsili, M., Vaupel, J.W., Wachter, K.W. 2018. The plateau of human mortality: demography of longevity pioneers. *Science* **360**, 1459–1461 (doi:10.1126/science.aat3119).
86. Belzile, L.R., Davison, A.C., Rootzén, H., Zholud, D. 2021. Human mortality at extreme age. *R. Soc. Open Sci.* **8**, 202097 (doi:10.1098/rsos.202097).
87. Gampe, J. 2010. Mortality of supercentenarians: estimates from the updated IDL. In *Supercentenarians* (eds H. Maier, J. Gampe, B. Jeune, J.-M. Robine, J. W. Vaupel), pp. 219–230. Springer-Verlag.
88. Pearce, E., Raftery, A. 2021. Probabilistic forecasting of maximum human lifespan by 2100 using Bayesian population projections. *Demogr. Res.* **44**, 1271–1294 (doi:10.4054/DemRes.2021.44.52).
89. Rootzén, H., Zholud, D. 2017. Human life is unlimited – but short. *Extremes* **20**, 713–728 (doi:10.1007/s10687-017-0305-5).
90. Beltrán-Sánchez, H., Austad, S.N., Finch, C.E. 2018. Comment on 'The plateau of human mortality: demography of longevity pioneers'. *Science* **361**, eaav1200 (doi:10.1126/science.aav1200).
91. Camarda, C.G. 2022. The curse of the plateau: measuring confidence in human mortality estimates at extreme ages. *Theor. Popul. Biol.* **144**, 24–36 (doi:10.1016/j.tpb.2022.01.002).
92. Einmahl, J.J., Einmahl, J.H.J., de Haan, L. 2019. Limits to human life span through extreme value theory. *J. Am. Stat. Assoc.* **114** 1075–1080 (doi:10.2139/ssrn.3089594).

93. Newman, S.J. 2018. Errors as a primary cause of late-life mortality deceleration and plateaus. *PLoS Biol.* **16**, e2006776 (doi:10.1371/journal.pbio.2006776).
94. Dang, L., Camarda, C.G., Meslé, F., Ouellette, N., Robine, J.-M., Vallin, J. 2023. The question of the mortality plateau in humans: when longevity pioneers give contrasting insights. *Demogr. Res.* **128**, 321–338 (doi:10.4054/DemRes.2023.48.11).
95. Gavrilova, N.S., Gavrilov, L.A. 2020. Are we approaching a biological limit to human longevity? *J. Gerontol. Ser. A* **75**, 1061–1067 (doi:10.1093/gerona/glz164).
96. Gavrilova, N.S., Gavrilov, L.A., Krut'ko, V.N. 2017. Mortality trajectories at exceptionally high ages: a study of supercentenarians. In *Proceedings of the 2017 Living to 100 Monograph*, pp. 1–17. Society of Actuaries.
97. Human Mortality Database. 2023. Max Planck Institute for Demographic Research (MPIDR, Germany), University of California, Berkeley (USA), and French Institute for Demographic Studies (INED, France). www.mortality.org.
98. International Database on Longevity. 2023. French Institute for Demographic Studies (INED, France). www.supercentenarians.org.
99. Maier, H., Gampe, J., Jeune, B., Robine, J.-M., Vaupel, J.W. 2010. *Supercentenarians.* Demographic Research Monographs. Springer.
100. Maier, H., Jeune, B., Vaupel, J.W. 2021. *Exceptional Lifespans.* Demographic Research Monographs. Springer.
101. Farr, W. 1885. *Vital Statistics: A Memorial Volume of Selections and Writings of William Farr (Posth.)* (ed. N.A. Humphreys). The Sanitary Institute of Great Britain.
102. Eyler, J.M. 1979. *Victorian Social Medicine: The Ideas and Methods of William Farr.* Johns Hopkins University Press.
103. Kannisto, V. 2000. Measuring the compression of mortality. *Demogr. Res.* **3**, 1–24 (doi:10.4054/demres.2000.3.6).
104. Kannisto, V. 2001. Mode and dispersion of the length of life. *Popul. Engl. Sel.* **13**, 159–171.
105. Canudas-Romo, V. 2008. The modal age at death and the shifting mortality hypothesis. *Demogr. Res.* **19**, 1179–1204 (doi:10.4054/demres.2008.19.30).
106. Canudas-Romo, V. 2010. Three measures of longevity: time trends and record values. *Demography* **47**, 299–312 (doi:10.1353/dem.0.0098).
107. Cheung, S.L.K., Robine, J.-M., Tu, E.J.C., Caselli, G. 2005. Three dimensions of the survival curve: horizontalization, verticalization, and longevity extension. *Demography* **42**, 243–258 (doi:10.1353/dem.2005.0012).
108. Diaconu, V., Ouellette, N., Camarda, C. G., Bourbeau, R. 2016. Insight on 'typical' longevity: an analysis of the modal lifespan by leading causes of death in Canada. *Demogr. Res.* **35**, 1–37 (doi:10.4054/demres.2016.35.17).
109. Horiuchi, S., Ouellette, N., Cheung, S.L.K., Robine, J.-M. 2013. Modal age at death: lifespan indicator in the era of longevity extension. *Vienna Yearb. Popul. Res.* **11**, 37–69 (doi:10.1553/populationyearbook2013s37).
110. Ouellette, N., Bourbeau, R. 2011. Changes in the age-at-death distribution in four low mortality countries: a nonparametric approach. *Demogr. Res.* **25**, 595–628 (doi:10.4054/demres.2011.25.19).
111. Ouellette, N., Bourbeau, R., Camarda, C.G. 2012. Regional disparities in Canadian adult and old-age mortality: a comparative study based on smoothed mortality ratio surfaces and age-at-death distributions. *Can. Stud. Popul.* **39**, 79–106 (doi:10.25336/P61P53).
112. Thatcher, A.R., Cheung, S.L.K., Horiuchi, S., Robine, J.-M. 2010. The compression of deaths above the mode. *Demogr. Res.* **22**, 505–538 (doi:10.4054/demres.2010.22.17).

113. Cheung, S.L.K., Robine, J.-M. 2007. Increase in common longevity and the compression of mortality: the case of Japan. *Popul. Stud.* **61**, 85–97 (doi:10.1080/00324720601103833).
114. Diaconu, V., van Raalte, A., Martikainen, P. 2022. Why we should monitor disparities in old-age mortality with the modal age at death. *PLoS ONE* **17**, e0263626 (doi:10.1371/journal.pone.0263626).
115. Robine, J.M., Herrmann, F.R. 2020. Maximal human lifespan. In *Encyclopedia of Biomedical Gerontology* (eds Suresh I.S. Rattan), pp. 385–399. Elsevier Science.
116. Wilmoth, J.R., Deegan, L.J., Lundström, H., Horiuchi, S. 2000. Increase of maximum life-span in Sweden, 1861–1999. *Science* **289**, 2366–2368 (doi:10.1126/science.289.5488.2366).
117. Wilmoth, J.R., Lundström, H. 1996. Extreme longevity in five countries: presentation of trends with special attention to issues of data quality. *Eur. J. Popul. Rev. Eur. Démographie* **12**, 63–93 (doi:10.1007/BF01797166).
118. de Beer, J., Bardoutsos, A., Janssen, F. 2017. Maximum human lifespan may increase to 125 years. *Nature* **546**, E16–E17 (doi:10.1038/nature22792).
119. Dong, X., Milholland, B., Vijg, J. 2016. Evidence for a limit to human lifespan. *Nature* **538**, 257–259 (doi:10.1038/nature19793).
120. Jeune, B. 2002. Living longer – but better? *Aging Clin. Exp. Res.* **14**, 72–93 (doi:10.1007/BF03324421).
121. Medford, A., Vaupel, J.W. 2019. Human lifespan records are not remarkable but their durations are. *PLoS ONE* **14**, e0212345 (doi:10.1371/journal.pone.0212345).
122. Strulik, H., Vollmer, S. 2013. Long-run trends of human aging and longevity. *J. Popul. Econ.* **26**, 1303–1323 (doi: 10.1007/s00148-012-0459-z).
123. Vijg, J., Le Bourg, E. 2017. Aging and the inevitable limit to human life span. *Gerontology* **63**, 432–434.
124. Weon, B.M., Je, J.H. 2009. Theoretical estimation of maximum human lifespan. *Biogerontology* **10**, 65–71 (doi: 10.1007/s10522-008-9156-4).
125. Wilmoth, J.R., Robine, J.-M. 2003. The world trend in maximum life span. *Popul. Dev. Rev.* **29**, 239–257.
126. Andreev, E.M., Shkolnikov, V.M., Begun, A.Z. 2002. Algorithm for decomposition of differences between aggregate demographic measures and its application to life expectancies, healthy life expectancies, parity-progression ratios and total fertility rates. *Demogr. Res.* **7**, 499–522 (doi:10.4054/demres.2002.7.14).
127. Arriaga, E.E. 1989. Changing trends in mortality decline during the last decades. In *Differential Mortality: Methodological Issues and Biological Factors* (eds L. Ruzicka, G. Wunsch, P. Kane), pp. 105–129. Clarendon Press.
128. Horiuchi, S., Wilmoth, J.R., Pletcher, S.D. 2008. A decomposition method based on a model of continuous change. *Demography* **45**, 785–801 (doi:10.1353/dem.0.0033).
129. Robine, J.-M., Paccaud, F. 2004. La démographie des nonagénaires et des centenaires en suisse. *Cah. Qué. Démographie* **33**, 51–81 (doi:10.7202/010852ar).
130. Vaupel, J.W., Jeune, B. 1995. The emergence and proliferation of centenarians. In *Exceptional Longevity: From Prehistory to the Present.* Monographs on Population Aging. (eds B. Jeune, J. W. Vaupel), pp. 109–116. Odense University Press.
131. Wood, S.N. 2017. *Generalized Additive Models: An Introduction with R* (2nd ed.). Chapman & Hall/CRC.

15 Mortality Modelling at the Oldest Ages in Human Populations

A Brief Overview

Linh Hoang Khanh Dang, Nadine Ouellette, Carlo Giovanni Camarda and France Meslé

Since the 1950s, the share of people aged 90 and above has increased steadily (see Chapters 12 and 14). The decline in fertility rates was first identified as the main driver for this demographic trend, until the unexpected and rapid decline in old-age death rates became a large contributor [1]. As the share of the older population grows ever larger, concerns over the sustainability of the pension systems and the necessary arrangements for ageing societies call for better knowledge on the age pattern of mortality at the highest ages, which remains largely debated. In this chapter, we review the main elements that have shaped research about mortality trajectory at the oldest ages over time.

15.1 History of Mortality Laws at the Oldest Ages

Mortality models have been essential for characterizing the shape of the mortality curve at very old ages [2]. Some models have been raised to the level of 'mortality laws' thanks to their ability to depict trends in empirical observations, their capacity to express a given mechanism of senescence and their elegance in application. In general, we can define a mortality law as a mathematical function that expresses a relationship between the instantaneous death rate (also known as the 'force of mortality') and age by a set of parameters. Therefore, mortality laws are often referred to as parametric mortality models. When applied to empirical data, these functions translate a hypothesis of a given ageing mechanism into a statistical distribution for the number of deaths, and the reliability of mortality models can thereby be assessed using standard statistical tools. Table 15.1 presents some of the mathematical details for the most widely used mortality models that are introduced in the forthcoming text.

A very first example of a mortality model was suggested by Abraham de Moivre in 1725 [3]. However, the most renowned contribution is unquestionably the law of mortality proposed by Benjamin Gompertz in 1825 [4]. Stating that 'this law of geometric progression pervades, in an approximate degree, large portions of different tables of mortality (…)' ([4], p. 514), the validity of this law over a large part of the age range remains unchanged after nearly 200 years from its publication. Today, because of mortality decline among the elderly, Gompertz's law may even extend to a broader

Table 15.1 Widely used mathematical mortality models at older ages

Model name	Functional form	Underlying hypotheses
Gompertz	$\mu(x) = a\exp(bx)$	Exponential increase of mortality with age Initial damage at starting age $x = 0$
Makeham	$\mu(x) = a\exp(bx) + c$	Exponential increase of mortality with age 'Chance' factor unrelated to age Initial damage at starting age $x = 0$
Perks	$\mu(x) = \dfrac{A + Bc^x}{1 + D(c-k)^x}$	Mortality deceleration
Beard	$\mu(x) = \dfrac{A + Bc^x}{1 + \lambda\exp(bx)}$	Mortality deceleration Upper asymptote at estimated value
Kannisto	$\mu(x) = \dfrac{a\exp(bx)}{1 + a\exp(bx)}$	Mortality deceleration Upper asymptote equal to 1
Weibull	$\mu(x) = ax^b$	Non-exponential increase of mortality with age Without damage at starting age $x = 0$
Log-quadratic	$\mu(x) = \exp(ax^2 + bx + c)$	Diverse age-patterns of mortality: increase, deceleration or decline with age

Note: The gamma-Gompertz model proposed by Vaupel and colleagues in 1979 [12], with functional form $\mu(x) = \dfrac{a\exp(bx)}{1 + \frac{a\gamma}{b}(\exp(bx) - 1)}$, is also frequently used in mortality modelling at advanced ages. This model ultimately leads to the same logistic trajectory as described by the Beard model, where $\lambda = (1 - \exp(bx))^{a\gamma}$ and γ is the coefficient of variation of the gamma distribution used to describe the unobserved heterogeneity or 'frailty' within the population.

age range than originally proposed (from ages 20 to 80 according to Gompertz [5, 6]), to as far as age 100 [7].

Gompertz also introduced a possible explanation for an underlying mechanism of human death, consisting of the two following main components: a chance factor that affects all individuals regardless of age and the 'deterioration, or an increased inability to withstand destruction' ([4], p. 517), described by an age-dependent exponential mortality rise. However, it took a few decades for the age-independent part of the equation accounting for chance – or background mortality – to be added to Gompertz's initial proposal by William M. Makeham [8]. Widely known as the 'Makeham term', it is often negligible at higher ages. Meanwhile, the age-dependent component became subject to finer and more diverse mobilizations of the senescence process after Gompertz.

In the 1930s, the deviation between observed death rates and the age trajectory of mortality predicted by Gompertz's law became increasingly apparent at the highest ages, leading researchers to focus on mathematical alternatives to Gompertz's exponential model. The leading work came from Wilfred Perks (1932), who introduced the family of logistic models as a result of his attempts to generalize Makeham's equation [9]. Logistic models showed remarkably good performance on mortality data at advanced ages, prompting the introduction of a simplified version (two-parameters model) of the original form by Väinö Kannisto [10, 22], most commonly known as the Kannisto model, which is still widely used to describe mortality at very old ages in various application settings.

The good performance of logistic models is not Perks' unique contribution. He was also among the first to further develop Gompertz's idea about 'the deterioration to withstand destruction' [4]. According to Perks, a given population can be stratified into (smaller) homogeneous groups of individuals with identical shapes for the force of mortality, but different levels of withstanding destruction. Consequently, overall population mortality is the average aggregated mortality of all stratified groups over age. The notion of individual 'deterioration to withstand destruction' also appears in later works but under different names, such as 'longevity' by Robert E. Beard and 'frailty' by James W. Vaupel and colleagues [11, 12]. By assuming it to be gamma distributed, Beard proved that if the force of mortality within each homogeneous group follows a Gompertz law, then the average population force of mortality has the logistic form proposed by Perks [9]. In other words, while Perks provided the arithmetic form for the population-level force of mortality and the idea of introducing a factor of stratification, Beard suggested the statistical distribution for the stratifying factor and the mortality function for practical use [13, 14]. Vaupel and colleagues went one step further and embedded this idea into a proportional hazard framework, by assuming a gamma distribution for the individual 'frailty' component and a Gompertz model for standard mortality (i.e. individual with mean frailty) [12]. This led to what became known as the gamma-Gompertz model. In practice, this approach eventually leads to the logistic form proposed by Perks and Beard. However, the contribution of Vaupel and colleagues' work lies in the fact that the authors went beyond the final trajectory of mortality at the highest ages [12]. Through a proportional hazard framework, they introduced another way of accounting for heterogeneity into the mortality model, and this has paved the way for many theoretical and empirical works to follow [15, 16].

It is noteworthy that Perks was aware, at the time of his publication, that a mortality plateau was inherent to the logistic model he proposed. He wrote:

> However, should anybody view the new curves with disfavor on account of the limit which they impose on μ_x, let him make a minor alteration to the formula, substituting $D(c-k)^x$ for Dc^x, where k may be as small a quantity as he pleases, so small indeed as to have no effect on μ_x throughout the whole practical change. Then, so long as k is finite, μ_x will tend to infinity as x tends to infinity. ([9], p. 39)

This statement, rarely acknowledged in the later literature, shows Perks' strong interest for capturing a change in the rate of mortality increase rather than focusing on an upper limit of mortality.

Another mortality model worth mentioning is the log-quadratic model. Popularized by Ansley J. Coale and Ellen E. Kisker (1990) with the objective to correct defective US data at the highest ages, the log-quadratic model is capable of portraying diverse patterns, from steep increase to deceleration and even a decline in human mortality with advancing age [17]. A trade-off for flexibility is, however, that the log-quadratic model is more a descriptive than an explanatory model.

In a broader perspective, death can be seen as cessation of all body functions and such a process can apply to living beings as much as machines. From engineering, Waloddi Weibull's 1951 proposition makes an analogy between the deterioration of a machine's components until its failure and that of human organs until a person's death [18]. Alongside the Gompertz model and its generalizations, the Weibull model offers an explanation for the mechanism of human mortality through degradation of its components.

15.2 Empirical Results

In this section, we present a selection of key studies that shaped research on the mortality trajectory at the oldest ages through parametric mortality models. Existing empirical studies can be divided into two main streams: one brings support to the validation of a decelerating pattern of mortality at the highest ages, while the other advocates for an exponential growth of mortality.

One of the earliest studies providing evidence of a mortality deceleration pattern at very old ages was done by John C. Barrett [19]. At the time, Arthur Roger Thatcher (1981) had just published new mortality data on centenarians that was collected in England and Wales [20]. Barrett fitted a quadratic function to these data in an attempt to detect possible deviations from an exponential growth pattern in death rates beyond age 100. He found that mortality decelerated for men after age 103, but not for women. A few years later, in 1988, Kannisto revisited Barrett's work, taking out three male deaths after age 110, and found that male mortality kept increasing with age [21]. This suggests that conclusions about the shape of the mortality curve at the highest ages are very sensitive to data availability and reliability. For a small number of old individuals, different conclusions can be drawn, not because of a lack of robustness in the statistical analysis, but because of the limited data at hand. It should be noted that so far, this is still a concern, especially at ages beyond 110 years old; an age which very few humans reach.

At the end of the twentieth century, in 1998, Thatcher and colleagues published a study that inspired many of the subsequent empirical works on the mortality trajectory at the oldest ages [22]. A set of six mortality models seen as the most pertinent ones to describe late-life trajectories were selected (see Table 15.1, except for the Perks model). The models were fitted to observed mortality data at ages 85 and above for 13 countries with long-running and reliable historical series. This database was specifically and carefully designed to allow for analyses by periods (1960–1970, 1970–1980, 1980–1990) and by birth cohorts (1871–1880). The authors found that the logistic (both the complete and the simplified Kannisto model) and quadratic models provided a better description of the data than those yielding exponential or logarithmic mortality increase over age (e.g. the Gompertz and Weibull models).

However, formal goodness-of-fit tests showed that between the logistic and quadratic models, no single one was systematically better than the others. Shortly after, in 1999, Thatcher extended the previous study to mortality data for England and Wales, going back as far as age 30 and as early in time as the year 1640 [23]. With this longer series, Thatcher was able to refine the earlier findings: the logistic model was found to be the best-fitting model for describing human mortality at older ages.

Later on, in 2018, Feehan ran a thorough investigation of a more extensive set of mortality models fitted to data on 360 birth cohorts at ages 80 and above [24]. Feehan's results show a better performance of the models, allowing for mortality deceleration for women and men in all study countries. However, among these models, the optimal one varied from a given population to another, and, considered systematically, no model worked best in all datasets. The author recommended using the log-quadratic model since it stood out as being the optimal one most often, while caution was advised in using the Kannisto model. The latter performed well overall, but fitted the data more poorly for some cohorts. These remarks are in line with recent findings by Dang [7], based on thoroughly validated data for French, Belgian and Quebec birth cohorts, and where the Kannisto model systematically underestimated mortality at very old ages. In contrast, the log-quadratic model performed well in each of the French and Belgian female datasets, but when data for all three populations were pooled together, the Beard model was the optimal model. More generally, mortality deceleration was found to be the most plausible trajectory at the highest ages for all female populations under various assumptions regarding the probability distribution of death counts at a given age (i.e. Poisson, binomial and negative binomial distribution). Deceleration started globally around age 100 in this study, meaning at relatively later ages compared to previous estimates made by Rau and colleagues at 92.82 [25]. As for male populations, an exponential growth of mortality with age remains the most plausible trajectory.

In 2011, Gavrilov and Gavrilova defended the idea of a continuous exponential growth of mortality following an analysis of US data [26]. They offered a critical overview of existing literature on the mortality trajectory at the highest ages, pointing out several factors that could possibly interfere with the detection of true underlying mortality patterns. Namely, they referred to the practice of pooling data for several birth cohorts (or countries) to standard assumptions such as constant mortality rates within single-year age intervals (e.g. the assumption of a uniform distribution of deaths, including at the highest ages) and to low-quality mortality data. Using data on deaths for individuals born between 1881 and 1895 extracted from the US Social Security Administration, the authors found an unabated exponential growth of mortality rates up to age 105. From age 106 onwards, the data quality was deemed unfit to draw a definite conclusion. A similar study was carried out in 2019 by the same authors [27]. In this paper, they based their analysis on data for US cohorts born between 1880 and 1899 taken from the Human Mortality Database (HMD) [28], limiting again the analysis to ages below 105 for data quality reasons. It was found that while the Kannisto model most often proved more suitable for older birth cohorts, the Gompertz model was a better choice for younger cohorts. The tipping point occurred around birth cohort 1886–1887, and the authors suggested that the questionable data quality of older cohorts may be the reason for the observed decelerating mortality pattern. An exponential growth of mortality at the oldest ages was also observed for Australia, Canada and the United States in a systematic investigation by Bebbington and colleagues [29], also using cohort data from the HMD.

15.3 Possible Limits to Human Lifespan

While studying the mortality trajectory at the oldest ages, the query on its natural limit becomes a subject of interest. In human longevity, one can identify various types of limits. The most commonly used definitions are the limit to life expectancy (i.e. mean lifespan), the limit of longevity (i.e. maximum lifespan) and the limit to the age trajectory of mortality (also commonly interpreted in literature as the mortality plateau). Although these inform us about specific metrics of human lifespan and make use of different statistical and demographic methods, they describe different characteristics of the same underlying variable: the length of human life or the human lifespan.

The limit to human life expectancy, more precisely life expectancy at birth, has been subject to many estimations throughout the twentieth century. Among others are estimates such as 64.7 years [30], 77.2 years [31], and 85 years [32] (readers interested in historical perspectives on this topic may refer to Chapter 14). The various authors who proposed those limits relied on the general idea that progress in life expectancy would necessarily come to an end someday. However, these limits were all exceeded shortly after being proposed as what Oeppen and Vaupel unveiled in their work in 2002 [33]. Furthermore, the authors showed that in fact, life expectancy in the record-holding country year after year had followed a very steady linear upward trend since 1840, illustrating that the belief that the average lifespan in human species might be close to reaching a limit is a misconception. Their observation, though remarkable, was challenged few years later by Vallin and Meslé [34]. Using higher-quality data, the authors demonstrated that although life expectancy generally increased over time, the pace of increase varied throughout history, revealing a segmented line. Some periods experienced faster growth than others, but this upward trend had slowed down in recent times, because most progress is now occurring at older ages. Although mortality improvements at older ages are promising, they contribute less to overall increase in life expectancy than declines in death rates at younger ages. Nonetheless, life expectancy at birth remains an important indicator of central tendency of the distribution of individual life durations, but its capacity to reflect the real dynamics in mortality improvement is reduced as populations age and changes in mortality rates at older ages become more significant. Even if decreases in death rates at older ages are not sufficient to lead to a big shift in life expectancy, such as was observed in the past, these changes are still significant. Other indicators, such as life expectancy at ages 50 and 65 [32], or the (adult) modal age at death [35, 36, 37], have been suggested as better candidates.

Moving on to the limit of longevity, or the highest age that a human could ever attain, various estimations have been proposed, such as 107 years old [38], and 115 years old [39, 40], among others. The question of whether this limit is an intrinsic human trait has been the subject of debate. Some scholars suggest that human lifespan is a variant characteristic that could evolve over time [41], while others argue that the maximum lifespan can be considered a stable characteristic of a given species [40]. However, regardless of whether the limit of longevity is a human intrinsic trait, according to actuary Steffesen and demographer Wilmoth [42, 43], the idea that there would be a maximum age to which individuals can survive, but that they would inevitably pass away in the moment (or in a 'fraction of a second') that follows is unthinkable.

The debate over the limits to human maximum lifespan has resurfaced with the publication of a paper by Dong and colleagues [44]. The authors argued in favour of such a limit, based on the lack of new maximum age at death records since 1995. They estimated this limit to be reached at 114.9 years old (with a 95% confidence interval: 113.1–116.7) and consider the age reached by Jeanne Calment (122 years and 164 days) to be an exceptional case. However, as previously shown, it is important to remember that human development is a complex and irregular process marked by periods of rapid and slow progress, stagnation and occasional temporary setbacks. Furthermore, when it comes to the highest attained ages, as more individuals live to increasingly older ages, the higher the chance we may observe a new record of maximum human lifespan.

Another way to approach the question of limit of longevity is through statistical modelling, specifically by applying extreme value theory (EVT) [45]. In 1994, Aarssen and de Haan were the first to explore this direction, using data from the Netherlands, and suggested the existence of a limit between ages 113 and 124 (with a 95% confidence level) [46]. The method was subsequently applied by others, including Barbi and colleagues and Rootzén and Zholud [47, 48], who found no evidence of a limit using French data and data on supercentenarians from the International Database on Longevity, respectively. Using Belgian data, Gbari and colleagues found supporting evidence for an ultimate age of 114.82 for males and 122.73 for females [49]. Similarly, Einmahl and colleagues presented compelling statistical evidence for an upper limit to the lifespan of both male and female Dutch residents [50]. These findings highlight that results can vary and be contradictory based on the population studied and the observation period, despite applying fundamentally similar methodologies.

Lastly, the mortality plateau at oldest ages is also a subject of great interest. Explanation for such hypothesis is well illustrated in Steffesen's 1931 lecture: 'The scanty observations available at extreme old age tend to show that the mortality in these distant regions of life is, at least for certain classes of lives, rather lower than generally assumed, perhaps even with a tendency to remain constant' ([42], p. 102). With gradually greater availability of data at extreme old ages, increasing number of studies have been conducted to verify such hypothesis using both parametric and non-parametric methods. Among parametric approaches, Rau and colleagues compared the performance of models predicting a mortality plateau and the Gompertz model that predicts a continuing exponential increase of mortality on aggregate data

[51]. Based on data from 16 low-mortality countries for years 1960–2010, they found a better performance of the gamma-Gompertz model compared to Gompertz model and estimated a plateau for the force of mortality at a level of 0.8 for females and 1.2 for males, which corresponds to an estimated probability of dying at 0.6 for females and 0.7 for males. Using individual data, Barbi and colleagues tested a constant baseline against a Gompertz baseline in a survival analysis setting [52]. Based on 3 836 Italian semi- and supercentenarians (i.e. individuals who died at age 105–109 and above 110, respectively), these authors claimed evidence for the existence of a mortality plateau. A few remarks should be made, however, on both strategies: first, a better performance of one model does not imply that all inherent mathematical properties of that model are valid. As such, a better performance of the gamma-Gompertz model might come solely from its flexibility to allow an inflection point and much less from its horizontal asymptote (also interpreted as the mortality plateau). Second, conclusions drawn on one country, or a set of countries, should not be generalized to populations outside the data used for analysis. Indeed, a replication study by Dang and colleagues proved that no evidence of mortality plateau can be found in French population [53]. Within a non-parametric setting, Gampe estimated a constant mortality rate around 0.7 between age of 110 and 114 [54]. They then replicated their approach on a larger set of data starting from age 105 and found again a constant mortality rate, but at a level slightly higher than 0.7, with a non-negligible variability in the estimates ([55], p. 30). After age 114, data are too scarce to obtain any reliable estimation. Therefore, even within a non-parametric setting that is free from all functional form hypotheses, there is still a lot of uncertainty around estimates at these extreme ages, and little is known after age 114. In sum, regardless of the approach used, a reasonable consensus on the existence of a limit to human longevity, and its level, has not yet been reached.

15.4 Concluding Remarks

Recent empirical works on the late-age trajectory of human mortality showed how models implying deceleration are preferable to the simple Gompertz trajectory at the highest ages. However, the onset of the mortality deceleration is being progressively delayed towards higher ages, approximately after age 100. Due to the small number of reliable data currently available, both in terms of number of survivors at the most advanced ages and of numbers of population whom data validation at these ages is duly done, complete certainty about these findings is not achievable yet. Each empirical observation made so far can thus be seen as one of many compelling possibilities. Any additional properly validated survivors at extreme ages, any new population integrated in a trustworthy source of data, such as the International Database on Longevity [56], as well as any new analysis have the potential to improve our understanding of human mortality at the oldest ages and/or provide guidance for new research directions. On the one hand, it is necessary to put more efforts in collecting reliable data, which is essential to extract more insights from this once data-starved field. On the other hand, it is critical to map out our ignorance by thoroughly quantifying the degree of our uncertainty when dealing with this relevant topic.

Furthermore, it is increasingly important to understand the source of deceleration patterns in human mortality at the oldest ages. At the time this chapter was written, the selection hypothesis by Vaupel and colleagues might be the most elaborated explanation for late-life mortality deceleration [12], but it is certainly not the only one [57, 58]. In addition, Salinari ran a test on female birth cohort data drawn from the HMD and found that selection is not the unique mechanism responsible for mortality deceleration [59]. However, it is also noteworthy that testing the validity of these hypotheses might be more challenging than expected. Not only there is a lack of properly verified age-at-death data, but the most commonly used mathematical mortality models in this area of research are not sufficiently advanced to test our target hypothesis [7], namely whether a deceleration of mortality at the population level could be due to a deceleration in individual mortality. Also, it is fundamental to remember that these hypotheses are not necessarily competing, but they could complement each other in providing us with insights to better understand the mortality process at the oldest ages. Further developments might be needed before analysing the deceleration of mortality in its finer components (e.g. the endogenous mortality, the background mortality) becomes possible.

Lastly, the number of survivors at very old ages has grown at an unprecedented rate in recent decades, notably at ages 90 and above. This was accompanied by an equally strong development in demographic research on mortality at the highest ages, a field of research that is rapidly expanding and ripe with possibilities, but not without challenges.

References

1. Horiuchi, S. 1991. Assessing the effects of mortality reduction on population ageing. *Popul. Bull. U. N.* **31/32**, 38–51.
2. Forfar, D., McCutcheon, J., Wilkie, A. 1988. On graduation by mathematical formula. *J. Inst. Actuar.* **115**, 1–149.
3. de Moivre, A. 1725. *Treatise on Annuities*. A. Millar.
4. Gompertz, B. 1825. On the nature of the function expressive of the law of human mortality. *Philos. Trans. R. Soc.* **115**, 513–585.
5. Gompertz, B. 1862. A supplement to two papers published in the transactions of the Royal Society on the science connected with human mortality. *Philos. Trans. R. Soc.* **152**, 511–559.
6. Gompertz, B. 1872. On one uniform law of mortality from birth to extreme old age, and on the law of sickness. *J. Inst. Actuar.* **16**, 329–344.
7. Dang, L. H. K. 2022. Risques de décès aux grands âges. PhD thesis, Université Paris Nanterre.
8. Makeham, W. M. 1860. On the law of mortality. *J. Inst. Actuar.* **13**, 283–287.
9. Perks, W. 1932. On some experiments in the graduation of mortality statistics. *J. Inst. Actuar.* **63**, 12–57.
10. Kannisto, V. 1992. Presentation at a Workshop on Old Age Mortality at Odense University, June.

11. Beard, R.E. 1952. Some further experiments in the use of the incomplete gamma function for the calculation of actuarial functions. *J. Institute Actuar.* **78**, 341–353.
12. Vaupel, J.W., Manton, K. G., Stallard, E. 1979. The impact of heterogeneity in individual frailty on the dynamics of mortality. *Demography* **16**, 439–454.
13. Beard, R. E. 1951. Some notes on graduation. *J. Inst. Actuar.* **77**, 382–431.
14. Beard, R. E. 1971. Some aspects of theories of mortality, cause of death analysis, forecasting and stochastic processes. In *Biological Aspect of Demography* (ed. W. Brass). 57–68 Taylor & Francis.
15. Salinari, G., De Santis, G. 2019. One or more rates of ageing? The extended gamma-Gompertz model (EGG). *Stat. Methods Appl.* **29**, 211–236 (doi:10.1007/s10260-019-00471-z).
16. Patricio, S.C., Missov, T.I. 2023. Makeham Mortality Models as Mixtures. arXiv:2304.08920 [stat.AP].
17. Coale, A.J., Kisker, E.E. 1990. Defect in data on old age mortality in the United States: new procedures for calculating approximately accurate mortality schedules and life table at the highest ages. *Asian Pac. Popul. Forum* **4**, 1–31.
18. Weibull, W. 1951. A statistical distribution function of wide applicability. *J. Appl. Mech.* **18**, 293–297.
19. Barrett, J. 1985. The mortality of centenarians in England and Wales. *Arch. Gerontol. Geriatr.* **4**, 211–218.
20. Thatcher, A.R. 1981. Centenarians. *Popul. Trends* **25**, 11–14.
21. Kannisto, V. 1988. On the survival of centenarians and the span of life. *Popul. Stud.* **42**, 389–406.
22. Thatcher, A.R., Kannisto, V., Vaupel, J.W. 1998. *The Force of Mortality at Ages 80 to 120*, vol. 5. Monographs on Population Aging. Odense University Press.
23. Thatcher, A.R. 1999. The long-term pattern of adult mortality and the highest attained age (with discussion). *J. R. Stat. Soc. Ser. A* **127**, 5–43.
24. Feehan, D.M. 2018. Separating the signal from the noise: evidence for deceleration in old-age death rates. *Demography* **55**, 2025–2044 (doi:10.1007/s13524-018-0728-x).
25. Rau, R., Muszynska, M., Vaupel, J.W., Baudisch, A. 2009. At What Age Does Mortality Start to Decelerate? Paper presented at the Annual Meeting of the Population Association of America, 1 May 2009, Detroit, USA.
26. Gavrilov, L.A., Gavrilova, N.S. 2011. Mortality measurement at advanced ages. *North Am. Actuar. J.* **15**, 432–447 (doi:10.1080/10920277.2011.10597629).
27. Gavrilov, L.A., Gavrilova, N.S. 2019. New trend in old-age mortality: Gompertzialization of mortality trajectory. *Gerontology* **65**, 451–457 (doi:10.1159/000500141).
28. Human Mortality Database. 2022. Max Planck Institute for Demographic Research (MPIDR, Germany), University of California, Berkeley (USA), and French Institute for Demographic Studies (INED, France). https://mortality.org.
29. Bebbington, M., Green, R., Lai, C.D., Zitikis, R. 2014. Beyond the Gompertz law: exploring the late-life mortality deceleration phenomenon. *Scand. Actuar. J.* **2014**, 189–207.
30. Dublin, L. 1928. *Health and Wealth.* Harper.
31. Bourgeois-Pichat, J. 1952. Essai sur la mortalité 'biologique' de l'homme. *Population*, 381–394.
32. Olshansky, J.S., Carnes, B.A., Cassel, C. 1990. In search of Methuselah: estimating the upper limits to human longevity. *Science* **250**, 634–640.
33. Oeppen, J., Vaupel, J.W. 2002. Broken limits to life expectancy. *Science* **296**, 1029–1031.
34. Vallin, J., Meslé, F. 2009. The segmented trend line of highest life expectancies. *Popul. Dev. Rev.* **35**, 159–187 (doi:10.1111/j.1728-4457.2009.00264.x).

35. Horiuchi, S., Ouellette, N., Cheung, S.L.K., Robine, J.-M. 2013. Modal age at death: lifespan indicator in the era of longevity extension. *Vienna Yearb. Popul. Res.* **11**, 37–69.
36. Kannisto, V. 2001. Mode and dispersion of the length of life. *Popul. Engl. Sel.* **13**, 159–171.
37. Ouellette, N., Bourbeau, R. 2011. Changes in the age-at-death distribution in four low mortality countries: a nonparametric approach. *Demogr. Res.* **25**, 595–628.
38. Vincent, P. 1951. La mortalité des vieillards. *Population* **6**, 181–204 (doi:10.2307/1524149).
39. Depoid, F. 1973. La mortalité des grands vieillards. *Population* **28**, 755–792 (doi:10.2307/1531256).
40. Vijg, J., Le Bourg, E. 2017. Aging and the inevitable limit to human life span. *Gerontology* **63**, 432–434.
41. Gavrilov, L.A., Krut'ko, V.N., Gavrilova, N.S. 2017. The future of human longevity. *Gerontology* **63**, 524–526.
42. Steffesen, J. 1931. Notes on the life table and the limits of life. *J. Inst. Actuar.* **62**, 99–108.
43. Wilmoth, J.R. 1997. In search of limits. In *Between Zeus and the Salmon: The Biodemography of Longevity* (eds K.W. Wachter, C.E. Finch), pp. 38–64. The National Academies Press.
44. Dong, X., Milholland, B., Vijg, J. 2016. Evidence for a limit to human lifespan. *Nature* **538**, 257–259.
45. Gumbel, E.J. 1937. *La durée extrême de la vie humaine*. Hermann.
46. Aarssen, K., de Haan, L. 1994. On the maximal lifespan of humans. *Math. Popul. Stud.* **4**, 259–281.
47. Barbi, E., Caselli, G., Vallin, J. 2003. Trajectories of extreme survival in heterogeneous populations. *Population* **58**, 43–65.
48. Rootzén, H., Zholud, D. 2017. Human life is unlimited – but short. *Extremes* **20**, 713–728.
49. Gbari, S., Poulain, M., Dal, L., Denuit, M. 2017. Extreme value analysis of mortality at the oldest ages: a case study based on individual ages at death. In *2017 Living to 100 Monograph*, pp. 1–28. Society of Actuaries.
50. Einmahl, J.J., Einmahl, J.H.J., de Haan, L. 2019. Limits to human life span through extreme value theory. *J. Am. Stat. Assoc.* **114**, 1075–1080.
51. Rau, R., Ebeling, M., Peters, F., Bohk-Ewald, C., Missov, T.I. 2017. Where is the level of the mortality plateau? In *2017 Living to 100 Monograph*, pp. 1–22. Society of Actuaries.
52. Barbi, E., Lagona, F., Marsili, M., Vaupel, J. W., Wachter, K. W. 2018. The plateau of human mortality: demography of longevity pioneers. *Science* **360**, 1459–1461.
53. Dang, L. H. K., Carmada, G. C., Ouellette, N., Meslé, F., Robine, J.-M., Vallin, J. 2023. The question of human mortality plateau: contrasting insights by longevity pioneers. *Demogr. Res.* **48**, 321–338.
54. Gampe, J. 2010. Human mortality beyond age 110. In *Supercentenarians* (eds H. Maier, J. Gampe, B. Jeune, J.-M. Robine, J.W. Vaupel), pp. 219–230. Springer-Verlag.
55. Gampe, J. 2021. Mortality of supercentenarians: estimates from the updated IDL. In *Exceptional Lifespans* (eds H. Maier, B. Jeune, J.W. Vaupel), pp. 29–36. Springer-Verlag.
56. International Database on Longevity. 2023. French Institute for Demo-Graphic Studies (INED, France). wwwsupercentenarians.org.
57. Greenwood, M., Irwin, J. O. 1939. The biostatistics of senility. *Hum. Biol.* **11**, 1–23.
58. Horiuchi, S., Wilmoth, J.R. 1998. Deceleration in the age pattern of mortality at older ages. *Demography* **35**, 391–412.
59. Salinari, G. 2018. Rethinking mortality deceleration. *Biodemography Soc. Biol.* **64**, 127–138.

16 Lessons from Exceptionally Long-Lived Individuals and Long-Living Families

Implications for Medical Research on Ageing and Age-Related Diseases

Larissa Smulders and Joris Deelen

16.1 Introduction

Average human life expectancy has, overall, been continuously increasing over the last two centuries [1], although recent data from the United States suggests that, now, we may have reached a plateau [2] (see Chapters 12 and 14). The worldwide increase in life expectancy has also resulted in an increased number of centenarians, at least in developed countries (i.e. those >100 years of age, see Figure 16.1). The currently observed increase in life expectancy is mainly attributable to changes in environmental conditions, such as improved hygiene, medical care and nutrition. Therefore, it is expected that the majority of the individuals who live to an age above 90 years (from here on referred to as long-lived individuals) are so-called phenocopies. Phenocopies are individuals who live long lives as a result of improved environmental factors as opposed to individuals who benefit from a favourable genetic background. In addition, the increase in life expectancy of individuals who have reached adulthood has not been accompanied by a similar increase in healthspan (i.e. number of years lived without major chronic diseases) [3]. The significant increase in both life expectancy and, to a lesser extent, healthspan is attributed to the rise of antibiotics and vaccines in the twentieth century, resulting in a substantial reduction of infectious conditions that led to early life mortality. In recent years, cardiovascular disease and cancer superseded infectious diseases as the leading causes of death. The more recent increase in life expectancy is attributable to medical interventions that reduce the risk of dying from chronic age-related diseases, such as cardiovascular disease, cancer, type 2 diabetes and Alzheimer's disease. Consequently, many long-lived individuals will still suffer from these diseases, so their healthspan is not increased [3]. Hence, the current focus of ageing research is on the prevention and compression of late-life morbidity, that is, reducing the time between the onset of age-related diseases and death [4].

Although the large majority of the increase in life expectancy is attributable to improved environmental conditions, including better medical care, increasing evidence suggests that survival to an age above 90 years, that is, longevity, is partly

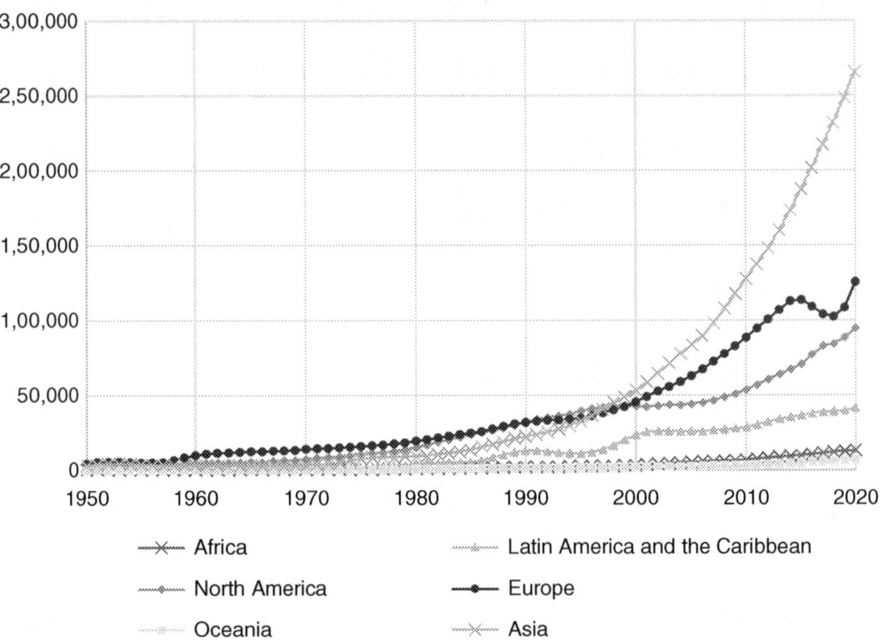

Figure 16.1 Increase in the number of centenarians worldwide, stratified by continent, since the mid-1950s.
Source: United Nations, Data Portal, Population division. From United Nations World Population Prospects. Copyright © 2022 United Nations. Used with the permission of the United Nations.

heritable and is accompanied by a lower prevalence of age-related diseases [5, 6]. These findings indicate that studies of the most exceptionally long-lived individuals, that is, individuals who reached an age above 100 years, and long-living families, that is, families with multiple long-lived individuals, may provide insights into the secrets of living a long and healthy life encoded in their genomes. This chapter will focus on the findings from studies of exceptionally long-lived individuals and long-living families, both at the phenotypic (with a focus on metabolic parameters) and genetic levels.

16.2 Studies of Long-Lived Individuals

The first study of exceptionally long-lived individuals, the Okinawa Centenarian Study, was established in Japan in 1975 (https://orcls.org/ocs) [7] (see also Chapter 17). Since then, many studies of both exceptionally long-lived individuals and long-living families have been established (see Table 16.1 for some representative examples). The most famous and most extensive population-based study of exceptionally long-lived individuals is the New England Centenarian Study, which

Table 16.1 Representative examples of population- and family-based studies of long-lived individuals

Study	Type of individuals	Reference
Population-based studies		
100-Plus Study	Cognitively healthy centenarians	[108]
Chinese Longitudinal Healthy Longevity Study	Centenarians	[78]
Danish Birth Cohort Studies	Nonagenarians and centenarians	[109]
Finnish Centenarians Study	Centenarians	[110]
Five-Country Oldest Old Project	Centenarians	[111]
French Long-Lived Individuals	Centenarians	[112]
German Long-Lived Individuals	Nonagenarians and centenarians	[113]
Japanese Semi-Supercentenarian Study	Centenarians	[114]
Longevity Genes Project	Centenarians	[115]
New England Centenarian Study	Centenarians	[8]
Okinawa Centenarian Study	Centenarians	[7]
Southern Italian Centenarian Study	Nonagenarians and centenarians	[116]
Swedish Centenarian Study	Centenarians	[117]
Tokyo Centenarian Study	Centenarians	[118]
Family-based studies		
European Challenge for Healthy Ageing Study	Centenarian siblings	[119]
Genetics of Healthy Ageing	Nonagenarian siblings	[120]
Leiden Longevity Study	Nonagenarian siblings	[42]
Long Life Family Study	Nonagenarians and their siblings	[10]

Note: This table provides some examples of well-known population- and family-based studies of long-lived individuals and should not be seen as a complete overview of all available studies.

consists of >1 800 centenarians (www.bumc.bu.edu/centenarian/) [8]. In addition to population-based studies focusing on unrelated individuals (i.e. singletons), several family-based studies have been established that focus on recruiting long-lived individuals and their (long-lived) siblings, often together with their offspring and the spouses thereof. In comparison to population-based studies, family-based studies often focus on 'younger' individuals, that is, those above 90 years of age (nonagenarians). Nonetheless, family-based studies have the advantage that they are likely enriched for genetic effects on longevity (on top of shared environmental effects). The family-based design allows for specific genetic analyses, such as linkage analysis, that are not possible using singletons. The most thoroughly investigated long-living families are those that are taking part in the Leiden Longevity Study (www.leidenlangleven.nl/) [9], comprised of nonagenarian siblings, and the Long Life Family Study (longlifefamilystudy.com/) [10], containing nonagenarians and their siblings. These studies are unique in their design and in the amount of data collected from both the long-lived individuals and their offspring.

In addition to these studies, the International Database on Longevity (IDL, www.supercentenarians.org/) keeps track of all (validated) individuals who have reached an age above 105 years or higher, so-called semi-supercentenarians (ages 105–109) and supercentenarians (ages >110 years), in Europe, Canada and the United States. This database is primarily used for demographic research in relation to mortality at the highest ages so it will not be discussed further in this chapter.

16.3 General Health and Morbidity of Long-Lived Individuals

One of the most striking observations made from analysing exceptionally long-lived individuals from different studies is the observed late onset of age-related diseases, including cancer, cardiovascular disease and type 2 diabetes, and of disabilities [6, 11]. In centenarians, the onset of age-related diseases is delayed by roughly 20 years, with fluctuations in this number resulting from differences in the gender of the studied individuals as well as methodological aspects of the studies. The same is observed in individuals from long-living families. It is clear that a decreased disease burden is transmittable to the next generation [12, 13]. The extent to which this impacts the heritability of longevity is still unknown, although it is evident from these studies that exceptionally long-lived individuals and individuals from long-living families are able to compress late-life morbidity (i.e. they spend a reduced proportion of time being diseased or disabled before dying). Although this does not rule out long-lived individuals developing age-related diseases – only a minority of long-lived individuals is able to complete escape them completely [14] – they have a longer lifespan free of disease, that is, an increased healthspan [12]. This characteristic makes such individuals and families especially interesting to study, considering that the mechanisms that allowed them to live a long and healthy life may offer clues on how to extend healthspan in individuals belonging to the general population.

16.4 Phenotypes in Offspring of Long-Lived Individuals

16.4.1 Methodological Issues

Many phenotypes change with age, yet it is challenging to study and quantify these changes in long-lived individuals because of the lack of an appropriate control group. Due to the nature of ageing, the perfect controls, that is, individuals from the same birth cohort who did not reach an exceptional old age, are often already deceased at the time of sample collection. To overcome this problem, one would require a large population that is studied over a very long time frame and to collect data before it becomes apparent which individuals will eventually become exceptionally long-lived. Since such studies are not (yet) available, researchers often turn to the middle-aged offspring of long-lived individuals; they can be compared to age-matched population controls and/or their partners. Although these offspring share only half of their genome with their long-lived parent, they also show a decreased prevalence of age-related diseases and a lower risk of mortality (e.g. [10, 13, 15, 16]). For example, the prevalence of stroke and heart diseases is decreased by (more than) twofold compared to age-matched controls. Moreover, their physical function and cognition seem to be improved (e.g. see [10, 17]). In the following subsections, we therefore decided to focus our attention on parameters that have been studied in offspring of long-lived individuals. We acknowledge that we thereby likely omit some parameters that may be relevant for longevity, such as the content of the gut microbiome [18].

16.4.2 Metabolic Parameters

Metabolites are the intermediates and end products of cellular regulatory processes. Examples of metabolites are amino acids, lipids and carbohydrates. Changes in the composition and amounts of different metabolites regulate the organism's response to the environment within the restrains of the organism's genetics. As blood is the most easily accessible tissue in human cohort studies, metabolites in the blood are the most widely studied. In comparison to the general population, the offspring of long-lived individuals show better metabolic health, as indicated by healthier levels of parameters involved in, among others, glycaemic control, and lipid and thyroid metabolism (Table 16.2). The phenotypes that are considered most robust are decreased levels of (fasting) glucose and triglycerides, increased levels of adiponectin, high-density lipoprotein (HDL) cholesterol and thyrotropin, and an increased low-density lipoprotein particle size, given that these have been observed in multiple independent populations. Other metabolic parameters, such as insulin-like growth factor 1 (IGF-1) levels, have also been reported to be altered in offspring of long-lived individuals, although findings between studies have been contradictory [19, 20, 21]. The observed differences are in line with prospective studies of individuals across the lifespan, which showed that these phenotypes (i.e. decreased (fasting) glucose and triglyceride levels and increased HDL cholesterol and thyrotropin levels) are associated with a decreased mortality risk and/or increased healthspan [22, 23]).

16.4.3 Immune-Related Parameters

Advanced age decreases the efficiency of the immune systems to mount an immune response upon exposure to antigens, which is termed immunosenescence. The critical role of the immune system in regulating an individual's health supports the notion that many age-related diseases, including immune disorders, cancer, type 2 diabetes and cardiovascular, neurodegenerative and chronic inflammatory diseases, originate from a dysregulation of the immune system with increasing age [24, 25]. Interestingly, the offspring of exceptionally long-lived individuals generally show a more youthful immune profile when compared to age-matched control groups (Table 16.2). By contrast, long-lived individuals also show an increase in pro- and anti-inflammatory cytokines [26], while these effects seem to be absent in their offspring [15, 27].

Table 16.2 Overview of metabolic phenotypes indicative of improved health in offspring of long-lived individuals

Metabolic phenotype	Effect in offspring of long-lived individuals	Link with health/disease	References
Glycaemic control			
Adiponectin	Increased levels	Higher levels are associated with a protective effect on the heart and prevention of age-related traits, such as insulin resistance and atherosclerosis.	[27, 121]
(Fasting) glucose	Decreased levels	Lower levels are associated with a decreased risk of developing type 2 diabetes.	[10, 122]
Lipid metabolism			
HDL cholesterol	Increased levels	Higher levels are associated with a decreased risk of developing cardiovascular disease.	[10, 123]
Triglycerides	Decreased levels	Lower levels are associated with a decreased risk of developing cardiovascular disease and atherosclerotic plaques.	[10, 124]
HDL particle size	Increased size	Larger particle sizes are associated with a decreased risk of developing cardiovascular disease and hypertension.	[10, 125]
Low-density lipoprotein particle size	Increased size	Larger particle sizes are associated with a decreased risk of developing cardiovascular disease and hypertension.	[10, 124]
Thyroid metabolism			
Thyrotropin (TSH)	Increased levels	Higher levels are associated with a decrease in oxidative stress and an increase in the tissue turnover rate.	[125, 126]
Free triiodothyronine (fT3)	Decreased levels	Lower levels are associated with lower blood pressure and a decreased risk of developing cardiovascular disease.	[127]
Free thyroxin (fT4)	Decreased levels	Extremely low levels are associated with hypothyroidism.	[15]
Immune response			
T-cell profile	More youthful profile	An accumulation of differentiated T cells causes a shift in the ratio of early and late stage T cells, reducing the immune system's capacity.	[128]
B-cell profile	More youthful profile	A higher percentage of naive B cells to differentiated memory B cell subtypes is critical in fighting off new infections.	[129, 130]
Immunosenescence	Decrease in hallmark features	A decrease in cell proliferation and an increase of late differentiated T cells reduce the response capacity of the immune system.	[131]
Autophagic activity (in T-cells)	Improved activity	Autophagy is critical for a cell to cope with stress and maintain cell homeostasis, mainly by degrading damaged proteins.	[132]

16.4.4 Omics-Based Profiles

Recent advances have made it possible to use omics-based approaches to identify molecular profiles indicative of longevity. Omics-based methodologies provide information on the entire set of biological components within a particular tissue, for example proteins, glycans or metabolites, while less technologically advanced methods normally only look at a narrow set of parameters. So far, these approaches have been used to look at differences in the glycome [28, 29], transcriptome [30, 31], proteome [32], metabolome [33, 34], and epigenome [35, 36, 37] of offspring of long-lived individuals. As expected, the offspring of long-lived individuals show a younger metabolic profile (reflected at multiple omics levels) in comparison to age-matched controls. They, for example, consistently show a lower epigenetic age, which is associated with a decreased mortality risk [38], and higher levels of ether phosphocholine species, indicating better antioxidant capacity and a healthier lipid profile. The same applies to the long-lived individuals themselves, that is, they often show an omics profile that matches that of a younger age group.

16.5 Genetics of Longevity

16.5.1 Heritability of Longevity

Survival to an exceptional old age is known to cluster in families, that is, first-degree relatives of long-lived individuals show a lifelong survival advantage compared to the general population [39, 40, 41, 42]. The presence of a genetic component contributing to longevity is further supported by the fact that Ashkenazi Jewish centenarians are not distinct in lifestyle factors, including obesity, alcohol consumption and smoking, compared to the general population [43]. However, the heritability of longevity has not yet been established, that is, only the heritability of related phenotypes, such as age at death, has been studied [44]. A recent study has shown that longevity, when defined as belonging to the 10% longest lived of a birth cohort, can be transmitted as a heritable trait [5]. This indicates that studying the genome of exceptionally long-lived individuals may enable the identification of the biological mechanisms underlying their longevity. To acquire a more accurate estimate of the heritability of longevity it will be crucial to establish extensive studies with exceptionally long-lived twins, ideally reared apart, to be able to separate genetic and shared environmental effects. However, such studies are not (yet) available.

16.5.2 Frequency of Disease Risk Alleles

The genome of long-lived individuals is assumed to contain a lower number of disease-associated genetic variants and/or a higher number of variants associated with protection against disease in comparison to individuals from the general population. However, multiple studies have shown that long-lived individuals have a similar burden of disease-associated variants, identified using genome-wide association studies (GWAS) [45, 46, 47], and carry specific disease risk alleles [48], indicating that their longevity is primarily determined by variants promoting health. This concept is further supported by the fact that the few genomes of exceptionally long-lived individuals that have been sequenced, using whole-genome or exome sequencing approaches, also show the presence of many pathogenic variants [49, 50].

16.5.3 Candidate Gene Studies

The first genetic studies of longevity were performed using a candidate gene approach. These studies focused on common genetic variation (minor allele frequency (MAF) >1%) in genes that had previously been implicated in lifespan regulation in model organisms or age-related diseases in humans. The most prominent and well-replicated loci that were identified using this approach are *apolipoprotein E* (*APOE*) and *forkhead box O3* (*FOXO3*). The *APOE* gene contains two longevity-associated alleles, ApoE ε2, which has been associated with increased odds of becoming long-lived, and ApoE ε4, which has shown the opposite effect [51]. The same alleles have also been associated with an increased (ApoE ε4) or decreased (ApoE ε2) risk for developing cardiovascular and Alzheimer's disease [51], which could explain the observed effects on longevity. Genetic variation in *FOXO3* has consistently been associated with increased odds of becoming long-lived across a variety of populations [52, 53, 54], and the most robust variant in this gene, rs2802292, has been functionally validated [55]. This study showed that the presence of rs2802292 creates a binding site for heat shock transcription factor 1 (HSF1), which induces the expression of *FOXO3* in response to stress. In addition to these two well-replicated loci, candidate gene studies have identified many population-specific genetic variants (i.e. variants that could not be replicated between different studies). An overview of these variants is provided in the LongevityMap database (genomics.senescence.info/longevity/) [56]. This database currently contains 3 144 variants that are located in 884 genes (as of March 2024). However, this database is susceptible to change due to ongoing sequencing efforts and the resulting increases in sample size that will enable the identification of additional variants.

Besides studying single genetic variants, several studies looked at the combined effect of common genetic variation and epistatic interactions in specific pathways on longevity. These studies highlight a role for genetic variation in insulin/insulin-like growth factor 1 signalling (IIS), mammalian target of rapamycin (mTOR) signalling, telomere maintenance and DNA damage/repair in longevity [57, 58, 59, 60].

16.5.4 Linkage Analyses

When it became technologically possible, the focus of the field shifted to linkage analyses and GWAS. These hypothesis-free approaches do not focus on specific genetic variants, genes or pathways, but rather on genetic variation in the whole genome. In a linkage approach, one looks for regions in the genome that are shared between long-lived siblings and/or exceptionally long-lived family members within extended pedigrees. Most of the linkage studies for longevity have been small scale, that is, they only contained a maximum of a couple of hundred families, each with a small number of long-lived siblings. These studies identified several interesting genomic regions [61, 62, 63, 64, 65], but most of these turned out to be population-specific. The most comprehensive linkage study of longevity to date is that by Beekman and colleagues [66], which studied >2000 European sibpairs that are part of the Genetics of Healthy Aging (GEHA) study. This study found four linkage regions (on chromosome 14q11.2, 17q12–22, 19p13.3–13.11 and 19q13.11–13.32). The region on chromosome 19q13.11–13.32 could be explained by the presence of the ApoE ε2 and ApoE ε4 alleles. In contrast, the other regions could not be explained by the association of common genetic variants, implying a contribution of rare variants with unknown effect sizes (i.e. MAF <1%).

16.5.5 Genome-Wide Association Studies

To determine the effect of single common genetic variants on longevity, GWAS can be performed. The first GWAS of longevity started in about the mid-2000s. The initial studies had relatively small sample sizes and focused on up to 500000 genetic variants that were directly measured using genotyping arrays (e.g. [67, 68, 69, 70, 71 72]). After it became possible to predict the presence of missing variants based on reference panels, so-called imputation, the number of studied common genetic variants increased immensely (up to ~10 million). However, this increase in the number of studied genetic variants, as well as in the sample sizes of the performed studies, did not lead to a dramatic increase in the number of longevity-associated variants [52, 73, 74, 75, 76, 77, 78]. With an increase in the sample size, it was expected that many more common variants would be identified, as is the case for most age-related diseases and traits, but this has not (yet) happened. The only locus for which genetic variation has consistently been associated with longevity is *APOE*, which was already known from candidate gene studies. In addition, several other variants have been identified in large meta-analyses combining the data from different studies [74, 76, 77], that is, in or near *ubiquitin carboxyl-terminal hydrolase 42* (*USP42*), *transmembrane O-mannosyltransferase targeting cadherins 2* (*TMTC2*), *G protein-coupled receptor 78* (*GPR78*), *bone morphogenetic protein-binding endothelial cell precursor-derived regulator* (*BMPER*) and *transmembrane protein 43* (*TMEM43*). However, most of these variants could not be replicated across meta-analyses, likely due to differences in the genetic background (e.g. ancestry) and the phenotypic definitions used for long-lived cases and controls. GWAS in the earlier days mainly focused on using a single age cut-off selection criterion for long-lived individuals from different countries, that is, survival to an age above 85, 90 or 100 years. However, to better consider non-genetic effects influencing the longevity phenotype, more recent GWAS used country-, sex- and birth cohort-specific life tables to define the age threshold for cases and controls. Although this stringent selection of cases and controls resulted in a more homogeneous phenotype, it did not results in the identification of many new variants

[74, 77]. The continuous development of methods to better take into account demographic information will hopefully further improve the power of GWAS of longevity [79, 80, 81]. Altogether, these findings indicate that longevity is more likely determined by rare genetic variants that cannot be detected through GWAS.

16.5.6 Whole-Genome/Exome Sequencing

To detect rare genetic variants contributing to longevity, which are usually not analysed in GWAS due to their low frequency, whole-genome or exome sequencing of exceptionally long-lived individuals and long-living families can be applied. A disadvantage of studying rare variants is that they cannot be directly linked to the longevity of the individuals in which they were identified; that is, it is not possible to determine their causality in humans as each long-lived individual will carry multiple rare variants in different genes. Moreover, the low frequency of rare variants limits the power to associate them with a specific phenotype [82]. Hence, it is crucial to perform functional characterisation of such variants to provide robust evidence that they are indeed involved in the successful ageing of their carriers. Thus far, this approach has identified rare functional genetic variants in insulin-like growth factor 1 receptor (*IGF1R*) and *FOXO3* [20, 83, 84], which are both involved in IIS, and in *nuclear factor of kappa light polypeptide gene enhancer in B-cells inhibitor, alpha* (*NFKBIA*), *clusterin* (*CLU*) and protein kinase C eta (*PRKCH*)[85], which are involved in the protein kinase C and nuclear factor κB signalling pathways. In addition, a recent exome sequencing study showed that long-lived individuals show enrichment of rare, coding longevity-associated variants in certain pathways, including IIS. Some of these variants, specifically those in genes involved in wingless-related integration site signalling (a pathway that controls embryonic development and tissue homeostasis in adults), seem to protect the long-lived individuals against the damaging effects of ApoE ε4 [86]. This finding provides additional evidence supporting the theory that long-lived individuals are likely less influenced by age-related diseases than by the presence of rare protective genetic variants in their genome. Combining the data for published (e.g. [87, 88, 89, 90, 91, 92, 93, 94]) and ongoing sequencing studies of exceptionally long-lived individuals and long-living families will hopefully result in the identification of additional rare variants contributing to longevity.

16.6 Conclusion

Exceptionally long-lived individuals and long-living families seem to be characterized by a compression of age-related diseases and disabilities and improved physical and cognitive functioning. Metabolically, this is reflected by a favourable immune-metabolic profile that is the opposite of metabolic syndrome, which could explain how some of the metabolic phenotypes are linked (Figure 16.2). The main processes influenced seem to be glycaemic control, lipid and thyroid metabolism, and immunity. The recent development in omics-based measurements will hopefully improve the identification

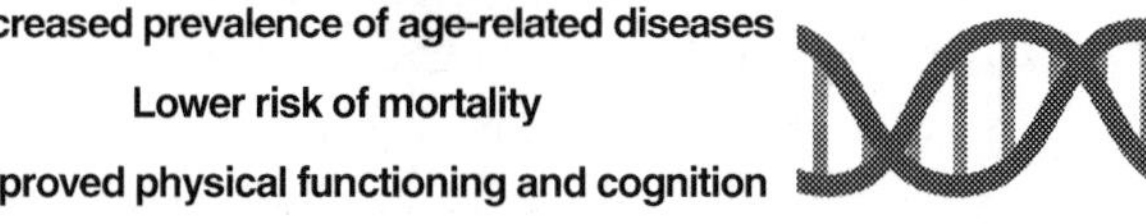

Figure 16.2 Summary of general health and morbidity-related phenotypes, metabolic and immune-related parameters and profiles, and genetic factors linked to longevity. Created with BioRender.com.

of the exact biological mechanisms that underlie longevity [95]. Most of the metabolic phenotypes associated with increased longevity are in line with what is observed in studies that look at so-called biomarkers of ageing in population-based cohorts, including prospective studies [22, 96, 97]. This indicates that the metabolic phenotypes that mark long-lived individuals and their families can be used to inform clinicians about the general health status of their patients. These findings have already resulted in a shift in the consensus on the treatment of age-related disease, that is, instead of treating each disease separately we should target specific metabolic phenotypes to treat multiple age-related diseases, including multi-morbidity, at once [98]. Moreover, some of these observed metabolic phenotypes are already used as biomarkers to estimate the effectiveness of specific lifestyle interventions [99, 100].

From a genetic point of view, it seems likely that longevity is mainly explained by many rare variants that each have a small compound additive effect on the phenotype. The burden of variants making an individual susceptible to age-related diseases is the same in long-lived individuals compared to normal ageing individuals, yet the former show a significantly improved healthspan. This discrepancy indicates that the genome of these long-lived individuals likely contains variants that not only promote longevity, but additively offer protection against disease, making these individuals healthier throughout their lives. This is coherent with a recent evolutionary study showing that natural selection may purify deleterious variants involved in susceptibility to late-onset diseases throughout adult life (even up to old ages) as well as favour the increase in frequency of variants promoting longevity [101]. Large GWAS successfully identified a handful of common genetic variants related to longevity (Figure 16.2), but they only explain a minor fraction of the heritable component. The recent advances that resulted in affordable sequencing make it possible to identify rare variants, which can subsequently be studied using functional approaches. The first studies using this approach have indicated that IIS/mTOR signalling plays a crucial role [20, 83, 84, 86], which is in line with studies that have been performed in model organisms [102]. Moreover, this agrees with the mentioned improved glycaemic control and decreased incidence of type 2 diabetes observed in the offspring of long-lived individuals (Figure 16.2). The link between the other metabolic phenotypes observed in long-lived individuals and their families and genetically encoded mechanisms is less clear. Yet we expect that future genetic studies of long-lived individuals, using whole genome/exome sequencing approaches, will be able to elucidate these missing links.

In conclusion, the findings presented in this chapter indicate that by targeting ageing we may be able to prevent or delay the onset of different diseases and disabilities. Thus, we should try to mimic the immune-metabolic profile of exceptionally long-lived individuals and long-living families using lifestyle and/or pharmacological interventions that, for example, target the genes/pathways that harbour rare protective genetic variants identified in long-lived individuals. In line with this idea, research in model organisms has already shown that such interventions, especially those targeting IIS/mTOR signalling, successfully extend both lifespan and healthspan in many different model organisms [103, 104]. The first intervention studies and clinical trials in humans applying lifestyle and/or pharmacological interventions, aiming to mimic the immune-metabolic profile observed in exceptionally long-lived individuals and long-living families and thereby improve healthspan, are ongoing [100, 105, 106, 107].

References

1. Oeppen, J., Vaupel, J.W. 2002. Demography: broken limits to life expectancy. *Science* **296**, 1029–1031.
2. Woolf, S.H., Schoomaker, H. 2019. Life expectancy and mortality rates in the United States, 1959–2017. *JAMA* **322**, 1996–2016.
3. Crimmins, E.M. 2015. Lifespan and healthspan: past, present, and promise. *Gerontologist* **55**, 901–911.
4. Fries, J.F. 2005. The compression of morbidity. *Milbank Q.* **83**, 801–823.
5. Van den Berg, N., Rodriguez-Girondo, M., Dijk, I.K., Mourits, R.J., Mandemakers, K., Janssens, A. 2019. Longevity defined as top 10% survivors and beyond is transmitted as a quantitative genetic trait. *Nat. Commun.* **10**, 35.
6. Andersen, S.L., Sebastiani, P., Dworkis, D.A., Feldman, L., Perls, T.T. 2012. Health span approximates life span among many supercentenarians: compression of morbidity at the approximate limit of life span. *J. Gerontol. Biol. Sci. Med. Sci.* **67**, 395–405.
7. Sanabe, E., Ashitomi, I., Suzuki, M. 1977. Social and medical survey of centenarians. *Okinawa J. Pub. Health* **9**, 98–106.

8. Perls, T.T., Bochen, K., Freeman, M., Alpert, L., Silver, M.H. 1999. Validity of reported age and centenarian prevalence in New England. *Age Ageing* **28**, 193–197.
9. Schoenmaker, M., Craen, A.J., Meijer, P.H., Beekman, M., Blauw, G.J., Slagboom, P.E. 2006. Evidence of genetic enrichment for exceptional survival using a family approach: the Leiden Longevity Study. *Eur. J. Hum. Genet.* **14**, 79–84.
10. Newman, A.B., Glynn, N.W., Taylor, C.A., Sebastiani, P., Perls, T.T., Mayeux, R. 2011. Health and function of participants in the Long Life Family Study: a comparison with other cohorts. *Aging* **3**, 63–76.
11. Ismail, K., Nussbaum, L., Sebastiani, P., Andersen, S., Perls, T., Barzilai, N. 2016. Compression of morbidity is observed across cohorts with exceptional longevity. *J. Am. Geriatr. Soc.* **64**, 1583–1591.
12. Sebastiani, P., Sun, F.X., Andersen, S.L., Lee, J.H., Wojczynski, M.K., Sanders, J.L. 2013. Families enriched for exceptional longevity also have increased health-span: Findings from the long life family study. *Front Public Health* **1**, 38.
13. Westendorp, R.G., Heemst, D., Rozing, M.P., Frolich, M., Mooijaart, S.P., Blauw, G.J. 2009. Nonagenarian siblings and their offspring display lower risk of mortality and morbidity than sporadic nonagenarians: The Leiden Longevity Study. *J. Am. Geriatr. Soc.* **57**, 1634–1637.
14. Evert, J., Lawler, E., Bogan, H., Perls, T. 2003. Morbidity profiles of centenarians: survivors, delayers, and escapers. *J. Gerontol. Biol. Sci. Med. Sci.* **58**, 232–237.
15. Bucci, L., Ostan, R., Cevenini, E., Pini, E., Scurti, M., Vitale, G. 2016. Centenarians' offspring as a model of healthy aging: a reappraisal of the data on Italian subjects and a comprehensive overview. *Aging* **8**, 510–519.
16. Terry, D.F., Wilcox, M.A., McCormick, M.A., Pennington, J.Y., Schoenhofen, E.A., Andersen, S.L. 2004. Lower all-cause, cardiovascular, and cancer mortality in centenarians' offspring. *J. Am. Geriatr. Soc.* **52**, 2074–2076.
17. Andersen, S.L., Sweigart, B., Sebastiani, P., Drury, J., Sidlowski, S., Perls, T.T. 2019. Reduced prevalence and incidence of cognitive impairment among centenarian offspring. *J. Gerontol. Biol. Sci. Med. Sci.* **74**, 108–113.
18. Sato, Y., Atarashi, K., Plichta, D.R., Arai, Y., Sasajima, S., Kearney, S.M. 2021. Novel bile acid biosynthetic pathways are enriched in the microbiome of centenarians. *Nature* **599**, 458–464.
19. Rozing, M.P., Westendorp, R.G., Frolich, M., Craen, A.J., Beekman, M., Heijmans, B.T. 2009. Human insulin/IGF-1 and familial longevity at middle age. *Aging* **1**, 714–722.
20. Suh, Y., Atzmon, G., Cho, M.O., Hwang, D., Liu, B., Leahy, D.J. 2008. Functionally significant insulin-like growth factor I receptor mutations in centenarians. *Proc. Natl. Acad. Sci. U. A* **105**, 3438–3442.
21. Vitale, G., Brugts, M.P., Ogliari, G., Castaldi, D., Fatti, L.M., Varewijck, A.J. 2012. Low circulating IGF-I bioactivity is associated with human longevity: findings in centenarians' offspring. *Aging* **4**, 580–589.
22. Deelen, J., Kettunen, J., Fischer, K., Spek, A., Trompet, S., Kastenmuller, G. 2019. A metabolic profile of all-cause mortality risk identified in an observational study of 44,168 individuals. *Nat. Commun.* **10**, 3346.
23. Li, X., Ploner, A., Wang, Y., Zhan, Y., Pedersen, N.L., Magnusson, P.K. 2021. Clinical biomarkers and associations with healthspan and lifespan: Evidence from observational and genetic data. *EBioMedicine* **66**, 103318.
24. Barbe-Tuana, F., Funchal, G., Schmitz, C.R.R., Maurmann, R.M., Bauer, M.E. 2020. The interplay between immunosenescence and age-related diseases. *Semin. Immunopathol.* **42**, 545–557.

25. Fulop, T., Larbi, A., Dupuis, G., Le Page, A., Frost, E.H., Cohen, A.A. 2017. Immunosenescence and inflamm-aging as two sides of the same coin: friends or foes? *Front Immunol.* **8**, 1960.
26. Rea, I.M., Gibson, D.S., McGilligan, V., McNerlan, S.E., Alexander, H.D., Ross, O.A. 2018. Age and age-related diseases: role of inflammation triggers and cytokines. *Front Immunol.* **9**, 586.
27. Deelen, J., Akker, E.B., Trompet, S., Heemst, D., Mooijaart, S.P., Slagboom, P.E. 2016. Employing biomarkers of healthy ageing for leveraging genetic studies into human longevity. *Exp. Gerontol.* **82**, 166–174.
28. Ruhaak, L.R., Uh, H.W., Beekman, M., Hokke, C.H., Westendorp, R.G., Houwing-Duistermaat, J. 2011. Plasma protein N-glycan profiles are associated with calendar age, familial longevity and health. *J. Proteome Res.* **10**, 1667–1674.
29. Ruhaak, L.R., Uh, H.W., Beekman, M., Koeleman, C.A., Hokke, C.H., Westendorp, R.G. 2010. Decreased levels of bisecting GlcNAc glycoforms of IgG are associated with human longevity. *PLoS ONE* **5**, 12566.
30. Passtoors, W.M., Boer, J.M., Goeman, J.J., Akker, E.B., Deelen, J., Zwaan, B.J. 2012. Transcriptional profiling of human familial longevity indicates a role for ASF1A and IL7R. *PLoS ONE* **7**, 27759.
31. Xiao, F.H., Chen, X.Q., Yu, Q., Ye, Y., Liu, Y.W., Yan, D. 2018. Transcriptome evidence reveals enhanced autophagy-lysosomal function in centenarians. *Genome Res.* **28**, 1601–1610.
32. Sebastiani, P., Federico, A., Morris, M., Gurinovich, A., Tanaka, T., Chandler, K.B. 2021. Protein signatures of centenarians and their offspring suggest centenarians age slower than other humans. *Aging Cell* **20**, 13290.
33. Collino, S., Montoliu, I., Martin, F.P., Scherer, M., Mari, D., Salvioli, S. 2013. Metabolic signatures of extreme longevity in northern Italian centenarians reveal a complex remodeling of lipids, amino acids, and gut microbiota metabolism. *PLoS ONE* **8**, 56564.
34. Gonzalez-Covarrubias, V., Beekman, M., Uh, H.W., Dane, A., Troost, J., Paliukhovich, I. 2013. Lipidomics of familial longevity. *Aging Cell* **12**, 426–434.
35. Gentilini, D., Mari, D., Castaldi, D., Remondini, D., Ogliari, G., Ostan, R. 2013. Role of epigenetics in human aging and longevity: genome-wide DNA methylation profile in centenarians and centenarians' offspring. *Age* **35**, 1961–1973.
36. Gutman, D., Rivkin, E., Fadida, A., Sharvit, L., Hermush, V., Rubin, E. 2020. Exceptionally long-lived individuals (ELLI) demonstrate slower aging calculated by DNA methylation clocks as possible modulators for healthy longevity. *Int. J. Mol. Sci.* **21**, 615.
37. Horvath, S., Pirazzini, C., Bacalini, M.G., Gentilini, D., Blasio, A.M., Delledonne, M. 2015. *Aging.* **7**, 1159–1170.
38. Marioni, R.E., Shah, S., McRae, A.F., Chen, B.H., Colicino, E., Harris, S.E. 2015. DNA methylation age of blood predicts all-cause mortality in later life. *Genome Biol.* **16**, 25.
39. Van den Berg, D., Rodriguez-Girondo, M., Craen, A.J.M., Houwing-Duistermaat, J.J., Beekman, M., Slagboom, P.E. 2018. Longevity around the turn of the 20th century: Life-long sustained survival advantage for parents of Today's Nonagenarians. *J. Gerontol. Biol. Sci. Med. Sci.* **73**, 1295–1302.
40. Pedersen, J.K., Elo, I.T., Schupf, N., Perls, T.T., Stallard, E., Yashin, A.I. 2017. The survival of spouses marrying into longevity-enriched families. *J. Gerontol. Biol. Sci. Med. Sci.* **72**, 109–114.

41. Perls, T.T., Wilmoth, J., Levenson, R., Drinkwater, M., Cohen, M., Bogan, H. 2002. Life-long sustained mortality advantage of siblings of centenarians. *Proc. Natl. Acad. Sci. U. A* **99**, 8442–8447.
42. Schoenmaker, M., Craen, A.J., Meijer, P.H., Beekman, M., Blauw, G.J., Slagboom, P.E. 2006. Evidence of genetic enrichment for exceptional survival using a family approach: the Leiden Longevity Study. *Eur. J. Hum. Genet.* **14**, 79–84.
43. Rajpathak, S.N., Liu, Y., Ben-David, O., Reddy, S., Atzmon, G., Crandall, J. 2011. Lifestyle factors of people with exceptional longevity. *J. Am. Geriatr. Soc.* **59**, 1509–1512.
44. Van den Berg, N., Beekman, M., Smith, K.R., Janssens, A., Slagboom, P.E. 2017. Historical demography and longevity genetics: Back to the future. *Ageing Res. Rev.* **38**, 28–39.
45. Beekman, M., Nederstigt, C., Suchiman, H.E., Kremer, D., Breggen, R., Lakenberg, N. 2010. Genome-wide association study (GWAS)-identified disease risk alleles do not compromise human longevity. *Proc. Natl. Acad. Sci. U. A* **107**, 18046–18049.
46. Gutman, D., Lidzbarsky, G., Milman, S., Gao, T., Sin-Chan, P., Gonzaga-Jauregui, C. 2020. Similar burden of pathogenic coding variants in exceptionally long-lived individuals and individuals without exceptional longevity. *Aging Cell* **19**, 13216.
47. Stevenson, M., Bae, H., Schupf, N., Andersen, S., Zhang, Q., Perls, T. 2015. Burden of disease variants in participants of the Long Life Family Study. *Aging* **7**, 123–132.
48. Bonafe, M., Olivieri, F., Mari, D., Baggio, G., Mattace, R., Sansoni, P. 1999. p53 variants predisposing to cancer are present in healthy centenarians. *Am. J. Hum. Genet.* **64**, 292–295.
49. Garagnani, P., Marquis, J., Delledonne, M., Pirazzini, C., Marasco, E., Kwiatkowska, K.M. 2021. Whole-genome sequencing analysis of semi-supercentenarians. *eLife* **10**, e57849.
50. Lin, J.-R., Sin-Chan, P., Napolioni, V., Torres, G.G., Mitra, J., Zhang, Q. 2021. Rare genetic coding variants associated with human longevity and protection against age-related diseases. *Nat. Aging* **1**, 783–794.
51. Mahley, R.W., Rall, S.C.J. 2000. Apolipoprotein E: far more than a lipid transport protein. *Annu. Rev. Genomics Hum. Genet.* **1**, 507–537.
52. Broer, L., Buchman, A.S., Deelen, J., Evans, D.S., Faul, J.D., Lunetta, K.L. 2015. GWAS of longevity in CHARGE consortium confirms APOE and FOXO3 candidacy. *J. Gerontol. Biol. Sci. Med. Sci.* **70**, 110–118.
53. Flachsbart, F., Caliebe, A., Kleindorp, R., Blanche, H., Eller-Eberstein, H., Nikolaus, S. 2009. Association of FOXO3A variation with human longevity confirmed in German centenarians. *Proc. Natl. Acad. Sci. U. A* **106**, 2700–2705.
54. Willcox, B.J., Donlon, T.A., He, Q., Chen, R., Grove, J.S., Yano, K. 2008. FOXO3A genotype is strongly associated with human longevity. *Proc. Natl. Acad. Sci. U. A* **105**, 13987–13982.
55. Grossi, V., Forte, G., Sanese, P., Peserico, A., Tezil, T., Lepore Signorile, M. 2018. The longevity SNP rs2802292 uncovered: HSF1 activates stress-dependent expression of FOXO3 through an intronic enhancer. *Nucleic Acids Res.* **46**, 5587–5600.
56. Budovsky, A., Craig, T., Wang, J., Tacutu, R., Csordas, A., Lourenco, J. 2013. LongevityMap: a database of human genetic variants associated with longevity. *Trends Genet.* **29**, 559–560.
57. Dato, S., Soerensen, M., Rango, F., Rose, G., Christensen, K., Christiansen, L. 2018. The genetic component of human longevity: New insights from the analysis of pathway-based SNP-SNP interactions. *Aging Cell* **17**, 12755.

58. Debrabant, B., Soerensen, M., Flachsbart, F., Dato, S., Mengel-From, J., Stevnsner, T. 2014. Human longevity and variation in DNA damage response and repair: study of the contribution of sub-processes using competitive gene-set analysis. *Eur. J. Hum. Genet.* **22**, 1131–1136.
59. Deelen, J., Uh, H.W., Monajemi, R., Heemst, D., Thijssen, P.E., Bohringer, S. 2013. Gene set analysis of GWAS data for human longevity highlights the relevance of the insulin/IGF-1 signaling and telomere maintenance pathways. *Age* **35**, 235–249.
60. Passtoors, W.M., Beekman, M., Deelen, J., Breggen, R., Maier, A.B., Guigas, B. 2013. Gene expression analysis of mTOR pathway: association with human longevity. *Aging Cell* **12**, 24–31.
61. Boyden, S.E., Kunkel, L.M. 2010. High-density genomewide linkage analysis of exceptional human longevity identifies multiple novel loci. *PLoS ONE* **5**, 12432.
62. Beekman, M., Blauw, G.J., Houwing-Duistermaat, J.J., Brandt, B.W., Westendorp, R.G., Slagboom, P.E. 2006. Chromosome 4q25, microsomal transfer protein gene, and human longevity: novel data and a meta-analysis of association studies. *J. Gerontol. Biol. Sci. Med. Sci.* **61**, 355–362.
63. Geesaman, B.J., Benson, E., Brewster, S.J., Kunkel, L.M., Blanche, H., Thomas, G. 2003. Haplotype-based identification of a microsomal transfer protein marker associated with the human lifespan. *Proc. Natl. Acad. Sci. U. A* **100**, 14115–14120.
64. Kerber, R.A., O'Brien, E., Boucher, K.M., Smith, K.R., Cawthon, R.M. 2012. A genome-wide study replicates linkage of 3p22-24 to extreme longevity in humans and identifies possible additional loci. *PLoS ONE* **7**, 34746.
65. Puca, A.A., Daly, M.J., Brewster, S.J., Matise, T.C., Barrett, J., Shea-Drinkwater, M. 2001. A genome-wide scan for linkage to human exceptional longevity identifies a locus on chromosome 4. *Proc. Natl. Acad. Sci. U. A* **98**, 10505–10508.
66. Beekman, M., Blanche, H., Perola, M., Hervonen, A., Bezrukov, V., Sikora, E. 2013. Genome-wide linkage analysis for human longevity: genetics of Healthy Aging Study. *Aging Cell* **12**, 184–193.
67. Deelen, J., Beekman, M., Uh, H.W., Helmer, Q., Kuningas, M., Christiansen, L. 2011. Genome-wide association study identifies a single major locus contributing to survival into old age; the APOE locus revisited. *Aging Cell* **10**, 686–698.
68. Flachsbart, F., Ellinghaus, D., Gentschew, L., Heinsen, F.A., Caliebe, A., Christiansen, L. 2016. Immunochip analysis identifies association of the RAD50/IL13 region with human longevity. *Aging Cell* **15**, 585–588.
69. Nebel, A., Kleindorp, R., Caliebe, A., Nothnagel, M., Blanche, H., Junge, O. 2011. A genome-wide association study confirms APOE as the major gene influencing survival in long-lived individuals. *Mech. Ageing Dev.* **132**, 324–330.
70. Newman, A.B., Walter, S., Lunetta, K.L., Garcia, M.E., Slagboom, P.E., Christensen, K. 2010. A meta-analysis of four genome-wide association studies of survival to age 90 years or older: the Cohorts for Heart and Aging Research in Genomic Epidemiology Consortium. *J. Gerontol. Biol. Sci. Med. Sci.* **65**, 478–487.
71. Sebastiani, P., Solovieff, N., Dewan, A.T., Walsh, K.M., Puca, A., Hartley, S.W. 2012. Genetic signatures of exceptional longevity in humans. *PLoS ONE* **7**, 29848.
72. Torres, G.G., Nygaard, M., Caliebe, A., Blanche, H., Chantalat, S., Galan, P. 2021. Exome-wide association study identifies FN3KRP and PGP as new candidate longevity genes. *J. Gerontol. Biol. Sci. Med. Sci.* **76**, 786-795.
73. Deelen, J., Beekman, M., Uh, H.W., Broer, L., Ayers, K.L., Tan, Q. 2014. Genome-wide association meta-analysis of human longevity identifies a novel locus conferring survival beyond 90 years of age. *Hum. Mol. Genet.* **23**, 4420–4432.

74. Deelen, J., Evans, D.S., Arking, D.E., Tesi, N., Nygaard, M., Liu, X. 2019. A meta-analysis of genome-wide association studies identifies multiple longevity genes. *Nat. Commun.* **10**, 3669.
75. Gurinovich, A., Song, Z., Zhang, W., Federico, A., Monti, S., Andersen, S.L. 2021. Effect of longevity genetic variants on the molecular aging rate. *Geroscience* **43**, 1237–1251.
76. Liu, X., Song, Z., Li, Y., Yao, Y., Fang, M., Bai, C. 2021. Integrated genetic analyses revealed novel human longevity loci and reduced risks of multiple diseases in a cohort study of 15,651 Chinese individuals. *Aging Cell* **20**, 13323.
77. Sebastiani, P., Gurinovich, A., Bae, H., Andersen, S., Malovini, A., Atzmon, G. 2017. Four genome-wide association studies identify new extreme longevity variants. *J. Gerontol. Biol. Sci. Med. Sci.* **72**, 1453–1464.
78. Zeng, Y., Nie, C., Min, J., Liu, X., Li, M., Chen, H. 2016. Novel loci and pathways significantly associated with longevity. *Sci. Rep.* **6**, 21243.
79. Rodriguez-Girondo, M., Berg, N., Hof, M.H., Beekman, M., Slagboom, E. 2021. Improved selection of participants in genetic longevity studies: family scores revisited. *BMC Med. Res. Methodol.* **21**, 7.
80. Sebastiani, P., Hadley, E.C., Province, M., Christensen, K., Rossi, W., Perls, T.T. 2009. A family longevity selection score: ranking sibships by their longevity, size, and availability for study. *Am. J. Epidemiol.* **170**, 1555–1562.
81. Yashin, A.I., Benedictis, G., Vaupel, J.W., Tan, Q., Andreev, K.F., Iachine, I.A. 1999. Genes, demography, and life span: the contribution of demographic data in genetic studies on aging and longevity. *Am. J. Hum. Genet.* **65**, 1178–1193.
82. Lee, S., Abecasis, G.R., Boehnke, M., Lin, X. 2014. Rare-variant association analysis: study designs and statistical tests. *Am. J. Hum. Genet.* **95**, 5–23.
83. Flachsbart, F., Dose, J., Gentschew, L., Geismann, C., Caliebe, A., Knecht, C. 2017 Identification and characterization of two functional variants in the human longevity gene FOXO3. *Nat Commun* **8**.
84. Tazearslan, C., Huang, J., Barzilai, N., Suh, Y. 2011. Impaired IGF1R signaling in cells expressing longevity-associated human IGF1R alleles. *Aging Cell* **10**, 551–554.
85. Ryu, S., Han, J., Norden-Krichmar, T.M., Zhang, Q., Lee, S., Zhang, Z. 2021. Genetic signature of human longevity in PKC and NF-kappaB signaling. *Aging Cell* **20**, 13362.
86. Lin, J.-R., Sin-Chan, P., Napolioni, V., Torres, G.G., Mitra, J., Zhang, Q. 2021. Rare genetic coding variants associated with human longevity and protection against age-related diseases. *Nat. Aging* **1**, 783–794.
87. Garagnani, P., Marquis, J., Delledonne, M., Pirazzini, C., Marasco, E., Kwiatkowska, K.M. 2021. Whole-genome sequencing analysis of semi-supercentenarians. *eLife* **10**.
88. Gierman, H.J., Fortney, K., Roach, J.C., Coles, N.S., Li, H., Glusman, G. 2014. Whole-genome sequencing of the world's oldest people. *PLoS ONE* **9**, 112430.
89. Gutman, D., Lidzbarsky, G., Milman, S., Gao, T., Sin-Chan, P., Gonzaga-Jauregui, C. 2020. Similar burden of pathogenic coding variants in exceptionally long-lived individuals and individuals without exceptional longevity. *Aging Cell* **19**, 13216.
90. Holstege, H., Pfeiffer, W., Sie, D., Hulsman, M., Nicholas, T.J., Lee, C.C. 2014. Somatic mutations found in the healthy blood compartment of a 115-yr-old woman demonstrate oligoclonal hematopoiesis. *Genome Res.* **24**, 733–742.
91. Nygaard, H.B., Erson-Omay, E.Z., Wu, X., Kent, B.A., Bernales, C.Q., Evans, D.M. 2019. Whole-exome sequencing of an exceptional longevity cohort. *J. Gerontol. Biol. Sci. Med. Sci.* **74**, 1386–1390.

92. Sebastiani, P., Riva, A., Montano, M., Pham, P., Torkamani, A., Scherba, E. 2011. Whole genome sequences of a male and female supercentenarian, ages greater than 114 years. *Front Genet.* **2**, 90.
93. Shen, S., Li, C., Xiao, L., Wang, X., Lv, H., Shi, Y. 2020. Whole-genome sequencing of Chinese centenarians reveals important genetic variants in aging WGS of centenarian for genetic analysis of aging. *Hum. Genomics* **14**, 23.
94. Van den Akker, E.B., Pitts, S.J., Deelen, J., Moed, M.H., Potluri, S., Rooij, J. 2016. Uncompromised 10-year survival of oldest old carrying somatic mutations in DNMT3A and TET2. *Blood* **127**, 1512–1515.
95. Ahadi, S., Zhou, W., Schussler-Fiorenza Rose, S.M., Sailani, M.R., Contrepois, K., Avina, M. 2020. Personal aging markers and ageotypes revealed by deep longitudinal profiling. *Nat. Med.* **26**, 83–90.
96. Kudryashova, K.S., Burka, K., Kulaga, A.Y., Vorobyeva, N.S., Kennedy, B.K. 2020. Aging biomarkers: From functional tests to multi-omics approaches. *Proteomics* **20**, 1900408.
97. Lara, J., Cooper, R., Nissan, J., Ginty, A.T., Khaw, K.T., Deary, I.J. 2015. A proposed panel of biomarkers of healthy ageing. *BMC Med.* **13**, 222.
98. Fabbri, E., Zoli, M., Gonzalez-Freire, M., Salive, M.E., Studenski, S.A., Ferrucci, L. 2015. Aging and multimorbidity: New tasks, priorities, and frontiers for integrated gerontological and clinical research. *J. Am. Med. Dir. Assoc.* **16**, 640–647.
99. Heilbronn, L.K., Jonge, L., Frisard, M.I., DeLany, J.P., Larson-Meyer, D.E., Rood, J. 2006. Effect of 6-month calorie restriction on biomarkers of longevity, metabolic adaptation, and oxidative stress in overweight individuals: a randomized controlled trial. *JAMA* **295**, 1539–1548.
100. Van de Rest, O., Schutte, B.A., Deelen, J., Stassen, S.A., Akker, E.B., Heemst, D. 2016. Metabolic effects of a 13-weeks lifestyle intervention in older adults: the Growing Old Together Study. *Aging* **8**, 111–126.
101. Pavard, S., Coste, C.F.D. 2021. Evolutionary demographic models reveal the strength of purifying selection on susceptibility alleles to late-onset diseases. *Nat. Ecol. Evol.* **5**, 392–400 (doi:10.1038/s41559-020-01355-2).
102. Fontana, L., Partridge, L., Longo, V.D. 2010. Extending healthy life span: from yeast to humans. *Science* **328**, 321–326.
103. Fontana, L., Partridge, L. 2015. Promoting health and longevity through diet: from model organisms to humans. *Cell* **161**, 106–118.
104. Johnson, S.C., Kaeberlein, M. 2016. Rapamycin in aging and disease: maximizing efficacy while minimizing side effects. *Oncotarget* **7**, 44876–44878.
105. Mannick, J.B., Del Giudice, G., Lattanzi, M., Valiante, N.M., Praestgaard, J., Huang, B. 2014. mTOR inhibition improves immune function in the elderly. *Sci. Transl. Med.* **6**, 268179.
106. Mannick, J.B., Morris, M., Hockey, H.P., Roma, G., Beibel, M., Kulmatycki, K. 2018. TORC1 inhibition enhances immune function and reduces infections in the elderly. *Sci. Transl. Med.* **10**, eaaq1564.
107. Most, J., Tosti, V., Redman, L.M., Fontana, L. 2017. Calorie restriction in humans: an update. *Ageing Res. Rev.* **39**, 36–45.
108. Holstege, H., Beker, N., Dijkstra, T., Pieterse, K., Wemmenhove, E., Schouten, K. 2018. The 100-plus Study of cognitively healthy centenarians: rationale, design and cohort description. *Eur. J. Epidemiol.* **33**, 1229–1249.
109. Rasmussen, S.H., Andersen-Ranberg, K., Thinggaard, M., Jeune, B., Skytthe, A., Christiansen, L. 2017. Cohort profile: the 1895, 1905, 1910 and 1915 Danish Birth Cohort

Studies – secular trends in the health and functioning of the very old. *Int. J. Epidemiol.* **46**, 1746–1746.
110. Frisoni, G.B., Louhija, J., Geroldi, C., Trabucchi, M. 2001. Longevity and the epsilon2 allele of apolipoprotein E: the Finnish Centenarians Study. *J. Gerontol. Biol. Sci. Med. Sci.* **56**, 75–78.
111. Robine, J.M., Cheung, S.L., Saito, Y., Jeune, B., Parker, M.G., Herrmann, F.R. 2010. Centenarians today: new insights on selection from the 5-COOP Study. *Curr. Gerontol. Geriatr. Res.*, **2010**, 120354.
112. Blanche, H., Cabanne, L., Sahbatou, M., Thomas, G. 2001. A study of French centenarians: are ACE and APOE associated with longevity? *C. R. Acad. Sci. III* **324**, 129–135.
113. Nebel, A., Croucher, P.J., Stiegeler, R., Nikolaus, S., Krawczak, M., Schreiber, S. 2005. No association between microsomal triglyceride transfer protein (MTP) haplotype and longevity in humans. *Proc. Natl. Acad. Sci. U. A* **102**, 7906–7909.
114. Arai, Y., Inagaki, H., Takayama, M., Abe, Y., Saito, Y., Takebayashi, T. 2014. Physical independence and mortality at the extreme limit of life span: supercentenarians study in Japan. *J. Gerontol. Biol. Sci. Med. Sci.* **69**, 486–494.
115. Barzilai, N., Gabriely, I., Gabriely, M., Iankowitz, N., Sorkin, J.D. 2001. Offspring of centenarians have a favorable lipid profile. *J. Am. Geriatr. Soc.* **49**, 76–79.
116. Anselmi, C.V., Malovini, A., Roncarati, R., Novelli, V., Villa, F., Condorelli, G. 2009. Association of the FOXO3A locus with extreme longevity in a southern Italian centenarian study. *Rejuvenation Res.* **12**, 95–104.
117. Samuelsson, S.M., Alfredson, B.B., Hagberg, B., Samuelsson, G., Nordbeck, B., Brun, A. 1997. The Swedish Centenarian Study: a multidisciplinary study of five consecutive cohorts at the age of 100. *Int. J. Aging Hum. Dev.* **45**, 223–253.
118. Gondo, Y., Hirose, N., Arai, Y., Inagaki, H., Masui, Y., Yamamura, K. 2006. Functional status of centenarians in Tokyo, Japan: developing better phenotypes of exceptional longevity. *J. Gerontol. Biol. Sci. Med. Sci.* **61**, 305–310.
119. De Rango, F., Dato, S., Bellizzi, D., Rose, G., Marzi, E., Cavallone, L. 2008. A novel sampling design to explore gene-longevity associations: the ECHA Study. *Eur. J. Hum. Genet.* **16**, 236–242.
120. Skytthe, A., Valensin, S., Jeune, B., Cevenini, E., Balard, F., Beekman, M. 2011. Design, recruitment, logistics, and data management of the GEHA (Genetics of Healthy Ageing) project. *Exp. Gerontol.* **46**, 934–945.
121. Atzmon, G., Pollin, T.I., Crandall, J., Tanner, K., Schechter, C.B., Scherer, P.E. 2008. Adiponectin levels and genotype: a potential regulator of life span in humans. *J. Gerontol. Biol. Sci. Med. Sci.* **63**, 447–453.
122. Wijsman, C.A., Rozing, M.P., Streefland, T.C., Cessie, S., Mooijaart, S.P., Slagboom, P.E. 2011. Familial longevity is marked by enhanced insulin sensitivity. *Aging Cell* **10**, 114–121.
123. Barzilai, N., Atzmon, G., Schechter, C., Schaefer, E.J., Cupples, A.L., Lipton, R. 2003. Unique lipoprotein phenotype and genotype associated with exceptional longevity. *JAMA* **290**, 2030–2040.
124. Vaarhorst, A.A., Beekman, M., Suchiman, E.H., Heemst, D., Houwing-Duistermaat, J.J., Westendorp, R.G. 2011. Lipid metabolism in long-lived families: the Leiden Longevity Study. *Age* **33**, 219–227.
125. Atzmon, G., Barzilai, N., Surks, M.I., Gabriely, I. 2009. Genetic predisposition to elevated serum thyrotropin is associated with exceptional longevity. *J. Clin. Endocrinol. Metab.* **94**, 4768–4775.

126. Jansen, S.W., Akintola, A.A., Roelfsema, F., Spoel, E., Cobbaert, C.M., Ballieux, B.E. 2015. Human longevity is characterised by high thyroid stimulating hormone secretion without altered energy metabolism. *Sci. Rep.* **5**, 11525.
127. Rozing, M.P., Westendorp, R.G., Craen, A.J., Frolich, M., Heijmans, B.T., Beekman, M. 2010. Low serum free triiodothyronine levels mark familial longevity: the Leiden Longevity Study. *J. Gerontol. Biol. Sci. Med. Sci.* **65**, 365–368.
128. Pellicano, M., Buffa, S., Goldeck, D., Bulati, M., Martorana, A., Caruso, C. 2014. Evidence for less marked potential signs of T-cell immunosenescence in centenarian offspring than in the general age-matched population. *J. Gerontol. Biol. Sci. Med. Sci.* **69**, 495–504.
129. Buffa, S., Pellicano, M., Bulati, M., Martorana, A., Goldeck, D., Caruso, C. 2013. A novel B cell population revealed by a CD38/CD24 gating strategy: CD38(-)CD24 (-) B cells in centenarian offspring and elderly people. *Age* **35**, 2009–2024.
130. Colonna-Romano, G., Buffa, S., Bulati, M., Candore, G., Lio, D., Pellicano, M. 2010. B cells compartment in centenarian offspring and old people. *Curr. Pharm. Des.* **16**, 604–608.
131. Derhovanessian, E., Maier, A.B., Beck, R., Jahn, G., Hahnel, K., Slagboom, P.E. 2010. Hallmark features of immunosenescence are absent in familial longevity. *J. Immunol.* **185**, 4618–4624.
132. Raz, Y., Guerrero-Ros, I., Maier, A., Slagboom, P.E., Atzmon, G., Barzilai, N. 2017. Activation-induced autophagy is preserved in CD4+ T-cells in familial longevity. *J. Gerontol. Biol. Sci. Med. Sci.* **72**, 1201–1206.

17 Human Populations with Extreme Longevities

Gianni Pes and Michel Poulain

17.1 Introduction

The interest in individuals with exceptional longevity, documented since antiquity, grew in the seventeenth and eighteenth centuries, reaching its peak in the nineteenth century. However, the potential causal factors of longevity began to attract interest only at the beginning of the twentieth century, when the hygienic, dietary and lifestyle factors that enabled a few exceptional individuals to potentially slow down the speed of the ageing process were considered. This research soon benefited from disciplines such as genetics, cultural anthropology, sociology and psychology, which tried to disentangle the seemingly complex set of factors that make it possible to maintain a good health state, even in old age [1], and, indirectly, lead to life extension. More recently, researchers, especially demographers, have questioned the validity of the age of the alleged exceptionally old centenarians who appeared worldwide at the end of the past century. Therefore, in the context of the development of the International Database on Longevity (IDL, see Chapters 14 and 15), strict rules for age validation were adopted, and the age of significantly older people was carefully checked as a prerequisite to any investigation aiming to identify potential extreme longevity factors. By contrast to individual longevity, the interest in long-lived populations arose in the early 1970s when the American doctor Alexander Leaf became known for his travel reports to countries renowned for the longevity of their inhabitants, such as the Vilcabamba valley in Ecuador [2]. In subsequent years, however, the alleged longevity of this community was invalidated by anthropologists who questioned the reliability of the demographic data [3]. Only the exceptional longevity recorded on the Japanese island of Okinawa has withstood criticism, owing to the support of a rigorous scientific methodology [4]. Thus, the scientific research aimed at identifying the existence of communities experiencing exceptional longevity is a relatively recent paradigm proposed mainly by demographers [5].

In this context, the discovery of well-documented, long-lived populations around the globe, later called *blue zones* (BZ) (Table 17.1), can be traced back to a seminar on longevity held in Montpellier, in October 1999, during which one of the authors of this chapter (G. Pes) presented, for the first time, data indicating the existence of a disproportionate number of centenarians in the male population of the highlands on the Mediterranean island of Sardinia [6]. This announcement attracted a lot of suspicious reactions from the audience of demographers, expressing the need for a thorough validation of the extreme ages of the oldest Sardinians. This stimulated the interest of the other author of this chapter (M. Poulain) and led to the establishment of a working plan to validate such exceptional longevity. From the demographers' viewpoint, the validation process of individual longevity was successful [7]. Even more interesting, during this investigation a new perspective emerged, as it was possible to confirm the existence of a relatively small and homogeneous geographical area in the east-central part of Sardinia, located between the historical subregions of Ogliastra and Barbagia, where the number of centenarians and the probability of reaching 100 years of age was demonstrably higher than elsewhere in Sardinia and mainland Italy [8]. The methods used for this characterization are summarized in Box 17.1. This area was initially called the 'blue zone' (BZ) due to the colour used to draw the first geographic map of the top long-lived villages.

Table 17.1 General characteristics of the four BZs

	Okinawa BZ	Ikaria BZ	Sardinia BZ	Nicoya BZ
Population	1 285 000	8 300	56 000	161 000
Survival probability[1]	11.4	6.0	13.5	4.5
Life expectancy at birth (years)	83.8	82.2	82.1	79.8
Male-to-female ratio at age 100	1:15	1:1.2	1:1	1:1.1
Traditional occupation	Agriculture	Animal husbandry	Animal husbandry	Forestry work
Climate	Subtropical [10]	Mediterranean [10]	Mediterranean [10]	Tropical [10]
Local income per capita	~$21 000 [10]	~$19 000 [10]	~$26 000 [10]	~$8 700 [10]
Daily energy intake per capita (kcal)	<2 000 [146]	<1 500 [131]	2 600 [84]	2 392 [147]
Smoking rate (%)	4.6	82 [131]	~20	40.3
Obesity rate (%, BMI ≥ 30 kg/m²)	10–40 [147]	12.5 [131]	9.8 [148]	23.6 [102]

[1] Probability of surviving between 1970 and 2000 for people who were aged 60–69 in 1970 and 90–99 in 2000 (based on census data).

Box 17.1 Identification of the Sardinian BZ

The shaping of the Sardinian BZ is based on information collected from death registers of all municipalities on the place of birth of all centenarians who died during the years 1980–2000 or were alive in 2001. The extreme population index (ELI) is computed by dividing these numbers of centenarians by the total number of newborns in each municipality, for the years 1880–1900.

To test the hypothesis that the geographical distribution of centenarians in Sardinia is non-random and that a peculiar geographical area could be identified where ELI is consistently and significantly higher than in the whole of Sardinia, we used a spatial deterministic model.

In practice, we estimated the difference between the number of centenarians born in the municipality and the expected theoretical number of centenarians for the same municipality, the latter being obtained by applying to the same number of newborns the average ELI for the whole Sardinian population (0.00219). The meaningful positive difference between specific municipalities and the whole of Sardinia were identified using the chi-square distance between theoretical and observed numbers of centenarians. Nevertheless, this test is valid assuming that the expected number of centenarians in the municipality is larger than five and the number of newborns is above 2283. None of the 377 Sardinian municipalities met these two conditions. To overcome this problem, we used a multi-scalar smoothing method based on Gaussian neighbourhood distribution that allows the study of the spatial concentration of centenarians as a continuum and is not limited by the division of the Sardinian territory into 377 municipalities (Figure 17.1). In the first

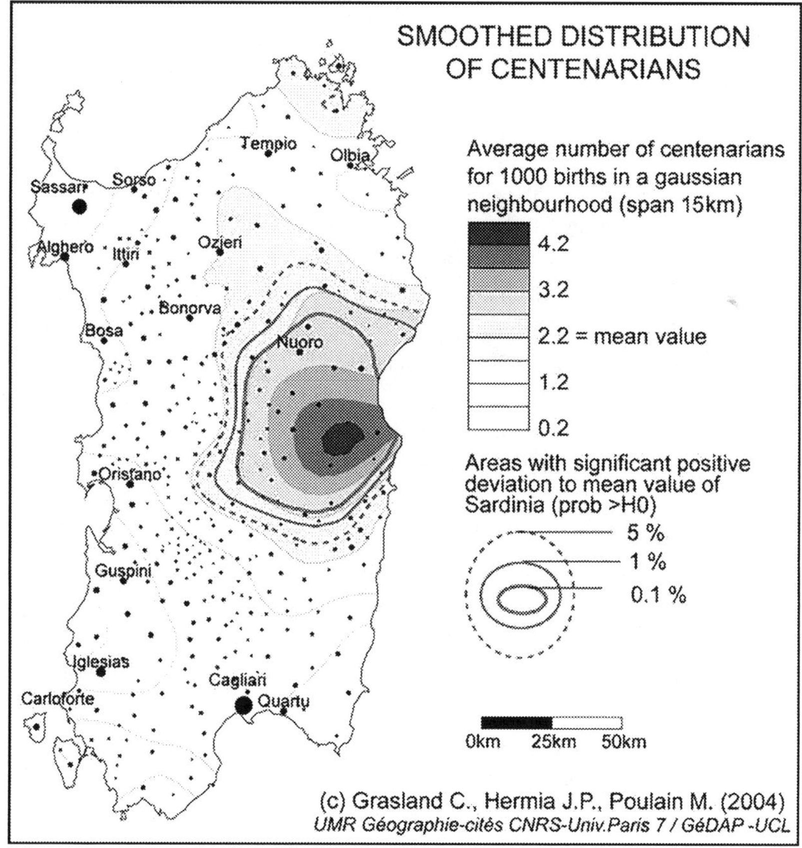

Figure 17.1 Map of Sardinia indicating the geographic distribution of the average number of centenarians, as well as isoclines if there are positive differences between the observed number of centenarians and the predicted numbers. Reprinted from [8]; copyright (2004) with permission from Elsevier.

step, we found that a 15 km Gaussian smoothing was large enough to capture the required minimum number of newborns without losing too much information on the spatial variation. Using this method, we produced a map where the concentration of centenarians is depicted as an area shaded in different colours, similar to the isobars in a weather map. On this map, the intensity of shades of grey is proportional to the value of ELI and the map appears continuous. Additionally, the black isoclines are related to a chi-square test of the assumption that the observed number of centenarians born in the neighbourhood of a point is not significantly higher than the expected value according to the average value for Sardinia (Figure 17.1). In other words, we verify if the number of births located in the 15 km neighbourhood and the number of centenarians born in the same neighbourhood are sufficiently high to reject this assumption.

The great advantages of the Gaussian smoothing method (compared to direct observation of municipality values or a simple aggregation of municipalities) are that the biases related to the shape and size of territorial divisions are reduced and it allows for the selection of the best compromise between a sample size of the neighbourhoods and variability of the spatial distribution.

In the following years, the term of BZ was popularized worldwide by American journalist Dan Buettner in an article published in November 2005 in *National Geographic* [9]. Besides the first BZs identified in Sardinia in 2000 and Okinawa in 2001, the second author of this chapter (M. Poulain) identified two other potential BZs in Costa Rica, the Nicoya Peninsula, and Greece, Ikaria Island. With the support of *National Geographic*, both authors participated in expeditions to the two places and validated these two additional long-lived BZs [10]. Due to their isolated locations, the genetic background and lifestyle of these populations are remarkably homogeneous, making it easier to gain better insight into human longevity determinants than from the more heterogeneous populations in developed countries. Therefore, investigating populations in the BZs may be regarded as a new and promising tool in studying human longevity. The comparative investigation of the different BZ populations allows us to test some hypotheses on longevity-promoting factors in communities which, despite differing from each other in several historical, anthropological, genetic and environmental aspects, maintain a striking internal homogeneity rarely found elsewhere. We have therefore endeavoured to offer an overview of the studies carried out since the early 2000s on the populations living in the BZs and to summarize the evidence for and against factors potentially affecting the longevity of these populations.

17.2 Genetic and Epigenetic Factors

Longevity is often assumed to be a heritable trait. Numerous studies conducted in model organisms, from yeast to primates, have suggested the possibility that lifespan is transmitted to offspring [11, 12, 13, 14]. In the case of humans, early studies on twins indicated that genes explain nearly 25% of lifespan variability [15]. Later, two studies conducted on family trees of over 13 million people have concluded that, after adjusting estimates to consider assortative mating, the heritability of human longevity does not exceed 10–16% [16, 17] (see also Chapter 16). Therefore, geneticists currently acknowledge the low heritability of the human lifespan [18, 19]. Nonetheless, the endless search for 'longevity genes' since the early 2000s, using sporadic and familial human longevity models, screened hundreds of genes (e.g. [20, 21, 22]). However, few results have been replicated in independent studies. The only two genes that proved to be 'robust' are the APOE gene and the FOXO3A gene [23, 24], both of which are involved in oxidative stress, inflammation, apoptosis and autophagy [24, 25, 26]. Lately, researchers have turned to the study of BZ populations in the hope that the search for population-wide genetic advantages there will be less altered than in more genetically heterogeneous populations [27, 28, 29, 30]. In general, the BZ populations have a long history of isolation due to geographical or cultural barriers [31], particularly evident in the case of Sardinia [32], which may have shaped their peculiar genetic makeup.

In the population of Okinawa, which is genetically more closely related to the Neolithic Jōmonese hunter-gatherers than to the Yamato Japanese widespread in the rest of Japan [33], several studies aimed at seeking longevity genes have been carried out, mainly addressing the human leukocyte antigens (HLA) system [34]. In 2008, Bradley John Willcox and colleagues, as part of the Hawaii Lifespan Study, provided evidence in favour of a role for the longevity of the FOXO3A gene, which is implicated in the insulin/insulin-like growth factor 1 (IGF-1) signalling pathway [24]. In particular, a variant of this gene (the rs2802292 G allele) protects the Okinawan male population against the shortening of telomeres, which are segments of DNA controlling a cell's entry into a senescence state [35, 36]. However, this association has not been confirmed in the populations of the Sardinia and Ikaria BZs [29], whereas its role in the population of the Nicoya BZ remains to be investigated. Therefore, the alleged pro-longevity effect of the FOXO3A gene appears to be exclusively limited to the Okinawa BZ population.

The other longevity-associated gene (APOE), involved in lipid transport, especially in the brain, has been investigated in the Sardinia and Ikaria BZs, where a slight non-significant reduction of the ε4 allele frequency was noted [29], already known for being deleterious to longevity in other populations [23]. This modest discrepancy can be attributed to an unbalanced allele frequency resulting from the small size of these populations and consequent genetic drift. So far, in the Nicoya BZ, the distribution of the APOE polymorphism has not been reported. A 2013 study investigated a different genetic trait, namely the length of telomeres (short strands of DNA that regulate cell replication) in relation to Nicoya longevity [37]. It revealed that Nicoyans have telomeres that are, on average, longer than other Costa Ricans, which could constitute a possible genetic basis for the exceptional survival of this community [30].

Overall, the impact of genetic factors on the longevity of the BZs is currently based on partial and unsystematic data, hampering the drawing of a definitive conclusion.

Therefore, future studies must be carefully planned, considering the singularity of these populations, the certain risk of underpowered statistics and the challenges related to the different genetic histories of these populations. For example, situations peculiar to a single BZ, such as the high rate of inbreeding and endogamy in Sardinia [38], admixture with Amerindian ancestry in Nicoya [39], and the smallness of the population of Ikaria [40, 41], should be carefully weighed.

Epigenetic research appears to be more promising. Evidence suggests that genomic DNA undergoes a methylation process throughout the lifespan, preserving a global memory of life-course exposures [42]. The analysis of the methylation pattern in several genomic sites allows the construction of a sort of epigenetic clock that accurately measures the biological age of an individual [43]. Therefore, it is now technically possible to evaluate the effect of various exposures (diet, physical activity and lifestyle) on the speed of biological ageing. Currently, these studies are ongoing in some of the BZs, and some results have already been published regarding the population of Nicoya. In particular, DNA methylation variability was found to be lower among Nicoyans than other Costa Ricans, suggesting a younger biological profile of the latter based on epigenetic drift [44]. In addition, regions of the genome were identified where differential methylation is greatest between the inhabitants of Nicoya and other Costa Ricans, providing the possibility of identifying novel DNA regions involved in longevity. Recently, DNA methylation-based predictors have been investigated in the Sardinia and Ikaria BZ, revealing that males of both communities are biologically older than females despite their equal survival to older age [45]. However, to what extent such methylation signatures can significantly impact the ageing phenotype remains an unanswered question.

17.3 Hormones

The abundant inter-species literature on hormones and ageing provides compelling evolutionary evidence implicating the endocrine system as a major regulator of ageing and longevity. However, these investigations are intrinsically difficult as the endocrine system is highly integrated, and the age-related alteration of one pathway is likely reflected in other distant hormonal pathways, making it difficult to differentiate between primary and secondary changes.

The hormones secreted by the pituitary gland affect the lifespan of mammals, including humans, as hypophysectomy research in rats has shown since the 1960s [46]. Research on animal models has established that the IGF-1 signalling pathway plays a crucial role in ageing and longevity [47, 48]. Laron's syndrome is a condition in which mutations in the growth hormone (GH) receptor gene result in a congenital deficiency of IGF-1 [49]. Despite evident metabolic changes, untreated patients are particularly resistant to the development of cancer [50]. These data support the concept that the GH/IGF-1 axis is part of an evolutionarily conserved network that regulates lifespan and longevity in many species. However, due to the complexity of the GH/IGF-1 system, data in humans are contradictory, with both beneficial and adverse effects of GH/IGF-1 downregulation reported [21, 51].

A rare loss-of-function mutation (R61X) of the GH receptor gene associated with Laron's syndrome and responsible for short stature has been identified in the Sardinia BZ [52]. However, it is not yet known whether it also entails increased longevity. Preliminary unpublished data reveal that plasma IGF-1 levels among the Sardinia BZ elderly are lower than those of middle-aged controls in both sexes (60.16 vs 184.5 ng/mL). However, whether this has causal significance or is simply due to age-dependent dysregulation of the hypothalamus–pituitary axis is uncertain.

Also, thyroid hormones fundamentally affect the growth and development of almost all tissues. They stimulate metabolic rate, lipolysis and protein catabolism, as well as central nervous and cardiovascular system activity. However, the long-term effect of thyroid hormones is likely an increased production of reactive oxygen species that ultimately damage the cell macromolecules and accelerate the ageing process. In animal models, the inhibition of thyroid function leads to life extension [53]. Mice carriers of gene variants associated with hypothyroidism generally have a longer lifespan than mice with normal thyroid hormone levels [54]. In humans, reduced thyroid function, usually as subclinical hypothyroidism revealed by high thyroid-stimulating hormone levels, is associated with increased longevity in Ashkenazi Jewish centenarians [55, 56].

In the Sardinia BZ, some historical data indicate a higher prevalence of thyroid hypofunction, probably derived from iodine deficiency, which can be cautiously interpreted as a contributor to longevity [57, 58]. In addition to Sardinia, this association was found in the other BZs as well. The Nicoya Peninsula, since the 1950s, has been considered a hotspot of endemic goitre [59], perhaps aggravated by an excess of calcium in drinking water [60]; even today, the inhabitants of this BZ show higher levels of thyroid-stimulating hormone (3.33 mU L for men and 2.61 mU/L for women) compared with the rest of the country [61], indicating a widespread condition of glandular hypofunction. Also, in the Ikaria BZ, iodine levels in drinking water are particularly low [62], which could explain the high prevalence of thyroid hypofunction on this island and in the North Aegean area in general [63]. The adverse health effect of high thyroid hormone levels is demonstrated in the population of Ikaria by the existence of a trend between thyroid-stimulating hormone levels and all-cause mortality after correction for possible confounders [64]. Interestingly, a non-linear relationship between thyroid hormone levels and psycho-affective status has been reported in the Sardinian BZ – individuals in the lowest and highest hormone quintiles were more depressed than those with values in the middle [65].

Sex hormones are steroid molecules involved in reproductive functions that show different blood levels in men and women. In men, the most important hormone is testosterone, produced mainly by the Leydig cells of the testicles, while in women it is oestrogen, produced by the ovaries. Both are involved in the susceptibility to age-related diseases and, indirectly, in longevity. The daily production of testosterone in men after age 40 tends to decrease progressively by 1% each year [66], reaching minimum values in old age; while in long-lived males, the levels remain elevated. A study conducted in the Ikaria BZ found that males with testosterone levels in the upper tertile had a lower prevalence of metabolic syndrome and likely a reduced cardiovascular risk [67]. A study in 70-year-old males from the Okinawa BZ demonstrated higher mean testosterone levels than American age-matched males, while no differences were found in females [68]. Similarly, a study that recruited centenarians from all around Sardinia reported that centenarian males have even greater total testosterone values than those of 60-year-old males and significantly higher than those of the centenarians of the Okinawa BZ [69]. In Nicoyan male centenarians, mean free testosterone levels were slightly lower than those recorded in Sardinia and Okinawa [61]. These data, taken together, indicate that, in general, among males in these BZs, androgen levels do not decline rapidly but remain in the range of younger elders. However, the direct role of these hormones in the longevity of BZ populations remains to be demonstrated, and the results of ongoing research in these areas are awaited.

Studies concerning the relationship between oestrogen and longevity are even more contentious than those concerning androgens. In mammals, including humans, females are known to live longer than males (see Chapter 9). It has been conjectured that females produce fewer free radicals than males, and oestrogen may enhance the antioxidant defence necessary for healthy ageing [70]. The oestradiol levels in a cohort of Sardinian female centenarians were comparable to those of younger women [69]. However, demographic studies suggest that the age of Sardinian women at the birth of their last child is significantly older than on the rest of the island [71], implying longer preserved oestrogenic function. Okinawan elderly males have significantly higher levels of oestradiol than Americans of the same age [68], a picture seemingly mirroring that of the elderly in the Sardinia BZ, where male centenarians had significantly higher oestradiol levels compared to middle-aged males [69]. It should be emphasized that oestrogen is produced in men through aromatization, a biochemical reaction that converts testosterone into oestradiol [72]. We must await the results of further endocrinological studies in the BZs to determine the precise significance of the increased ability of males from these areas to produce oestrogen hormones for a longer period of time.

17.4 Nutritional Factors

Diet is a prominent factor among the significant determinants that may prolong life. Taking a glimpse at the literature on this subject, one can be astonished by the myriad of foods, dietary behaviours and eating habits that are claimed to increase longevity and supported by more or less convincing scientific evidence [73]. The characteristics of the BZs' diet have been the subject of a recent publication [74], and in the following, we will focus mainly on the role of calorie restriction and the type of diet.

17.4.1 Calorie Restriction

In model organisms, a 30–60% reduction in energy intake without inducing malnutrition prolongs their lifespan. This fact was first described in 1934 [75], and it has since been confirmed by numerous studies [11, 76, 77]. The molecular underpinnings by which a lower energy intake expands lifespan are still not completely elucidated. However, it has been suggested that energy restriction induces a reconfiguration of energy metabolism [78], thereby abating overall free radical production in cells [79], a mechanism mediated by sirtuin-deacetylated transcription factors. These are responsible for the upregulation of key antioxidant enzymes [80]. Since the effects of caloric restriction are observed across the entire evolutionary history, from yeasts to non-human primates, it is assumed that it can also act on humans. In the Okinawa BZ, some studies seem to support the positive effect of calorie restriction on the metabolic state of individuals [81]. According to Willcox and colleagues [68], the Okinawan diet in pre-Second World War cohorts had characteristics very similar to the energy-restricted diet used in experimental calorie restriction models. These authors estimated caloric intake in septuagenarian Okinawans at less than 2 000 kcal per day (i.e. significantly lower than that of Americans of the same chronological age), which could explain why elderly Okinawans have shorter stature than other age-matched Japanese. However, the Okinawan cohorts after the Second World War were likely no longer subjected to an energy-restricted regime. A certain impact of calorie restriction has also been hypothesized for the inhabitants of the Sardinia BZ [82], which would explain their short stature [83]. However, a 1928 article reported that the average daily caloric intake of the Sardinian population at that time was 2 400 kcal per day, compared to a national average of 2 600 kcal per day, a difference of only 7% [84]. An article published in 1943 estimated the average daily caloric intake of Sardinian herders as 2 608 kcal per day, excluding alcoholic beverages, and 2 719 kcal per day with alcohol included [85]. These data seem to cast doubt on the existence of an overt calorie restriction in the Sardinia BZ population, although the traditional Sardinian diet has always been described as rather frugal [86]. In theory, the situation could be slightly different for the Ikaria BZ population. Many islanders are members of the Greek Orthodox Church, which formally prescribes up to 200 days of fasting each year, consisting of a diet based exclusively on fish and seafood [87]. However, a 2019 study found that at least 54.9% of residents of Ikaria do not follow Greek Orthodox fasting [88]. Therefore, the role played by calorie restriction on longevity is likely to be considered rather marginal, even for the population of Ikaria. The inhabitants of the Nicoya BZ consume slightly more calories than the general population of Costa Rica [89]; therefore, the diet of this community cannot be regarded as calorie-restricted. From a more general point of view, the hypothesis of a causal relationship between calorie restriction and longevity has been criticized [90, 91], and the most recent line of research is more focused on the potential life-prolonging effect of limiting the daily hours of food intake (time-restricted feeding) rather than implementing a global reduction of dietary energy [92].

17.4.2 Type of Diet

In general, the dietary regimes of the four BZs differ significantly, as one would expect given the different historical vicissitudes and cultural traditions of these populations [74]. However, the Mediterranean diet, a regime considered to be associated with a prolonged lifespan [93, 94], prevails in two out of the four BZs (Sardinia and Ikaria). This fuelled the idea that the Mediterranean diet, or some local variant, was (at least in part) responsible for their longevity. Sardinians and Ikarians, being rural populations, continue to consume large quantities of self-produced food, including bread and dairy products [95]. The role of a diet based primarily on plant-based foods in promoting longevity has been highlighted by popular publications on the BZs [96]. The consumption of fibre-rich whole grain products is associated with improved metabolic status and a reduced prevalence of chronic diseases, such as diabetes and cancer. In the Sardinia BZ, whole wheat bread was a substantial part of the staple diet of the population for a long time, with a daily consumption that could exceed 1 000 g [95]. Another basic food in Sardinia was fresh vegetable soup (minestrone with onion, fennel and carrots) and legumes (beans, broad beans and peas), as well as potatoes and pork broth in the mountainous area. However, it would be incorrect to claim that the diets in the Sardinia BZ and that of Ikaria are primarily 'vegetarian'. The food choices of the pastoral communities of Ogliastra and Ikaria were traditionally conditioned by the abundance of livestock products, although, in the traditional Ogliastra diet, the consumption of meat was moderate. Before the nutrition transition in the 1950s and 1960s, meat was consumed on average only two to four times a month, and protein intake depended mainly on dairy products [95]. After this period, the consumption of beef gradually increased as did, to a lesser extent, the consumption of fish, fruit and vegetables and olive oil [97]. The diets adopted by the two Mediterranean BZs diverged from the traditional Mediterranean diet in various ways but did not contradict the NHANES III (Third National Health and Nutrition Examination Survey) data by Levine et al., which suggest that a high intake of animal proteins after the age of 65 does not raise the mortality rate [98]. In both Mediterranean BZs, consumption of fish and seafood was surprisingly low. In Sardinia, it was traditionally limited almost exclusively to riverside villages [95]. Similarly, Ikarians preferred the practice of animal husbandry over fishing (historically they were wary of the sea and associated it with pirate attacks), and therefore, their fish consumption was lower than that of the other islands of the Aegean Sea [99]. Nonetheless, Chrysohoou and colleagues reported that fish consumption in Ikaria is associated with a lower prevalence of depression in the elderly [100].

With regard to the other BZs, a study on the eating habits of the centenarians from Nicoya reported only 28% with a good nutritional status [61]. A study on residents from this BZ between 90 and 109 years of age revealed that 74% ate fruit one to three times a day, and the remainder ate fruit two to six times a week [101]. The consumption of tubers (potatoes, sweet potatoes and cassava) ranged from two times a week to three times a day in 88% of the participants, which was comparable to Sardinia and Ikaria. Despite this, the carbohydrates consumed by Nicoyans in the diet had a low glycaemic index [88], which may have delayed the onset of metabolic disorders. Similarly to what is observed in other BZs, the consumption of dairy products was very high in Nicoya. In a survey, 79% of the participants consumed dairy products from one to three times a day, and a further 12% consumed dairy products two to six times a week [102]. Moreover, half of Nicoyans ate no less than three to five servings of meat per week, and at least a quarter ate meat almost every day. A comparison with the habits of Sardinians revealed that the most consumed type of meat was pork, albeit in moderation [74, 102]. Therefore, as in the two first BZs, the habits of the Nicoyans cannot be defined as purely vegetarian. The eating habits of Okinawans have been extensively discussed in various studies [68, 74, 103, 104], therefore, it is necessary to highlight that they differ significantly from those of the rest of Japan. It has been observed that throughout the Tokugawa era, from the sixteenth century to the mid-nineteenth century, the absence of conditioning by the Buddhist tradition meant that the diet contained a non-negligible consumption of meat, especially pork [103]. Even the traditional consumption of Kombu seaweed and Goya bitter gourds, which would play a decisive role in combating diabetes and atherosclerosis [105], can be attributed to cultural and historical influences.

17.5 Physical Activity

Undoubtedly, physical activity plays a prominent role among the modifiable factors associated with longer survival [106]. Various epidemiological studies have shown that regular exercise can improve motor fitness, reduce mortality and prolong survival [107, 108, 109, 110, 111]. BZ populations, without exception, are characterized by the practice of moderate to high levels of physical activity, even in old age, which can be partly explained by the fact that they are often rural communities, mainly agricultural, mostly resident in mountainous regions, entailing high levels of outdoor physical activity up to advanced age [112]. In the Sardinia BZ, before the advent of mechanization, most of the male population was engaged in seminomadic pastoralism, which involved long-distance walking. An ecological study highlighted a significant relationship between longevity and some variables assumed as proxies of physical activity, including: (1) traditional male employment (agricultural or pastoral), implying constant physical activity over time; (2) the average daily distance to reach the workplace on foot outside the village area; and (3) the average inclination of the land, obtained through the geographical data of the geographic information system (GIS), based on the hypothesis that walking on steep terrain involves, on average, a higher energy expenditure than if walking on flat ground [113]. Assessment of physical activity in a 90-year-old free-living cohort from the Sardinia BZ using wearable accelerometers revealed that moderate activity (≥3 metabolic equivalents, METs) exceeded 40% in most of the participants [114], when compared to the highest reported in 90-year-old cohorts [109, 111, 115]. Aside from the metabolic benefits, regular physical activity by elderly people in the Sardinia BZ appears to be associated with high levels of subjective psychological well-being, which is especially pronounced in men [116].

Physical activity has also been thoroughly investigated in the population of the Ikaria BZ. A study revealed that only 12% of men and 18% of women on the island fell within the range of low self-reported physical activity; moreover, physical activity conferred certain anti-arrhythmic protection, albeit limited to women [117, 118], and better endothelial function, especially in middle-aged subjects, and therefore, indirectly, protection from the premature onset of atherosclerosis [119]. Part of the physical activity consists of the social rituals of Panegyris [120], which includes a circular dance reminiscent of the traditional collective dance in the Sardinia BZ [121], as well as dancing in the other BZs. In Okinawa, agricultural activity is often integrated with traditional dance [81]. In the centenarians of the Nicoya Peninsula, a study reported that over 44% of respondents were very active, with an average gait speed of 13.04 seconds [61]. Regular, moderate physical activity until advanced age, common to all BZs, may have helped to improve the metabolic state and to delay the onset of chronic age-related diseases, as well as to improve mental health [122].

17.6 Mental Health

The relationship between individual mental health and longevity is particularly complex. Unlike biological and behavioural aspects, which are measurable with relative ease, an individual's mental world is more elusive and sometimes hardly quantifiable. The tools used by psychologists to evaluate the mental characteristics of individuals are often unreliable in the face of the frequent tendency of respondents to provide complaisant answers under the pressure of social desirability [123]. It is even more difficult to perform cross-cultural psychological comparisons due to the obvious heterogeneity of cultural factors across populations, including the average level of education, early life experiences, frequency of positive and negative life events, care received from a spouse or children and so on. Much literature has recently been produced on the relationship between an 'optimistic attitude' and longevity. In general, optimism is considered a psychological trait that promotes longevity [124, 125].

The psychology of the BZ inhabitants is often oriented in a positive sense. Their 'optimism' is counted among the causal factors for their longevity [126, 127, 128]. In the Sardinia BZ, the elderly have lower levels of depressive symptoms, fewer cognitive deficits and better working memory performance [129]. However, one study found that the average cognitive level of nonagenarians in the Sardinia BZ was lower than that of nonagenarian cohorts elsewhere [130], although this may have been biased by the low level of education typical of this population. The average psycho-affective status is very good in the oldest population of Ikaria (Geriatric Depression Scale, GDS, score 3.1 and 4.9 in men and women, respectively) [131], and the frequency of depression is lower than in age-matched peers from other Greek islands or Cyprus [132]. According to the data from the Costa Rican Longevity and Healthy Aging Study (CRELES), the inhabitants of the Nicoya BZ have, in general, fewer disabilities (physical and mental) compared to other Costa Ricans [89], in whom the burden of mental disability is considerable [133]. In Okinawa BZ, mental health may depend on a particular cultural tradition. Alongside 'Western' psychiatry, there is still a strong shamanic component, called *yuta* [134]. Many inhabitants of the island, therefore, seek spiritual comfort from this traditional approach, which is still thriving nowadays. In addition, shamans have renewed themselves by adopting a modern style of communication. There is no direct evidence that this approach is a significant element of Okinawan longevity, but it may have reinforced the resilience and emotional balance of many islanders [135].

However, there are also psychologically negative aspects to some of the BZ populations. Suicide is common in the BZ of Sardinia and Okinawa [136, 137, 138]. It primarily affects rural communities and appears to be the result of the recent perception by the oldest inhabitants that the traditional identity of the community is threatened by overwhelming powers and a strong sense of cultural isolation [137]. In any case, it is a recent phenomenon that was likely rare in the past. Also, the women living in Nicoya, but not men, have a higher prevalence of geriatric depression than their matched peers in the rest of the country, for which an explanation is currently lacking [89]. It is clear that the inhabitants of the BZs are not exempt from elements that may potentially cause mental disturbances, they are only able to cope with them by maintaining a remarkable level of autonomy through individual and social resources [139, 140].

17.7 Conclusion

The emergence of long-lived populations scattered across three of the four continents of the world poses a hard challenge to epidemiologists and demographers who try to discover the underlying factors of their existence. The close link between economic well-being and longevity was repeatedly underlined until the end of the last century by scholars such as Albert Evans, with his theory of the social gradient of life expectancy, according to which the longevity of a given population would grow in direct proportion to its economic prosperity, owing to the reduction of anxiety and stress levels [141]. In reality, it is precisely the study of the BZs that seems to question this conceptual framework and rather suggest that a certain degree of underdevelopment linked to a modernization process that began later and took place more slowly, as well as the relative maintenance of traditional lifestyles and eating habits, may have favoured longevity, as William C. Cockerham aptly argued for the Okinawa case [142, 143]. This probably allowed these populations not only to achieve relative well-being without the trauma of the forced industrialization of the early twentieth century in most Western populations but also decreased social class disparities, sources of competition and individual stress [144]. This is evidenced by the popularity of the Panegyris, a traditional festival of Ikaria, through which the inhabitants reaffirm their cultural identity and their belonging to the community beyond social differences [117]. These aspects have created favourable health conditions in large sections of the population that were unavailable in the more competitive context of industrialized countries, both in the West and in the East. In any case, it is clear that BZs did not exist in the past but are likely the result of a 'longevity transition' that appears driven by a combination of contextual transitions in environmental, nutritional and behavioural factors. The latter was fuelled by the strong economic transition around the middle of the last century, which heralded the advent of our postmodern society. Another crucial aspect is that, in various respects, these populations have benefited from some peculiar aspects of modernity, as is clear from both the quick improvement of access to better health services and the evolution of their diet, which results from the combination of traditional foods and new opportunities brought about by the epidemiological transition, which are not necessarily negative. The condition of long isolation for geographical or cultural reasons to which these populations have been exposed has deeply affected the psychology of the members of these communities, nourishing in them a strong perception of being different from their neighbours, and consequently strengthening their attachment to traditional social norms with a clear defensive significance. The 'premodern' trait of almost sacred respect for the elderly within the community follows the same vein and may represent one of the reasons why the elderly survive longer, owing to the attention shown to them by family members and, in general, by the social clan [130]. Poised between modernity and postmodernity, these BZs appear as intrinsically fragile communities, likely unable to withstand the further changes imposed by societal evolution. Nonetheless, as long as they last, they deserve to be regarded as a great asset for the study of human longevity from a community point of view and may teach our post-industrial societies how to capitalize on the opportunities presented by the acquired new degree of longevity [145].

References

1. Rowe, J.W., Kahn, R.L. 1987. Human aging: usual and successful. *Science* **237**, 143–149 (doi:10.1126/science.3299702).
2. Leaf, A. 1975. *Youth in Old Age*. McGraw-Hill.
3. Mazess, R.B., Forman, S.H. 1979. Longevity and age exaggeration in Vilcabamba, Ecuador. *J. Gerontol.* **34**, 94–98 (doi:10.1093/geronj/34.1.94).
4. Suzuki, M., Mori, H., Asato, T., Sakugawa, H., Ishii, T., Hosoda, Y. 1985. Medical researches upon centenarians in Okinawa (1): case controlled study of family history as hereditary influence on longevity. *Nihon Ronen Igakkai Zasshi* **22**, 457–467 (doi:10.3143/geriatrics.22.457).
5. Poulain, M. 2019. Individual longevity versus population longevity. In *Centenarians: An Example of Positive Biology* (ed. C. Caruso), pp. 53–70. Springer Nature.
6. Pes, G.M. 1999. *The Sardinian Centenarian Study*. Max-Planck-Gesellschaft.
7. Poulain, M., Pes, G.M., Carru, C., Ferrucci, L., Baggio, G., Franceschi, C., Deiana, L. 2007. The validation of exceptional male longevity in Sardinia. In *Human Longevity, Individual Life Duration, and the Growth of the Oldest-Old Population* (eds J.-M. Robine, E.M. Crimmins, S. Horiuchi, Z. Yi), pp. 147–166. Springer Netherlands.
8. Poulain, M., Pes, G.M., Grasland, C., Carru, C., Ferrucci, L., Baggio, G., Franceschi, C., Deiana, L. 2004. Identification of a geographic area characterized by extreme longevity in the Sardinia island: the AKEA Study. *Exp. Gerontol.* **39**, 1423–1429 (doi:10.1016/j.exger.2004.06.016).
9. Buettner, D. 2005. The secrets of long life. *Natl. Geogr.* **2005**, 2–26.
10. Poulain, M., Herm, A., Pes, G.M. 2013. The blue zones: areas of exceptional longevity around the world. *Vienna Yearb. Popul. Res.* **11**, 87–108 (doi:10.1553/populationyearbook2013s87).
11. Fontana, L., Partridge, L., Longo, V.D. 2010. Extending healthy life span: from yeast to humans. *Science* **328**, 321–326 (doi:10.1126/science.1172539).

12. Martin, L.J., Mahaney, M.C., Bronikowski, A.M., Carey, K.D., Dyke, B., Comuzzie, A.G. 2002. Lifespan in captive baboons is heritable. *Mech. Ageing Dev.* **123**, 1461–1467 (doi:10.1016/s0047-6374(02)00083-0).
13. Fabrizio, P., Pozza, F., Pletcher, S.D., Gendron, C.M., Longo, V.D. 2001. Regulation of longevity and stress resistance by Sch9 in yeast. *Science* **292**, 288–290 (doi:10.1126/science.1059497).
14. Longo, V.D., Fabrizio, P. 2002. Regulation of longevity and stress resistance: a molecular strategy conserved from yeast to humans? *Cell Mol. Life Sci.* **59**, 903–908 (doi:10.1007/s00018-002-8477-8).
15. Herskind, A.M., McGue, M., Iachine, I.A., Holm, N., Sorensen, T.I., Harvald, B., Vaupel, J.W. 1996. Untangling genetic influences on smoking, body mass index and longevity: a multivariate study of 2464 Danish twins followed for 28 years. *Hum. Genet.* **98**, 467–475 (doi:10.1007/s004390050241).
16. Kaplanis, J. et al. 2018. Quantitative analysis of population-scale family trees with millions of relatives. *Science* **360**, 171–175 (doi:10.1126/science.aam9309).
17. Ruby, J.G. et al. 2018. Estimates of the heritability of human longevity are substantially inflated due to assortative mating. *Genetics* **210**, 1109–1124 (doi:10.1534/genetics.118.301613).
18. van den Berg, N., Rodriguez-Girondo, M., van Dijk, I.K., Mourits, R.J., Mandemakers, K., Janssens, A., Beekman, M., Smith, K.R., Slagboom, P.E. 2019. Longevity defined as top 10% survivors and beyond is transmitted as a quantitative genetic trait. *Nat. Commun.* **10**, 35 (doi:10.1038/s41467-018-07925-0).
19. Garagnani, P. et al. 2021. Whole-genome sequencing analysis of semi-supercentenarians. *eLife* **10**, e57849 (doi:10.7554/eLife.57849).
20. Bonafe, M. et al. 2002. Genetic analysis of paraoxonase (PON1) locus reveals an increased frequency of Arg192 allele in centenarians. *Eur. J. Hum. Genet.* **10**, 292–296 (doi:10.1038/sj.ejhg.5200806).
21. Bonafe, M., Olivieri, F. 2009. Genetic polymorphism in long-lived people: cues for the presence of an insulin/IGF-pathway-dependent network affecting human longevity. *Mol. Cell Endocrinol.* **299**, 118–123 (doi:10.1016/j.mce.2008.10.038).
22. Lescai, F., Marchegiani, F., Franceschi, C. 2009. PON1 is a longevity gene: results of a meta-analysis. *Ageing Res. Rev.* **8**, 277–284 (doi:10.1016/j.arr.2009.04.001).
23. Schächter, F., Faure-Delanef, L., Guenot, F., Rouger, H., Froguel, P., Lesueur-Ginot, L., Cohen, D. 1994. Genetic associations with human longevity at the APOE and ACE loci. *Nat. Genet.* **6**, 29–32 (doi:10.1038/ng0194-29).
24. Willcox, B.J., Donlon, T.A., He, Q., Chen, R., Grove, J.S., Yano, K., Masaki, K.H., Willcox, D.C., Rodriguez, B., Curb, J.D. 2008. FOXO3A genotype is strongly associated with human longevity. *Proc. Natl. Acad. Sci. U. A* **105**, 13987–13992 (doi:10.1073/pnas.0801030105).
25. Morris, B.J., Willcox, D.C., Donlon, T.A., Willcox, B.J. 2015. FOXO3: a major gene for human longevity: a mini-review. *Gerontology* **61**, 515–525 (doi:10.1159/000375235).
26. Jofre-Monseny, L., Minihane, A.-M., Rimbach, G. 2008. Impact of ApoE genotype on oxidative stress, inflammation and disease risk. *Mol. Nutr. Food Res.* **52**, 131–145 (doi:10.1002/mnfr.200700322).
27. Pes, G.M. et al. 2004. Association between longevity and cytokine gene polymorphisms: a study in Sardinian centenarians. *Aging Clin. Exp. Res.* **16**, 244–248 (doi:10.1007/BF03327391).
28. Georgiopoulos, G. et al. 2017. Arterial aging mediates the effect of TNF-alpha and ACE polymorphisms on mental health in elderly individuals: insights from IKARIA Study. *QJM* **110**, 551–557 (doi:10.1093/qjmed/hcx074).

29. Poulain, M. et al. 2021. Specific features of the oldest old from the longevity blue zones in Ikaria and Sardinia. *Mech. Ageing Dev.* **198**, 111543 (doi:10.1016/j.mad.2021.111543).
30. Rehkopf, D.H., Dow, W.H., Rosero-Bixby, L., Lin, J., Epel, E.S., Blackburn, E.H. 2013. Longer leukocyte telomere length in Costa Rica's Nicoya Peninsula: a population-based study. *Exp. Gerontol.* **48**, 1266–1273 (doi:10.1016/j.exger.2013.08.005).
31. Pes, G., Poulain, M., Pachana, N. 2016. *Encyclopedia of Geropsychology*. Springer.
32. Vona, G. 1997. The peopling of Sardinia (Italy): history and effects. *Int. J. Anthropol.* **12**, 71–87.
33. Hanihara, K. 1991. Dual structure model for the population history of Japanese. *Anthropol. Sci.* **2**, 1–33.
34. Akisaka, M., Suzuki, M. 1998. Okinawa Longevity Study: molecular genetic analysis of HLA genes in the very old. *Nihon Ronen Igakkai Zasshi* **35**, 294–298 (doi:10.3143/geriatrics.35.294).
35. Bao, J.M., Song, X.L., Hong, Y.Q., Zhu, H.L., Li, C., Zhang, T., Chen, W., Zhao, S.C., Chen, Q. 2014. Association between FOXO3A gene polymorphisms and human longevity: a meta-analysis. *Asian J. Androl.* **16**, 446–452 (doi:10.4103/1008-682X.123673).
36. Allsopp, R., Willcox, D.C., Torigoe, T., Gerschenson, M., Chen, R., Donlon, T., Willcox, B., Shimabukuro, M. 2021. Effects of FOXO3 on markers of aging in blood: an Okinawan longevity cohort study. *Innov. Aging* **5**, 667–667 (doi:10.1093/geroni/igab046.2500).
37. Banks, D.A., Fossel, M. 1997. Telomeres, cancer, and aging: altering the human life span. *JAMA* **278**, 1345–1348.
38. Portas, L. et al. 2010. History, geography and population structure influence the distribution and heritability of blood and anthropometric quantitative traits in nine Sardinian genetic isolates. *Genet. Res. Camb.* **92**, 199–208 (doi:10.1017/S001667231000025X).
39. Azofeifa, J., Ruiz-Narvaez, E.A., Leal, A., Gerlovin, H., Rosero-Bixby, L. 2018. Amerindian Ancestry and extended longevity in Nicoya, Costa Rica. *Am. J. Hum. Biol.* **30** (doi:10.1002/ajhb.23055).
40. Stefanadis, C.I. 2013. Aging, genes and environment: lessons from the Ikaria Study. *Hell. J. Cardiol.* **54**, 237–238.
41. Stefanadis, C.I. 2011. Unveiling the secrets of longevity: the Ikaria Study. *Hell. J. Cardiol.* **52**, 479–480.
42. Szyf, M., Bick, J. 2013. DNA methylation: a mechanism for embedding early life experiences in the genome. *Child Dev.* **84**, 49–57 (doi:10.1111/j.1467-8624.2012.01793.x).
43. Horvath, S. et al. 2016. An epigenetic clock analysis of race/ethnicity, sex, and coronary heart disease. *Genome Biol.* **17**, 171 (doi:10.1186/s13059-016-1030-0).
44. McEwen, L.M., Morin, A.M., Edgar, R.D., MacIsaac, J.L., Jones, M.J., Dow, W.H., Rosero-Bixby, L., Kobor, M.S., Rehkopf, D.H. 2017. Differential DNA methylation and lymphocyte proportions in a Costa Rican high longevity region. *Epigenetics Chromatin.* **10**, 21 (doi:10.1186/s13072-017-0128-2).
45. Engelbrecht, H.-R., Merrill, S.M., Gladish, N., MacIsaac, J.L., Lin, D.T.S., Ecker, S., Chrysohoou, C.A., Pes, G.M., Kobor, M.S., Rehkopf, D.H. 2022. Sex differences in epigenetic age in Mediterranean high longevity regions. *Front. Aging* **3**, 1007098 (doi: 10.3389/fragi.2022.1007098).
46. Everitt, A.V., Olsen, G.G., Burrows, G.R. 1968. The effect of hypophysectomy on the aging of collagen fibers in the tail tendon of the rat. *J. Gerontol.* **23**, 333–336 (doi:10.1093/geronj/23.3.333).
47. Ewald, C.Y., Landis, J.N., Porter Abate, J., Murphy, C.T., Blackwell, T.K. 2015. Dauer-independent insulin/IGF-1-signalling implicates collagen remodelling in longevity. *Nature* **519**, 97–101 (doi:10.1038/nature14021).

48. Kimura, K.D., Tissenbaum, H.A., Liu, Y., Ruvkun, G. 1997. Daf-2, an insulin receptor-like gene that regulates longevity and diapause in *Caenorhabditis elegans*. *Science* **277**, 942–946 (doi:10.1126/science.277.5328.942).
49. Shevah, O., Laron, Z. 2006. Genetic analysis of the pedigrees and molecular defects of the GH-receptor gene in the Israeli cohort of patients with Laron syndrome. *Pediatr. Endocrinol. Rev.* **3** (Suppl. 3), 489–497.
50. Laron, Z., Werner, H. 2021. Laron syndrome: a historical perspective. *Rev. Endocr. Metab. Disord.* **22**, 31–41 (doi:10.1007/s11154-020-09595-0).
51. Verhelst, J., Abs, R. 2009. Cardiovascular risk factors in hypopituitary GH-deficient adults. *Eur. J. Endocrinol.* **161**(Suppl. 1), S41–S49 (doi:10.1530/EJE-09-0291).
52. Zoledziewska, M. et al. 2015. Height-reducing variants and selection for short stature in Sardinia. *Nat. Genet.* **47**, 1352–1356 (doi:10.1038/ng.3403).
53. Ooka, H., Fujita, S., Yoshimoto, E. 1983. Pituitary-thyroid activity and longevity in neonatally thyroxine-treated rats. *Mech. Ageing Dev.* **22**, 113–120 (doi:10.1016/0047-6374(83)90104-5).
54. Buffenstein, R., Pinto, M. 2009. Endocrine function in naturally long-living small mammals. *Mol. Cell Endocrinol.* **299**, 101–111 (doi:10.1016/j.mce.2008.04.021).
55. Atzmon, G., Barzilai, N., Hollowell, J.G., Surks, M.I., Gabriely, I. 2009. Extreme longevity is associated with increased serum thyrotropin. *J. Clin. Endocrinol Metab.* **94**, 1251–1254 (doi:10.1210/jc.2008-2325).
56. Atzmon, G., Barzilai, N., Surks, M.I., Gabriely, I. 2009. Genetic predisposition to elevated serum thyrotropin is associated with exceptional longevity. *J. Clin. Endocrinol. Metab.* **94**, 4768–4775.
57. Martino, E. et al. 1994. Endemic goiter and thyroid function in central-southern Sardinia: report on an extensive epidemiological survey. *J. Endocrinol. Invest.* **17**, 653–657 (doi:10.1007/BF03349681).
58. Tolu, F., Palermo, M., Dore, M.P., Errigo, A., Canelada, A., Poulain, M., Pes, G.M. 2019. Association of endemic goitre and exceptional longevity in Sardinia: evidence from an ecological study. *Eur. J. Ageing* **16**, 405–414 (doi:10.1007/s10433-019-00510-4).
59. Perez, C., Salazar-Baldioceda, A., Scrimshaw, N.S., Tandon, O.B. 1956. Endemic goiter in Costa Rican school children. *Am. J. Public Health Nations Health* **46**, 1283–1286 (doi:10.2105/ajph.46.10.1283).
60. Mora-Alvarado, D.A., Portuguez-Barquero, C.F., Alfaro-Herrera, N., Hernández-Miraulth, M. 2015. Diferencias de dureza del agua y las tasas de longevidad en la península de nicoya y los otros distritos de guanacaste. *Rev. Tecnol. en Marcha* **28**, 3–14.
61. Madrigal-Leer, F., Martinez-Montandon, A., Solis-Umana, M., Helo-Guzman, F., Alfaro-Salas, K., Barrientos-Calvo, I., Camacho-Mora, Z., Jimenez-Porras, V., Estrada-Montero, S., Morales-Martinez, F. 2020. Clinical, functional, mental and social profile of the Nicoya Peninsula centenarians, Costa Rica, 2017. *Aging Clin. Exp. Res.* **32**, 313–321 (doi:10.1007/s40520-019-01176-9).
62. Karakatsanis, S., D'Alessandro, W., Kyriakopoulos, K., Voudouris, K. 2011. Chemical characterization of the thermal springs along the South Aegean Volcanic Arc and Ikaria Island. In *Advances in the Research of Aquatic Environment. Environmental Earth Sciences* (eds N. Lambrakis, G. Stournaras, K. Katsanou), pp. 239–247. Springer.
63. Giassa, T., Mamali, I., Gaki, E., Kaltsas, G., Kouraklis, G., Markou, K., Karatzas, T. 2018. Iodine intake and chronic autoimmune thyroiditis: a comparative study between coastal and mainland regions in Greece. *Horm. Athens* **17**, 565–571 (doi:10.1007/s42000-018-0057-x).
64. Chrysohoou, C., Pitsavos, C., Lazaros, G., Skoumas, J., Tousoulis, D., Stefanadis, C., Ikaria Study, I. 2016. Determinants of all-cause mortality and incidence of cardiovascular

disease (2009 to 2013) in older adults: the Ikaria Study of the blue zones. *Angiology* **67**, 541–548 (doi:10.1177/0003319715603185).

65. Delitala, A.P., Terracciano, A., Fiorillo, E., Orru, V., Schlessinger, D., Cucca, F. 2016. Depressive symptoms, thyroid hormone and autoimmunity in a population-based cohort from Sardinia. *J. Affect. Disord.* **191**, 82–87 (doi:10.1016/j.jad.2015.11.019).
66. Golan, R., Scovell, J.M., Ramasamy, R. 2015. Age-related testosterone decline is due to waning of both testicular and hypothalamic-pituitary function. *Aging Male* **18**, 201–204 (doi:10.3109/13685538.2015.1052392).
67. Chrysohoou, C. et al. 2013. Low total testosterone levels are associated with the metabolic syndrome in elderly men: the role of body weight, lipids, insulin resistance, and inflammation; the Ikaria Study. *Rev. Diabet. Stud.* **10**, 27–38 (doi:10.1900/RDS.2013.10.27).
68. Willcox, B.J., Willcox, D.C., Todoriki, H., Yano, K., Curb, J.D., Suzuki, M. 2007. Caloric restriction, energy balance and healthy aging in Okinawans and Americans: biomarker differences in septuagenarians. *Okinawan J. Am. Stud.* **4**, 60–72.
69. Delitala, G., Sanciu, F., Fanciulli, G., Carru, C., Maglione, F., Pes, G.M., Riccardi, E., Baggio, G., Deiana, L. 2006. Gonadal hormones and adrenal steroidogenesis in centenarians. *Biochim. Clin.* **30**, S37–S37.
70. Viña, J., Borras, C., Gambini, J., Sastre, J., Pallardo, F.V. 2005. Why females live longer than males? Importance of the upregulation of longevity-associated genes by oestrogenic compounds. *FEBS Lett.* **579**, 2541–2545 (doi:10.1016/j.febslet.2005.03.090).
71. Poulain, M., Herm, A., Chambre, D., Pes, G. 2016. Fertility history, children's gender, and post-reproductive survival in a longevous population. *Biodemography Soc. Biol.* **62**, 262–274 (doi:10.1080/19485565.2016.1207502).
72. Wu, A., Shi, Z., Martin, S., Vincent, A., Heilbronn, L., Wittert, G. 2018. Age-related changes in estradiol and longitudinal associations with fat mass in men. *PLoS ONE* **13**, e0201912 (doi:10.1371/journal.pone.0201912).
73. Dhalaria, R., Verma, R., Kumar, D., Puri, S., Tapwal, A., Kumar, V., Nepovimova, E., Kuca, K. 2020. Bioactive compounds of edible fruits with their anti-aging properties: a comprehensive review to prolong human life. *Antioxid. Basel* **9**, 1123 (doi:10.3390/antiox9111123).
74. Pes, G.M., Dore, M.P., Tsofliou, F., Poulain, M. 2022. Diet and longevity in the blue zones: a set-and-forget issue? *Maturitas* **164**, 31–37 (doi:10.1016/j.maturitas.2022.06.004).
75. McCay, C.M., Crowel, M.F. 1934. Prolonging the life span. *Sci. Mon.* **39**, 405–414.
76. Colman, R.J., Anderson, R.M. 2011. Nonhuman primate calorie restriction. *Antioxid. Redox Signal* **14**, 229–239 (doi:10.1089/ars.2010.3224).
77. Libert, S., Guarente, L. 2013. Metabolic and neuropsychiatric effects of calorie restriction and sirtuins. *Annu. Rev. Physiol.* **75**, 669–684 (doi:10.1146/annurev-physiol-030212-183800).
78. Anderson, R.M., Weindruch, R. 2006. Calorie restriction: progress during mid-2005-mid-2006. *Exp. Gerontol.* **41**, 1247–1249 (doi:10.1016/j.exger.2006.10.019).
79. Drew, B., Phaneuf, S., Dirks, A., Selman, C., Gredilla, R., Lezza, A., Barja, G., Leeuwenburgh, C. 2003. Effects of aging and caloric restriction on mitochondrial energy production in gastrocnemius muscle and heart. *Am. J. Physiol. Regul. Integr. Comp. Physiol.* **284**, R474–480 (doi:10.1152/ajpregu.00455.2002).
80. Guarente, L. 2012. Sirtuins and calorie restriction. *Nat. Rev. Mol. Cell Biol.* **13**, 207 (doi:10.1038/nrm3308).
81. Suzuki, M. 2001. Cultural climate and social custom for longevity region, Okinawa. *Nihon Ronen Igakkai Zasshi* **38**, 163–165.

82. Salaris, L., Poulain, M., Samaras, T.T. 2012. Height and survival at older ages among men born in an inland village in Sardinia (Italy), 1866–2006. *Biodemography Soc. Biol.* **58**, 1–13 (doi:10.1080/19485565.2012.666118).
83. Pes, G.M., Tognotti, E., Poulain, M., Chambre, D., Dore, M.P. 2017. Why were Sardinians the shortest Europeans? A journey through genes, infections, nutrition, and sex. *Am. J. Phys. Anthr.* **163**, 3–13 (doi:10.1002/ajpa.23177).
84. Tivaroni, J. 1928. Intorno alle condizioni alimentari della popolazione Sarda. *Riv. Polit. Econ.* **2**, 1–11.
85. Peretti, G. 1943. Rapporti tra alimentazione e caratteri antropometrici. studio statistico-biometrico in Sardegna. *Quad. Della Nutr.* **9**, 69–130.
86. Le Lannou, M. 1941. *Pâtres et paysans de la sardaigne*. Tours Arrault.
87. Koufakis, T., Karras, S., Antonopoulou, V., Angeloudi, E., Zebekakis, P., Kotsa, K. 2017. Effects of orthodox religious fasting on human health: a systematic review. *Eur. J. Nutr.* **56**, 2439–2455 (doi:10.1007/s00394-017-1534-8).
88. Legrand, R., Manckoundia, P., Nuemi, G., Poulain, M. 2019. Assessment of the health status of the oldest olds living on the Greek island of Ikaria: a population based-study in a blue zone. *Curr. Gerontol. Geriatr. Res.* **2019**, 8194310 (doi:10.1155/2019/8194310).
89. Rosero-Bixby, L., Dow, W.H., Rehkopf, D.H. 2013. The Nicoya region of Costa Rica: a high longevity island for elderly males. *Vienna. Yearb. Popul. Res.* **11**, 109–136 (doi:10.1553/populationyearbook2013s109).
90. Le Bourg, E. 2018. Is it time to state that diet restriction does not increase life span in primates? *J. Gerontol. Biol. Sci. Med. Sci.* **73**, 308–309 (doi:10.1093/gerona/glx159).
91. Le Bourg, E., Redman, L.M. 2018. Do-it-yourself calorie restriction: the risks of simplistically translating findings in animal models to humans. *Bioessays* **40**, e1800087 (doi:10.1002/bies.201800087).
92. Donato, A., Pietrangeli, F., Serafini, M. 2022. Early dinner time and caloric restriction lapse contribute to the longevity of nonagenarians and centenarians of the Italian Abruzzo Region: a cross-sectional study. *Front. Nutr.* **9**, 863106.
93. Trichopoulou, A., Critselis, E. 2004. Mediterranean diet and longevity. *Eur. J. Cancer Prev.* **13**, 453–456 (doi:10.1097/00008469-200410000-00014).
94. Trichopoulou, A., Vasilopoulou, E. 2000. Mediterranean diet and longevity. *Br. J. Nutr.* **84** (Suppl. 2), S205–S209 (doi:10.1079/096582197388554).
95. Pes, G.M., Tolu, F., Dore, M.P., Sechi, G.P., Errigo, A., Canelada, A., Poulain, M. 2015. Male longevity in Sardinia: a review of historical sources supporting a causal link with dietary factors. *Eur. J. Clin. Nutr.* **69**, 411–418 (doi:10.1038/ejcn.2014.230).
96. Buettner, D., Skemp, S. 2016. Blue Zones: Lessons From the World's Longest Lived. *Am. J. Lifestyle Med.* **10**, 318–321 (doi:10.1177/1559827616637066).
97. Tessier, S., Gerber, M. 2005. Factors determining the nutrition transition in two Mediterranean islands: Sardinia and Malta. *Public Health Nutr.* **8**, 1286–1292 (doi:10.1079/phn2005747).
98. Levine, M.E. et al. 2014. Low protein intake is associated with a major reduction in IGF-1, cancer, and overall mortality in the 65 and younger but not older population. *Cell Metab.* **19**, 407–417 (doi:10.1016/j.cmet.2014.02.006).
99. Foscolou, A., Polychronopoulos, E., Paka, E., Tyrovolas, S., Bountziouka, V., Zeimbekis, A., Tyrovola, D., Ural, D., Panagiotakos, D. 2016. Lifestyle and health determinants of cardiovascular disease among Greek older adults living in Eastern Aegean Islands: an adventure within the MEDIS Study. *Hell. J. Cardiol.* **57**, 407–414 (doi:10.1016/j.hjc.2016.11.021).

100. Chrysohoou, C. et al. 2010. Fish consumption moderates depressive symptomatology in elderly men and women from the IKARIA Study. *Cardiol. Res. Pr.* **2011**, 219578 (doi:10.4061/2011/219578).
101. Momi-Chacón, A., Capitán-Jiménez, C., Campos, H. 2017. Dietary habits and lifestyle among long-lived residents from the Nicoya Peninsula of Costa Rica. *Rev. Hisp. Cienc. Salud.* **3**, 53–60.
102. Nieddu, A., Vindas, L., Errigo, A., Vindas, J., Pes, G.M., Dore, M.P. 2020. Dietary habits, anthropometric features and daily performance in two independent long-lived populations from Nicoya Peninsula (Costa Rica) and Ogliastra (Sardinia). *Nutrients* **12** 1621 (doi:10.3390/nu12061621).
103. Sho, H. 2001. History and characteristics of Okinawan longevity food. *Asia Pac. J. Clin. Nutr.* **10**, 159–164 (doi:10.1111/j.1440-6047.2001.00235.x).
104. Todoriki, H., Willcox, D.C., Willcox, B.J. 2004. The effect of post-war dietary change on longevity and health in Okinawa. *Okinawan J. Am. Stud.* **1**, 52–61.
105. Alam, M.A., Uddin, R., Subhan, N., Rahman, M.M., Jain, P., Reza, H.M. 2015. Beneficial role of bitter melon supplementation in obesity and related complications in metabolic syndrome. *J. Lipids* **2015**, 496169 (doi:10.1155/2015/496169).
106. Lissner, L., Bengtsson, C., Bjorkelund, C., Wedel, H. 1996. Physical activity levels and changes in relation to longevity: a prospective study of Swedish women. *Am. J. Epidemiol.* **143**, 54–62 (doi:10.1093/oxfordjournals.aje.a008657).
107. Paffenbarger, R.S., Jr., Hyde, R.T., Hsieh, C.C., Wing, A.L. 1986. Physical activity, other life-style patterns, cardiovascular disease and longevity. *Acta Med. Scand. Suppl.* **711**, 85–91 (doi:10.1111/j.0954-6820.1986.tb08936.x).
108. Yates, L.B., Djousse, L., Kurth, T., Buring, J.E., Gaziano, J.M. 2008. Exceptional longevity in men: modifiable factors associated with survival and function to age 90 years. *Arch. Intern. Med.* **168**, 284–290 (doi:10.1001/archinternmed.2007.77).
109. Frisard, M.I., Fabre, J.M., Russell, R.D., King, C.M., DeLany, J.P., Wood, R.H., Ravussin, E., Louisiana Healthy Aging, S. 2007. Physical activity level and physical functionality in nonagenarians compared to individuals aged 60–74 years. *J. Gerontol. Biol. Sci. Med. Sci.* **62**, 783–788 (doi:10.1093/gerona/62.7.783).
110. Venturelli, M., Schena, F., Richardson, R.S. 2012. The role of exercise capacity in the health and longevity of centenarians. *Maturitas* **73**, 115–120 (doi:10.1016/j.maturitas.2012.07.009).
111. Pancani, S. et al. 2022. 12-month survival in nonagenarians inside the Mugello Study: on the way to live a century. *BMC Geriatr.* **22** (doi:10.1186/s12877-022-02908-9).
112. Herbert, C., House, M., Dietzman, R., Climstein, M., Furness, J., Kemp-Smith, K. 2022. Blue zones: centenarian modes of physical activity: a scoping review. *J. Popul. Ageing* (doi:10.1007/s12062-022-09396-0).
113. Pes, G.M., Tolu, F., Poulain, M., Errigo, A., Masala, S., Pietrobelli, A., Battistini, N.C., Maioli, M. 2013. Lifestyle and nutrition related to male longevity in Sardinia: an ecological study. *Nutr. Metab. Cardiovasc. Dis.* **23**, 212–219 (doi:10.1016/j.numecd.2011.05.004).
114. Pes, G.M., Dore, M.P., Errigo, A., Poulain, M. 2018. Analysis of physical activity among free-living nonagenarians from a Sardinian longevous population. *J. Aging Phys. Act.* **26**, 254–258 (doi:10.1123/japa.2017-0088).
115. Tiainen, K., Raitanen, J., Vaara, E., Hervonen, A., Jylha, M. 2015. Longitudinal changes in mobility among nonagenarians: the Vitality 90+ Study. *BMC Geriatr.* **15**, 124 (doi:10.1186/s12877-015-0116-y).

116. Fastame, M.C., Mulas, I., Pau, M. 2020. Mental health and motor efficiency of older adults living in the Sardinia's blue zone: a follow-up study. *Int. Psychogeriatr.* **33**, 1–12 (doi:10.1017/S1041610220001659).
117. Legrand, R., Nuemi, G., Poulain, M., Manckoundia, P. 2021. Description of lifestyle, including social life, diet and physical activity, of people >/=90 years living in Ikaria, a longevity blue zone. *Int. J. Env. Res. Public Health* **18**, 6602 (doi:10.3390/ijerph18126602).
118. Oikonomou, E. et al. 2011. Gender variation of exercise-induced anti-arrhythmic protection: the Ikaria Study. *QJM* **104**, 1035–1043 (doi:10.1093/qjmed/hcr112).
119. Siasos, G. et al. 2013. The impact of physical activity on endothelial function in middle-aged and elderly subjects: the Ikaria Study. *Hell. J. Cardiol.* **54**, 94–101.
120. Fatourou, K. 2016. Experiencing dance as social process: a case study of the summer Paniyiri in Ikaria Island. *Cord. Proc.*, 148–155 (doi:10.1017/cor.2016.21).
121. Cugusi, L., Massidda, M., Matta, D., Garau, E., Di Cesare, R., Deidda, M., Satta, G., Chiappori, P., Solla, P., Mercuro, G. 2015. A new type of physical activity from an ancient tradition: the Sardinian folk dance 'Ballu Sardu'. *J. Dance Med. Sci.* **19**, 118–123 (doi:10.12678/1089-313X.19.3.118).
122. Pes, G.M., Dore, M.P. 2021. Understanding the impact of motor activity on the mental well-being of older people. *Int. Psychogeriatr.* **33**, 1237–1239 (doi:10.1017/S1041610220003701).
123. Latkin, C.A., Edwards, C., Davey-Rothwell, M.A., Tobin, K.E. 2017. The relationship between social desirability bias and self-reports of health, substance use, and social network factors among urban substance users in Baltimore, Maryland. *Addict. Behav.* **73**, 133–136 (doi:10.1016/j.addbeh.2017.05.005).
124. Jacobs, J.M., Maaravi, Y., Stessman, J. 2021. Optimism and longevity beyond age 85. *J. Gerontol. Biol. Sci. Med. Sci.* **76**, 1806–1813 (doi:10.1093/gerona/glab051).
125. Lee, L.O., James, P., Zevon, E.S., Kim, E.S., Trudel-Fitzgerald, C., Spiro, A., III, Grodstein, F., Kubzansky, L.D. 2019. Optimism is associated with exceptional longevity in 2 epidemiologic cohorts of men and women. *Proc. Natl. Acad. Sci. U. A* **116**, 18357–18362 (doi:10.1073/pnas.1900712116).
126. Fastame, M.C., Hitchcott, P.K., Penna, M.P. 2015. Do self-referent metacognition and residential context predict depressive symptoms across late-life span? A developmental study in an Italian sample. *Aging Ment. Health* **19**, 698–704 (doi:10.1080/13607863.2014.962003).
127. Fastame, M.C., Ruiu, M., Mulas, I. 2021. Mental health and religiosity in the Sardinian blue zone: life satisfaction and optimism for aging well. *J. Relig. Health* **60**, 2450–2462 (doi:10.1007/s10943-021-01261-2).
128. Ruiu, M., Carta, V., Deiana, C., Fastame, M.C. 2022. Is the Sardinian blue zone the new Shangri-La for mental health? Evidence on depressive symptoms and its correlates in late adult life span. *Aging Clin. Exp. Res.* **34**, 1315–1322 (doi:10.1007/s40520-021-02068-7).
129. Fastame, M.C., Penna, M.P. 2014. Psychological well-being and metacognition in the fourth age: an explorative study in an Italian oldest old sample. *Aging Ment. Health* **18**, 648–652 (doi:10.1080/13607863.2013.866635).
130. Pes, G.M., Errigo, A., Tedde, P., Dore, M.P. 2020. Sociodemographic, clinical and functional profile of nonagenarians from two areas of Sardinia characterized by distinct longevity levels. *Rejuvenation Res.* **23**, 341–348 (doi:10.1089/rej.2018.2129).
131. Panagiotakos, D.B., Chrysohoou, C., Siasos, G., Zisimos, K., Skoumas, J., Pitsavos, C., Stefanadis, C. 2011. Sociodemographic and lifestyle statistics of oldest old people

(>80 years) living in Ikaria Island: the Ikaria Study. *Cardiol. Res. Pr.* **2011**, 679187 (doi:10.4061/2011/679187).

132. Tourlouki, E. et al. 2010. The 'secrets' of the long livers in Mediterranean islands: the MEDIS Study. *Eur. J. Public Health* **20**, 659–664 (doi:10.1093/eurpub/ckp192).
133. Rosero-Bixby, L., Modrek, S., Domino, M.E., Dow, W.H. 2017. Aging and mental health in a longitudinal study of elderly Costa Ricans. In *Encyclopedia of Geropsychology* (ed. N.A. Pachana), pp. 185–196. Springer Singapore.
134. Naka, K., Toguchi, S., Takaishi, T., Ishizu, H., Sasaki, Y. 1985. Yuta (shaman) and community mental health on Okinawa. *Int. J. Soc. Psychiatry* **31**, 267–274 (doi:10.1177/002076408503100404).
135. Allen, M. 2002. Therapies of resistance? Yuta, help-seeking, and identity in Okinawa. *Crit. Asian Stud.* **34**, 221–242 (doi:10.1080/14672710220146215).
136. Azcueta, R., Pinna, M., Manchia, M., Simbula, S., Tondo, L., Baldessarini, R.J. 2021. Suicidal risks in rural versus urban populations in Sardinia. *J. Affect. Disord.* **295**, 1449–1455 (doi:10.1016/j.jad.2021.09.024).
137. Lorettu, L., Nivoli, A., Bellizzi, S., Piu, D., Meloni, R., Dore, M.P., Pes, G.M. 2021. Geospatial clustering of suicide mortality in Sardinia. *Curr. Psychol.* **42**, 11556–11564 (doi: 10.1007/s12144-021-02448-2).
138. Kageyama, T., Naka, K. 1996. Longitudinal change in youth suicide mortality in Okinawa after World War II: a comparative study with mainland Japan. *Psychiatry Clin. Neurosci.* **50**, 239–242 (doi:10.1111/j.1440-1819.1996.tb00556.x).
139. Fastame, M.C., Hitchcott, P.K., Mulas, I., Ruiu, M., Penna, M.P. 2018. Resilience in elders of the Sardinian blue zone: an explorative study. *Behav. Sci. Basel* **8**, 30 (doi:10.3390/bs8030030).
140. Fastame, M.C., Hitchcott, P.K., Penna, M.P. 2018. The impact of leisure on mental health of Sardinian elderly from the 'blue zone': evidence for ageing well. *Aging Clin. Exp. Res.* **30**, 169–180 (doi:10.1007/s40520-017-0768-x).
141. Evans, R.G. 1994. *Why Are Some People Healthy and Others Not? The Determinants of Health of Populations*. A. de Gruyter.
142. Cockerham, W.C., Hattori, H., Yamori, Y. 2000. The social gradient in life expectancy: the contrary case of Okinawa in Japan. *Soc. Sci. Med.* **51**, 115–122 (doi:10.1016/s0277-9536(99)00444-x).
143. Cockerham, W.C., Yamori, Y. 2001. Okinawa: an exception to the social gradient of life expectancy in Japan. *Asia Pac. J. Clin. Nutr.* **10**, 154–158 (doi:10.1111/j.1440-6047.2001.00232.x).
144. Poland, B., Coburn, D., Robertson, A., Eakin, J. 1998. Wealth, equity and health care: a critique of a 'population health' perspective on the determinants of health. Critical Social Science Group. *Soc. Sci. Med.* **46**, 785–798 (doi:10.1016/s0277-9536(97)00197-4).
145. Scott, A.J. 2021. The longevity society. *The Lancet* **2**, E820–E827.
146. Willcox, B.J., Willcox, D.C., Todoriki, H., Fujiyoshi, A., Yano, K., He, Q., Curb, J.D., Suzuki, M. 2007. Caloric restriction, the traditional Okinawan diet, and healthy aging: the diet of the world's longest-lived people and its potential impact on morbidity and life span. *Ann. N. Acad. Sci.* **1114**, 434–455 (doi:10.1196/annals.1396.037).
147. Kabagambe, E.K., Baylin, A., Siles, X., Campos, H. 2002. Comparison of dietary intakes of micro- and macronutrients in rural, suburban and urban populations in Costa Rica. *Public Health Nutr.* **5**, 835–842 (doi:10.1079/PHN2002372).
148. Loviselli, A., Ghiani, M.E., Velluzzi, F., Piras, I.S., Minerba, L., Vona, G., Calo, C.M. 2010. Prevalence and trend of overweight and obesity among Sardinian conscripts (Italy) of 1969 and 1998. *J. Biosoc. Sci.* **42**, 201–211 (doi:10.1017/S0021932009990411).

18 Socio-Economic Consequences of Increased Longevity in Contemporary Populations

Miguel Sánchez-Romero and Alexia Prskawetz

18.1 Introduction

Rising life expectancy will have a pronounced effect on the age structure of most developed countries in the next decades. While individual longevity, if combined with gains in healthy years of life, is regarded as an important improvement, at the aggregate level, however, longer lives constitute a challenge as the share of elderly dependent people will increase.

Important determinants of the overall socio-economic effects of increasing life expectancy are individual-level economic behaviour (such as consumption, savings, labour force participation etc.) together with the prevailing public and private transfer systems. To quantify age-specific economic behaviour and transfers, we draw on a rich dataset of age-structured economic data (National Transfer Accounts, NTA) that also includes private and public transfers by age. Based on these data we first explain the impact of increasing life expectancy on the components of economic growth (Section 18.2). We start by reviewing the literature on life expectancy improvements on human capital (Section 18.2.2) and on financial wealth (Section 18.2.3). To account for behavioural reactions induced by changes in the economic and demographic environment we apply a multi-country dynamic general equilibrium model populated by overlapping generations (OLG) (Section 18.2.4), previously developed in earlier work by Sánchez-Romero [1]. From now on we name the implemented economic model the OLG-NTA model. Our model extends [2] by accounting not only for the changing age structure of the population in quantifying future economic growth but also for the changing age profiles of economic characteristics. To disentangle the role of life expectancy from other demographic factors (e.g. fertility and education) on consumption and output per capita and the evolution of inequality across generations within societies at a specific point of time, we run several counterfactual experiments in which we alternately fix each demographic component at the level of 2020.

We present our results for two groups of 35 European countries that differ by income level based on Organisation for Economic Co-operation and Development (OECD) income-level grouping: (1) high-income countries (HICs: Austria, Belgium, Croatia,

We would like to thank two anonymous reviewers and Samuel Pavard for valuable comments and suggestions. This project has received funding from the Austrian National Bank (OeNB) under Grant No. 18744.

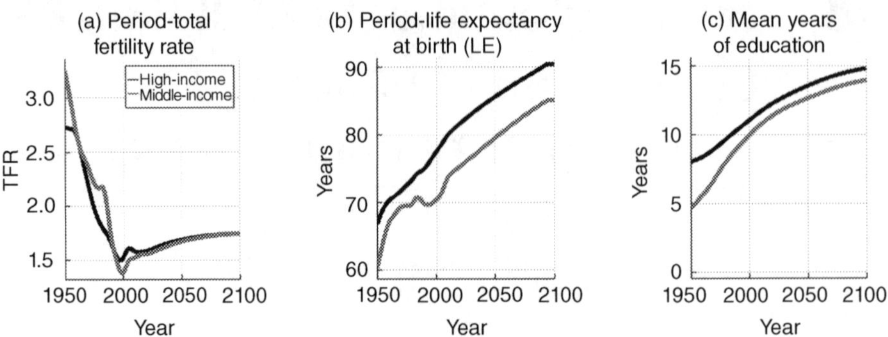

Figure 18.1 Time path of the period: (a) total fertility rate, (b) life expectancy and (c) mean years of education in European countries from 1950 to 2100 by OECD income level. Sources: [31, 40].

Denmark, Estonia, Finland, France, Germany, Greece, Hungary, Iceland, Ireland, Italy, Latvia, Lithuania, Luxembourg, Malta, the Netherlands, Norway, Poland, Portugal, Slovakia, Slovenia, Spain, Sweden and Switzerland); (2) middle-income countries (MICs: Belarus, Bosnia and Herzegovina, Bulgaria, Moldova, Montenegro, Romania, the Russian Federation, Serbia and Ukraine). We have opted for this setup since these two country groups also represent different demographic developments in the past and projected future.

Figure 18.1 plots the past and projected trends after year 2019 for fertility, life expectancy and education for the two European country groups we consider. These constitute the demographic developments for our baseline simulations. In all countries a pronounced decline in fertility until 2000 is followed by a slow recovery thereafter with the MICs having the lowest fertility rate. All countries will experience further gains in life expectancy with the differences between HICs and MICs being persistent. The fall in life expectancy in MICs during the 1990s is related to the fall of the Soviet Union and the Balkan Wars [3, 4]. Education, as measured by the mean years of education is assumed to further increase nearly 15 years in the HICs, and lag behind for one year for MICs [5].

18.2 Impact of Increased Life Expectancy on Standards of Living

In this section we discuss how increasing life expectancy impacts the standard of living as measured by consumption per capita (C/N). We decompose consumption per capita (C/N) into its macroeconomic components and explain, based on previous research, how improvements in survival at different stages in life affect each of these components. Based on empirical age profiles of the NTA project we illustrate how these components differ across the two country groups we introduced previously. To simulate the role of future demographic change on consumption per capita we apply our multi-country OLG-NTA model.

18.2.1 Growth Rate of Consumption per Capita

Total output (Y) in an economy can be spent on public and private consumption (C) and on savings: $Y = C + sY$, where s denotes the propensity to save. Consumption per capita is therefore proportional to output per capita (Y/N):

$$(\mathrm{C}/\mathrm{N}) = (1-s)(Y/N) = (1-s)(Y/L)(L/N), \tag{18.1}$$

with $(1-s)$ denoting the marginal propensity to consume out of total output. Output per capita can be decomposed into output per hour worked (Y/L) and per capita hours worked by the population (L/N). By taking logs in Eq. (18.1) and differentiation with respect to time we obtain

$$(C/N)_{\mathrm{gr}} = (1-s)_{\mathrm{gr}} + (Y/L)_{\mathrm{gr}} + (L/N)_{\mathrm{gr}}. \tag{18.2}$$

The growth rate of consumption per capita is the sum of the growth rate of the marginal propensity to consume, $(1-s)_{\mathrm{gr}}$, the growth rate of output per hour worked, $(Y/L)_{gr}$, which is known as the productivity component and the growth rate of the ratio of workers to the total population $(L/N)_{gr}$ [6], which represents the first demographic dividend [7]. The first demographic dividend is positive (resp. negative) when the working-age population is growing faster (resp. slower) as compared to the overall population. Hence, the first demographic dividend is expected to become negative when the baby boomers retire, since the total number of hours worked will decline faster than the growth rate of the population.

In an open economy, the growth rate of the marginal propensity to consume is given by (see the derivation in Appendix 18.1):

$$(1-s)_{\mathrm{gr}} = \left(1-\alpha+\left(r-N_{\mathrm{gr}}-Z_{\mathrm{gr}}\right)(A/Y)\right)_{\mathrm{gr}}, \tag{18.3}$$

Where α is the labour share in total output, r is the interest rate, N_{gr} is the growth rate of the population, Z_{gr} is the growth rate of technological progress and A is the total financial wealth of households. Substituting (18.3) into (18.2) results in the following expression of the growth rate of consumption per capita:

$$(\mathrm{C}/\mathrm{N})_{\mathrm{gr}} = (\mathrm{Y}/\mathrm{L})_{\mathrm{gr}} + (\mathrm{L}/\mathrm{N})_{\mathrm{gr}} + \left(1-\alpha+\left(r-N_{\mathrm{gr}}-Z_{\mathrm{gr}}\right)(\mathrm{A}/\mathrm{Y})\right)_{\mathrm{gr}}. \tag{18.4}$$

In the face of population ageing, future increases in income per capita will mainly come through the productivity component, as represented by the growth rate of the output per hour worked. Using a standard Cobb–Douglas production function (i.e. $Y = K^{\alpha}(ZH)^{1-\alpha}$, where α is the capital share, K is the stock of productive capital, Z is exogenous technology and H is the stock of human capital), and dividing the output by the total number of hours worked (L), we have that the growth rate of the output per hour worked is (see the derivation in Appendix 18.2).

$$(Y/L)_{gr} = \frac{\alpha}{1-\alpha}(K/Y)_{gr} + Z_{gr} + (H/L)_{gr}. \tag{18.5}$$

The first term on the right-hand side of Eq. (18.5) is the capital-to-output ratio, which is determined by the world capital market in open economies. The second term is the growth rate of the exogenous technological progress (Z_{gr}) and the last term on the right-hand side is the growth rate of human capital per hour worked.

Combining Eqs. (18.4) and (18.5), the growth rate of consumption per capita can be decomposed into the following five components:

$$\left(\frac{C}{N}\right)_{gr} = \frac{\alpha}{1-\alpha}\left(\frac{K}{Y}\right)_{gr} + Z_{gr} + \left(\frac{H}{L}\right)_{gr} + \left(\frac{L}{N}\right)_{gr} + \left(1-\alpha+\left(r-N_{gr}-Z_{gr}\right)\frac{A}{Y}\right)_{gr}. \tag{18.6}$$

Given that $(\alpha/(1-\alpha))(K/Y)_{gr}$ is the same across all countries (since we assumed an open economy setup) and Z_{gr} is assumed to be exogenously given, Eq. (18.6) implies that the impact of life expectancy improvements on the growth rate of consumption per capita is determined by the following three components: the human capital per hour worked (i.e. $(H/L)_{gr}$), the first demographic dividend (i.e. $(L/N)_{gr}$) and the evolution of the fraction of asset income not devoted to investment (i.e. $(1-\alpha+(r-N_{gr}-Z_{gr})(A/Y))_{\mathrm{gr}}$), which is a function of the financial wealth-to-income ratio (A/Y).

All of these three terms are influenced by changes in the population age structure together with changes in the economic behaviour of individuals along their life cycle. In a first step we discuss, based on previous literature, the effect of life expectancy improvements on human capital (H) and aggregate financial wealth (A) in Sections 18.2.2 and 18.2.3. We explain in Section 18.2.3 how to empirically calculate the aggregate financial wealth (A), using NTA data, and the role that the transfer system plays on the aggregate financial wealth. To assess the total impact of the expected future increases in life expectancy on consumption per capita in countries with different transfer systems we apply the OLG-NTA model in Section 18.2.4 [1].

18.2.2 Survival Improvements and the Stock of Human Capital

An essential factor for production and an important driver of per capita consumption growth is the stock of (productive) human capital, which measures the total economic value of the skills and quality of a country's workforce. The stock of human capital in year t, $H(t)$, is calculated as the total effective labour supplied by the population:

$$H(t) = \int_0^{\omega} N(x,t)h(x,t)dx, \tag{18.7}$$

where ω is the maximum lifespan, $N(x,t)$ is the population size of age x in year t and $h(x,t)$ is the per capita effective labour supply at age x at time t. Per capita effective labour supply reflects factors such as individuals' behaviour, individuals' characteristics like education and health, as well as institutional settings that incentivize or discourage labour at specific ages.

To calculate $h(x,t)$, the per capita age profile of labour income is frequently used as a good proxy, since both measures are proportional from a cross-sectional

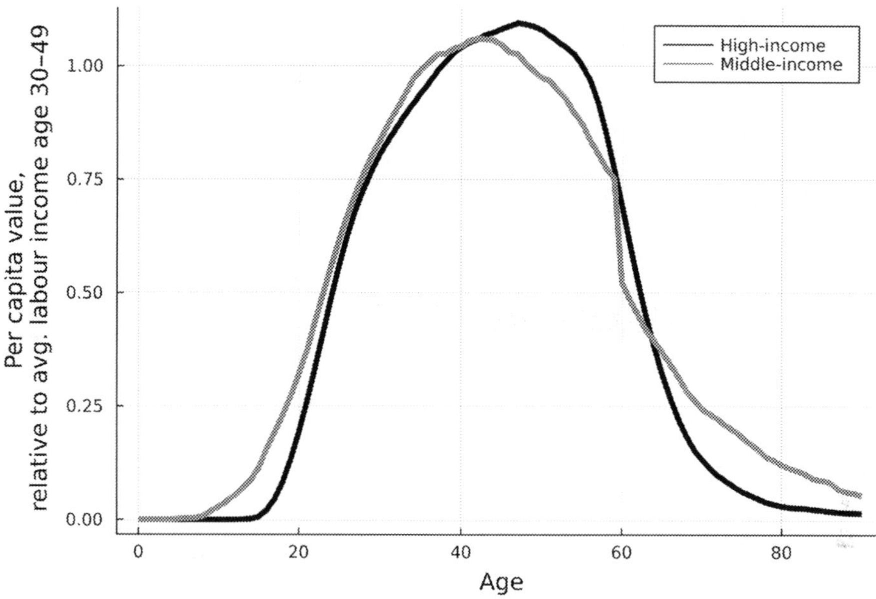

Figure 18.2 Average per capita labour income profiles by OECD income level, year 2007. Source: https://ntaccounts.org/web/nta/show/NTA%20data%20visualization data visualization, [22].

perspective.[1] Figure 18.2 shows the average per capita age profiles of labour income for our two country groups that differ by their income level. All age profiles are taken from the NTA database (www.ntaccounts.org), as, for instance, presented in Lee and Ogawa [8]. These profiles show two different patterns that are consistent with studies on the effects of pension and tax systems on labour supply [9]. For instance, many HICs (black line), that frequently have a more aged population, have already implemented policies that promote delaying the retirement age [10]. Their labour income profiles are frequently higher at older ages because of retirement policies that incentivize late retirement, the existence of seniority-wage schemes and a more educated labour force. In MICs, the age profile of labour income (grey line) reflects: (1) that young workers are better educated than older workers; and (2) that many retirement policies still create incentives for retiring at specific ages for some workers; while (3) other workers need to continue working until very old ages. This is probably due to the lack of a generous welfare state or a universal pension system that provides enough support at old age to all retirees.

Many factors can shape the evolution of the stock of human capital (e.g. mortality, fertility, migration, education, economic shocks, changes in institutions etc.). In the following we focus on explaining how improvements in survival affect the stock of human capital.

[1] Per capita labour income earned at age x in year t $y(x, t)$ is equivalent to $w(t)h(x, t)$, where $w(t)$ is the wage rate per effective unit of labour supply in year t.

Table 18.1 Direct and indirect effects of survival improvements at different stages over the life cycle on the stock of human capital

		Survival improvements		
Effects		Before work	At working ages	After retirement
Direct: *Demographic*				
$N(x,t)$	Short run		+	
	Long run	+	+	
Indirect: *Economic*				
$h(x,t)$	Retirement age		–	+
	Hours worked			
	Education	+	+	+
Aggregate				
$H(t)$	Short run		±	+
	Long run	+	±	+

Note: (+) positive effect, (–) negative effect, blank spaces imply no clear influence in either direction.

An increase in survival leads to direct (demographic) and indirect (economic) effects on $H(t)$. A direct effect is related to a change in the population size $N(x, t)$. An indirect effect relates to a change in the economic characteristic $h(x,t)$ itself. These effects will depend on the age at which survival improves (see Table 18.1). Given that demographic changes may have persistent effects over time, we further distinguish in Table 18.1 between effects in the short and in the long run. In the short run, only survival improvements at working ages have a positive effect on Nh, since this increases the number of workers. In the long run and in a one-sex female model, only survival improvements until menopause, which raises the number of women who survive to fertile years, guarantee a persistent increase in the population growth rate, *ceteris paribus* the fertility pattern.[2] Survival improvements after retirement do not influence the number of workers, neither in the short nor in the long run. The positive sign of the direct impact of survival improvements on the direct demographic effect contrasts with the non-monotonicity of the impact of survival improvements on economic variables (the indirect effect). This is because survival improvements at different stages in the life cycle create different incentives on economic variables such as retirement age, hours worked and education. If we start from the end of life, survival improvements after retirement increase the willingness to postpone retirement and to continue investing in education [11, 12, 13], since individuals need more wealth to finance the additional years lived in retirement. Survival improvements at working ages, however, lead to two opposite effects on $h(.)$. On the one hand, they

[2] Notice that survival improvements at retirement ages increase the population at old age, which will not raise the total number of workers unless that there is a significant fraction of labour income earners at old age.

lead to early retirement [11, 14], since individuals have a higher expected wealth – due to the increasing odds of receiving a future income – that allows them to buy more leisure time in the form of retirement. On the other hand, the higher expected wealth increases the returns to education, known as the Ben-Porath mechanism [15], which incentivizes continued studying [13]. Thus, if survival is not considered in the calculation of the expected total number of years worked, these two effects tend to shorten this later. However, interestingly, if survival is taken into consideration, Lee found that the expected number of years worked – that is, survival weighted – remained constant [16]. Survival improvements in early life have a positive effect on educational attainment [17, 18], which raises $h(.)$. However, a clear causal relationship between survival improvements and hours worked has not been found [12].

The aggregate effects shown in Table 18.1 are the result of combining the direct demographic and indirect economic effects. At the aggregate level survival improvements at retirement increase $H(t)$ both in the short and in the long run, while survival improvements during childhood only have a positive effect on $H(t)$ in the long run. The gains in life expectancy at working ages have an ambiguous effect on $H(t)$ in the short as well as in the long run, since the positive demographic effects can be offset by the reduction in labour supply due to later entrance into the labour market and earlier retirement.

18.2.3 Survival Improvements and Financial Wealth

Another potential driver of per capita consumption growth is aggregate financial wealth. Financial wealth positively affects per capita consumption because it generates asset income to the owners of assets and it raises the return to complementary production factors such as the stock of human capital.[3] Financial wealth is accumulated by individuals to primarily shift resources to older ages (when consumption exceeds labour income) or to leave bequests to descendants [19]. Moreover, financial wealth is affected by the public transfer system since the latter can complement or substitute how consumption is financed at different stages in life.

The aggregate financial wealth in year t, $A(t)$, is given by the total financial wealth held by all individuals alive in year t:

$$A(t)=\int_0^{\omega}N(x,t)a(x,t)dx, \tag{18.8}$$

where $a(x,t)$ are the per capita assets held at age x in year t.

Survival improvements also affect $A(t)$, as summarized in Table 18.2. As for $H(t)$ we can differentiate between direct demographic and indirect economic effects on $A(t)$ and distinguish between survival improvements at different stages over the life cycle. In this case, a direct effect is related to a change in the population size at a specific age where the economic characteristic, $a(x)$, is not equal to zero, while an indirect effect relates to a change in the economic characteristic $a(x)$ itself. The direct

[3] Notice that by increasing the capital per worker, workers become more productive and their wages rise.

Table 18.2 Direct and indirect effects of survival improvements at different stages over the lifecycle on the stock of financial wealth

Effects		Survival improvements		
		Before work	At working ages	After retirement
Direct: *Demographic*				
$N(x,t)$	Short run		+	+
	Long run	+	+	
Indirect: *Economic*				
$a(x,t)$	Savings	–		+
Aggregate				
$A(t)$	Short run	–	+	+
	Long run	±	+	+

Note: (+) positive effect, (–) negative effect, blank spaces imply no clear influence in either direction.

effect of survival improvements on $N(x)$ are always non negative. In the short run, only increases in survival of adult individuals (workers and retirees) have positive effects, since children do not hold assets. In the long run, only survival improvements until menopause, which raises the number of women who survive to fertile years, guarantee a persistent increase in the population growth rate, *ceteris paribus* the fertility pattern. However, the impact of survival improvements on per capita financial wealth $a(x,t)$ is more complex. To determine financial wealth involves having knowledge about individuals' preferences, future interest rates, financial constraints, the state of the economy, all the expected public and private transfers and even the demographic characteristics of the household, among others.

To simplify the analysis, we consider a simple savings model without public transfers (e.g. publicly provided childcare and education, public health care or public pensions). The most frequently used model to explain savings at the micro level is the life-cycle theory of saving. The model predicts that survival improvements during childhood reduce life-cycle savings, because parents need to take care of more surviving children [20]. Survival improvements after retirement, however, induce individuals to increase their life-cycle savings [21], since they need to finance the additional years in retirement, *ceteris paribus* other transfers and the state of the economy. At the aggregate level, the last two rows of Table 18.2 summarize how survival improvements affect $A(t)$ in the short and in the long run. The last two rows are obtained combining the direct demographic effects and the indirect economic effects. They show that survival improvements at working ages and at retirement ages increase $A(t)$ in the short and in long run. In contrast, survival improvements during childhood reduce $A(t)$ in the short run, while the effect is ambiguous in the long run. This is because parents will need to reduce their savings in order to take care of more surviving children, which has a negative effect on $A(t)$, while there will be an increasing number of adults as a result of the higher population growth rate, which has a positive effect on $A(t)$.

However, in reality, since future needs can be financed both by assets and by transfers, per capita financial wealth is also influenced by all kinds of public and private transfers. To make explicit all the transfers that individuals have at each age, the NTA framework allows per capita financial wealth at age x to be computed as the difference between life cycle and transfer wealth [22]:

$$a(x,t) = W(x,t) - T(x,t), \tag{18.9}$$

where $W(x,t)$ and $T(x,t)$ are, respectively, the life-cycle wealth and the transfer wealth of an individual at age x in year t [23–25], which is the net balance between all the expected received and given transfers. The life-cycle wealth is defined as the present value of the expected difference between the future per capita consumption and the future per capita labour income:

$$W(x,t) = \int_x^{\infty} \left(c(a,t+a-x) - y(a,t+a-x)\right) e^{-\int_x^a r(t+p-x)+\mu(p,t+p-x)dp} da, \tag{18.10}$$

where c denotes per capita consumption (public and private), y is per capita labour income, which is given by the product of the wage rate w and the per capita effective labour supply h, r is the interest rate and μ is the mortality rate. Life-cycle wealth reflects the amount of wealth demanded by an individual to finance their remaining consumption given expected future earnings [26]. As Eq. (18.9) shows, the life-cycle wealth demanded by an individual can be financed by assets and by transfers.

Using data from the Global NTA database [22], Figure 18.3(a) shows the age profiles of per capita consumption (public and private) and per capita labour income (y) for our two country groupings (high- and middle-income groups). Notice from Eq. (18.10) how these two profiles are key to calculate life-cycle wealth. We can extract two important results from this figure. First, per capita consumption is higher than labour income for dependent children and old-age groups (known as life-cycle deficit), while per capita consumption is lower than the labour income for prime-age workers (known as life-cycle surplus). Second, per capita consumption (relative to the average labour income between ages 30–50) is higher for dependent children and old-age groups in HICs as compared to MICs.

In comparison to life-cycle wealth, transfer wealth is defined as the present value of the expected net transfers to be received in the future:

$$T(x,t) = \int_x^{\infty} \left(\tau^+(a,t+a-x) - \tau^-(a,t+a-x)\right) e^{-\int_x^a r(t+p-x)+\mu(p,t+p-x)dp} da, \tag{18.11}$$

where τ^+ are per capita transfers received (public and private transfer inflows) and τ^- are per capita transfers given (public and private transfer outflows). An example of a well-known transfer wealth is social security wealth or, equivalently, the present value of expected future benefits minus remaining social contributions. Hence, Eqs. (18.10) and (18.11) are important because they explicitly show the necessity of using an economic model that simulates the future evolution of the NTA flows.

Figure 18.3(b) shows age profiles of transfers: per capita transfer inflows (TI or τ^+), which are transfers received from other generations, and per capita transfer outflows (TO or τ^-), or transfers given to other generations, for our two country groupings by

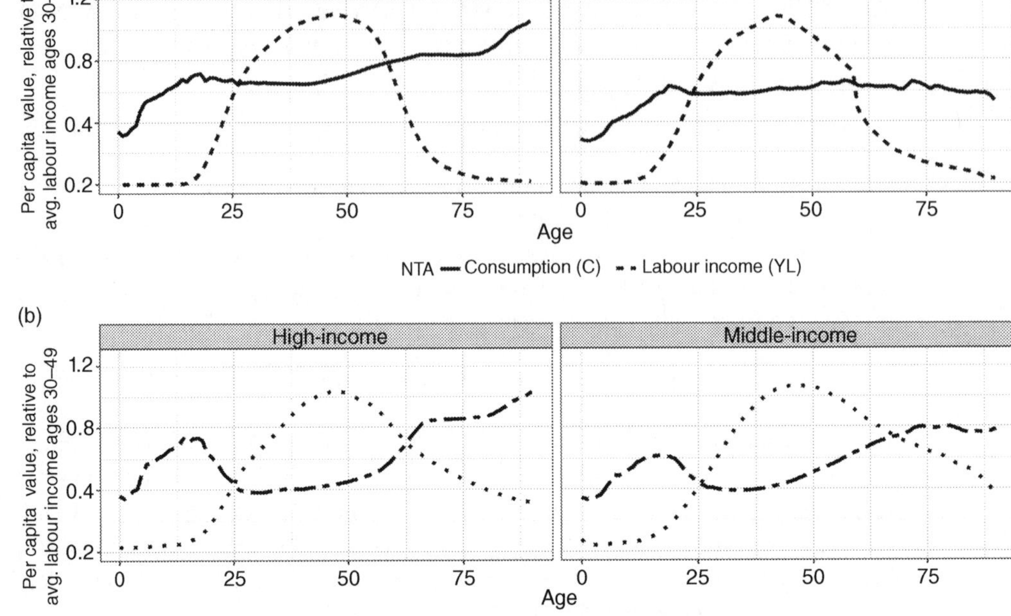

Figure 18.3 Economic life cycle (a) and transfers inflows (received) and outflows (given) (b) in per capita values: average of European countries around 2007.
Note: C denotes total consumption (public and private), YL is labour income (labour earnings plus mixed income), TI denotes private and public transfer inflows (transfers received from other generations of the family and transfers received from the public sector), TO denotes private and public transfer outflows (transfers given to other generations of the family and transfers paid to the public sector).
Source: https://ntaccounts.org/web/nta/show/NTA data visualization, [22].

income level. This figure shows that transfers inflows for dependent children, which are mostly related to educational expenditures, are higher in HICs than in MICs.

Using (18.9) in (18.8) gives a key identity in the NTA methodology [23]. The aggregate financial wealth at time t is the difference between the aggregate life-cycle wealth and the aggregate transfer wealth at time t

$$A(t)=\int_0^{\omega}N(x,t)\big(W(x,t)-T(x,t)\big)dx=W(t)-T(t). \tag{18.12}$$

In countries where the aggregate financial wealth exceeds the aggregate life-cycle wealth, Eq. (18.12) implies that $T<0$. According to Eq. (18.11), negative transfer wealth occurs when the remaining transfers to be given (outflows) exceed the remaining transfers to be received (inflows), which, according to Figure 18.3(b), occurs early in the life cycle of individuals. Thus, negative aggregate transfer wealth is associated with an economy in which transfers go from prime-age workers to dependent children. In contrast, in countries where aggregate financial wealth falls short of aggregate life-cycle wealth, transfers go from prime-age workers to old-age groups

$(T > 0)$. Notice in Figure 18.3(b) that when individuals approach old age the remaining transfers to be received exceed the remaining transfers to be given. Thus, positive aggregate transfer wealth is associated with an economy in which transfers go from prime-age workers to old-age groups. Hence, in countries with generous pension systems (i.e. high transfer inflows at old age), it is expected that with population ageing T will increase and the difference between W and A will be reduced.

From Eq. (18.12) we obtain that changes in life expectancy will not only have an effect on the age distribution of the population (through N) but also on the demand for life-cycle wealth and on the evolution of the transfer system.

18.2.4 Simulation Results of the OLG-NTA Model

The calculation of Eq. (18.12) requires future NTA profiles. Unfortunately, since NTAs are based on existing data, we cannot use them for analysing the future impact of the increasing life expectancy on per capita consumption growth. To cope with the lack of information for each country on future NTA profiles, in this section we use the multi-country OLG-NTA model developed by Sánchez-Romero to construct for each country the future evolution of the NTA profiles [1] (see Box 18.1).

Box 18.1 The OLG-NTA Model

The OLG-NTA model takes as given all per capita public transfer profiles as well as per capita private transfers on education and health relative to the labour income from the Global NTA database [29]. We scale all per capita transfers received each year according to the evolution of the average labour income between ages 30 to 49. The OLG-NTA model endogenously calculates the private consumption of the household and the asset-based reallocation for all cohorts in each country.

The multi-country OLG-NTA model has the following additional characteristics. Each country is comprised by a neoclassical firm, a government and households. Firms produce a single good, which can be stored or consumed by households. To produce the single good, firms combine capital and labour using Cobb–Douglas technology under constant returns to scale (with the production elasticities being one-third for capital and two-thirds for labour). The interest rate and the wage rate are determined in international markets. For the sake of comparability with the European Commission [30], we assume Z_{gr} is constant over time and equal to 1.5% per year for our simulations. Households are comprised of a household head and dependent children. Household heads supply labour, taking the number of hours worked as given (which is indirectly calculated using the labour income profile), optimize consumption for all household members and save for retirement. Individuals face exogenous mortality risks that differ by age. The labour income received by the household head is determined in competitive markets and based on the educational attainment and experience of the household head. Over the life cycle of household heads, the number of dependent children varies and is determined by fertility, mortality and the number of children who had already left the parental home by the age of 20. In addition, we assume all governments run a balanced budget. Demography – that is, age-specific fertility and mortality, population size and age structure – is considered exogenous.

Our population projections for each country rely on the medium variant from World Population Prospects 2019 by the United Nations (UN) Population Division. Population is divided into single-year age groups and the time interval is a year. The mean years of education of each birth cohort for each country are taken from Wittgenstein Centre for Demography and Global Human Capital (2018) database [31]. For further details about the model see Sánchez-Romero [1].

Based on the OLG-NTA model, Table 18.3 presents simulated average per capita growth rates for selected macroeconomic variables from 2020 to 2100 for 35 European countries grouped by OECD income level. We chose the year 2100 as the final date in order to account for the contribution to per capita income growth of the cohort born in 2020 until death. The table is divided in two panels and contains results in four columns. Each panel summarizes the results for European countries belonging to the same OECD income level (either high or middle income). The first column reports growth rates for the baseline simulation (see the evolution of the total fertility rate, life expectancy at birth and the mean years of schooling for the baseline simulation in our two income groups in Figure 18.1). The last three columns show the growth rates of three simulated counterfactuals, in which we cancel the effect of one demographic component (i.e. life expectancy, fertility and education) by keeping its rates at the level of 2020 from that year onwards. Thus, through counterfactuals, we can estimate the total contribution of each demographic factor to the growth rate of consumption per capita, since we not only account for direct demographic effects, but also for indirect demographic effects (see Tables 18.1 and 18.2). To assess the total contribution of each demographic factor to the growth rate, we follow Sánchez-Romero and colleagues by comparing the baseline simulation to each counterfactual [27, 28].

Rows 1 and 2 in the HICs and, respectively, in the MICs report the average yearly growth rates for the period 2020–2100 of consumption per capita, $\left(C/N\right)_{gr}$, and of income per capita, $\left(Y/N\right)_{gr}$. In the baseline simulation (or status quo) we obtain that the average yearly growth rate of consumption per capita is 1.35% in HICs and 1.44% in MICs. In both panels (for HICs and MICs countries), the average yearly growth rates of income per capita are smaller than the assumed exogenous labour productivity of 1.5% per year, which is mainly explained by the negative first demographic dividend, $\left(L/N\right)_{gr}$ (see row 7) of −0.24% in both country groups. Moreover, it is worth noticing two facts. First, the average yearly growth rates of income per capita are quite similar across both income groups. Second, the average yearly growth rate of consumption per capita is lower in HICs (1.35%) than in MICs (1.44%). According to Eq. (18.7), this difference is explained by the moderated decline in the financial

Table 18.3 Source of average yearly growth rates from 2020 to 2100 in European countries by income level and demographic component (annual average logarithmic rate in %)

		Counterfactual		
	Baseline	I: Without increase in life expectancy	II: Without increase in fertility	III: Without increase in education
High-income countries (HICs)†				
1. $(C/N)_{gr}$	1.35	1.43	1.38	1.28
2. $(Y/N)_{gr}$	1.44	1.51	1.36	1.39
Wealth accounts				
3. $(W/Y)_{gr}$	−0.91	−1.10	−0.41	−0.97
4. $(A/Y)_{gr}$	−0.73	−0.57	−0.25	−0.82
5. $(K/Y)_{gr}$	0.24	0.17	0.10	0.27
Productivity component				
6. $(Y/L)_{gr}$	1.68	1.64	1.62	1.63
Translation component				
7. $(L/N)_{gr}$	−0.24	−0.12	−0.26	−0.24
Middle-income countries (MICs)†				
1. $(C/N)_{gr}$	1.44	1.51	1.51	1.34
2. $(Y/N)_{gr}$	1.46	1.55	1.38	1.40
Wealth accounts				
3. $(W/Y)_{gr}$	−0.41	−0.76	0.28	−0.52
4. $(A/Y)_{gr}$	−0.15	−0.22	0.26	−0.21
5. $(K/Y)_{gr}$	0.24	0.17	0.10	0.27
Productivity component				
6. $(Y/L)_{gr}$	1.70	1.65	1.65	1.63
Translation component				
7. $(L/N)_{gr}$	−0.24	−0.11	−0.28	−0.24

Note: †All simulation results are obtained with a growth rate of the labour-augmented technological progress of 1.5% per year, which corresponds to an annual total factor productivity (TFP) growth rate with a capital share of 1/3 of 1%. In the baseline we follow the UN medium variant demographic assumptions for the period 2020–2100, which implies an average percentage increase in life expectancy of 10.9 in HICs and of 13.1 in MICs, in the total fertility rate of 10.1 in HICs and of 12.1 in MICs, and, using the Wittgenstein Centre (WIC) Human Capital Data Explorer assumption, an increase in the mean years of schooling of 21.1 in HICs and of 22.7 in MICs.
Source: Authors' calculations based on the OLG-NTA model.

wealth-to-output ratio (A/Y) of −0.15% in MICs compared to the reduction in (A/Y) of −0.73% in HICs.

Rows 3–5 in Table 18.3 report the average yearly growth rate of life-cycle wealth, financial wealth and the stock of productive capital-to-output ratio. Three important results can be highlighted. First, the average yearly growth rate of the stock of

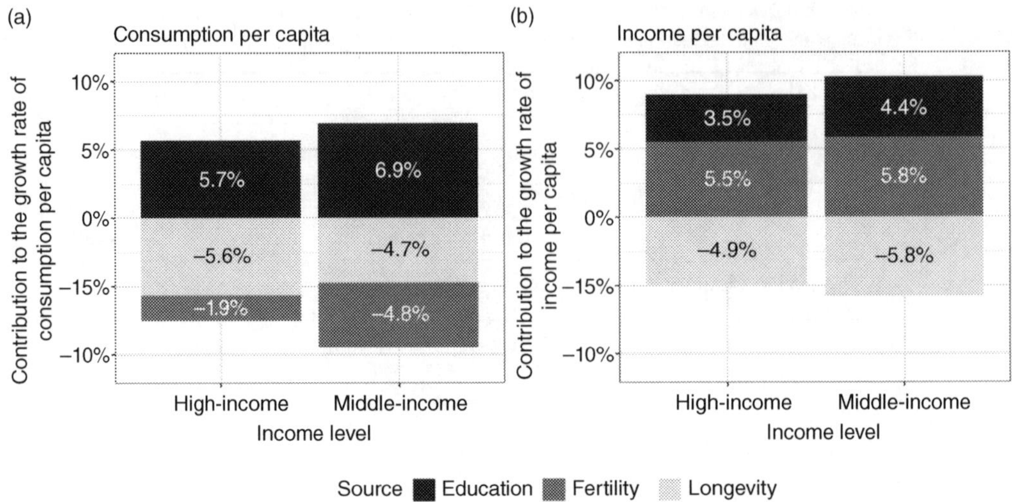

Figure 18.4 Contribution to the growth rate of consumption per capita (a) and income per capita (b) from 2020 to 2100 in European countries by demographic component and income level.
Note: The contribution is calculated as $\left((X)_{gr}^{Baseline} - (X)_{gr}^{Counterfactual}\right) / (X)_{gr}^{Baseline}$.
Source: Authors' simulations.

productive capital-to-output ratio (i.e. 0.24%) implies a future increase in the capital per effective worker, known as capital deepening [32, 33], and a small decline in asset prices. Thus, like other multi-country general equilibrium models we find there that world population ageing has a small impact on asset prices [34]. Second, looking at HICs, we obtain in the baseline simulation that the life-cycle wealth-to-income ratio declines faster than the growth rate of financial wealth-to-income ratio, $(W/Y)_{gr} = -0.91\%$ versus $(A/Y)_{gr} = -0.73\%$. According to Eq. (18.10), this difference implies that consumption per capita will be increasingly financed by transfers to the elderly in HICs. Notice that since the aggregate transfer wealth is, in general, negative, a lower growth rate implies an increase in T. In other words, there will be more upward transfers from young generations to older generations, *ceteris paribus* the private and public transfer system. Similar results are also obtained for MICs, since $(A/Y)_{gr} > (W/Y)_{gr}$, with the difference that the increase in T declines in MICs. Third, given that $r > N_{i,gr} + Z_{gr}$ [35], the difference in the growth rate of consumption per capita across income groups is explained by the evolution of the aggregate financial wealth-to-income ratio (see row 4 (A/Y) in Table 18.3). In HICs, A/Y declines faster than in MICs (−0.73% vs −0.15%).

To assess the relative importance of life expectancy improvements on $(C/N)_{gr}$ and $(Y/N)_{gr}$, we compare their effects in relation to other demographic components. Figure 18.4 depicts the contribution of life expectancy, fertility and education on the growth rate of consumption per capita (Figure 18.4(a)) and income per capita (Figure 18.4(b)) from 2020 to 2100. The plot is based on data from Table 18.3. Figure 18.4 shows a negative impact of life expectancy gains on $(C/N)_{gr}$ and $(Y/N)_{gr}$ from 2020 to 2100. The impact of life expectancy on $(C / N)_{gr}$ is –5.6% in HICs and –4.7% in MICs. The impact of life expectancy on $(Y / N)_{gr}$ ranges between –4.9% in HICs and –5.8% in MICs. The negative impact of life expectancy on $(Y / N)_{gr}$ is explained by the increasing age of the workforce. This is explained by the lower productivity growth of older workers as compared with younger workers. However, it is worth keeping in mind that the negative effect of life expectancy could be mitigated or even reversed when life expectancy positively influences education, which is known as the Ben-Porath mechanism, or would lead to postponing the retirement age [11, 13, 36].

Comparing the contribution of life expectancy on $\left(C / N\right)_{gr}$ and $\left(Y / N\right)_{gr}$ to other demographic components, we see that higher education has a positive effect on $\left(C / N\right)_{gr}$ and $\left(Y / N\right)_{gr}$. Higher fertility positively influences $\left(Y / N\right)_{gr}$, since the workforce is rejuvenated, but it has a negative effect on $\left(C / N\right)_{gr}$, which is consistent with the fact that lower fertility improves the standard of living, or $\left(C / N\right)_{gr}$, in ageing populations [37].

18.3 Impact of Increasing Life Expectancy on Intergenerational Income Inequality

Life expectancy gains affect not only economic growth but also intergenerational income inequality. From an intergenerational perspective, life expectancy gains coupled with decreasing fertility rates raise the old-age dependency ratio (the proportion of people above age 64 relative to the population between ages 15 and 64). On the one side, if old-age transfers remain unchanged, a higher old-age dependency ratio implies that the cost borne by workers to support the transfer system will increase, their disposable income will decline and the risk of poverty among workers will rise. On the other hand, if a reduction in old-age transfers is implemented to keep the current cost on workers unchanged, old-age transfers will decline and the risk of poverty among the elderly will increase. Thus, survival improvements and lower fertility rates can raise income inequality across age groups (i.e. intergenerational inequality) in ageing populations.

To assess how population ageing may affect intergenerational inequality we apply the Gini coefficient calculated across ages. The Gini coefficient is a measure of the degree of inequality that ranges between zero and one. A value of zero means perfect equality, while a value of one corresponds to the maximum inequality. We calculate the Gini coefficient using our simulated per capita profiles of labour income, asset income and net public transfers obtained through the OLG-NTA model. We exclude ages from 0 to 20 in our calculation of the Gini coefficient, since during this stage of life individuals generally neither earn labour income nor asset income.

Figure 18.5 shows the time path of the average Gini coefficient across European countries for our two groups of countries (high-income and middle-income countries). Figure 18.5(a) depicts the Gini coefficient without net (in-cash) public transfers, while in Figure 18.5(b) the Gini coefficient includes net public transfers. Figure 18.5 depicts several interesting results. First, Figure 18.5(a) shows a higher Gini

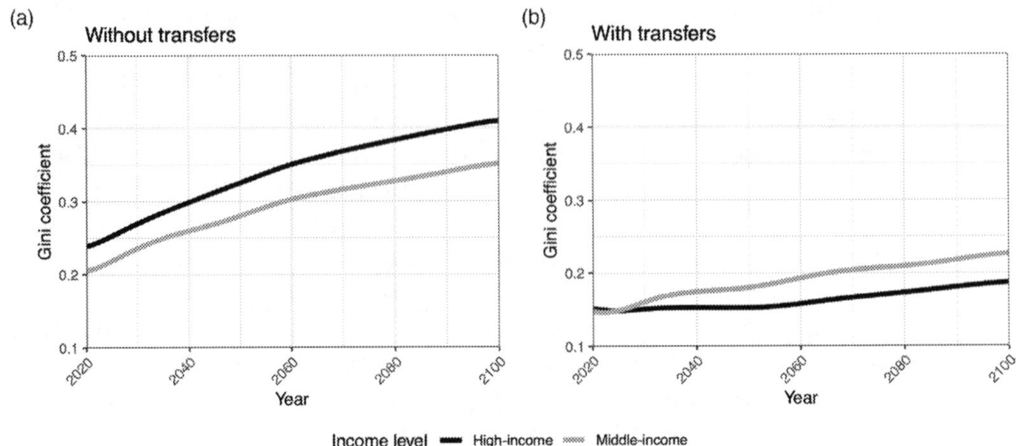

Figure 18.5 Time path of the Gini coefficient without transfers (a) and with transfers (b) in European countries from 2020 to 2100 by OECD income level.
Source: Authors' simulations.

coefficient for HICs compared to MICs if we neglect transfers. For both country groups the Gini coefficient is projected to increase in the twenty-first century with a slightly higher increase in HICs (from 0.24 in 2020 to 0.41 in 2100 compared to an increase from 0.20 in 2020 to 0.35 in 2100 in MICs). Second, Figure 18.5(b) shows that once public transfers are considered, the Gini coefficient is significantly lower. Moreover, inequality becomes lower in HICs compared to MICs, due to the more generous pension systems in the former relative to the latter. Indeed, in HICs inequality is projected to be rather flat during the period 2020–2050 and slightly increases from 0.16 in 2050 to 0.19 in 2100. Instead, in MICs, inequality increases from 0.16 in 2020 to 0.23 in 2100.

To gain insight into the main demographic drivers of the increasing inequality during the twenty-first century, we consider the growth rates of the Gini coefficient. In particular, we assess the contribution of each demographic component on inequality by comparing the results from the baseline simulation (demography based on the medium variant from World Population Prospects 2019 by the UN Population Division) with three different counterfactuals, in which we keep one of the demographic components (life expectancy, fertility) and education constant each time.[4] The results are presented in Table 18.4.

Four main results follow from Table 18.4 when looking at the contribution of each demographic factor on the Gini coefficient without transfers. First, higher life expectancy and fertility rates induce higher-income inequality. This holds for both country groups as well as for both variants of the Gini coefficient (with and without transfers). See Figure 18.1(a) and (b) and the note after Table 18.4. Life expectancy

[4] We exclude from the analysis net migration, since in the simulations we consider that migrants have the same economic life cycle as nationals.

Table 18.4 Source of the growth rate of the Gini coefficient across age groups from 2020 to 2100 in European countries by demographic component and income level (annual average logarithmic rate in %)

		Counterfactual					
	Baseline	I: Life expectancy		II: Fertility		III: Education	
Income level	Growth	Growth	(contribution)	Growth	(contribution)	Growth	(contribution)
High							
without transfers	0.69	0.47	(31.0)	0.59	(14.3)	0.71	(−3.8)
with transfers	0.25	0.23	(9.8)	0.11	(54.7)	0.30	(−19.9)
Middle							
without transfers	0.68	0.45	(33.6)	0.55	(19.1)	0.72	(−5.7)
with transfers	0.53	0.39	(25.3)	0.37	(30.4)	0.58	(−10.0)

Note: Contribution values are calculated as $\left(\text{Growth}^{\text{Baseline}} - \text{Growth}^{\text{Counterfactual}}\right) / \text{Growth}^{\text{Baseline}}$. The average percentage increase in life expectancy (counterfactual I) is 10.9 in high- and 13.1 in middle-income countries. In the baseline we follow the UN medium variant demographic assumptions for the period 2020–2100, which implies an average percentage increase in life expectancy of 10.9 in HICs and of 13.1 in MICs, in the total fertility rate of 10.1 in HICs and of 12.1 in MICs, and, using the Wittgenstein Centre for Global Human Capital Data Explorer assumption, an increase in the mean years of schooling of 21.1 in HICs and of 22.7 in MICs. The increase in life expectancy is not considered in counterfactual I, the increase in fertility is not considered in counterfactual II and the increase in education is not considered in counterfactual III.
Source: Authors' calculations based on the OLG-NTA model.

gains contribute to one-third of the increase in the growth rate of inequality without considering transfers (see column I in Table 18.4), while fertility contributes between 14.3% (for HICs) and 19.1% (for MICs) of the increase in the growth rate of inequality without considering transfers (see counterfactual II in Table 18.4).[5] Higher life expectancy increases inequality since higher life expectancy raises the number of elderly people relative to prime-age workers, which increases the cost of public old-age support systems and reduces the savings of prime-age workers and hence the asset income of retirees. As a result, inequality increases because the difference between the gross labour income, which workers receive, and the asset income, mainly received by retirees, widens.

Second, when the Gini coefficient includes net public transfers (second and fourth row in column counterfactual I in Table 18.4), the contribution of life expectancy gains on inequality is lower in countries with a more generous old-age public transfer system. This is because the income gap between the disposable income of workers

[5] The large contribution of demographic factors to the growth rate of the Gini coefficient are related to the measure we apply. When using a relative deviation, small baseline growth rates that are in the denominator exaggerate the contribution values, while absolute differences are not affected by the baseline growth rate. The reported effects of demographic factors on inequality are in line with those shown in the literature [38, 39].

and the pension benefit of retirees is reduced due to the increase in social contributions. For instance, in HICs, with more generous pension benefits, life expectancy gains only explain 9.8% of the increase in the growth rate of the Gini coefficient that includes transfers, while in MICs, with less generous pension systems, life expectancy gains explain 25.3% of the increase in the growth rate of the Gini coefficient with transfers.

Third, the counterfactual simulations show that an increase in fertility raises income inequality across cohorts. Similar results are theoretically shown at the population level by Sánchez-Romero and colleagues and Lam [38, 39]. The increases in income inequality when fertility increases across age groups is explained by the reduction in asset income at old age and the increase in labour income of workers. When the Gini coefficient includes net public transfers, the increase in inequality due to higher fertility is even higher (see, in Table 18.4, the contribution column in counterfactual II). This latter effect is opposite to life expectancy gains. Higher fertility reduces the old-age dependency ratio, which makes supporting pension benefits less costly and further increases the disposable income of workers. Consequently, the gap between disposable income, which workers earn, and pension benefits, which retirees receive, increases.

Fourth, education reduces both measures of inequality (see, in Table 18.4, the contribution column in counterfactual III). An increase in human capital raises labour income, life-cycle savings and asset income. Since all European countries have a financial wealth-to-output ratio above the world capital-to-output ratio, the investment in education yields, in these countries, an additional asset income at old age that further reduces the intergenerational income inequality.

18.4 Conclusion

To assess the effect of increasing life expectancy on economic growth and hence consumption growth, it is important to understand at which life stage survival increases as well as the mechanisms through which longer lives will influence the components of economic growth and the transfers that shift resources across generations. Based on a decomposition analysis of per capita consumption growth and previous literature we discussed the direct demographic and indirect economic effects of life expectancy increases on the aggregate human capital and financial wealth of a country. To also account for behavioural effects of changing demographic and economic environments in the future, we applied a multi-country dynamic general equilibrium economic model, which also allowed us to investigate the projected income inequality across generations. Besides life expectancy improvements we also investigated the role of changes in fertility and education for economic growth. To account for the heterogeneity in demographic developments in Europe we grouped the countries by income level (high income and middle income).

We obtain that in contemporary European countries life expectancy gains negatively impact on consumption per capita and output per capita from 2020 up to 2100. The lower increase in the total number of hours worked relative to the population growth rate implies that the average growth rate of consumption and output per capita will fall short of the growth rate of technological progress, which was set at 1.5% annually [30]. This negative effect is more pronounced in MICs compared to HICs. Thus, the negative effect of increasing life expectancy on output per capita growth will be smaller in HICs compared to that in MICs. This result can, however, be mitigated or even reversed if life expectancy induces higher education [15], which is known as the Ben-Porath mechanism, or if retirement ages increase. In case of per capita consumption growth, the order is reversed, with HICs having a more pronounced decline compared to MICs. Life expectancy gains in HICs will have a more pronounced negative effect on the accumulation of assets than in MICs, since the former have more generous transfers to the old-age population.

Life expectancy gains will also shape the income inequality across generations in the next decades. During the twenty-first century for all country groups the Gini coefficient of income inequality across generations (without considering transfer flows) will increase, with a more pronounced increase in HICs. This is explained by the increasing reduction in asset income at old age relative to the gross income of prime-age workers. Since the proportion of elderly people relative to prime-age workers will increase due to life expectancy gains, and hence the cost of public old-age support systems will also rise, retirees will have accumulated a lower level of savings for retirement and therefore their asset income will be lower. When the Gini coefficient includes public transfers, we observe a significant reduction in income inequality across generations. Inequality is reduced more in HICs compared to MICs, due to the more generous pension systems in HICs relative to that in MICs. Life expectancy gains contribute to one-third of the increase in the growth rate of such inequality without considering transfers. The increase in inequality is based on the widening of the difference between the gross labour income, which workers receive, and the asset income, mainly received by retirees. A more generous old-age public transfer system reduces the impact of life expectancy gains on inequality.

To continue experiencing economic growth and avoid intergenerational inequality, while enjoying increases in life expectancy, our results suggest the necessity of increasing the labour supply and reducing the cost of the transfer system to old age. Several policies can be implemented, which have not been analysed in this chapter. These include increasing the labour force at the intensive and extensive margin by also fostering labour-force participation for females, the unemployed and migrants. At the same time, transfer systems need to be reformed to cope with the changing age structure and reduce inequality. A multi-pillar system that not only focuses on state pensions but also supports pension programmes at the firm level and individual savings for old age is an important further development that also will help to diversify demographic and economic risks.

Appendix 18.1 Growth Rate of the Marginal Propensity to Consume

In an open economy the aggregate consumption satisfies

$$C = (1-s)Y = Y_l + rA - \dot{A}, \tag{A.1}$$

where s is the saving rate, Y is total output, C is the total consumption, which includes public and private consumption, Y_l is the total labour income A is the financial wealth, and $\dot{A}$ is the change in financial wealth with respect to time or household savings. The total labour income Y_l can be expressed as a fraction $(1-\alpha)$ of the total output, that is, $Y_l = (1-\alpha)Y$, where, under constant returns to scale, α is the capital share. Taking Y as common factor in (A.1) we have

$$(1-s)Y = \left(1-\alpha + rA/Y - \dot{A}/Y\right)Y. \tag{A.2}$$

Dividing both sides of (A.2) by the total output, provided that $\dot{A}/Y = (A/Y)\left((A/Y)_{\mathrm{gr}} + Y_{\mathrm{gr}}\right)$, and if the economy is assumed to be in a balanced growth path (i.e. $(A/Y)_{\mathrm{gr}} = 0$ and $Y_{\mathrm{gr}} = \mathrm{N}_{\mathrm{gr}} + \mathrm{Z}_{\mathrm{gr}}$), with Z_{gr} denoting the growth rate of the technological progress, the growth rate of the marginal propensity to consume is:

$$(1-s)_{\mathrm{gr}} = \left(1-\alpha + \left(r - N_{\mathrm{gr}} - Z_{\mathrm{gr}}\right)(\mathrm{A}/\mathrm{Y})\right)_{\mathrm{gr}}. \tag{A.3}$$

Appendix 18.2 Growth Rate of the Output per Hour Worked

Dividing both sides of the Cobb–Douglas production function by Y^{α}, solving for Y and dividing both sides by the total number of hours worked L gives

$$Y/L = (K/Y)^{\frac{\alpha}{1-\alpha}} ZH/L. \tag{A.4}$$

Taking logarithms in both sides of (A.4) and differentiating with respect to time, we find that the growth rate of output per hour worked equals the growth rate of the capital-output ratio, the growth rate of the labour-augmenting technological progress and the growth rate of human capital per hour worked.

References

1. Sánchez-Romero, M. 2022. Assessing the generational impact of COVID-19 Using National Transfer Accounts (NTAs). *Vienna Yearb. Popul. Res.* **20**, 1–35 (doi:10.1553/population-yearbook2022.res1.2).
2. Mason, A., Lee, R., Members of the NTA Network. 2022. Six ways population change will affect the global economy. *Popul. Dev. Rev.* **48**, 51–73 (doi:10.1111/padr.12469).
3. Leon, D.A. 2011. Trends in European life expectancy: a salutary view. *Int. J. Epidemiol.* **40**, 271–277.

4. Mackenbach, J.P., Looman, C.W. 2013. Life expectancy and national income in Europe, 1900–2008: an update of Preston's analysis. *Int. J. Epidemiol.* **42**, 1100–1110.
5. Lutz, W., Goujon, A., Stonawski, M., Stilianakis, N., Samir, K. 2018. *Demographic and Human Capital Scenarios for the 21st Century: 2018 Assessment for 201 Countries* Publications Office of the European Union.
6. Kelley, A.C., Schmidt, R.M. 2005. Evolution of recent economic-demographic modeling: a synthesis. *J. Popul. Econ.* **18**, 275–300.
7. Bloom, D.E., Williamson, J.G. 1998. Demographic transitions and economic miracles in emerging Asia. *World Bank Econ. Rev.* **12**(3), 419–455.
8. Lee, S.H., Ogawa, N. 2011. Labour income over the lifecycle. *Popul. Aging Gener. Econ.* 109–135.
9. Gruber, J., Wise, D.A. 1999. *Social Security and Retirement around the World.* University of Chicago Press.
10. OECD (Organisation for Economic Co-operation and Development. 2019. *Pensions at a Glance 2019: OECD and G20 Indicators.* OECD Publishing.
11. d'Albis, H., Lau, S.-H.P., Sánchez-Romero, M. 2012. Mortality transition and differential incentives for early retirement. *J. Econ. Theory* **147**, 261–283.
12. Restuccia, D., Vandenbroucke, G. 2013. A century of human capital and hours. *Econ. Inq.* **51**, 1849–1866.
13. Sánchez-Romero, M., d'Albis, H., Prskawetz, A. 2016. Education, lifetime labour supply, and longevity improvements. *J. Econ. Dyn. Control* **73**, 118–141.
14. Kalemli-Ozcan, S., Weil, D. 2010. Mortality change, the uncertainty effect, and retirement. *J. Econ. Growth* **15**, 65–91.
15. Ben-Porath, Y. 1967. The production of human capital and the life cycle of earnings. *J. Polit. Econ.* **75**(4), 352–365.
16. Lee, C. 2001. The expected length of male retirement in the United States, 1850–1990. *J. Popul. Econ.* **14**, 641–650 (doi:10.1007/s001480100064).
17. Bleakley, H. 2007. Disease and development: evidence from hookworm eradication in the American South. *Q. J. Econ.* **122**(1), 73–117.
18. Field, E., Robles, O., Torero, M. 2009. Iodine deficiency and schooling attainment in Tanzania. *Am. Econ. J. Appl. Econ.* **1**, 140–169.
19. Mason, A., Ogawa, N., Chawla, A., Matsukura, R. 2011. Asset-based flows from a generational perspective. In *Population Aging and the Generational Economy* (eds R. Lee, A. Mason), pp. 209–236. Edward Elgar.
20. Coale, A.J., Hoover, E. 1958. *Population Growth and Economic Development in Low-Income Countries.* Princeton University Press.
21. Modigliani, F., Ando, A.K. 1957. Tests of the life cycle hypothesis of savings: comments and suggestions 1. *Bull. Oxf. Univ. Inst. Econ. Stat.* **19**, 99–124.
22. Lee, R.D., Mason, A. 2011. *Population Aging and the Generational Economy: A Global Perspective.* Edward Elgar.
23. Willis, R.J. 1988. Life cycles, institutions and population growth: a theory of the equilibrium interest rate in an overlapping-generations model. In *Economics of Changing Age Distributions in Developed Countries* (eds R.D. Lee, W.B. Arthur, G. Rogers), pp. 106–138. Oxford University Press.
24. Lee, R.D. 1994. Age structure, intergenerational transfer, and wealth: a new approach, with applications to the United States. *J. Hum. Resour.* **29**, 1027–1063.

25. Bommier, A., Lee, R.D. 2003. Overlapping generations models with realistic demography. *J. Popul. Econ.* **16**(1), 135–160.
26. Lee, R.D. 2016. Macroeconomics, aging, and growth. In *Handbook of the Economics of Population Aging*, vol. 1, pp. 59–118. North-Holland. Elsevier.
27. Sánchez-Romero, M. 2013. The role of demography on per capita output growth and saving rates. *J. Popul. Econ.* **26**, 1347–1377 (doi:10.1007/s00148-012-0447-3).
28. Sánchez-Romero, M., Abio, G., Patxot, C., Souto, G. 2018. Contribution of demography to economic growth. *SERIEs* **9**, 27–64 (doi:10.1007/s13209-017-0164-y).
29. Lee, R.D., Mason, A. 2011. *Population Aging and the Generational Economy: A Global Perspective*. Edward Elgar.
30. European Commission. 2020. The 2021 ageing report: underlying assumptions and projection methodologies. European Commission Institutional Paper 142 (doi:10.2765/733565).
31. Wittgenstein Centre for Demography and Global Human Capita 2018. Wittgenstein Centre Data Explorer Version 2.0. www.wittgensteincentre.org/dataexplorer.
32. Mason, A., Lee, R.D. 2007. Transfers, capital, and consumption over the demographic transition. In *Population Aging, Intergenerational Transfers and the Macroeconomy* (eds R. Clark, A. Mason, N. Ogawa), pp. 128–162. Edward Elgar.
33. Lee, R., Mason, A. 2010. Fertility, human capital, and economic growth over the demographic transition. *Eur. J. Popul. Rev. Eur. Démographie* **26**, 159–182 (doi:10.1007/s10680-009-9186-x).
34. Börsch-Supan, A., Ludwig, A., Winter, J. 2006. Ageing pension reform and capital flows: a multi-country simulation model. *Economica* **73**(292), 625–658.
35. Piketty, T. 2018. *Capital in the Twenty-First Century*. Harvard University Press.
36. Cervellati, M., Sunde, U. 2013. Life expectancy, schooling, and lifetime labour supply: theory and evidence revisited. *Econometrica* **81**(5), 2055–2086.
37. Lee, R.D., Mason, A. 2014. Is low fertility really a problem? Population aging, dependency, and consumption. *Science* **346**, 229–234.
38. Sánchez-Romero, M., Wrzaczek, S., Prskawetz, A., Feichtinger, G. 2018. Does demography change wealth inequality? In *Control Systems and Mathematical Methods in Economics* (eds G. Feichtinger, R. Kovacevic, G. Tragler), pp. 349–375. Springer.
39. Lam, D. 1986. The dynamics of population growth, differential fertility, and inequality. *Am. Econ. Rev.* **76**, 1103–1116.
40. United Nations Department of Economic and Social Affairs, Population Division. 2019. World Population Prospects 2019 (online ed. rev. 1). https://population.un.org/wpp2019.

Index

Printed in the United States
by Baker & Taylor Publisher Services